Electricity & Controls for

HVAC/R

7th Edition

Electricity & Controls for HVAC/R

Stephen L. Herman
Ron Sparkman

7th
Edition

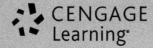

CENGAGE
Learning®

Australia • Brazil • Japan • Korea • Mexico • Singapore • Spain • United Kingdom • United States

CENGAGE
Learning®

Electricity and Controls for HVAC-R, Seventh Edition

Stephen L. Herman, Ron Sparkman

Vice President, Careers & Computing:
 Dave Garza
Director of Learning Solutions:
 Sandy Clark
Senior Acquisitions Editor: James DeVoe
Director, Development-Career and
 Computing: Marah Bellegarde
Managing Editor: Larry Main
Senior Product Manager: John Fisher
Editorial Assistant: Aviva Ariel
Brand Manager: Kristin McNary
Market Development Manager:
 Erin Brennan
Senior Production Director:
 Wendy Troeger
Production Manager: Mark Bernard
Content Project Manager:
 Barbara LeFleur
Senior Art Director: David Arsenault
Technology Project Manager: Joe Pliss
Media Editor: Deborah Bordeaux
Cover Image(s):
INSET IMAGES (top to bottom):
 © iStockphoto/Dmitriy Kalinin
 © iStockphoto/grafvision
 © iStockphoto/David Lee
Interior Design Image(s):
Title (background) and opener pages:
 © Sven Hoppe/www.Shutterstock.com
Title Page (right side, top to bottom):
 © iStockphoto/David Lee
 © David Lee/www.Shutterstock.com
 © iStockphoto/Sharon Meredith

Library of Congress Control Number: 2013932695
ISBN-13: 978-1-1332-7820-7
ISBN-10: 1-1332-7820-5

Cengage Learning
200 First Stamford Place, 4th Floor
Stamford, CT 06902
USA

Cengage Learning is a leading provider of customized learning solutions with office locations around the globe, including Singapore, the United Kingdom, Australia, Mexico, Brazil, and Japan. Locate your local office at: **international.cengage.com/region**

Cengage Learning products are represented in Canada by Nelson Education, Ltd.

To learn more about Delmar, visit **www.cengage.com/delmar**

Purchase any of our products at your local college store or at our preferred online store **www.cengagebrain.com**

Notice to the Reader

Publisher does not warrant or guarantee any of the products described herein or perform any independent analysis in connection with any of the product information contained herein. Publisher does not assume, and expressly disclaims, any obligation to obtain and include information other than that provided to it by the manufacturer. The reader is expressly warned to consider and adopt all safety precautions that might be indicated by the activities described herein and to avoid all potential hazards. By following the instructions contained herein, the reader willingly assumes all risks in connection with such instructions. The publisher makes no representations or warranties of any kind, including but not limited to, the warranties of fitness for particular purpose or merchantability, nor are any such representations implied with respect to the material set forth herein, and the publisher takes no responsibility with respect to such material. The publisher shall not be liable for any special, consequential, or exemplary damages resulting, in whole or part, from the readers' use of, or reliance upon, this material.

Printed in the United States of America
7 8 9 10 11 12 13 21 20 19 18 17

CONTENTS

PREFACE

Electricity and Controls for HVAC/R is written with the assumption that the student has no prior knowledge of electricity or control systems. Basic electrical theory is presented in a practical, straightforward manner. Mathematical explanations are used only when necessary to explain certain concepts of electricity. Each unit starts with the objectives of the unit and ends with a summary of important facts.

The text begins with the study of **basic electrical theory** and progresses to **series circuits, parallel circuits, alternating current, inductive circuits,** and **capacitive circuits.** The text also includes information on different types of three-phase services found in industrial and commercial locations as well as single-phase residential services. Individual devices and components common to the air-conditioning, heating, and refrigeration field are presented in a practical manner. Devices are explained from a standpoint of how they operate and how they are used. The text contains testing procedures for many of the devices covered. The practical presentation of these devices makes this text a *must-have* reference book for the service technician working in the field.

Electricity and Controls for HVAC/R, seventh edition, includes information on isolation transformers, autotransformers, and current transformers. The three major types of three-phase motors—squirrel-cage induction, wound rotor, and synchronous—are also covered. Coverage of single-phase motors includes split-phase motors, resistance-start induction-run motors, capacitor-start induction-run motors, and permanent-split capacitor motors. Shaded-pole induction motors and multispeed motors are also covered. The seventh edition also provides information on variable frequency drives.

Control circuits are developed using the components in the text. The text assumes that the student has no prior knowledge of control systems. **Operation of the manufacturer's control schematics is explained to aid the student in understanding how a control system operates and how to troubleshoot the system.**

Electricity and Controls for HVAC/R, seventh edition, includes information on **household and commercial icemaker controls.** These circuits are explained in a step-by-step procedure to ensure that students have a thorough working knowledge of these units.

Solid-state devices common to the HVAC/R field are covered in a straightforward manner. The devices covered are: **diode, transistor, SCR, diac, triac,** and **operational amplifier.** The last section of the text covers **programmable logic controllers,** which are becoming more and more common in the field.

NEW FOR THE SEVENTH EDITION

- Expanded information concerning the differences between direct current and alternating current
- Expanded coverage of resistors
- Added information concerning the voltage rating of capacitors
- Additional information on diodes
- Explanation of zener diodes
- Explanation of the silicon bilateral switch
- Extended coverage of rectifiers
- Updated to the 2011 National Electrical Code
- Expanded coverage of current transformers
- Explanation of the gate-turnoff SCR
- Extended information on short cycle timers

FEATURES OF THIS BOOK

Electricity and Controls for HVAC/R, seventh edition, contains many features to help enhance learning for the student:

- A special section covering **Safety** rules at the front of the book reminds students to follow correct procedures and take the necessary precautions when working around electricity.
- **Step-by-Step Procedures** are integrated throughout the text where applicable and provide students with a thorough working knowledge of the HVAC systems.
- **Troubleshooting Questions** present situations in which students must develop critical-thinking and problem-solving skills to prepare them for the field.
- **Review Questions** and **Practice Problems** are included at the end of each unit to allow students to evaluate their comprehension of the material and apply what they have learned from the information presented in the unit.
- **An extensive art program** includes schematics, line drawings, and up-to-date photos that help to reinforce the information presented in the text.

SUPPLEMENT TO THIS BOOK

An online Instructor Companion website contains an Instructor Guide with answers to end-of-chapter review questions, testbanks, and chapter presentations done in PowerPoint.

Accessing an Instructor Companion Web Site from SSO Front Door

1. Go to http://login.cengage.com and log in using the Instructor email address and password.
2. Enter author, title or ISBN in the Add a title to your bookshelf search box, and click the Search button.
3. Click Add to My Bookshelf to add Instructor Resources.
4. At the Product page, click the Instructor Companion site link

New Users

If you're new to Cengage.com and do not have a password, contact your sales representative.

ACKNOWLEDGMENTS

The author and Delmar, Cengage Learning gratefully acknowledge the time and effort put forth by the review panel of this revision. Our special thanks to:

Dan Burris
Eastfield College
Mesquite, TX

John A. Chemsak
Technical College of the Low Country
Beaufort, SC

Phil Coulter
Durham College
Whitby, Ontario, Canada

Mark Davis
New Castle School of Trades
Sharon, PA

Eugene C. Dickson
Indian River Community College
Fort Pierce, FL

Malcolm Evett
San Jose City College
San Jose, CA

James Hutchinson
Roanoke-Chowan CC
Ahoskie NC

Robert S. Kish
Belmont Technical College
Wheeling, WV

Doug Lacey
Grayson County College
Denison, TX

Reed Lovell
Lanier Technical Institute
Oakwood, GA

William Matthews
Linn State Technical College
Linn, MO

Marvin Maziarz
Niagara County Community College
Niagara Falls, NY

Peter McCann
New Britain, CT

Joseph Moravek
Lee College
Houston, TX

Tom Nieson
Gateway Technical College
Kenosha, WI

Randy Paul Smith
Altamaha Technical Institute
Jesup, GA

Frank M. Sylvester
Erie Community College
Buffalo, NY

Timothy Tiegs
St. Clair College
Windsor, ON Canada

Richard Wirtz
Columbus State Community College
Columbus, OH

SECTION 1

Basic Electricity

A SPECIAL NOTE ON SAFETY

The purpose of this textbook is to provide the air-conditioning and refrigeration technician with knowledge of electricity. Electricity is an extremely powerful force and should never be treated in a careless manner. The air-conditioning and refrigeration technician commonly works with voltages ranging from 24 volts to 480 volts. One mistake can lead to serious injury or death.

Never work on an energized circuit if it is possible to disconnect the power. When possible, use a three-step check to make certain that the power is turned off. The three-step check is as follows:

1. Test the meter on a known live circuit to make sure the meter is operating.
2. Test the circuit that is to be de-energized with the meter.
3. Test the meter on the known live circuit again to make certain that the meter is still operating.

Install a warning tag at the point of disconnection to warn people not to restore power to the circuit, as shown in Figure SF–1.

GENERAL SAFETY RULES

Think

Of all the rules concerning safety, this one is probably the most important: No amount of safeguarding or "idiot-proofing" a piece of equipment can protect a person as well as the person's taking time to think before acting. Many technicians have been killed by supposedly "dead" circuits. Do not depend on circuit breakers, fuses, or someone else to open a circuit. Test it yourself before you touch it. If you are working on high-voltage equipment, use insulated gloves and meter probes designed to be used on the voltage being tested. Your life is your own, so *think* before you touch something that can take it away.

▶ **Figure SF–1**
Warning tags warn people that the circuit should not be turned back on.

Certain pieces of equipment can be especially hazardous if you are not aware of them. Some central air-conditioning units use a main contactor that has only one set of contacts to disconnect a 240-volt circuit, as shown in Figure SF–2. The contactor operates on the principle that a complete circuit must exist for current to flow. If one line is broken or open, no current can flow to the compressor. The hazard lies in the fact that one of the 240-volt lines is still supplying power to the unit. If a technician should touch the unbroken line and ground, a 120-volt circuit is completed through his or her body. Other contactors employ two load contacts to break the circuit to the compressor, as shown in Figure SF–3. This type of contactor is much safer and can prevent a serious injury.

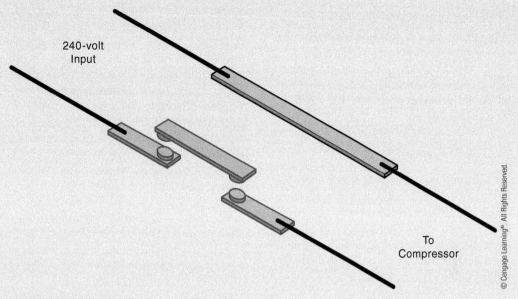

240-volt
Input

To
Compressor

▶ **Figure SF–2**
Some main contactors use one set of load contacts to break a 240-volt connection to the compressor.

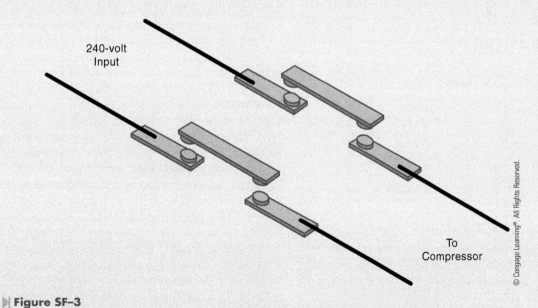

240-volt
Input

To
Compressor

▶ **Figure SF–3**
Contactors that employ two load contacts to break both sides of the 240-volt line are much safer and can prevent a serious injury.

Avoid Horseplay

Jokes and horseplay have a time and place, but the time and place are not when someone is working on an electric circuit or a piece of moving machinery. Do not be the cause of someone's being injured or killed, and do not let someone else be the cause of your being injured or killed.

Do Not Work Alone

Work with someone else, especially when working in a hazardous location or on a live circuit. Have someone with you to turn off the power or give artificial respiration or cardiopulmonary resuscitation (CPR). One of the effects of electrical shock is that it causes breathing difficulties and can cause the heart to go into fibrillation.

Work with One Hand When Possible

The worst case for electrical shock is when the current path is from one hand to the other. This path causes the current to pass directly through the heart. A person can survive a severe shock between the hand and one foot that would otherwise cause death if the current path were from one hand to the other.

Learn First Aid

Anyone working on electrical equipment should make an effort to learn first aid. Knowing first aid is especially important for anyone who must work with voltages above 50 volts. A knowledge of first aid and especially CPR may save your life or someone else's.

Effects of Electric Current on the Body

Most people have heard that it is not the voltage that kills but the current. Although this is a true statement, do not be misled into thinking voltage cannot harm you. Voltage is the force that pushes the current though the circuit. Voltage can be compared to the pressure that pushes water through a pipe. The more pressure available, the greater the volume of water flowing through a pipe. Students often ask how much current will flow through the body at a particular voltage. There is no easy answer to this question. The amount of current that can flow at a particular voltage is determined by the resistance of the current path. Different people have different resistances. A body will have less resistance on a hot day, when sweating, because salt water is a very good conductor. What you ate and drank for lunch can have an effect on your body resistance. The length of the current path can affect the resistance. Is the current path between two hands or from one hand to one foot? All of these factors affect body resistance.

The chart in Figure SF–4 illustrates the effects of different amounts of current on the body. This chart is general and shows the effects on most people. Some people may have less tolerance to electricity, and others may have greater tolerance.

A current of 2 to 3 milliamperes generally causes a slight tingling sensation. The tingling sensation increases as current increases, and becomes very noticeable at about 10 milliamperes. The tingling sensation is very painful at about 20 milliamperes. Currents between 20 and 30 milliamperes generally cause a person to seize the line and become unable to let go of the circuit. Currents between 30 and 40 milliamperes cause muscular paralysis, and currents between 40 and 60 milliamperes cause breathing difficulty. By the time the current increases to about 100 milliamperes, breathing is extremely difficult. Currents from 100 to 200 milliamperes generally cause death because the heart usually goes into fibrillation. Fibrillation is a condition in which the heart begins to "quiver" and the pumping action stops. Currents above 200 milliamperes generally cause the heart to squeeze shut. When the current is removed, the heart typically returns to a normal pumping action. This is the principle of operation of a defibrillator. It is often said that 120 volts is the most dangerous voltage to work with. The reason is that 120 volts generally cause a current flow between 100 and 200 milliamperes through the bodies of most people. Large amounts of current can cause severe electrical burns. Electrical burns are usually very serious because the burn occurs on the inside of the body. The exterior of the body may not look seriously burned, but the inside may be severely burned.

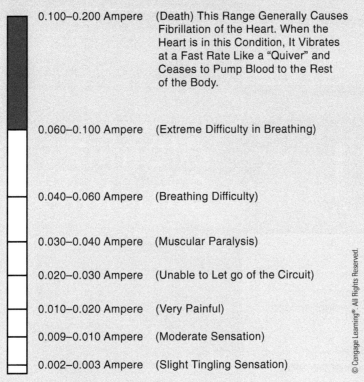

0.100–0.200 Ampere   (Death) This Range Generally Causes Fibrillation of the Heart. When the Heart is in this Condition, It Vibrates at a Fast Rate Like a "Quiver" and Ceases to Pump Blood to the Rest of the Body.

0.060–0.100 Ampere   (Extreme Difficulty in Breathing)

0.040–0.060 Ampere   (Breathing Difficulty)

0.030–0.040 Ampere   (Muscular Paralysis)

0.020–0.030 Ampere   (Unable to Let go of the Circuit)

0.010–0.020 Ampere   (Very Painful)

0.009–0.010 Ampere   (Moderate Sensation)

0.002–0.003 Ampere   (Slight Tingling Sensation)

▶ **Figure SF–4**
The effects of electric current on the body.

Atomic Structure

OBJECTIVES

After studying this unit, the student should be able to:

▷ Discuss basic atomic theory

▷ Name the principal parts of an atom

▷ Discuss the law of charges

▷ Define electricity

▷ Discuss the differences between conductors and insulators

To understand electricity, one should start with the study of atoms. The **atom** is the basic building block of the universe. All matter is composed of atoms. An atom is the smallest part of any element. Atoms are composed of three principal parts: the electron, the proton, and the neutron. Electrons exhibit a negative charge; protons exhibit a positive charge; and neutrons have no charge. Some theories suggest that neutrons are composed of both a proton and an electron. The positive charge of the proton and the negative charge of the electron cancel each other.

Protons and neutrons are extremely massive particles as compared to the electron. The latest scientific measurements suggest protons and neutrons weigh about 1838 times more than the electron and that the electron is approximately 1/1000 the size of a proton, Figure 1–1. Many scientists believe that

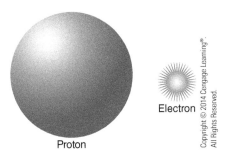

▶ **Figure 1–1**
The electron is only about 1/1000 the size of the proton, and the proton weighs about 1838 times more than the electron.

it is almost impossible to accurately measure the size of these particles and that it is not the actual size but the characteristics of the particles that is important.

THE LAW OF CHARGES

One of the basic laws of physics concerning atoms is the **law of charges**. This law states that opposite charges attract each other and like charges repel each other. If charged particles were suspended from a string, a positively charged particle and a negatively charged particle would attract each other, but two positively charged particles or two negatively charged particles would repel each other, as shown in Figure 1–2. Because **electrons** are negatively charged particles and **protons** are positively charged particles, they are attracted to each other.

STRUCTURE OF THE ATOM

The **nucleus** or center of the atom is composed of protons and **neutrons**. Electrons orbit the exterior of the atom in specific orbits or shells. The smallest of all atoms is hydrogen. Hydrogen does not contain a neutron in its nucleus. The hydrogen atom contains one proton and one electron, as shown in Figure 1–3. The smallest atom that contains both protons and neutrons in the nucleus is helium, shown in Figure 1–4. Helium contains two protons, two neutrons, and two electrons.

In 1808, a scientist named John Dalton proposed that all matter was composed of atoms. Although the assumptions that Dalton used to prove his theory

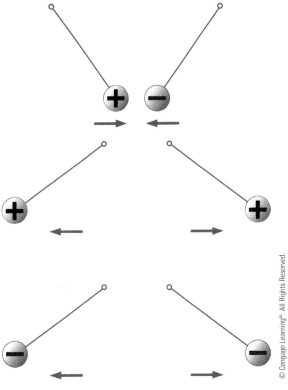

▶ **Figure 1–2**
The law of charges states that opposite charges attract and like charges repel.

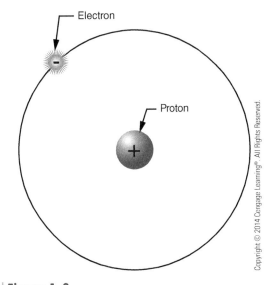

▶ **Figure 1–3**
An atom of hydrogen contains one proton and one electron.

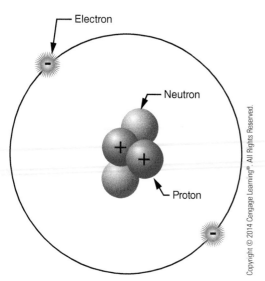

▶ **Figure 1–4**
A helium atom contains both protons and neutrons in the nucleus.

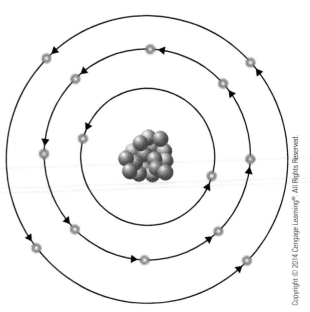

▶ **Figure 1–5**
Electrons exist in allowed orbits or shells.

were later found to be factually incorrect, the idea that all matter is composed of atoms was adopted by most of the scientific world. Then in 1897, J. J. Thompson discovered the electron. Thompson determined that electrons have a negative charge and that they have very little mass compared to the atom. He proposed that atoms have a large, positively charged, massive body with negatively charged electrons scattered throughout it. Thompson also proposed that the negative charge of the electrons exactly balanced the positive charge of the large mass, causing the atom to have a net charge of zero. Thompson's model of the atom proposed that electrons existed in a random manner within the atom, much like firing BBs from a BB gun into a slab of cheese. This was referred to as the *plum pudding model* of the atom.

In 1913, a Danish scientist named Niels Bohr presented the most accepted theory concerning the structure of an atom. In the Bohr model, electrons exist in specific, or "allowed," orbits around the nucleus, similar to the way that planets orbit the sun, as shown in Figure 1–5. The orbit in which the electron exists is determined by the electron's mass times its speed times the radius of the orbit. These

factors must equal the positive force of the nucleus. In theory there can be an infinite number of allowed orbits.

When an electron receives enough energy from some other source, it "quantum jumps" into a higher allowed orbit. Electrons, however, tend to return to a lower allowed orbit. When this occurs, the electron emits the excess energy as a single photon of electromagnetic energy.

Atoms have a set number of electrons that can be contained in one orbit or shell, as shown in Figure 1–6. The number of electrons that can be contained in any one shell is determined by the formula ($2N^2$). The letter N represents the number of the orbit or shell. For example, the first orbit can hold no more than two electrons.

$$2 \times (1)^2$$

$$2 \times 1 = 2$$

The second orbit can hold no more than eight electrons.

$$2 \times (2)^2$$

$$2 \times 4 = 8$$

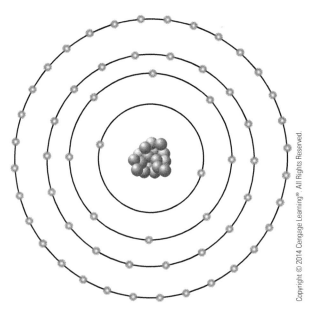

▶ **Figure 1–6**
    There is a maximum number of electrons that can be contained in one orbit or shell.

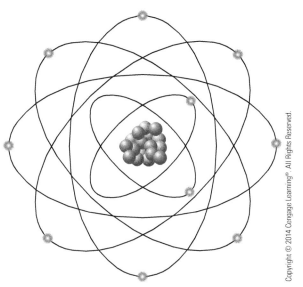

▶ **Figure 1–7**
    Electrons orbit the nucleus in a circular manner.

The third orbit can contain no more than 18 electrons.

$$2 \times (3)^2$$

$$2 \times 9 = 18$$

The fourth and fifth orbits cannot hold more than 32 electrons, which is the maximum number of electrons that can be contained in any orbit.

$$2 \times (4)^2$$

$$2 \times 16 = 32$$

Although atoms are often drawn flat, as illustrated in Figure 1–6, the electrons orbit around the nucleus in a circular fashion, as shown in Figure 1–7. The electrons travel at such a high rate of speed that they form a shell around the nucleus. It would be similar to a golf ball being surrounded by a tennis ball that is surrounded by a basketball. For this reason, electron orbits are often referred to as *shells*.

VALENCE ELECTRONS

The outermost shell of an atom is known as the **valence shell**. Any electrons located in the outer shell of an atom are known as **valence electrons**,

as shown in Figure 1–8. The valence shell of an atom cannot hold more than eight electrons. It is the valence electrons that are of primary concern in the study of **electricity**, because it is these electrons that explain much of electrical theory. A **conductor**, for instance, is made from a material that contains one or two valence electrons. Atoms with one or two valence electrons are unstable and will give up these electrons easily. Conductors are materials that permit electrons to flow through them easily. Silver, copper, and gold all contain one valence electron and are excellent conductors of electricity. Silver is the best natural conductor of electricity, followed by copper, gold, and aluminum. An atom of copper is shown in Figure 1–9. Although it is known that atoms containing few valence electrons are the best conductors, it is not known why some of these materials are better conductors than others. Copper, gold, platinum, and silver all contain only one valence electron. Silver, however will conduct electricity more readily than any of the others. Aluminum, which contains three valence electrons, is a better conductor of electricity than platinum, which contains only one valence electron.

Electricity is composed of electrons. If it were possible to connect a spigot to a wall outlet and

Valence Electron

Valence Shell

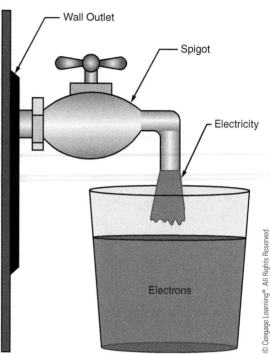

Figure 1–8
Valence electrons are located in the outermost orbit of an atom.

Wall Outlet

Spigot

Electricity

Electrons

Figure 1–10
Electricity is composed of electrons.

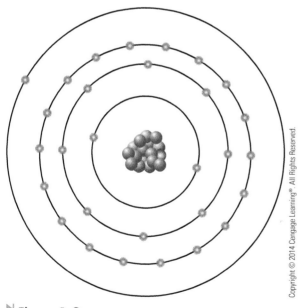

Figure 1–9
Copper atom.

catch electricity in a glass, the glass would be full of electrons, as shown in Figure 1–10. Electric current is the flow of electrons, which means that the electrons must be moving to produce current. It is the movement of the electrons that produces electrical energy. There are several theories concerning how electrons are made to flow through a conductor. One theory is generally referred to as the *bump theory*. It states that current flow is produced when an electron from one atom knocks electrons of another atom out of orbit, as shown in Figure 1–11. The striking electron gives its energy to the electron being struck. The striking electron settles into orbit around the nucleus, and the electron that was struck moves off to strike another electron. This action can often be seen in the game of pool. The moving cue ball gives it energy to the ball being struck, as shown in Figure 1–12. The stationary ball then moves off with most of the energy of the cue ball, and the cue ball stops moving. Although most of the energy was transferred to the stationary ball, some energy was dissipated as heat caused by the impact of the cue ball striking the stationary ball.

Other theories deal with the fact that all electric power sources produce a positive and negative

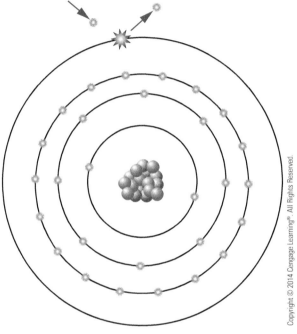

Figure 1–11
One electron knocks another electron out of orbit.

Copyright © 2014 Cengage Learning® All Rights Reserved.

terminal. The negative terminal is created by causing an excess of electrons to form at that terminal, and the positive terminal is created by removing a large number of electrons from that terminal, as shown in Figure 1–13. Different methods can be employed to produce the excess of electrons at one terminal and deficiency of electrons at the other, but when a circuit is completed between the two terminals, negative electrons are repelled away from the negative terminal and attracted to the positive terminal, as shown in Figure 1–14. The greater the difference in the number of electrons between the negative and positive terminals, the greater the force of repulsion and attraction.

Most of the world's electric power is produced by rotating machines called alternators. Alternators work on the principle of cutting magnetic lines of flux with a conductor. One of the basic principles of electricity is that anytime a conductors cuts magnetic lines of flux, a voltage is inducted into the conductor. Another common source of electricity is batteries. Batteries produce electricity by chemical action. Other methods include solar cells, thermocouples, and piezoelectric devices. Solar cells produce electricity when they receive photons of light that cause electrons to move. Thermocouples produce electricity by heating the junction of two different types of metal, and the piezoelectric effect produces current flow from pressure.

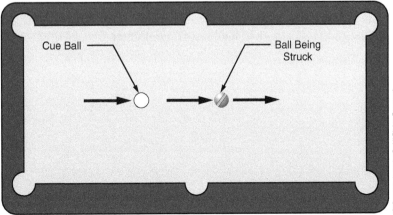

Figure 1–12
The cue ball gives its energy to the ball it strikes.

© Cengage Learning® All Rights Reserved.

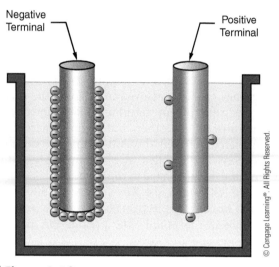

▶ **Figure 1–13**
The negative terminal of a power source has an excess of electrons, and the positive terminal has a lack of electrons.

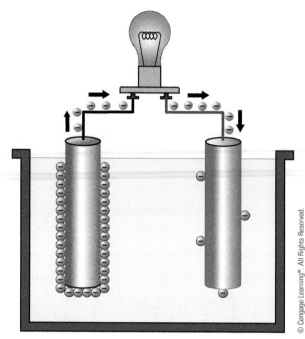

▶ **Figure 1–14**
Electrons flow from the negative terminal through the load to the positive terminal of the power source.

INSULATORS

Insulators are materials that do not permit electrons to flow through them easily. The atoms of an insulator generally contain seven or eight valence electrons. When an atom contains seven or eight valence electrons, they are tightly held by the atom and not easily given up. Insulator materials are generally made from compounds instead of a pure element. The molecules of the compounds are extremely stable and do not break

their bond easily. Some examples of insulators are wood, ceramic, mica, rubber, and thermoplastic. Insulator materials are generally rated by voltage. The voltage rating indicates the amount of voltage they can withstand without breaking down. A very common electric cable used in residential wiring is the nonmetallic two-conductor cable with ground, shown in Figure 1–15. This cable is generally referred to as NM or Romex. The insulation around the conductors has a voltage rating of 600 volts.

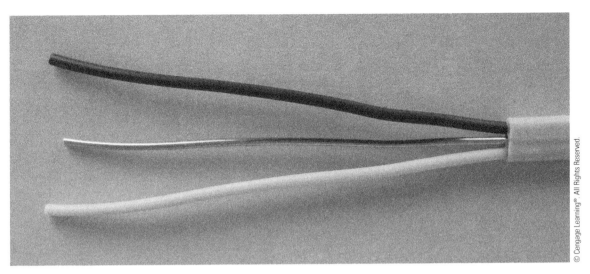

▶ **Figure 1–15**
Two conductor cables with ground are commonly used in residential wiring.

▶▶ SUMMARY

- ⊙ The three major parts of an atom are the electron, proton, and neutron.
- ⊙ Electrons have a negative charge; protons have a positive charge; and neutrons have no charge.
- ⊙ The nucleus of the atom contains protons and neutrons.
- ⊙ An electron is about 1000 times smaller than a proton, and the proton weighs about 1838 times more than an electron.
- ⊙ The law of charges states that opposite charges attract and like charges repel.
- ⊙ Electrons exist in allowed orbits around the nucleus of the atom.
- ⊙ Neutral atoms contain the same number of electrons and protons.
- ⊙ Valence electrons are the electrons located in the outermost orbit of an atom.
- ⊙ The best conductors of electricity generally contain one or two valence electrons.
- ⊙ Insulators are materials that do not conduct electricity easily.
- ⊙ Insulator materials are rated by voltage.

KEY TERMS

atom	insulator	proton
conductor	law of charges	valence electrons
electricity	neutron	valence shell
electron	nucleus	

REVIEW QUESTIONS

1. What are the three subatomic parts of an atom, and what charge does each carry?
2. How many times smaller is an electron than a proton?
3. The weight of a proton is how many times heavier than that of an electron?
4. State the law of charges.
5. What scientist presented the most accepted theory concerning the behavior of atoms?
6. Materials that make the best conductors generally contain how many valence electrons?
7. What are valence electrons?
8. What is the maximum number of electrons that can exist in the valence orbit?
9. How is most of the electric power in the world produced?
10. Insulator materials are generally rated by the amount of _____ they can withstand without breaking down.

Electrical Quantities and Ohm's Law

OBJECTIVES

After studying this unit, the student should be able to:

» Define a coulomb
» Define an ampere
» Define voltage
» Discuss resistance
» Define a watt
» Use Ohm's law to solve electrical problems

Electricity has a standard set of values. Before you can work with electricity, you must have a knowledge of these values and how to use them. Because the values of electrical measurement have been standardized, they are understood by everyone who uses them. For instance, carpenters use a standard system for measuring length, such as the inch or foot. Imagine what a house would look like that was constructed by two carpenters who used different lengths of measure for an inch or foot. The same holds true for people who work with electricity. A volt, ampere, or ohm is the same for everyone who uses them.

COULOMB

A **coulomb** is a quantity measurement of electrons. One coulomb contains 6.25×10^{18} electrons. The number shown in Figure 2–1 is the number of

6,250,000,000,000,000,000

▶ **Figure 2–1**
The number of electrons in a coulomb.

electrons in 1 coulomb. Because the coulomb is a unit of measurement, it is similar to a quart, gallon, or liter. It takes a certain amount of liquid to equal a quart, just as it takes a certain amount of electrons to equal a coulomb.

AMPERE

The **ampere**, or amp, is defined as 1 coulomb per second. Notice that the definition of an amp involves a quantity measurement (the coulomb) combined with a time measurement (the second). One amp of current flows through a conductor when 1 coulomb flows past a point in 1 second, Figure 2–2. Amperage is a measurement of the actual amount of electricity that is flowing through a circuit. In a water system, it would be comparable to gallons per minute or gallons per second, Figure 2–3. If 1 coulomb were to flow past a point in a half

second, there would be 2 amperes of current flow. If 1 coulomb were to flow past a point in 2 seconds, there would be a half amp of flow. Current is normally measured in amperes, milliamperes, and microamperes. A milliamp is one-thousandth of an amp (0.001), and a microamp is one-millionth of an amp (0.000,001).

VOLT

Voltage is actually defined as **electromotive force**, or EMF. It is the force that pushes the electrons through a wire and is often referred to as **electrical pressure**. Remember that voltage cannot flow. To say that voltage flows is like saying that pressure flows through a pipe. Pressure can push water through a pipe, and it is correct to say that water flows through a pipe, but it is not correct to say that pressure flows through a pipe. The same is true for voltage. The voltage pushes current through a wire, but voltage cannot flow through a wire. In a water system, the voltage could be compared to the pressure of the system, Figure 2–4.

▶ **Figure 2–2**
One coulomb flowing past a point in 1 second.

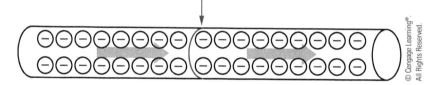

Pump

GPM

Battery

AMPS

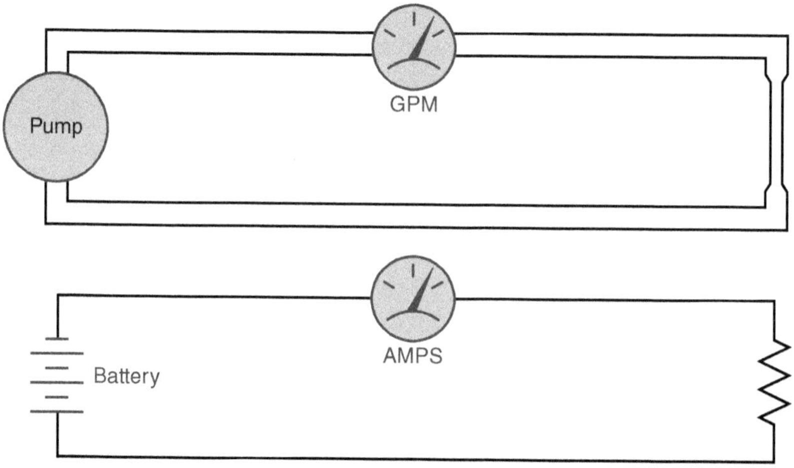

▶ **Figure 2–3**
Water flow is similar to current flow.

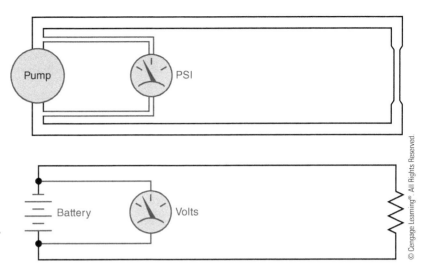

Figure 2–4
The pressure in a water system is comparable to the voltage in an electric circuit.

OHM

The **ohm** is the measure of the **resistance** to the flow of current. The voltage of the circuit must overcome the resistance before it can cause electrons to flow through it. Without resistance, every electrical circuit would be a short circuit. All electrical loads, such as heating elements, lamps, motors, transformers, and so forth, are measured in ohms. In a water system, a reducer is used to control the flow of water. In an electrical circuit, a resistor can be used to control the flow of electrons. Figure 2–5 illustrates this concept.

Electrical quantities other than resistance are also measured in ohms. Inductive reactance and capacitive reactance are current-limiting forces that exist in alternating current circuits and are discussed later in the text. Another term used to describe a current-limiting force is *impedance*. **Impedance** is a combination of all current-limiting forces in a circuit and can include resistance, inductive reactance, and capacitive reactance. Because resistance is only one of the current-limiting forces found in alternating current circuits, impedance is the quantity generally used to describe the current-limiting force in an alternating current circuit. Impedance is discussed in more detail later in the text.

WATT

Wattage is a measure of the amount of power that is being used in the circuit. It is proportional to the amount of voltage and the amount of current

Figure 2–5
A reducer reduces the flow of water through a water system just as a resistor reduces the flow of electrons in an electric circuit.

flow. To understand watts, return to the example of the water system. Assume that a water pump has a pressure of 120 psi (pounds per square inch) and causes a flow rate of 1 gallon per second. Now assume that this water is used to drive a waterwheel, as shown in Figure 2–6. Notice that the waterwheel has a radius of 1 foot from the center shaft to the rim of the wheel. Because water weighs 8.34 pounds per gallon and is being forced against the wheel at a pressure of 120 psi, the wheel could develop a torque of 1000.8 foot-pounds (120 × 8.34 × 1 = 1000.8). If the pressure were increased to 240 psi, but the water flow remained constant, the force against the wheel would double (240 × 8.34 × 1 = 2001.6). If the pressure remained at 120 psi, but the water flow were increased to 2 gallons per second, the force against the wheel would again double (120 ×16.68 × 1 = 2001.6). Notice that the amount of power developed by the water-wheel is determined by both the amount of pressure driving the water and the amount of flow.

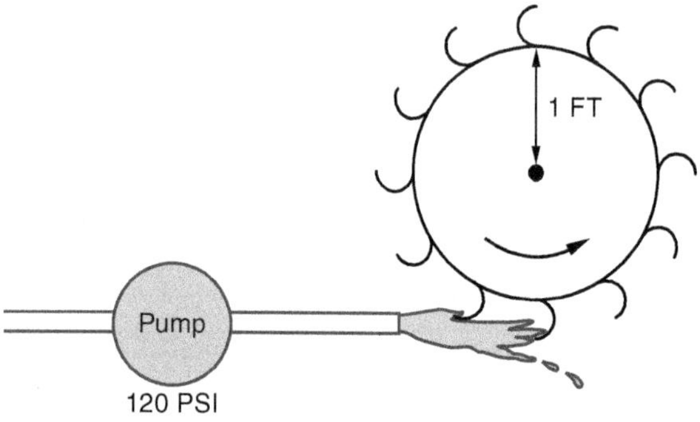

The power of an electrical circuit is very similar. Figure 2–7 shows a resistor connected to a circuit with a voltage of 120 volts and a current flow of 1 amp. The resistor shown represents an electric heating element. When 120 volts forces a current of 1 amp through it, the heating element produces 120 watts of heat (120 × 1 = 120 watts). If the voltage were increased to 240 volts, but the current remained constant, the element would produce 240 watts of heat (240 × 1 = 240 watts). If the voltage remained at 120 volts, but the current were increased to 2 amps, the heating element would again produce 240 watts (120 × 2 = 240). Notice that the amount of power used by the heating element is determined by the amount of current flow and the voltage driving it.

A good rule to remember concerning watts, or **true power**, is that before watts can exist in an electric circuit, electrical energy must be converted or changed into some other form. When current flows through a resistor, the resistor becomes hot and dissipates heat. Heat is a form of energy. The wattmeter connected in the circuit shown in Figure 2–7 measures the amount of electrical energy that is being converted into heat energy. If the resistor shown in Figure 2–7 were replaced with an electric motor, the wattmeter would measure the amount of electrical energy that was converted into mechanical energy.

OHM'S LAW

Ohm's law is named for the German scientist George S. Ohm. Ohm discovered that all electrical quantities are proportional to each other and can therefore be expressed as mathematical formulas. In its simplest form, Ohm's law states that *it takes 1 volt to push 1 amp through 1 ohm.*

1 FT

Pump

120 PSI

▶ **Figure 2–6**
Pump used to drive a waterwheel.

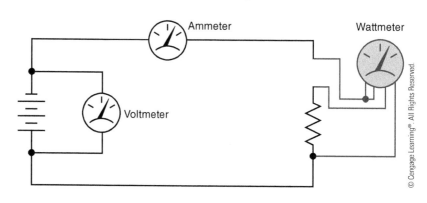

Ammeter

Wattmeter

Voltmeter

▶ **Figure 2–7**
The amount of voltage and current determine the power.

Figure 2–8 shows three basic Ohm's law formulas. In these formulas, E stands for electromotive force and is used to represent the voltage. The I stands for the intensity of the current and is used to represent the amount of current flow or amps. The letter R stands for resistance and is used to represent the ohms.

The first formula states that the voltage can be found if the current and resistance are known. Voltage is equal to amps multiplied by ohms. For example, assume a circuit has a resistance of 50 ohms and a current flow through it of 2 amps. The voltage connected to this circuit is 100 volts (2 amps × 50 ohms = 100 volts). The second formula indicates that if the voltage and resistance of the circuit are known, the amount of current flow can be found. Assume a 120-volt circuit is connected to a resistance of 30 ohms. The amount of current that will flow in the circuit is 4 amps (120 volts/30 ohms = 4 amps). The third formula states that if the voltage and current flow in a circuit are known, the resistance can be found. Assume a circuit has a voltage of 240 volts and a current flow of 10 amps. The resistance in the circuit is 24 ohms (240 volts/10 amps = 24 ohms).

Figure 2–9 shows a simple chart that can be a great help when trying to remember the Ohm's law formula. To use the chart, cover the quantity to be found. For example, if the voltage, E, is to be found, cover the E on the chart. The chart now shows the remaining letters IR. Therefore, $E = I \times R$. If the current is to be found, cover the I on the chart. The chart now shows E/R. Therefore, $I = E/R$. If the resistance of a circuit is to be found, cover the R on the chart. The chart now shows E/I. Therefore, $R = E/I$.

A larger chart that shows the formulas needed to find watts as well as the voltage, amperage, and resistance is shown in Figure 2–10. Because watt is the unit of electric power, the letter P is used to represent watts. The chart is divided into four quadrants. Each quadrant contains three formulas that can be used to find the electrical quantity represented in that quadrant.

EXAMPLE 1: An electric heating element is connected to 120 volts and has a resistance of 18 ohms. What is the power consumption of this element?

SOLUTION: The electrical quantity to be determined is power, or watts. The known electrical values are voltage and resistance. Using the formula chart in

$$E = I \times R \qquad I = \frac{E}{R} \qquad R = \frac{E}{I}$$

▶ **Figure 2–8**
Three Ohm's law formulas.

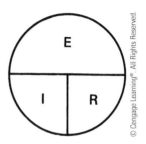

▶ **Figure 2–9**
Ohm's law formula chart.

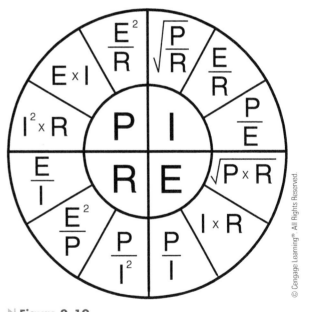

▶ **Figure 2–10**
Ohm's law chart.

Figure 2–10, choose the formula for finding power that includes the two known electrical quantities. The formulas for determining power are located in the first quadrant of the chart. The formula that

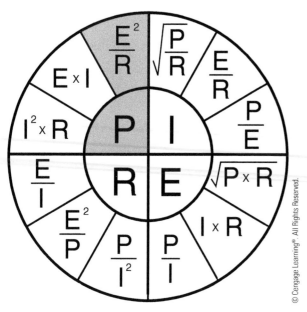

> **Figure 2–11**
> Power can be determined when the voltage and
> resistance are known.

> **Figure 2–12**
> Current can be determined when the power and
> voltage are known.

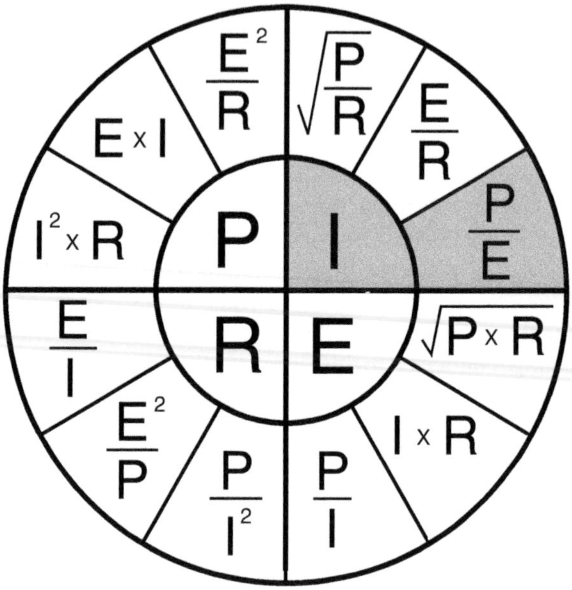

contains the known values of voltage and resistance
is $P = E^2/R$, as shown in Figure 2–11.

$$P = \frac{E^2}{R}$$

$$P = \frac{120 \times 120}{18}$$

$$P = 800 \text{ watts}$$

EXAMPLE 2: A 240-volt electric furnace has a power
rating of 15,000 watts, or 15 kW. How much current will this furnace draw when in operation?

SOLUTION: The quantity you are looking for is current or amps, represented by the letter I.

Formulas for determining current are located in
the second quadrant of the formula chart shown in
Figure 2–10. The two known quantities are power
(watts) and voltage. The correct formula is $I = P/E$,
as shown in Figure 2–12.

$$I = \frac{P}{E}$$

$$I = \frac{15,000}{240}$$

$$I = 62.5 \text{ amperes}$$

ALTERNATING CURRENT

The type of current discussed thus far in this unit
is direct current, or DC. Direct current is unidirec-
tional, meaning that it flows in only one direction,
from negative to positive. The purest form of direct
current is probably the battery. Current leaves
the negative terminal, flows through the load,
and returns to the positive terminal, Figure 2–13.

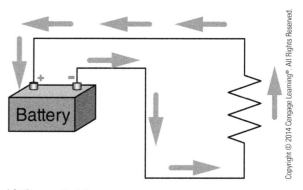

> **Figure 2–13**
> Direct current (DC) flows in only one direction.

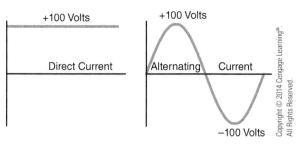

Figure 2–14
Direct current is shown as a straight line, and alternating current reverses polarity at regular intervals.

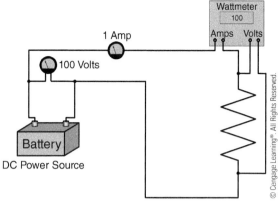

Figure 2–15
A 100 volt-battery connected to a 100-ohm resistor produces 100 watts of power.

Alternating current differs from direct current (DC) in that AC current reverses its direction of flow at periodic intervals. If an oscilloscope were connected to a 100-volt battery, you would see a straight line. If the oscilloscope were connected to a source of alternating current that reaches a maximum value of 100 volts, you would see a waveform that rises from zero to a peak value of 100 volts in the positive direction, returns to zero, rises to a value of 100 volts in the negative direction, and returns to zero, Figure 2–14. This is referred to as one cycle. Each time the voltage polarity changes, the direction of current flow changes. Frequency is the number of complete cycles that occur in 1 second and is measured in Hertz (Hz). The standard frequency used throughout the United States and Canada is 60 Hz.

Almost all air-conditioning and refrigeration units are powered by alternating current. Some differences exist between DC and AC currents. If a 100-volt battery were connected to a 100-ohm resistor, an ammeter would indicate a current flow of 1 ampere. A wattmeter connected in the circuit would indicate 100 watts of power, Figure 2–15.

$$P = E \times I \quad P = 100 \times 1 \quad P = 100 \text{ watts}$$

If the battery were replaced with an AC power source that produces a peak voltage of 100 volts, the wattmeter would indicate a power of approximately 50 watts, Figure 2–16. The reason for this difference is that, during the same time period, the DC voltage is constant, but the AC voltage reaches the peak value for only a short time and is less than maximum for most of the time, Figure 2–17. This

causes a problem with Ohm's law because volts multiplied by amps no longer equal watts.

$$P = E_{Peak} \times I_{Peak} \quad P = 100 \times 1 \quad P = 50 \text{ watts}$$

To compensate for this problem, the peak value of AC voltage is increased by the square root of 2, or 1.414—or in this example to a value of 141.4 volts (100 × 1.414). If an AC voltage with a peak value of 141.4 volts were connected to a 100-ohm

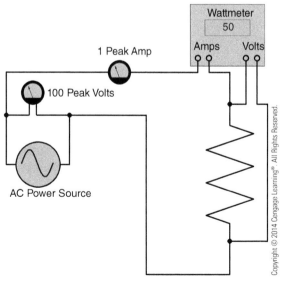

Figure 2–16
If an AC power source that produces 100 peak volts were connected to a 100-ohm resistor, it would produces about 50 watts of power.

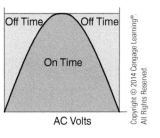

Figure 2–17
The AC waveform does not produce a constant voltage.

To Change	To	Multiply By
Peak	RMS	0.707
RMS	Peak	1.414

Figure 2–18
Conversion factors for peak and RMS voltages.

kilo	1000
hecto	100
deka	10
Base unit	1
deci	$\frac{1}{10}$, or 0.1
centi	$\frac{1}{100}$, or 0.01
milli	$\frac{1}{1000}$, or 0.001

Figure 2–19
Standard metric prefixes.

resistor, 100 watts would be produced. In this example, 141.4 volts is the peak value, and 100 volts is known as the **RMS (root mean square)** value. The RMS **peak** value of AC voltage produces the same amount of power as a like amount of DC voltage. Voltage ratings for equipment operated on alternating current lists the RMS value of voltage. If the nameplate on an air-conditioning unit lists a voltage of 240 volts, that is the RMS value of voltage. AC voltmeters indicate the RMS value of voltage unless the meter specifically states that it indicates peak voltage. An oscilloscope connected to an AC voltage, however, will indicate the peak value. The conversion factors for determining the RMS value if the peak value is known, or determining the peak value if the RMS value is known, are shown in the chart in Figure 2–18.

METRIC PREFIXES

Metric prefixes are used in the electrical field just as they are in most other scientific fields. In the English system of measure, different divisions are used for different measurements. A yard, for example, can be divided into 3 feet. A foot is generally divided into 12 inches, and an inch is often divided into sixteenths. In the metric system, 10 or a multiple of 10 is always the dividing factor. A kilometer can be divided into 10 hectometers. A hectometer can be divided into 10 dekameters, and a dekameter can be divided into 10 meters. A chart listing standard metric prefixes is shown in Figure 2–19.

In the electrical field, a different type of metric notation called *engineering notation* is used instead of the standard metric prefixes. Engineering notation is in steps of 1000 instead of 10. A chart listing common engineering notation prefixes and their symbols is shown in Figure 2–20. Starting with the base unit, or 1, the first engineering unit greater than 1 is kilo, or 1000. The next engineering unit greater than kilo is mega, or 1 million. A million is 1000 times larger than a thousand. The first engineering unit less than the base unit is milli, or 1 one thousandth. The next engineering unit less than milli is micro, or 1 one-millionth. A millionth is 1000 times smaller than a thousandth.

These prefixes are used to simplify standard electrical measurements. Five billion watts can be written as 5,000,000,000 W or as 5 GW. Twenty-five microamperes can be written as 0.000,025 A or as 25 μA.

ENGINEERING UNIT	SYMBOL	MULTIPLY BY:	
Tera	T	1,000,000,000,000	$\times 10^{12}$
Giga	G	1,000,000,000	$\times 10^{9}$
Meg	M	1,000,000	$\times 10^{6}$
kilo	k	1,000	$\times 10^{3}$
Base unit		1	
Milli	m	0.001	$\times 10^{-3}$
Micro	μ	0.000,001	$\times 10^{-6}$
Nano	n	0.000,000,001	$\times 10^{-9}$
Pico	p	0.000,000,000,001	$\times 10^{-12}$

▶| **Figure 2–20**
Standard prefixes used in engineering notation.

▶ SUMMARY

- ⊙ A coulomb is a quantity measurement of electrons that is equal to 6.25×10^{18} electrons.
- ⊙ An ampere is the amount of current that flows through a circuit.
- ⊙ An ampere is defined as 1 coulomb per second.
- ⊙ Voltage is the force that pushes the electrons through a circuit.
- ⊙ Voltage is defined as electromotive force (EMF).
- ⊙ Voltage is sometimes referred to as electrical pressure.
- ⊙ Resistance to the flow of electricity is measured in ohms.
- ⊙ Wattage is a measurement of electric power.
- ⊙ Before an electric circuit can have true power, or watts, electrical energy must be converted into some other form of energy.
- ⊙ Ohm's law can be used to mathematically find electrical values when at least two of these values are known.
- ⊙ The RMS value of AC voltage produces the same amount of power as an equivalent value of DC voltage.

KEY TERMS

ampere	ohm	true power
coulomb	Ohm's law	voltage
electrical pressure	peak	wattage
electromotive force	resistance	
impedance	RMS (root mean square)	

REVIEW QUESTIONS

1. What is a coulomb?

2. What is the definition of an amp?

3. Define the term *voltage*.

4. Define the term *ohm*.

5. Define the term *watt*.

6. An electric heating element has a resistance of 16 ohms and is connected to a voltage of 120 volts. How much current (amps) will flow in the circuit?

7. How many watts of heat are being produced by the heating element in question 6?

8. A 240-volt circuit has a current flow of 20 amps. How much resistance (ohms) is connected in the circuit?

9. An electric motor has an apparent resistance of 15 ohms. If a current of 8 amps is flowing through the motor, what is the connected voltage?

10. A 240-volt air-conditioning compressor has an apparent resistance of 8 ohms. How much current (amps) will flow in the circuit?

11. How much power (watts) is being used by the compressor in question 10?

12. A 5000-watt electric heating unit is connected to a 240-volt line. What is the current flow (amps) in the circuit?

13. If the voltage in question 12 were reduced to 120 volts, how much current (amps) would be needed to produce the same amount of power?

14. Is it less expensive to operate the electric heating unit in question 12 on 240 volts or on 120 volts? Explain your answer.

15. The nameplate on a heat pump indicates that the unit should be connected to 240 volts AC. Is this the peak or RMS value of AC voltage?

16. An oil-filled capacitor has a voltage rating of 300 peak volts. If this capacitor were connected to an RMS voltage of 240 volts, would it be damaged?

UNIT 3 ▷

Measuring Instruments

OBJECTIVES

After studying this unit the student should be able to:

- ≫ Discuss the operation and connection of a voltmeter
- ≫ Measure voltage, using a multirange voltmeter
- ≫ Discuss the operation of an ammeter
- ≫ Describe different types of ammeters
- ≫ Measure current with an ammeter

Anyone who wants to work in the air-conditioning and refrigeration field must become proficient with the common instruments used to measure electrical quantities. These instruments are the **voltmeter**, **ammeter**, and **ohmmeter**. In the air-conditioning and refrigeration field, the technician works almost exclusively with alternating current. For this reason, the meters covered in this unit are intended to be used in an alternating current system.

VOLTMETER

The voltmeter is designed to be connected directly across the source of power. Figure 3–1 shows a voltmeter being used to test the voltage of a panel box. Notice that the leads of the meter are connected directly across the source of voltage. The reason a voltmeter can be connected directly across the power

Figure 3–1
Voltmeter being used to test the voltage of a panel.

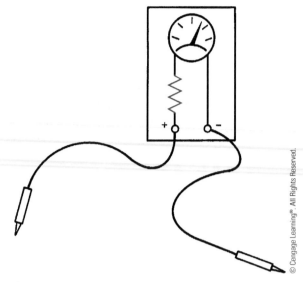

Figure 3–2
A voltmeter has high resistance connected in series with the meter movement.

line is because it has a very high resistance connected in series with the meter movement, Figure 3–2. A common resistance for a voltmeter is about 20,000 ohms per volt for DC and 5000 ohms per volt AC. Assume the voltmeter shown in Figure 3–2 is an AC meter and has a full-scale range of 300 volts. The resistor connected in series with the meter would, therefore, have a resistance of 1,500,000 ohms (300 volts × 5000 ohms per volt = 1,500,000 ohms).

Most voltmeters are **multiranged**, which means that they are designed to use one meter movement to measure several ranges of voltage. For example, one meter may have a selector switch that permits full-scale ranges to be selected. These ranges may be 3 volts full-scale, 12 volts full-scale, 30 volts full-scale, 60 volts full-scale, 120 volts full-scale, 300 volts full-scale, and 600 volts full-scale.

The reason for making a meter with this number of scales is to make the meter as versatile as possible. If it is necessary to check for a voltage of 480 volts, the meter can be set on the 600-volt range. If it becomes necessary to check a control voltage of 24 volts, however, it would be very difficult to do on the 600-volt range. If the meter is set on the 30-volt range, however, it becomes a simple matter to test for a voltage of 24 volts. The meter shown in Figure 3–3 has multirange selection for voltage.

When the selector switch of this meter is turned, steps of resistance are either inserted in the circuit to increase the range or removed from the circuit to decrease the range, Figure 3–4. Notice that when the higher voltage settings are selected, more resistance is inserted in the circuit.

Another type of voltmeter that is gaining popularity is the **digital meter**. A digital meter displays the voltage in digits instead of using a meter movement, Figure 3–5. Digital meters have several advantages over voltmeters that use a meter movement (commonly called **analog meters**). The greatest advantage is that the **input impedance**, or resistance, is higher. Analog meters commonly have a resistance of about 5000 ohms per volt. This means that on a 3-volt full-scale range, the meter movement has a resistance of 15,000 ohms connected in series with it (3 × 5000 = 15,000). On the 600-volt full-scale range, the meter movement has a resistance of 3,000,000 ohms connected in series with it (600 × 5000 = 3,000,000). Digital meters commonly have an input impedance of 10,000,000 ohms (10 megohms), regardless of the

range they are set on. The advantage of this high input impedance is that it does not interfere with a low-power circuit. The advantage of this may not be too clear at first, because most technicians are used to working with circuits that have more than enough power to operate the meter. However, many of the newer controls are electronic; these circuits may be greatly altered if tested with a low-impedance meter.

For example, assume an electronic control is operated on 5 volts and has a total current capacity of 100 microamps (0.000100 amps). Now assume that a 5000-ohm-per-volt meter is set on the 12-volt scale and is to be used to check the control circuit. The voltmeter has a resistance of 5000 × 12 = 60,000 ohms. If this meter were used to test the circuit, the meter would have a current draw of 83.3 microamps (5 volts/60,000 ohms = 0.0000833 amps). Because the control circuit only has a total current capacity of 100 microamps, the meter is using most of the current to operate. The circuit has been changed to such a degree that it can no longer operate.

If the digital meter were used to test this same circuit, it would have a current draw of 0.5 microamp (5 volts/10,000,000 = 0.00000005 amp). The circuit would be able to furnish the 0.5 microamp needed to operate the meter without a problem or altering the circuit.

Another advantage of the digital meter is that it is generally easier for an inexperienced person to learn to read. Analog meters can be used for about 99% of the measurements that must be taken, but it generally takes some time and practice to read them properly.

Figure 3–3
Multimeter with an analog scale.

Courtesy of Triplett Test Equipment and Tools

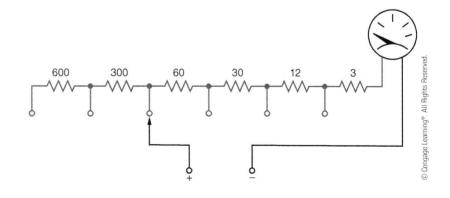

Figure 3–4
A multirange voltmeter.

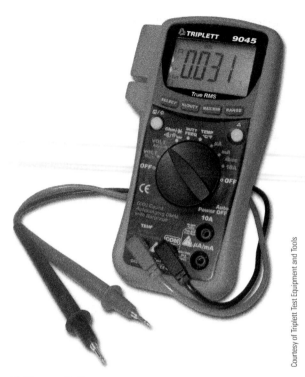

▶ **Figure 3–5**
Digital multimeter.

▶ **Figure 3–6**
Speedometer.

▶ **Figure 3–7**
Fuel gauge.

Learning to read the scale of a multimeter takes time and practice. Most people use meters everyday without thinking about it. A very common type of meter used daily by most people is shown in Figure 3–6. The meter illustrated is a speedometer similar to those seen in an automobile. This meter is designed to measure speed and is calibrated in miles per hour (mph). The speedometer shown has a full-scale value of 80 mph. If the pointer were positioned as shown in Figure 3–6, most people would know instantly that the speed of the automobile is 55 mph.

Figure 3–7 illustrates another common meter used by most people. This meter is used to measure the amount of fuel in the tank of an automobile. Most people can glance at the pointer of the meter and know that the meter is indicating that there is one-quarter of a tank of fuel remaining. Now assume that the tank has a capacity of 20 gallons. The meter is now indicating that there are a total of 5 gallons of fuel remaining in the tank.

Learning to read the scale of a multimeter is similar to learning to read a speedometer or fuel gauge.

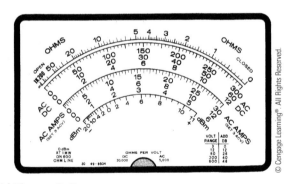

▶ **Figure 3–8**
Multimeter face.

The meter scale shown in Figure 3–8 has several scales used to measure different values and quantities. The very top of the scale is used to measure resistance or ohms. Notice that the scale begins at the left-hand side with infinity, and zero can be found at the far right-hand side. Ohmmeters are covered later in this unit. The second scale is labeled AC–DC and is used to measure voltage. Notice this scale has three different full-scale values. The top scale is 0 to 300, the second scale is 0 to 60, and the third scale is 0 to 12. The scale used is determined by the setting of the range control switch. The third set of scales is labeled AC AMPS. This scale is used with a clamp-on ammeter attachment that can be used with some meters. The last scale is labeled dbm and is seldom if ever used by the technician in the field.

Reading a Voltmeter

Notice that the three voltmeter scales use the primary numbers 3, 6, and 12 and are in multiples of 10 of these numbers. Because these numbers are in multiples of 10, it is an easy matter to multiply or divide the readings in your head by moving a decimal point. Remember that any number can be multiplied by 10 by moving the decimal point one place to the right, and any number can be divided by 10 by moving the decimal point one place to the left. For example, if the selector switch were set to permit the meter to indicate a voltage of 3 volts full-scale, the 300-volt scale would be used, and the reading divided by 100. The reading can be divided by 100 by moving the decimal point 2 places to the left. In Figure 3–9, the meter is indicating a voltage of 2.5 volts when the selector switch were set for 3 volts full-scale. The pointer is indicating a value of 250. Moving the decimal point 2 places to the left will give a reading of 2.5 volts. If the selector switch were set for a full-scale value of 30 volts, the meter shown in Figure 3–9 would be indicating a value of 25 volts. This reading is obtained by dividing the scale by 10 and moving the decimal point one place to the left.

Now assume that the meter has been set to have a full-scale value of 600 volts. The meter shown in Figure 3–10 is indicating a voltage of 440 volts. Because the full-scale value of the meter is set for 600 volts, use the 60-volt range and multiply the reading on the meter by 10. This can be done by moving the decimal point one place to the right. The pointer in Figure 3–10 indicates a value of 44. When this value is multiplied by 10, the correct voltage reading becomes 440 volts.

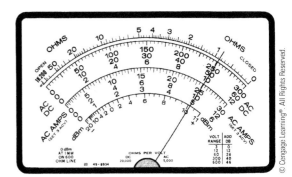

▶ **Figure 3–9**
Reading an analog multimeter.

▶ **Figure 3–10**
Reading an analog multimeter.

There are three distinct steps that should be followed when reading a meter. Following the steps is especially important for someone who has not had a great deal of experience reading a multimeter. These steps are as follows:

1. **Determine what the meter indicates.** Is the meter set to read a value of DC voltage, DC current, AC voltage, AC current, or ohms? It is impossible to read a meter if you do not know what the meter is measuring.

2. **Determine the full-scale value of the meter.** The advantage of a multimeter is that it has the ability to measure a wide range of values and quantities. After it has been determined what quantity the meter is set to measure, it must then be determined what the range of the meter is. There is a great deal of difference in readings when the meter is set to indicate a value of 600 volts full-scale and when it is set for 30 volts full-scale.

If you are not sure of the voltage value being tested, always set the meter on its highest range and then reduce the range until the indicator moves up scale to a value that can be easily read. A meter can never be damaged by placing it on a high range setting and then reducing it, but it can be damaged by placing it on a low range setting and then connecting it to a high voltage.

Another consideration is the position of the indicator. Analog meters are more accurate at the upper end of the scale than at the lower end. If possible, adjust the range setting so that the indicator is more than halfway up the scale.

3. **Read the meter.** The last step is to determine what the meter is indicating. It may be necessary to determine the value of the hatch marks on the meter face for the range the selector switch is set for. If the meter in Figure 3–8 were set for a value of 300 volts full-scale, each hatch mark would have a value of 5 volts. If the full-scale value of the meter were 60 volts, however, each hatch mark would have a value of 1 volt.

When reading a meter, always look straight at the indicator. Viewing the indicator from the side causes a parallax view, and the reading will not be accurate. It is the same effect as produced when trying to determine the speed of an automobile by looking at the speedometer from the passenger seat. The only way to accurately determine the speed is to view the speedometer straight on, not from the side. Some meters have a mirror section on the meter face. This permits the person using the meter to line up the indication with the reflection behind it to ensure an accurate reading.

AMMETER

There are several different types of ammeters used to measure electric current. Clamp-on types are generally used in the field for troubleshooting, and inline ammeters are panel mounted. The inline ammeter, unlike the voltmeter, is a very-low-impedance device. Inline ammeters must be connected in series with the load to permit the load to limit the current flow, Figure 3-11.

An ammeter has a typical impedance of less than 0.1 ohm. If this meter were connected in parallel with the power supply, the impedance of the ammeter would be the only thing to limit the amount of current flow in the circuit. Assume that an ammeter with an impedance of 0.1 ohm is connected across a 240-volt AC line. The current flow in this circuit would be 2400 amps (240/0.1 = 2400). The blinding flash of light would be followed by the destruction of the ammeter. Ammeters that are connected directly into the circuit, as shown in Figure 3–11, are referred to as **inline** ammeters. Figure 3–12 shows an inline ammeter.

Notice that the meter in Figure 3–12 has several ranges. AC ammeters use a current transformer to provide multiscale capability. The primary of the transformer is connected in series with the load, and the ammeter is connected to the secondary of

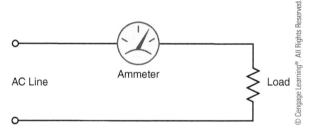

▶ **Figure 3–11**
An ammeter must be connected in series with the load.

▶ **Figure 3–12**
AC inline ammeter.

the transformer. Figure 3–13 illustrates this type of connection. Notice that the range of the meter is changed by selecting different taps on the secondary of the current transformer. The different taps on the transformer provide different turns ratios between the primary and secondary of the transformer.

When a large amount of AC current must be measured, a current transformer is connected in to the power line. The ammeter is then connected to the secondary of the transformer. The AC ammeters are designed to indicate a current of 5 amps, and the current transformer determines the value of line current that must flow to produce a current of 5 amps on the secondary of the transformer. The incoming line may be looped around the opening in the transformer several times to produce the proper turns ratio between the primary and the secondary windings. Figure 3–14 shows a transformer of this type. This type of connection is often used for panel meters mounted on large commercial units.

The type of ammeter used in the field by most air-conditioning service technicians is the **clamp-on** type of ammeter similar to the one shown in Figure 3–15. To use this meter, the jaw of the meter is clamped around one of the conductors supplying power to the load. Figure 3–16 shows this connection. Notice that the meter is clamped around only one of the power lines. When the meter is clamped around more than one line, the magnetic fields of the wires cancel each other and the meter indicates zero.

This type of meter also uses a current transformer to operate the meter. The jaw of the meter is part of the core material of the transformer. When the meter is connected around the current-carrying

▶ **Figure 3–14**
Current transformer used to meter large AC currents.

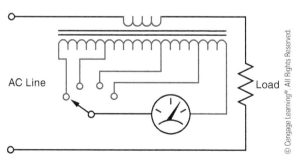

▶ **Figure 3–13**
A current transformer provides different scales.

▶ **Figure 3–15**
Multirange and clamp-on ammeter combination.

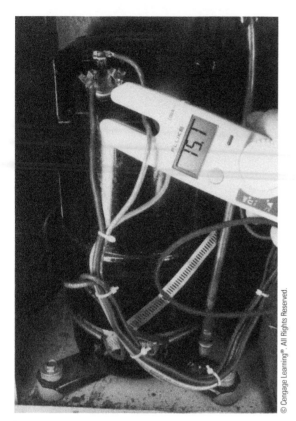

Figure 3–16
Clamp-on ammeter used to measure the running current of a compressor.

wire, the changing magnetic field produced by the AC current induces a voltage into the current transformer. The strength of the magnetic field and its frequency determine the amount of voltage induced in the current transformer. Because 60 Hz is a standard frequency throughout the country, the amount of induced voltage is proportional to the strength of the magnetic field.

The clamp-on type of ammeter can have different range settings by changing the **turns ratio** of the secondary of the transformer just as the inline ammeter does in Figure 3–13. The primary of the current transformer is the conductor the ammeter is connected around. When the ammeter is connected around one wire, as shown in Figure 3–16, the primary has one turn of wire as compared to the number of turns of wire in the secondary. When two turns of wire are wrapped around the

jaw of the ammeter, the primary winding now contains two turns instead of one, and the turns ratio of the transformer is changed, Figure 3–17. The ammeter will now indicate double the amount of current in the circuit. The reading on the scale of the meter would have to be divided by two to get the correct reading. For example, assume two turns of wire have been wrapped around the ammeter jaw, and the meter indicates a current of 3 amps. The actual current in this circuit is 1.5 amps ($3 \div 2 = 1.5$). The ability to change the turns ratio of a clamp-on ammeter can be very useful for measuring low currents such as those found in a control circuit. Changing the turns ratio of the transformer is not limited to wrapping two turns of wire around the jaw of the ammeter. Any number of turns of wire can be wrapped around the jaw of the ammeter, and the reading is then divided by that number. If three turns of wire were wrapped around the jaw of the meter, the reading would be divided by three.

A very handy device can be made by wrapping 10 turns of wire around some core of nonmagnetic material, such as a thin piece of plastic pipe. Because the device is intended to be used for low current readings, the wire size does not have to be large. A 20 or 18 American Wire Gauge (AWG) wire is large enough. Plastic tape is used to secure the turns of wire to the core material, and two alligator clips are connected to the ends of the wire. This device is shown in Figure 3–18. To use this device, break connection in the circuit to be tested, and insert the 10 turns of wire, using the alligator

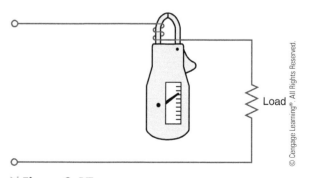

Figure 3–17
Two turns of wire change the turns ratio of the transformer.

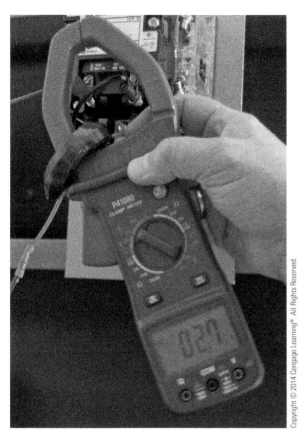

▶ **Figure 3–18**
Scale divider used with clamp-on ammeter. The
ammeter indicates a value of 2.7 amperes. If the scale
divider had a 10:1 ratio, the actual circuit current
would be 0.27 ampere.

▶ **Figure 3–19**
Digital clamp-on ammeter.

clips. The jaw of the ammeter is placed around the
plastic core. The primary of the transformer now
contains 10 turns of wire, and the scale factor
can now be divided by a factor of 10. The correct
ammeter reading is found by moving the decimal
point one place to the left. If the ammeter had a
low scale of 6 amps full-scale, it could now be used
to measure 0.6 amp full-scale. This can be a real
advantage when it is necessary to measure control
currents that may not be greater than 0.2 amp
under normal operating conditions.

Some clamp-on ammeters use a digital read-out
instead of a meter movement. A digital-type meter
is shown in Figure 3–19. The digital ammeters are

generally better for measuring low current values,
but the 10-turn scale divider can be used with these
ammeters also. Just remember to divide the reading
shown by a factor of 10. The clamp-on ammeters
discussed in this unit are intended to be used for
measuring AC currents only and will not operate
if connected to a DC line. There are clamp-on type
ammeters, however, that can be used to measure
DC current.

Many clamp-on ammeters are designed to measure AC volts and ohms as well as AC current. This
makes the meter a more versatile instrument. When
voltage or resistance is to be measured, a set of leads
is attached to the meter.

OHMMETER

The ohmmeter is used to measure resistance. The common **volt-ohm-milliammeter (VOM)** contains an ohmmeter. The ohmmeter must provide its own power supply to measure resistance. This is done with batteries located inside the instrument. When resistance is to be measured, the meter must first be zeroed. Zeroing is done with the ohms adjust control located on the front of the meter. To zero the meter, connect the leads together and adjust the ohms adjust knob until the meter indicates 0 at the far right end of the scale, Figure 3–20. When the leads are separated, the meter will again indicate infinity resistance at the far left side of the meter scale. When the leads are connected across a resistance, the meter will again indicate up scale. Figure 3–21 shows a meter indicating a resistance of 25 ohms, assuming the range setting is R × 10.

Ohmmeters can have different range settings such as R × 1, R × 100, R × 1000, or R × 10,000. On the R × 1 setting, the resistance is measured straight off the resistance scale located at the top of the meter. If the range were set for R × 1000, however, the reading would have to be multiplied by 1000. The ohmmeter reading shown in Figure 3–21 would be indicating a resistance of 2500 ohms if the range had been set for R × 1000. Notice that the ohmmeter scale is read backward from the other scales. Zero ohms is located on the far right side of the scale, and maximum ohms is located at the far left side. It generally takes a little time and practice to read the ohmmeter properly.

Digital ohmmeters display the resistance in figures instead of using a meter movement. When using a digital ohmmeter, care must be taken to notice the scale indication on the meter. For example, most digital meters display a K on the scale to indicate kilohms or an M to indicate megohms. (*Kilo* means 1000, and *mega* means 1,000,000.) When the meter is showing a resistance of (0.200 K), it means 0.2 × 1000, or 200 ohms. If the meter indicates (1.65 M), it means 1.65 × 1,000,000, or 1,650,000 ohms.

The ohmmeter must never be connected to a circuit with power on. Because the ohmmeter uses its own internal power supply, it has a very low operating voltage. If a meter is connected to power when it is set in the ohms position, it will probably damage or destroy the meter.

▶ **Figure 3–20**
The ohmmeter must be set at zero.

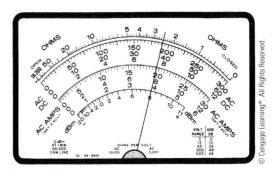

▶ **Figure 3–21**
Read the ohmmeter from right to left.

▶ SUMMARY

- ⊙ A voltmeter is a high-resistance device and is designed to be connected directly to the power line.
- ⊙ The resistance of an analog voltmeter, a voltmeter with a scale and a moving pointer, changes with the setting of the meter.

⊙ An analog voltmeter typically has a resistance of 5000 ohms per volt on the AC setting and 20,000 ohms per volt on the DC setting.

⊙ Digital voltmeters generally have a resistance of 10 million ohms on all range settings.

⊙ These are the three basic steps to reading an analog meter:

A. Determine what the meter indicates.

B. Determine the full-scale value of the meter.

C. Read the meter.

⊙ Ammeters are low-resistance devices and must be connected in series with the load.

⊙ AC ammeters generally use a current transformer to change scale values.

⊙ Current transformers designed to measure large amounts of current have a standard output current of 5 amps.

⊙ Most clamp-on ammeters can measure values of AC current only, because they depend on magnetic induction to operate.

⊙ The range setting of clamp-on ammeters can be changed by wrapping more turns of wire around the meter jaw.

⊙ Ohmmeters are used to measure the resistance of a circuit.

⊙ Ohmmeters should never be connected to a circuit that has power applied to it.

⊙ Ohmmeters must have their own source of power to operate.

KEY TERMS

ammeter	inline	turns ratio
analog meter	input impedance	voltmeter
clamp-on	multiranged	volt-ohm-milliammeter (VOM)
digital meter	ohmmeter	

REVIEW QUESTIONS

1. What type of meter has a high resistance connected in series with the meter movement?
2. How is a voltmeter connected into the circuit?
3. If a voltmeter has a resistance of 5000 ohms per volt, what is the resistance of the meter when it is set on the 300-volt range?
4. What is the advantage of using a voltmeter that has a high impedance as opposed to a low-impedance meter?
5. What is an analog meter?
6. Why must an ammeter be connected in series with the load?
7. What device is used to change the scale values of an AC ammeter?
8. What is meant by the term *inline* ammeter?

9. A clamp-on ammeter has three turns of wire wrapped around the movable jaw. If the meter is indicating a current of 15 amps, how much current is actually flowing in the circuit?

10. List the three steps for reading a meter.

11. What type of meter contains its own internal power supply?

12. What precaution must be taken when using an ohmmeter?

UNIT 4 ▷ Electrical Circuits

OBJECTIVES

After studying this unit the student should be able to:

▷ Define a series circuit
▷ Find Ohm's law values for series circuits
▷ Define a parallel circuit
▷ Find Ohm's law values for parallel circuits
▷ Discuss combination circuits
▷ Find Ohm's law values for combination circuits

Electrical circuits can be divided into three basic types. These are **series**, **parallel**, and **combination**. The simplest of these circuits is the series circuit, shown in Figure 4–1. A series circuit is characterized by the fact that it has only one path for current flow. If it is assumed that current must flow from point A to point B in the circuit, as shown in Figure 4–1; it will flow through each of the resistors. Therefore, *the current flow in a series circuit must be the same at any point in the circuit.* Another rule of series circuits states that *the sum of the voltage drops around the circuit must equal the applied voltage.* A third rule of series circuits states that *the total resistance is equal to the sum of the individual resistors.*

The circuit shown in Figure 4–2 shows the values of **current flow**, **voltage drop**, and resistance for each of the resistors. Notice that the total resistance of the circuit can be found by adding the

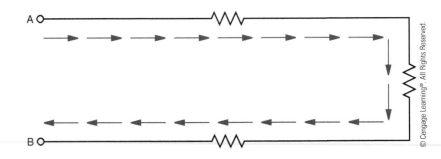

▶ **Figure 4–1**
A series circuit has only
one path for current flow.

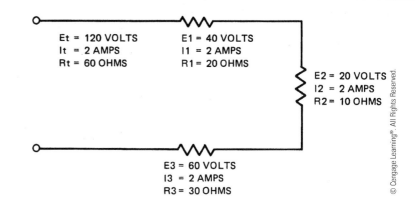

▶ **Figure 4–2**
Voltage, current and resistance
values for a series circuit.

values of each of the individual resistors (20 + 10 + 30 = 60 ohms). The amount of current flow in the circuit can be found by using Ohm's law.

$$I = \frac{E}{R}$$

$$I = \frac{120}{60}$$

$$I = 2 \text{ amps}$$

There is a current flow in the circuit of 2 amps. Notice that the same current flows through each of the resistors. The voltage drop across each resistor can be found using Ohm's law ($E = I \times R$). The voltage dropped across resistor R1 is 2 × 20 = 40 volts. This means that it takes a voltage of 40 volts to push 2 amps of current through 20 ohms of resistance. If a voltmeter is connected across resistor R1, it indicates a voltage drop of 40 volts. The voltage drop of resistor R2 can be found the same way ($E = I \times R$), (2 × 10 = 20). The voltage dropped across resistor R2 is 20 volts. The third resistor has a voltage drop of 2 × 30 = 60 volts. Notice that if the voltage drops are added together,

they will equal the voltage applied to the circuit (40 + 20 + 60 = 120).

Because a series circuit has only one path for current flow, if any point in the circuit should become open, current flow throughout the entire circuit will stop. Some strings of Christmas tree lights are wired in series. If any bulb in the string burns out, all of the lights will go out. When the defective bulb is replaced, all of the lights will operate. Because of this characteristic of series circuits, fuses and circuit breakers are connected in series with what they are intended to protect. Figure 4–3 shows a fuse used to protect an air-conditioning unit. If the fuse should open, current flow to the entire circuit will stop.

PARALLEL CIRCUITS

Parallel circuits are characterized by the fact that they have more than one path for current flow. The circuit shown in Figure 4–4 illustrates multiple paths of a parallel circuit. If the current in this

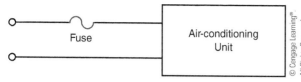

Figure 4–3
Fuses are connected in series.

circuit is assumed to flow from point A to point B, there are three separate paths through which it can flow. Current can flow from point A through resistor R1 to point B, or it can flow from point A through resistor R2 to point B, or from point A through resistor R3 to point B. Because current can flow through each of these resistors, the total current flow in the circuit is the sum of these individual currents. A rule for parallel circuits states that *the total current in a parallel circuit is the sum of the currents through the individual paths* ($It = I1 + I2 + I3$). Notice in Figure 4–4 that each of the resistors is connected directly across the incoming power line. Therefore, *all the components in a parallel circuit have the same voltage drop.*

Each time a new component is added to a parallel circuit, a new path for current flow is created. Because there is less opposition to current flow each time a component is added, the total resistance of the circuit is decreased. The total resistance of a parallel circuit can be found using either of three formulas. The first of these formulas is

$$Rt = \frac{R1 \times R2}{R1 + R2}$$

The second formula is

$$\frac{1}{Rt} = \frac{1}{R1} + \frac{1}{R2} + \frac{1}{R3} + \frac{1}{RN}$$

or

$$\frac{1}{Rt} = \frac{1}{R1} + \frac{1}{R2} + \frac{1}{R3} + \frac{1}{RN}$$

The third formula is

$$Rt = \frac{R}{N}$$

The third formula can be used only when all resistors connected in parallel are the same value. Assume that four resistors with a value of 100 ohms each are connected in parallel. The total resistance would be 25 ohms. To use this formula, divide the resistance of one resistor by the total number of resistors.

$$Rt = \frac{R}{N}$$
$$Rt = \frac{100}{4}$$
$$Rt = 25\ \Omega$$

Figure 4–5 shows a parallel circuit containing three resistors with the values of 15, 10, and 30 ohms. The total resistance of the circuit can be found by using either of the two formulas.

$$Rt = \frac{15 \times 10}{15 + 10} = \frac{150}{25} = 6$$

Notice that in this formula only two of the resistors can be found at a time. It is now necessary to use the total resistance of the first two resistors and use that value for R1 in the formula. Resistor R3 is used in the R2 position in the formula.

$$Rt = \frac{6 \times 30}{6 + 30}$$
$$Rt = \frac{180}{36}$$
$$Rt = 5\ ohms$$

The total resistance of this parallel circuit is 5 ohms. The second formula can be used to find the total

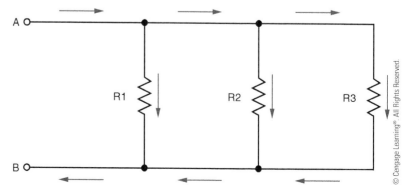

Figure 4–4
Parallel circuits have more than one path for current flow.

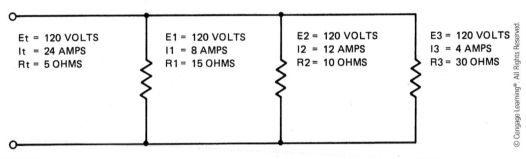

▶ **Figure 4–5**
Three parallel resistors.

resistance by plugging the values of resistance into the formula. With these values, it becomes a matter of adding fractions. When fractions are to be added, the first thing that must be done is to find some number that all the denominators will divide into. This is called finding a **common denominator**. For this problem, 30 is the common denominator.

$$\frac{1}{Rt} = \frac{1}{15} + \frac{1}{10} + \frac{1}{30}$$

$$\frac{1}{Rt} = \frac{2}{30} + \frac{3}{30} + \frac{1}{30}$$

$$\frac{1}{Rt} = \frac{6}{30}$$

$$\frac{Rt}{1} = \frac{30}{6}$$

$$Rt = 5 \text{ ohms}$$

Notice that the two formulas give the same answer.

Another method of finding the total resistance in a parallel circuit is to find the **reciprocal** of each individual resistor. A third rule for parallel circuits states that *total resistance is the reciprocal of the sum of the reciprocals of the individual resistors*. The problem can therefore be solved by finding the reciprocal of each individual resistor, adding them together, and finding the reciprocal of the sum. (The reciprocal of any number can be found by dividing that number into 1.) The problem can be solved as follows:

$$\frac{1}{Rt} = \frac{1}{R1} + \frac{1}{R2} + \frac{1}{R3}$$

$$\frac{1}{Rt} = \frac{1}{15} + \frac{1}{10} + \frac{1}{30}$$

$$\frac{1}{Rt} = 0.0667 + 0.1 + 0.0333$$

$$\frac{1}{Rt} = 0.02$$

$$Rt = 5 \text{ ohms}$$

When 120 volts is applied to the circuit, the values of voltage and current for the entire circuit can be found. Because each of the resistors is connected directly across the power line, each resistor will have the same voltage drop of 120 volts. The current flow through resistor R1 can be found using Ohm's law.

$$I = \frac{E}{R}$$

$$I = \frac{120}{15}$$

$$I = 8 \text{ amps}$$

The current flow through resistor R2 is (120/10 = 12 amps). The current flow through resistor R3 is (120/30 = 4 amps). The total current flow in the circuit can be found by using the formula

$$It = \frac{Et}{Rt}$$

$$It = 120/5$$

$$It = 24 \text{ amps}$$

Notice that the total current can also be found by adding the currents flowing through the individual resistors (8 + 12 + 4 = 24 amps).

Most circuits are connected in parallel. The lights and outlets in a house are connected in parallel. Because all of the lights and outlets are connected in parallel, each light has an applied voltage of 120 volts, and each outlet can supply 120 volts to whatever is connected to it. If the lights in a house were wired in series, all of the lights would have to be turned on before any of them would burn.

COMBINATION CIRCUITS

A combination circuit contains both series and parallel connections within the same circuit. In

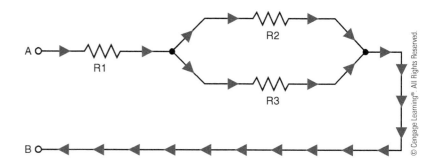

▶ **Figure 4–6**
Combination circuit.

Figure 4–6, resistor R1 is connected in series with resistors R2 and R3. Resistors R2 and R3 are connected in parallel with each other. If it is assumed that current flows from point A to point B, all of the current will have to flow through resistor R1. At the junction of resistors R2 and R3, however, the current divides and flows through separate paths. The amount of current that flows through each resistor is determined by its resistance. Notice that all of the circuit current must flow through R1. Because there is only one current path through R1, it is connected in series with the rest of the circuit. When the current reaches the junction of resistors R2 and R3, there is more than one path for current flow. These resistors are, therefore, connected in parallel.

Values will now be added to the example circuit shown in Figure 4–6. It will be assumed that resistor R1 has a value of 75 ohms; resistor R2 has a value of 100 ohms; and resistor R3 has a value of 125 ohms. It will also be assumed that a voltage of 24 volts is connected to the circuit, Figure 4–7. To find the missing values in this circuit, it is first necessary to determine the total resistance. The first step in determining

the total resistance is to calculate the resistance of the parallel block formed by resistors R2 and R3. This value will become RC (resistance of the combination).

$$Rc = \frac{1}{\frac{1}{R2} + \frac{1}{R3}}$$

$$Rc = \frac{1}{\frac{1}{100} + \frac{1}{125}}$$

$$Rc = \frac{1}{0.01} \times 0.008$$

$$Rc = \frac{1}{0.018}$$

$$Rc = 55.556 \ \Omega$$

Now that the combined resistance of resistors R2 and R3 is known, this value can be treated as one single resistor. The circuit will now be redrawn as shown in Figure 4–8. The circuit is now a simple series circuit containing resistors R1 and RC.

The total circuit resistance of the circuit can be determined by adding the values of resistors R1 and RC.

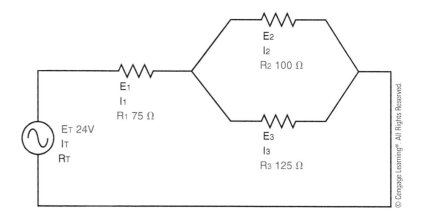

▶ **Figure 4–7**
Adding circuit values.

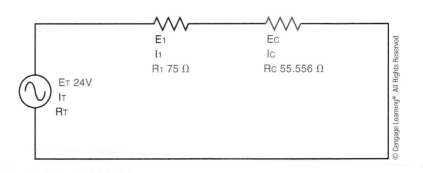

▶ **Figure 4–8**
Resistors R2 and R3 become RC.

$Rt = R1 + RC$

$Rt = 75 + 55.556$

$Rt = 130.556 \ \Omega$

Because the applied or total circuit voltage is known and the total resistance is known, the total circuit current can be determined using Ohm's law.

$It = \dfrac{Et}{Rt}$

$It = \dfrac{24}{130.556}$

$It = 0.184 \ amp$

In a series circuit, the current is the same at any point in the circuit. The voltage drop across resistors R1 and RC can now be determined.

$E1 = I1 \times R1$

$E1 = 0.184 \times 75$

$E1 = 13.8 \ volts$

$EC = IC \times RC$

$EC = 0.184 \times 55.556$

$EC = 10.2 \ volts$

The computed circuit values are shown in Figure 4–9.

Resistor RC in reality is the combined values of resistors R2 and R3. The values that apply to resistor RC, therefore, apply to the parallel block formed by resistors R2 and R3. In a parallel circuit, the voltage is the same across each branch. Therefore, the voltage dropped across resistor RC is dropped across both R2 and R3. Now that the voltage drop across each is known, the current flow through each can be determined using Ohm's law.

$I2 = \dfrac{E2}{R2}$

$I2 = \dfrac{10.2}{100}$

$I2 = 0.102 \ amp$

$I2 = \dfrac{E3}{R3}$

$I2 = \dfrac{10.2}{125}$

$I2 = 0.0816 \ amp$

The circuit with all calculated values is shown in Figure 4–10.

Another example of a combination circuit is shown in Figure 4–11. In this circuit, resistors R1 and R2 are connected in series with each other and resistors R3 and R4 are connected in series with each other. Resistors R1 and R2 are connected in

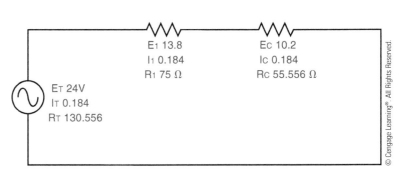

▶ **Figure 4–9**
Circuit values for the series circuit are determined.

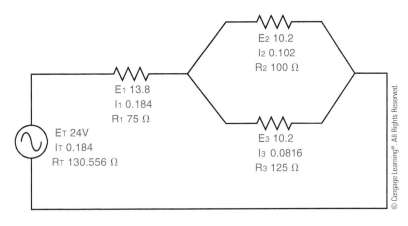

▶ **Figure 4–10**
Circuit with all computed values.

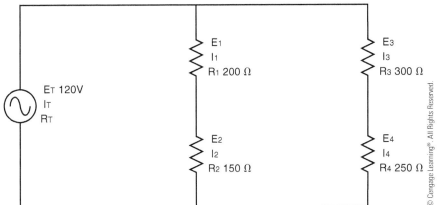

▶ **Figure 4–11**
Example combination
circuit 2.

parallel with resistors R3 and R4. It is assumed that resistor R1 has a value of 200 ohms, resistor R2 has a value of 150 ohms, resistor R3 has a value of 300 ohms, and resistor R4 has a value of 250 ohms. It is also assumed that 120 volts is applied to the circuit.

The first step in solving this circuit is to compute the total circuit resistance. This can be done by computing the total resistance of the branch containing resistors R1 and R2 and the branch containing resistors R3 and R4. Resistors R1 and R2 are connected in series. The total resistance of this branch can be computed by adding the values of resistors R1 and R2. This value will be RC1 (resistance of combination 1).

$$RC1 = R1 + R2$$
$$RC1 = 200 + 150$$
$$RC1 = 350 \ \Omega$$

The resistance of the second branch can be computed by adding the values of R3 and R4. This resistance value will be RC2 (resistance of combination 2).

$$RC2 = R3 + R4$$
$$RC2 = 300 + 250$$
$$RC2 = 550 \ \Omega$$

The circuit can now be reduced to a simple parallel circuit containing a 300-ohm and a 550-ohm resistor, Figure 4–12. The total resistance can now be computed using one of the parallel resistance formulas.

$$Rt = \frac{Rc1 \times Rc2}{Rc1 + Rc2}$$
$$Rt = \frac{350 \times 550}{350 + 550}$$
$$Rt = \frac{192.500}{900}$$
$$Rt = 213.889 \ \Omega$$

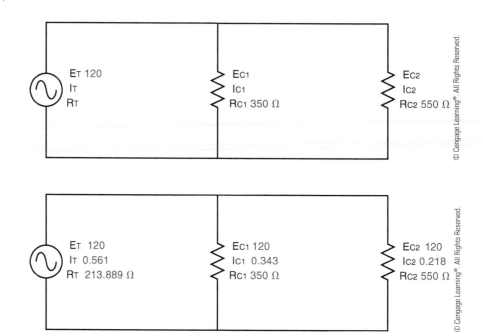

Figure 4–12
Reducing the circuit to a
simple parallel circuit.

Figure 4–13
The values for the
parallel circuit have been
determined.

The total current can now be computed using Ohm's law.

$$I_t = \frac{E_t}{R_t}$$

$$I_t = \frac{120}{213.889}$$

$$I_t = 0.561 \text{ amp}$$

Because resistors RC1 and RC2 are connected in parallel, the voltage across each is the same as the applied voltage of the circuit. The current through each branch can now be determined.

$$IC1 = \frac{120}{350}$$

$$IC1 = 0.343 \text{ amp}$$

$$IC2 = \frac{120}{350}$$

$$IC2 = 0.218 \text{ amp}$$

The complete values for the parallel circuit are shown in Figure 4–13.

The values of RC1 and RC2 can now be substituted in the original circuit. Resistor RC1 is a combination of resistors R1 and R2. The values of RC1 apply to the branch composed of R1 and R2. Resistors R1 and R2 are connected in series. In a series circuit, the current flow is the same

through any part of the circuit. Therefore, the current flowing through RC1 is the same current flowing through R1 and R2. Likewise, the current flowing through RC2 is the same current that flows through resistors R3 and R4, Figure 4–14. Now that the value of resistance and current are known for each of the resistors, the voltage drop across each can be computed.

$$E1 = I1 \times R1$$

$$E1 = 0.343 \times 200$$

$$E1 = 68.6 \text{ V}$$

$$E2 = I2 \times R2$$

$$E2 = 0.343 \times 150$$

$$E2 = 51.45 \text{ V}$$

$$E3 = I3 \times R3$$

$$E3 = 0.218 \times 300$$

$$E3 = 65.4 \text{ V}$$

$$E4 = I4 \times R4$$

$$E4 = 0.218 \times 250$$

$$E4 = 54.5 \text{ V}$$

All values for the circuit are shown in Figure 4–15.

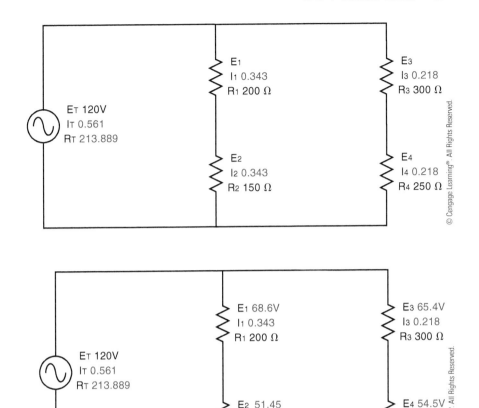

Figure 4–14
Determining the current
flow through each branch.

Figure 4–15
All missing values have
been determined.

SUMMARY

⊙ Three basic types of electric circuits are the series, parallel, and combination.

⊙ Series circuits contain only one path for current flow.

⊙ Three rules concerning the electrical values in a series circuit are these:

A. The current flow is the same in all parts of a series circuit.

B. The sum of the voltage drops across each element is equal to the applied voltage.

C. The total resistance can be determined by adding the resistance of each element in the circuit.

⊙ Parallel circuits contain more than one path for current flow.

⊙ There are three rules concerning the electrical values in a parallel circuit:

A. The voltage is the same across all branches in a parallel circuit.

B. The total current can be found by adding the current flow through each branch of the circuit.

C. The total resistance can be found by adding the reciprocal of the resistance of each branch and then taking the reciprocal of that sum.

⊙ Combination circuits contain both series and parallel branches.

KEY TERMS

combination
common denominator
current flow

parallel
reciprocal

series
voltage drop

REVIEW QUESTIONS

1. List the three basic types of electrical circuits.

2. What is the major characteristic of a series circuit?

3. List the three basic rules for series circuits.

4. What is the major characteristic of a parallel circuit?

5. List the three basic rules for the parallel circuit.

6. What type of circuit is used most often in industry and the home?

7. What type of circuit is used the least in industry and the home?

8. Three resistors valued at 300 ohms, 200 ohms, and 600 ohms are connected in series. What is their total resistance?

9. If the three resistors in question 8 were connected in parallel, what would be their total resistance?

10. How are fuses and circuit breakers connected in a circuit, and why are they connected this way?

PRACTICE PROBLEMS SET 1

1. Refer to the circuit shown in Figure 4–2. Find the missing values.

ET = 240	E1	E2	E3
IT	I1	I2	I3
RT	R1 = 1200 Ω	R2 = 2000 Ω	R3 = 1800 Ω

2.

ET = 48	E1	E2	E3 = 22.795
IT	I1	I2	I3
RT = 990 Ω	R1 = 220 Ω	R2 = 300 Ω	R3

3.

ET = 120	E1 = 24.75	E2	E3 = 59.4
IT	I1	I2	I3
RT	R1	R2 = 2.2 kΩ	R3

4. Refer to the circuit shown in Figure 4–5. Find the missing values.

ET	E1	E2	E3
IT = 26	I1	I2	I3
RT	R1 = 24 Ω	R2 = 18 Ω	R3 = 36 Ω

5.

ET	E1	E2	E3
IT	I1	I2	I3 = 1.5
RT = 8 Ω	R1 = 24 Ω	R2 = 48 Ω	R3

6.

ET	E1	E2	E3
IT = 0.289	I1 = 0.139	I2 = 0.1	I3
RT	R1 = 860 Ω	R2 = 1200 Ω	R3 = 2400 Ω

7. Refer to the circuit shown in Figure 4–7. Find all missing values.

ET = 120	E1	E2	E3
IT	I1	I2	I3
RT	R1 = 360 Ω	R2 = 1200 Ω	R3 = 1600 Ω

8.

ET	E1	E2	E3
IT	I1 = 0.080	I2	I3
RT = 2600 Ω	R1	R2 = 2400 Ω	R3 = 4800 Ω

9.

ET	E1	E2	E3 = 114.18
IT = 0.466	I1	I2	I3
RT	R1 = 270 Ω	R2 = 510 Ω	R3 = 470 Ω

10. Refer to the circuit shown in Figure 4–11. Find all missing values.

ET = 277	E1	E2	E3	E4
IT	I1	I2	I3	I4
RT	R1 = 3300 Ω	R2 = 4300 Ω	R3 = 2700 Ω	R4 = 5100 Ω

11.

ET	E1	E2	E3	E4
IT	I1 = 0.06	I2	I3	I4
RT = 960 Ω	R1	R2 = 1000 Ω	R3 = 650 Ω	R4 = 950 Ω

12.

ET	E1	E2	E3	E4
IT	I1 = 0.0444	I2	I3 = 0.0369	I4
RT	R1 = 3000 Ω	R2 = 2400 Ω	R3 = 1800 Ω	R4 = 4700 Ω

▷ PRACTICE PROBLEMS SET 2

Ohm's Law

1. E =
  I = 10 A
  R = 12 Ω

2. I =
  E = 220 V
  R = 10 Ω

3. R =
  E = 120 V
  I = 3 A

4. E =
  I = 1 A
  R = 240 Ω

5.

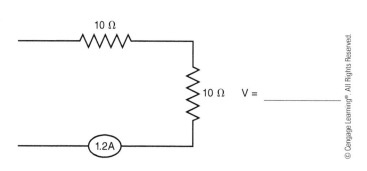

10 Ω

10 Ω   V = _____

1.2A

6.

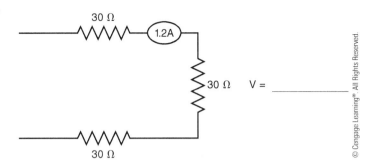

30 Ω

1.2A

30 Ω   V = _____

30 Ω

PRACTICE PROBLEMS SET 3

Ohm's Law

1.

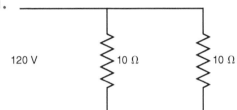

120 V   10 Ω   10 Ω I_T = _____

2.

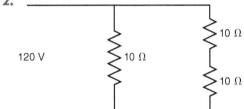

120 V   10 Ω   10 Ω

10 Ω I_T = _____

3.

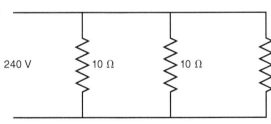

240 V   10 Ω   10 Ω   5 Ω I_T = _____

4.

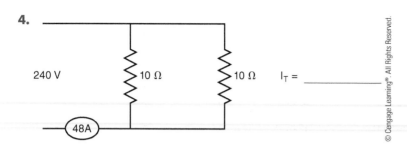

PRACTICE PROBLEMS SET 4

Ohm's Law

1. E =
   I = 1 A
   R = 12 Ω

2. I =
   E = 220 V
   R = 5 Ω

3. R =
   E = 120 V
   I = 5 A

4. E =
   I = 0.5 A
   R = 240 Ω

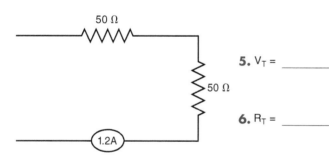

5. V_T = _____

6. R_T = _____

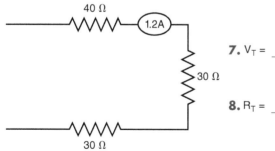

7. V_T = _____

8. R_T = _____

PRACTICE PROBLEMS SET 5

1.

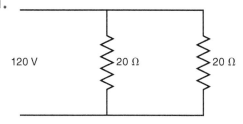

120 V   20 Ω   20 Ω $I_T =$ _____

2.

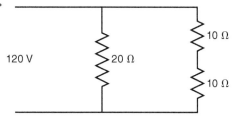

120 V   20 Ω   10 Ω   10 Ω $I_T =$ _____

3.

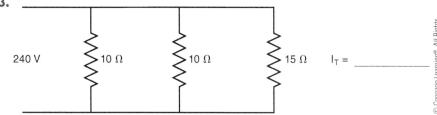

240 V   10 Ω   10 Ω   15 Ω $I_T =$ _____

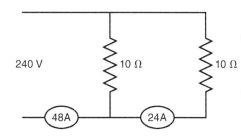

240 V   10 Ω   10 Ω   48A   24A

4. $I_T =$ _____

5. $R_T =$ _____

UNIT 5 ▷

Electrical Services

Air-conditioning equipment must be connected to an electrical service. The type of air-conditioning equipment used is generally determined by the type of electrical service available to operate it. The air-conditioning technician must have knowledge of different types of electrical services.

POWER GENERATION

In the United States and Canada, power is generated as a three-phase 60-hertz voltage. The term **hertz** means 60 cycles per second. This means that the voltage increases from zero to its maximum positive value, returns to zero, increases to its maximum negative value and returns to zero 60 times each second. Figure 5–1 shows one complete cycle of AC voltage.

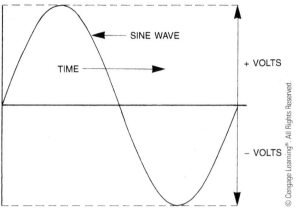

▶ **Figure 5–1**
AC sine wave.

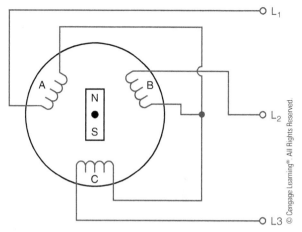

▶ **Figure 5–3**
Wye or star connection.

WYE CONNECTION

The wye connection is also referred to as the **star** connection. This connection is made by joining one end of each of the windings together as shown in Figure 5–3. The connection shown in Figure 5–4 is a wye connection that has been drawn schematically to make it easier to see and understand. Notice how one end of each of the windings is joined at the center point. The wye connection can be used to provide an increase in the output or line voltage. The phase voltage is the voltage produced across one of the windings. The line voltage is the voltage produced across the output points of the connection. Figure 5–5 shows a wye connection connected to a three-phase load bank. Ammeters and voltmeters are used to illustrate the differences between phase values and line values. Notice that the phase value of voltage is measured from the output of the winding, at point C, to the center point of the

The term **three phase** means that there are three separate voltage waveforms produced by the **alternator**. An alternator is a generator that produces AC voltage. For the alternator to produce the three phases, the internal windings of the alternator—called the **stator**—are wound 120° apart. Figure 5–2 illustrates the windings of an alternator. The moving part of the alternator, called the **rotor**, is actually a large electromagnet. When the magnet is turned, the magnetic field cuts through the windings of the stator and induces a voltage into them. The amount of voltage induced is controlled by the strength of the magnetic field, and the frequency or hertz is controlled by the speed of the rotation of the magnet. Because the windings of the stator are physically wound 120° apart, the three voltages are 120° out of phase with each other. The windings of the stator are connected to form one of the two basic three-phase connections. These connections are the **delta** and **wye**.

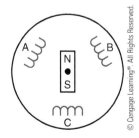

▶ **Figure 5–2**
The windings of an alternator are 120° apart.

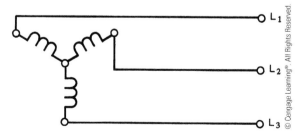

▶ **Figure 5–4**
Schematic of a wye connection.

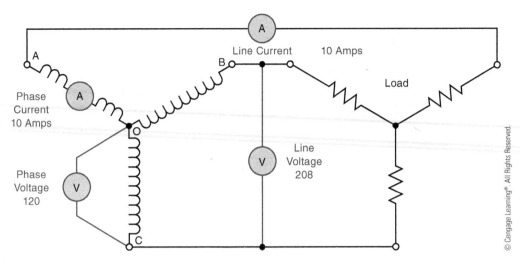

▶ Figure 5–5
Phase and line values of a wye connection.

wye connection at the point labeled O. The line value is measured across two of the output points of the connection (B&C). The phase current meter is inserted in the winding of the alternator, and the line current meter is inserted in the output line. Notice also that the two ammeters indicate the same value of current. *In a wye connection, phase current and line current are equal.* The voltages, however, are not. *The line voltage in a wye connection is 1.732 times greater than the phase voltage* (1.732 is the square root of 3). The reason for this voltage increase is that the voltages are 120° out of phase with each other. Figure 5–6 shows a diagram to illustrate this. Because the three voltages are out of phase with each other, they will be added. Vector addition must be used, however, because of the 120° phase shift. If three voltages are shown in a length that corresponds to 120 volts, and a resultant is drawn to the point of intersection, it will be found that the length of the resultant corresponds to 208 volts. The 120° phase shift between voltages is the reason the two 120-volt phases add to produce 208 volts instead of 240 volts.

Wye-connected systems often use a fourth conductor connected to the center of the connection. This conductor becomes the neutral, Figure 5–7. Notice in this connection that the voltage between any line and neutral is the phase voltage or 120 volts,

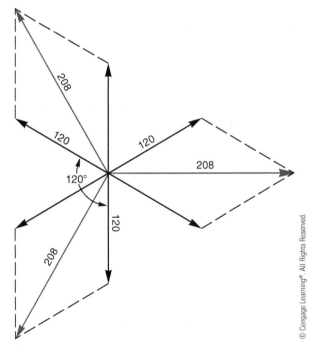

▶ Figure 5–6
Vector diagram of the phase and line values in a three-phase system.

and the voltage between any two of the lines is 208 volts. The 208/120-volt three-phase connection is very common in industry and commercial buildings. Another very common three-phase

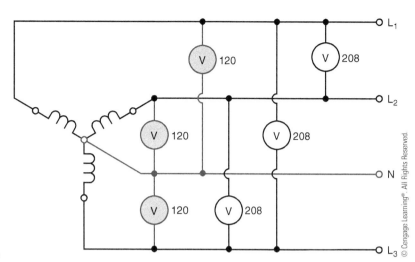

Figure 5–7
Fourth wire connected for a neutral.

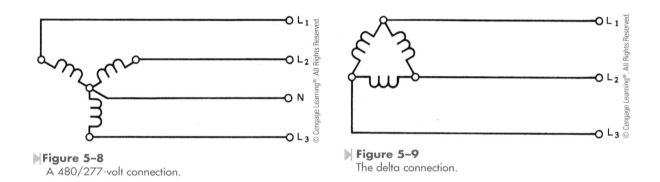

Figure 5–8
A 480/277-volt connection.

Figure 5–9
The delta connection.

four-wire connection is shown in Figure 5–8. This is a 480/277-volt connection. Two hundred seventy-seven volts is often used in large stores and office buildings to operate the fluorescent lights, whereas the 480-volt connection is used to operate large air-conditioning systems. The 120-volt connections are provided by transformers that step down the 480 volts to 120 volts.

DELTA CONNECTION

The next connection to be covered is the delta. A schematic diagram of a delta connection is shown in Figure 5–9. This connection gets its name from the fact that it looks like the Greek letter delta (Δ).

Figure 5–10 shows a delta system connected to a three-phase load bank. Ammeters and voltmeters are used to illustrate the differences in phase and line values of voltage and current. Notice that the values of phase voltage and line voltage are equal for the delta connection. One of the rules for three-phase systems states that *line voltage and phase voltage are equal in a delta connection*. The ammeters, however, are not equal. *In a delta connection, the line current is 1.732 times greater than the phase current*. This is the reason that the delta connection is so popular in industry. The current flow through the windings of a transformer are less than the line amps if the transformer bank is connected in delta.

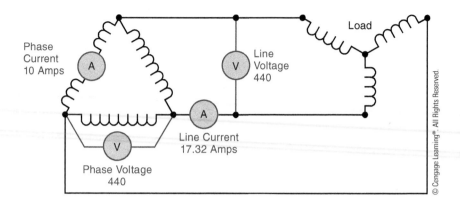

▶ **Figure 5–10**
Voltage and current relationships in a delta connection.

HIGH-LEG SYSTEM

Figure 5–11 illustrates another common type of transformer connection. This is a 240/120-volt system with a **high leg**. Three transformers are connected to form a delta connection. One of the transformers is larger than the other two, however, and is center tapped. The large transformer must be able to supply power for both three-phase and single-phase loads. The other two transformers supply power for the three-phase loads only. If the phase voltage of the transformers is 240 volts, the voltage between any two of the three lines is 240 volts. If the center-tap connection is used as a neutral conductor, however, the voltages between L2 and neutral, and L3 and neutral will be 120 volts. Therefore, L2,

L3, and neutral are used to supply 240/120 volts for single-phase loads. Care must be taken not to connect a 120-volt device across L1 and neutral. Line L1 is known as a high leg, and a voltage of about 208 volts exists between these two points.

OPEN-DELTA SYSTEM

Another type of three-phase service is known as the open delta. The **open-delta system** has the advantage of needing only two transformers to provide three-phase voltage. This connection is often used when the amount of three-phase power needed is low or when the power needs are expected to increase in the future. The open-delta connection, however, does have some disadvantages. The total output power is

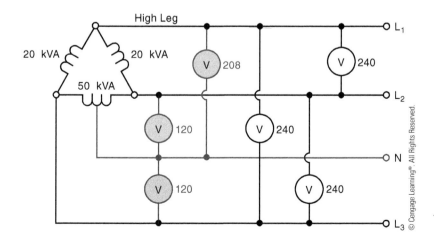

▶ **Figure 5–11**
High-leg system.

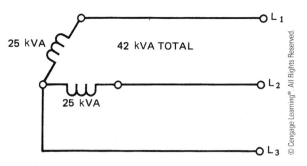

▶ **Figure 5-12**
Open-delta system.

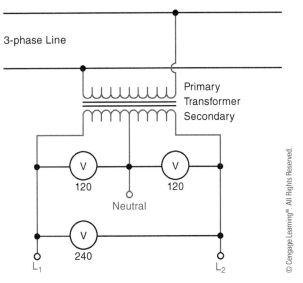

▶ **Figure 5-13**
Single-phase transformer.

only 84% of the combined rating of the transformers. If the two transformers shown in Figure 5–12 each have a power rating of 25 kVA (kilovolt amps), the total delivered power of this connection is only 42 kVA (25 + 25 = 50) (50 × 84% = 42). If at a later date the power requirements increase, a third transformer can be added to close the delta. The total output power of this connection is the combined rating of all three transformers. In this case, it is 75 kVA (25 + 25 + 25 = 75).

SINGLE-PHASE SERVICE

A **single-phase** 240/120-volt system can be obtained by connecting a single transformer to two lines of a three-phase system. The primary of the transformer shown in Figure 5–13 is connected to two of the three-phase lines of the power company.

The secondary voltage of the transformer is 240 volts. The secondary winding of the transformer is center tapped. The center tap is grounded and becomes the **neutral** conductor. If the voltage across the entire secondary is measured, it will be 240 volts. If the voltage between either of the secondary leads is measured to the center tap, it will be 120 volts. The reason this is true for single-phase and not three-phase transformers is that the voltages of the single-phase system are in phase with each other, Figure 5–14. Because the transformer center

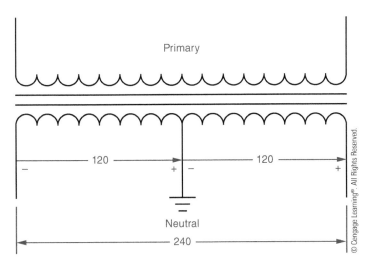

▶ **Figure 5-14**
The voltages across the winding of the secondary of the single-phase transformer are in phase with each other.

tap is the neutral conductor, it will be 120 volts more positive than one side of the secondary winding and 120 volts more negative than the other side of the secondary winding at a particular point in time. Because these two vectors are in phase or are in the same direction, they produce a total voltage of 240 volts.

PANEL BOX

Regardless of the type of service used, connection will be made at a **fuse** or **circuit-breaker** box. Figure 5–15 shows a 150-amp, single-phase circuit-breaker panel. Circuit breakers are made in different sizes and types. Figure 5–16 shows three different types of circuit breakers. The **single-pole breaker** is used for connecting a 120-volt circuit; the **two-pole breaker** is used for connecting a 240-volt single-phase circuit; and the **three-pole breaker** is used for connecting a three-phase circuit. The three-pole breaker must be used with a three-phase circuit-breaker panel and cannot be used in a single-phase panel.

▶ **Figure 5–16**
Single-pole, double-pole, and three-pole circuit breakers.

When a 120-volt connection is to be made, cable is brought into the panel. A **two-conductor romex** cable contains three wires—a black, a white, and a bare copper. The bare copper wire is the grounding wire or safety wire and is not considered a circuit conductor. Only the black and white wires are considered to be circuit conductors. The black wire is used as the "hot" conductor, and the white wire is used as the neutral. Figure 5–17 shows a 120-volt, single-phase circuit connected into the panel box. Notice that the black wire is connected to the circuit breaker, and the white wire is connected to the neutral bus. Notice also that the bare copper wire is connected to the neutral bus with the white wire.

When a 240-volt connection must be made, a two-pole circuit breaker is used. If the connection is to use only two-circuit conductors as shown in Figure 5–18, the black wire connects to one pole of the two-pole breaker. The *National Electrical Code* does not permit a white wire to be used as a "hot" circuit conductor. For this reason the wire must be identified by wrapping a piece of colored tape around it. The tape can be any color except white, gray, or green. Black or red tape is generally used. The identified conductor is then connected to the other pole of the two-pole breaker. The bare copper wire is connected to the neutral bus.

If a 240-volt, three-wire circuit is to be connected to the panel, a three-conductor cable is used. The three-conductor cable contains four wires—a black, a red, a white, and a green. The green is the grounding or safety wire and is not considered a circuit

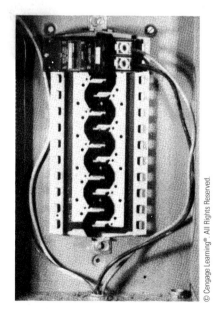

▶ **Figure 5–15**
150-amp single-phase panel.

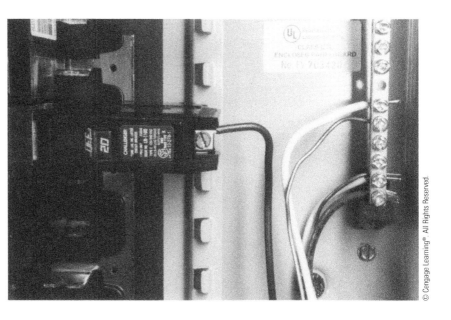

▶ **Figure 5–17**
120-volt single-phase
connection.

▶ **Figure 5–18**
240-volt single-phase two-wire
connection.

conductor. Figure 5–19 shows a 240-volt three-wire connection. The black and red wires are connected to the two poles of the circuit breaker. The white and green wires are connected to the neutral bus.

When a three-phase panel connection is made, a three-pole circuit breaker is used. There may or may not be a neutral, depending on the type of circuit. For example, a 208/120-volt connection would use a fourth wire connected to the neutral bus. A 440-volt straight, three-phase connection would use only three conductors connected to a three-pole breaker.

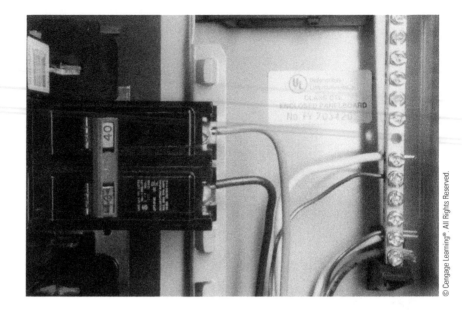

▶ **Figure 5–19**
240-volt single-phase,
three-wire connection.

FUSES

Circuit breakers are not the only means used to provide circuit protection. Fuses are still used to a great extent. Fuses are rated in two ways—by voltage and current. The voltage rating of a fuse indicates the amount of voltage the fuse is designed to interrupt without arcing across. Although fuses can be obtained that have ratings of several thousand volts, the most common fuses used in the air-conditioning field are 250 volt and 600 volt. The 600-volt fuse is longer to provide a greater distance between the two contact ends if the fuse link should blow. The extra length is needed at higher voltages to prevent **arc-over**.

Figure 5–20 shows a type of fuse that uses a replaceable link. When this fuse blows, the fuse cartridge can be taken apart and the fuse link replaced. This type of fuse is more expensive to purchase, but it could be a savings if the fuse has to be replaced frequently.

Fuses used for circuit protection are made in standard ampere ratings. Figure 5–21 shows these ratings as taken from the *National Electrical Code*. Fuses for air-conditioning and refrigeration equipment are normally sized at 175% of the rated full-load current of the motor. If this does not permit the motor to start, however, compressors can be

▶ **Figure 5–20**
Replaceable-link type of fuse.

fused as much as 225% of their full-load running current. If the fuse size needed does not correspond with one of the standard fuse sizes, the next smaller size fuse must be used. For example, assume it has been determined that a fuse rating of 130 amps is needed. The standard ratings chart for fuses shown in Figure 5–21 does not list a 130-amp fuse. Therefore, the closest standard rating less than 130 amps

240.6 Standard Ampere Ratings.
(A) Fuses and Fixed-Trip Circuit Breakers. The standard ampere ratings for fuses and inverse time circuit breakers shall be considered 15, 20, 25, 30, 35, 40, 45, 50, 60, 70, 80, 90, 100, 110, 125, 150, 175, 200, 225, 250, 300, 350, 400, 450, 500, 600, 700, 800, 1000, 1200, 1600, 2000, 2500, 3000, 4000, 5000, and 6000 amperes. Additional standard ampere ratings for fuses shall be 1, 3, 6, 10, and 601. The use of fuses and inverse time circuit breakers with nonstandard ampere ratings shall be permitted.

Figure 5–21
Standard fuse ratings.
Reprinted with permission from NFPA 70®-2011, National Electrical Code®. Copyright ©2010, National Fire Protection Association, Quincy, MA 02169. This reprinted material is not the complete and official position of the NFPA on the referenced subject, which is represented only by the standard in its entirety.

is 125 amps. A 125-amp fuse will be used. Notice that fuses can be sized as much as 225% of the full-load current of the compressor. Fuses are sized this much above the running current of the motor to permit the fuse the ability to withstand the starting current of the motor. Fuses are designed to protect the circuit against short circuits; they are not used to protect the motor from overloads.

Overload protection for the motor is provided by the overload relay, which will be covered later, or by dual-element time-delay fuses. Dual-element time-delay fuses are designed to provide both types of protection. Figure 5–22 illustrates a dual-element time-delay fuse. The fuse link is designed to open quickly in the event of a short circuit. Short-circuit currents are generally several hundred times the

rating of the fuse. The fuse link is also designed to allow some amount of overload for a short period of time. This time delay permits the motor to start without opening the fuse link. The overload protection is provided by a solder link. The solder is intended to melt at a specific temperature, and a spring is used to pull the link apart. Although the motor starting current is greater than the overload protection would permit, it takes time for the solder to melt, permitting the motor to start.

FUSED DISCONNECTS

Fused disconnects provide both a disconnect switch and fuse holders. Figure 5–23 shows a fused disconnect used for three-phase circuits. Fused disconnects,

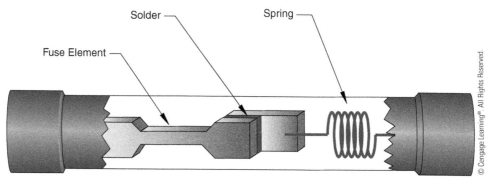

Solder — Spring —

Fuse Element —

Figure 5–22
Dual-element time-delay fuse.

© Cengage Learning®. All Rights Reserved.

STANDARD DISCONNECT SIZES
30 amp
60 amp
100 amp
200 amp
400 amp
600 amp
1000 amp
1200 amp
1600 amp
3000 amp
4000 amp
5000 amp
6000 amp

▶ **Figure 5–24**
Standard disconnect sizes.

▶ **Figure 5–23**
Three-phase fused disconnect.

like fuses, have standard ratings. The standard sizes for fused disconnects are shown in Figure 5–24. The rating of the disconnect indicates the maximum size of fuse that can be used in that enclosure. For example, assume the 125-amp fuse discussed earlier in this unit is to be mounted in a disconnect. Because the fuse size is greater than 100 amps, it cannot be mounted in a 100-amp enclosure. The next standard size enclosure is 200 amps. The 125-amp fuses will have to be mounted in a 200-amp disconnect.

When servicing equipment, it is often necessary to turn off the power to the equipment. When this is necessary, certain precautions should be taken by the service technician. Remember that your life is your own, and do not trust someone else not to turn the circuit back on while it is being serviced. Most industries provide a tag that is hung on the disconnect while it is being serviced. A paper tag, however, cannot stop someone from turning the power back on. For this reason, a small padlock should be used to lock the disconnect in the off position. If a lock is not available, the fuses should be removed with fuse pullers. There is no such thing as being too safe when working with high-voltage electricity.

▷ SUMMARY

- ⊙ The electric power in the United States and Canada is generated as three phase with a frequency of 60 Hz.
- ⊙ The voltages of a three-phase system are 120° out of phase with each other.
- ⊙ The two basic types of three-phase connections are the wye and delta.
- ⊙ In a wye-connected system, the line voltage is greater than the phase voltage by a factor of the $\sqrt{3}$ (1.732).

▶ In a wye-connected system, the line current and phase current are equal.

▶ In a delta-connected system, the line voltage and phase voltage are equal.

▶ In a delta-connected system, the line current is greater than the phase current by a factor of $\sqrt{3}$.

▶ Some delta-connected systems can produce a high leg that has one phase with a voltage that is 1.732 times greater than the other two-phase voltages.

▶ Open-delta systems use only two transformers.

▶ In an open-delta system, the total kVA capacity is only about 84% of the combined kVA capacity of the two transformers.

▶ Single-phase services produce voltages that are 180° out of phase with each other.

▶ Single-phase services generally provide a 120/240-volt connection by grounding the center tap of a transformer secondary and using it as a neutral conductor.

▶ A single-pole circuit breaker is used to provide power to a 230-volt circuit.

▶ A two-pole circuit breaker is used to provide power to a 240-volt circuit.

▶ A three-pole circuit breaker is used to provide power to a three-phase circuit.

▶ Fuses are sometimes used instead of circuit breakers.

▶ *Section 240.6* of the *National Electrical Code* lists standard sizes of fuses and circuit breakers.

KEY TERMS

alternator	neutral	three phase
arc-over	open-delta system	three-pole breaker
circuit breaker	rotor	two-conductor
delta	single-phase	romex
fuse	single-pole breaker	two-pole breaker
hertz	star	wye
high-leg	stator	

REVIEW QUESTIONS

1. What is an alternator?

2. What controls the output voltage of an alternator?

3. What controls the frequency of the alternator?

4. How many degrees out of phase with each other are the voltages of a three-phase system?

5. What are the two major types of three-phase connections?

6. List the rules concerning line and phase values of current and voltage in a wye connection.

7. List the rules concerning line and phase values of current and voltage in a delta connection.

8. In a high-leg delta-connected system, what is the voltage between the high leg and neutral?

9. What type of three-phase transformer connection uses only two transformers?

10. How many degrees out of phase are the voltages of a single-phase system?

11. A two-conductor romex cable contains three wires. Which wire is not counted, and why?

12. What type of circuit breaker is used to make a 240-volt connection?

13. Where does the grounding conductor connect in a panel?

14. In what two electrical units are fuses rated?

15. It has been calculated that a 290-amp fuse is needed to protect the circuit supplying an air-conditioning compressor. What standard rating of fuse should be used?

16. What size fuse disconnect will be used for the fuse in question 15?

17. What is a dual-element fuse?

Wire Size and Voltage Drop

OBJECTIVES

After studying this unit, the student should be able to:

≫ List factors that determine wire resistance

≫ Determine the resistance of a piece of wire

≫ Use the *National Electrical Code®* to determine wire size

≫ Test an installation of excessive voltage drop on the conductors

When installing air-conditioning equipment, it is important to use the proper size wire. If wire is used that is larger than needed, it is an unnecessary expense. If wire is used that is too small, it will cause excessive voltage drop and damage the equipment.

WIRE RESISTANCE

Most people think of wire as having zero resistance. In fact, many electrical calculations are made that assume the resistance of the wire is so little that it is negligible. In actual practice, however, all wires have resistance. There are four factors that determine the resistance of a piece of wire. The factors are listed here:

1. The diameter of the wire
2. The material the wire is made of
3. The length of the wire
4. The temperature of the wire

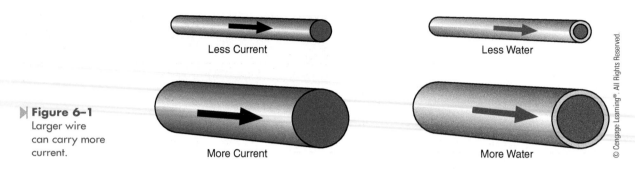

Less Current

Less Water

Figure 6–1
Larger wire can carry more current.

More Current

More Water

AREA

The cross-sectional area of wire is measured in **circular mils**. The circular mil area of a wire can be found by finding the diameter of the wire in thousandths of an inch (1 mil = $\frac{1}{1000}$ inch) and squaring that number. (To square a number means to multiply that number by itself.) For example, assume a piece of wire is measured with a micrometer and is found to have a diameter of 8 thousandths of an inch (0.008). The circular mil area of the wire is 64 (8 × 8 = 64). Notice that 64 is written as a whole number, not a decimal number. A wire that has a diameter of 0.064 inch has a circular mil area of 4096 (64 × 64 = 4096).

The circular mil area of stranded wire is determined by finding the area of one of the strands and then multiplying by the number of strands. For example, assume that a wire has 24 strands of wire that are 0.012 inch in diameter. The area of one wire is 144 CM (circular mils). The entire conductor has a circular mil area of 3456 CM (144 × 24 = 3456). Large wire is generally stranded to make it easier to bend.

The larger the diameter of a wire, the less resistance it will have and the more current it can carry, Figure 6–1. Current flowing through a wire is very similar to water flowing through a pipe. A large wire can carry more current at a specific voltage than a small wire. A large pipe can carry more water at a specific pressure than a small pipe.

MATERIAL

A standard measurement used for finding the resistance of wire is the **mil-foot**. A mil-foot of wire is a piece of wire 1 circular mil in diameter and 1 foot

long. If the resistance of a mil-foot of different types of wire is found, a mathematical formula can be used to determine the resistance of different types, sizes, and lengths of wire. This formula is

$R = K \times L/CM$

Where: K = the ohms per mil-foot of the wire.
L = the length of the wire in feet.
CM = the circular mil area of the wire.

The table in Figure 6–2 gives the resistance of different types of wire in ohms per mil-foot. Using the table shown in Figure 6–3, the diameter and circular mil area for different sizes of wire can be found.

EXAMPLE 1: What is the resistance of a piece of 18 **AWG (American Wire Gauge)** copper wire 400 feet long?

K = OHMS RESISTANCE PER MIL-FOOT (AT 70°F)	
Aluminum	17
Brass	42
Cadmium Bronze	12
Copper	10.4
Copperclad Aluminum (20% Cu)	15.2
Copperweld	26–34
Iron	60
Nichrome	600
Silver	9.6
Steel	75
Tungsten	33

Figure 6–2
Ohms per Mil-Foot for different materials.

B & S Gauge No.	Diam. in Mils	Area in Circular Mils	Ohms per 1000 Ft. (ohms per 100 meters)						Pounds per 1000 Ft. (kg per 100 meters)			
			Copper*		Copper*		Aluminum		Copper		Aluminum	
			68°F	(20°C)	167°F	(75°C)	68°F	(20°C)				
0000	460	211,600	0.049	(0.016)	0.0596	(0.0195)	0.0804	(0.0263)	640	(95.2)	195	(29.0)
000	410	167,800	0.0618	(0.020)	0.0752	(0.0246)	0.101	(0.033)	508	(75.5)	154	(22.9)
00	365	133,100	0.078	(0.026)	0.0948	(0.031)	0.128	(0.042)	403	(59.9)	122	(18.1)
0	325	105,500	0.0983	(0.032)	0.1195	(0.0392)	0.161	(0.053)	320	(47.6)	97	(14.4)
1	289	83,690	0.1239	(0.0406)	0.151	(0.049)	0.203	(0.066)	253	(37.6)	76.9	(11.4)
2	258	66,370	0.1563	(0.0512)	0.191	(0.062)	0.256	(0.084)	201	(29.9)	61.0	(9.07)
3	229	52,640	0.1970	(0.0646)	0.240	(0.079)	0.323	(0.106)	159	(23.6)	48.4	(7.20)
4	204	41,740	0.2485	(0.0815)	0.302	(0.099)	0.408	(0.134)	126	(18.7)	38.4	(5.71)
5	182	33,100	0.3133	(0.1027)	0.381	(0.125)	0.514	(0.168)	100	(14.9)	30.4	(4.52)
6	162	26,250	0.395	(0.129)	0.481	(0.158)	0.648	(0.212)	79.5	(11.8)	24.1	(3.58)
7	144	20,820	0.498	(0.163)	0.606	(0.199)	0.817	(0.268)	63.0	(9.37)	19.1	(2.84)
8	128	16,510	0.628	(0.206)	0.764	(0.250)	1.03	(0.338)	50.0	(7.43)	15.2	(2.26)
9	114	13,090	0.792	(0.260)	0.963	(0.316)	1.30	(0.426)	39.6	(5.89)	12.0	(1.78)
10	102	10,380	0.999	(0.327)	1.215	(0.398)	1.64	(0.538)	31.4	(4.67)	9.55	(1.42)
11	91	8,234	1.260	(0.413)	1.532	(0.502)	2.07	(0.678)	24.9	(3.70)	7.57	(1.13)
12	81	6,530	1.588	(0.520)	1.931	(0.633)	2.61	(0.856)	19.8	(2.94)	6.00	(0.89)
13	72	5,178	2.003	(0.657)	2.44	(0.80)	3.29	(1.08)	15.7	(2.33)	4.80	(0.71)
14	64	4,107	2.525	(0.828)	3.07	(1.01)	4.14	(1.36)	12.4	(1.84)	3.80	(0.56)
15	57	3,257	3.184	(1.044)	3.98	(1.27)	5.22	(1.71)	9.86	(1.47)	3.00	(0.45)
16	51	2,583	4.016	(1.317)	4.88	(1.60)	6.59	(2.16)	7.82	(1.16)	2.40	(0.36)
17	45.3	2,048	5.06	(1.66)	6.16	(2.02)	8.31	(2.72)	6.20	(0.922)	1.90	(0.28)
18	40.3	1,624	6.39	(2.09)	7.77	(2.55)	10.5	(3.44)	4.92	(0.713)	1.50	(0.22)
19	35.9	1,288	8.05	(2.64)	9.79	(3.21)	13.2	(4.33)	3.90	(0.580)	1.20	(0.18)
20	32	1,022	10.15	(3.33)	12.35	(4.05)	16.7	(5.47)	3.09	(0.459)	0.94	(0.14)
21	28.5	810	12.8	(4.2)	15.6	(5.11)	21.0	(6.88)	2.45	(0.364)	0.745	(0.11)
22	25.4	642	16.1	(5.3)	19.6	(6.42)	26.5	(8.69)	1.95	(0.290)	0.591	(0.09)
23	22.6	510	20.4	(6.7)	24.8	(8.13)	33.4	(10.9)	1.54	(0.229)	0.468	(0.07)
24	20.1	404	25.7	(8.4)	31.2	(10.2)	42.1	(13.8)	1.22	(0.181)	0.371	(0.05)
25	17.9	320	32.4	(10.6)	39.4	(12.9)	53.1	(17.4)	0.97	(0.14)	0.295	(0.04)
26	15.9	254	40.8	(13.4)	49.6	(16.3)	67.0	(22.0)	0.77	(0.11)	0.234	(0.03)
27	14.2	202	51.5	(16.9)	62.6	(20.5)	84.4	(27.7)	0.61	(0.09)	0.185	(0.03)
28	12.6	160	64.9	(21.3)	78.9	(25.9)	106	(34.7)	0.48	(0.07)	0.147	(0.02)
29	11.3	126.7	81.8	(26.8)	99.5	(32.6)	134	(43.9)	0.384	(0.06)	0.117	(0.02)
30	10	100.5	103.2	(33.8)	125.5	(41.1)	169	(55.4)	0.304	(0.04)	0.092	(0.01)
31	8.93	79.7	130.1	(42.6)	158.2	(51.9)	213	(69.8)	0.241	(0.04)	0.073	(0.01)
32	7.95	63.2	164.1	(53.8)	199.5	(65.4)	269	(88.2)	0.191	(0.03)	0.058	(0.01)
33	7.08	50.1	207	(68)	252	(82.6)	339	(111)	0.152	(0.02)	0.046	(0.01)
34	6.31	39.8	261	(86)	317	(104)	428	(140)	0.120	(0.02)	0.037	(0.01)
35	5.62	31.5	329	(108)	400	(131)	540	(177)	0.095	(0.01)	0.029	
36	5	25	415	(136)	505	(165)	681	(223)	0.076	(0.01)	0.023	
37	4.45	19.8	523	(171)	636	(208)	858	(281)	0.0600	(0.01)	0.0182	
38	3.96	15.7	660	(216)	802	(263)	1080	(354)	0.0476	(0.01)	0.0145	
39	3.53	12.5	832	(273)	1012	(332)	1360	(446)	0.0377	(0.01)	0.0115	
40	3.15	9.9	1049	(344)	1276	(418)	1720	(564)	0.0299	(0.01)	0.0091	
41	-											
42	2.5											
43	-											
44	1.97											

*Resistance figures are given for standard annealed copper. For hard-drawn copper, add 2%

▶ **Figure 6–3**
American Wire Gauge table.

SOLUTION: First, state the formula to be used.

$$R = \frac{K \times L}{CM}$$

Second, substitute known numeric values in the formula. The value of K can be found in the table shown in Figure 6–2. The K value for copper is 10.4. The CM area of 18 AWG wire can be found from the chart in Figure 6–3. The circular mil area of 18 AWG wire is 1624 CM. If these values are substituted in the formula for letters, the formula will be

$$R = \frac{10.4 \times 400}{1624}$$

$$R = \frac{4160}{1624}$$

$$R = 2.56 \text{ ohms}$$

EXAMPLE 2: What is the resistance of a piece of 12 AWG aluminum wire 250 feet long?

SOLUTION:

$$R = \frac{K \times L}{CM}$$

$$R = \frac{17 \times 250}{6530}$$

$$R = \frac{4250}{6530}$$

$$R = 0.6508 \text{ ohm}$$

Table 310.15(B)(2)(a) Ambient Temperature Correction Factors Based on 30°C (86°F)

For ambient temperatures other than 30°C (86°F), multiply the allowable ampacities specified in the ampacity tables by the appropriate correction factor shown below.

Ambient Temperature (°C)	Temperature Rating of Conductor			Ambient Temperature (°F)
	60°C	75°C	90°C	
10 or less	1.29	1.20	1.15	50 or less
11–15	1.22	1.15	1.12	51–59
16–20	1.15	1.11	1.08	60–68
21–25	1.08	1.05	1.04	69–77
26–30	1.00	1.00	1.00	78–86
31–35	0.91	0.94	0.96	87–95
36–40	0.82	0.88	0.91	96–104
41–45	0.71	0.82	0.87	105–113
46–50	0.58	0.75	0.82	114–122
51–55	0.41	0.67	0.76	123–131
56–60	—	0.58	0.71	132–140
61–65	—	0.47	0.65	141–149
66–70	—	0.33	0.58	150–158
71–75	—	—	0.50	159–167
76–80	—	—	0.41	168–176
81–85	—	—	0.29	177–185

▶ **Figure 6–4**
Ambient temperature correction factors.
NFPA 70®, National Electrical Code and NEC® are registered trademarks of the National Fire Protection Association, Quincy, MA.

TEMPERATURE

The resistance of a piece of wire is also affected by **temperature**. As the temperature increases, the resistance of wire increases also. Notice that the charts in Figure 6–2 and Figure 6–3 state the resistance of wire at a specific temperature. Most wire tables provide some means for determining the resistance of wire as temperature increases. Figure 6–4 shows *Table 310.(15)(B)(2)(a)* of the *National Electrical Code*. The table is divided into columns. The first column lists ambient temperature in Celsius degrees, and the last column lists ambient temperature in Fahrenheit degrees. The three center columns list the correction factors in accord with the temperature rating of the conductor found in *NEC Table 310.15(B)(16)* and *Table 310.15(B)(17)*.

INSULATION

The type of **insulation** around the wire also partly determines the amount of current the wire is permitted to carry. Some types of insulation can withstand more heat than others, and are

Table 310.15(B)(16) (formerly Table 310.16) **Allowable Ampacities of Insulated Conductors Rated Up to and Including 2000 Volts, 60°C Through 90°C (140°F Through 194°F), Not More Than Three Current-Carrying Conductors in Raceway, Cable, or Earth (Directly Buried), Based on Ambient Temperature of 30°C (86°F)***

Size AWG or kcmil	Temperature Rating of Conductor [See Table 310.104(A).]						Size AWG or kcmil
	60°C (140°F)	75°C (167°F)	90°C (194°F)	60°C (140°F)	75°C (167°F)	90°C (194°F)	
	Types TW, UF	Types RHW, THHW, THW, THWN, XHHW, USE, ZW	Types TBS, SA, SIS, FEP, FEPB, MI, RHH, RHW-2, THHN, THHW, THW-2, THWN-2, USE-2, XHH, XHHW, XHHW-2, ZW-2	Types TW, UF	Types RHW, THHW, THW, THWN, XHHW, USE	Types TBS, SA, SIS, THHN, THHW, THW-2, THWN-2, RHH, RHW-2, USE-2, XHH, XHHW, XHHW-2, ZW-2	
	COPPER			ALUMINUM OR COPPER-CLAD ALUMINUM			
18	—	—	14	—	—	—	—
16	—	—	18	—	—	—	—
14**	15	20	25	—	—	—	—
12**	20	25	30	15	20	25	12**
10**	30	35	40	25	30	35	10**
8	40	50	55	35	40	45	8
6	55	65	75	40	50	55	6
4	70	85	95	55	65	75	4
3	85	100	115	65	75	85	3
2	95	115	130	75	90	100	2
1	110	130	145	85	100	115	1
1/0	125	150	170	100	120	135	1/0
2/0	145	175	195	115	135	150	2/0
3/0	165	200	225	130	155	175	3/0
4/0	195	230	260	150	180	205	4/0
250	215	255	290	170	205	230	250
300	240	285	320	195	230	260	300
350	260	310	350	210	250	280	350
400	280	335	380	225	270	305	400
500	320	380	430	260	310	350	500
600	350	420	475	285	340	385	600
700	385	460	520	315	375	425	700
750	400	475	535	320	385	435	750
800	410	490	555	330	395	445	800
900	435	520	585	355	425	480	900
1000	455	545	615	375	445	500	1000
1250	495	590	665	405	485	545	1250
1500	525	625	705	435	520	585	1500
1750	545	650	735	455	545	615	1750
2000	555	665	750	470	560	630	2000

*Refer to 310.15(B)(2) for the ampacity correction factors where the ambient temperature is other than 30°C (86°F).
**Refer to 240.4(D) for conductor overcurrent protection limitations.

Figure 6–5
NFPA 70®, National Electrical Code and NEC® are registered trademarks of the National Fire Protection Association, Quincy, MA.

therefore permitted to carry more current. For example, in *Table 310.15(B)(16)*, the type of wire insulation is listed at the top of each column, Figure 6–5. Notice the different temperature ratings for the types of insulation listed. Also notice the amount of current a wire is permitted to carry for different types of insulation. Find a 2 AWG conductor on the far left-hand side of the wire table. Notice the different amounts of current this conductor is permitted to carry with different types of insulation.

VOLTAGE RATING

Wire also has a **voltage rating**. The voltage rating of wire has nothing to do with the type of material the wire is made of or its diameter, but rating is determined by the type of insulation. Most wire used in industry has a voltage rating of 600 volts. The amount of voltage the insulation can effectively hold off is determined by the material the insulation is made of and its thickness.

SIZING CONDUCTORS FOR HERMETICALLY SEALED COMPRESSORS

Requirements for the installation of hermetically sealed compressors are covered in *Article 440* of the *National Electrical Code*. Note that this section covers hermetically sealed units only and does not apply to separate motor and compressor units. *NEC Section 440.6(A)* states that the rated-load current marked on the nameplate of the equipment is to be used in determining the rating or ampacity of the disconnecting means, branch circuit conductors, controller, fuses or circuit breakers, ground fault protection equipment, and overload protection. If the rated-load current is not shown on the equipment nameplate, the rated load current shown on the compressor is to be used. Note the difference between the equipment nameplate and compressor nameplate. The equipment nameplate is generally shown on the unit itself, Figure 6–6. The compressor nameplate is located on the compressor.

NEC Section 440.6(A), Exception 1, states that where so marked, the branch circuit selection current shall be used instead of the rated load current to determine the rating or ampacity of the disconnecting means, branch circuit conductors, controller, short circuit protective device, and ground fault protective device. *NEC Section 440.32* states that for a single motor compressor, the branch circuit conductors shall have an ampacity not less than 125% of either the motor compressor rated load current or the branch circuit selection current, whichever is greater. As a general rule the branch circuit selection current

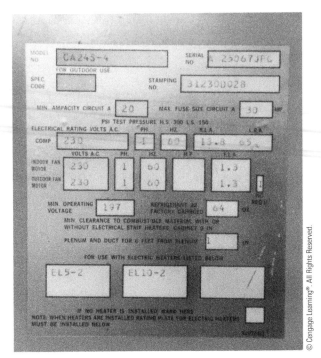

▶ **Figure 6–6**
Equipment nameplate.

will be greater than the compressor rated load amps because the selection current includes any indoor and/or outdoor fans.

Termination Temperature

Another factor that must be taken into consideration is the temperature rating of different terminations, or **termination temperature**. The termination is the point of attachment for conductors such as circuit breakers, disconnects, switches, and so forth. *NEC Section 110.14(C)* basically states that the temperature rating of the termination cannot be exceeded. Although a wire with a higher temperature rating may be used, the ampacity of the wire must be selected on the basis of the lowest temperature rating in the circuit. Type THHN insulation has a temperature rating of 90°C, but assume that it is connected to a device with a temperature rating of 75°C. The ampacity of the wire would have to be selected on the basis of 75°C, not 90°C.

Occasionally the temperature rating of a device is listed on the device or in the manufacturer's literature, but as a general rule it is not known. For this reason, the *National Electrical Code* states that conductors for circuits rated at 100 amperes or less should be selected from the 60°C column, and conductors rated over 100 amperes are to be selected from the 75°C, column. An exception to this is if the insulation type is rated less than 75°C, as is the case with type TW and UF. If either of these two types of insulation is employed, the ampere rating of the conductor must be determined from the 60°C column, regardless of the circuit current.

EXAMPLE

Assume that an equipment nameplate lists the rated load amps (RLA) of the compressor at 14.8 amps and the circuit selection current at 22 amps. To determine the correct conductor size for this unit, multiply the larger of the two rating by 125%.

$$22 \times 1.25 = 27.5 \text{ amps}$$

The next step is to determine the conductor size from *NEC Table 310.15(B)(16)*. It is assumed that copper conductors with type THWN insulation will be used for this installation. Although THWN insulation is located in the 75°C column, the conductor size should be selected from the 60°C column because the total circuit current is less than 100 amperes. *Table 310.15(B)(16)* indicates a 10 AWG conductor is the closest size without going under 27.5 amperes. A 10 AWG conductor should be used for this installation.

The *National Electrical Code* also requires that a disconnecting means be located within sight from and readily accessible to the air-conditioning or refrigeration equipment, Figure 6–7. The *NEC* does permit the disconnect to be installed on or within the equipment provided it is not installed on a panel that is designed to allow access to the equipment. The disconnect may be located away from the air-conditioning or refrigeration equipment provided there is a working clearance of at least 30 inches (76.2 cm) to allow service accessibility. This is to enable the service technician to disconnect power from the unit without having to enter the building to find the main power panel. A typical disconnect used for air-conditioning service is shown in Figure 6–8.

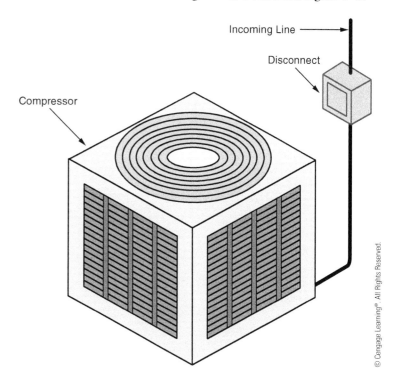

▶ **Figure 6–7**
A disconnecting means must be located in sight of the compressor.

▶ **Figure 6–8**
Disconnect switch used for air-conditioning equipment.

TESTING FOR EXCESSIVE VOLTAGE DROP

Testing a unit for excessive voltage can be done with a voltmeter. First, test the voltage at the panel with the unit turned off. Assume this voltage to be 240 volts. Next, start the air-conditioning unit and again check the voltage at the panel. If the voltage remains unchanged, it is an indication that there is no voltage drop at the panel and that all connections for the part of the circuit are good. If there is excessive voltage drop at the panel, it is an indication of bad connections, or the service entrance is too small for the load. For this example, assume the voltage remains at 240 volts when the unit is turned on. Next, check the voltage at the unit, Figure 6–9. If there is a significant voltage drop at the unit, it indicates that the wire size is too small and that too much voltage is being used to push current through the wire. This problem can be corrected by connecting larger wires from the panel to the unit.

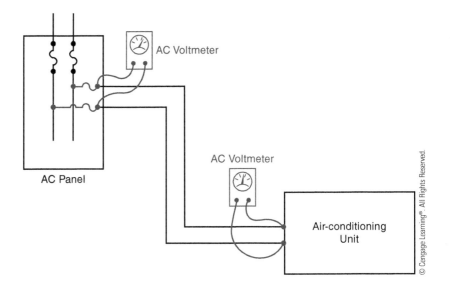

AC Voltmeter

AC Panel

AC Voltmeter

Air-conditioning Unit

▶ **Figure 6–9**
Testing for voltage drop.

SUMMARY

⊙ The resistance of wire is determined by four factors:

    A. The diameter of the wire

    B. The material the wire is made of

    C. The length of the wire

    D. The temperature of the wire

⊙ The cross-sectional area of wire is measured in circular mils.

⊙ Circular mil area is found by squaring the diameter of the wire.

⊙ The standard measurement for wire resistance in the English system is the mil-foot.

⊙ Different types of materials have different resistances per mil-foot.

⊙ The wire tables in the *National Electrical Code* are often used to determine the size of wire needed for a particular installation.

⊙ The voltage rating of wire is determined by the type of insulation.

⊙ Excessive voltage drop can be determined by measuring the voltage at the source and at the load when the unit is in operation.

KEY TERMS

AWG (American Wire Gauge)	**insulation**	**termination**
circular mils	**mil-foot**	**temperature**
	temperature	**voltage rating**

REVIEW QUESTIONS

1. Name four factors that determine the resistance of wire.

2. A wire has a diameter of 0.057 inch. What is its circular mil area?

3. What is a mil-foot of wire?

4. When the temperature of wire increases, does its resistance increase or decrease?

5. What determines the voltage rating of wire?

6. What two factors determine the amount of voltage rating a certain type of insulation will have?

7. How much resistance does 75 feet of 24 AWG wire have?

8. If a current of 4 amps flows through the wire in question 7, how much voltage will be dropped by the wire?

UNIT 7 ▷

Inductance

OBJECTIVES

After studying this unit, the student should be able to:

▷ Discuss the voltage and current relationship in an AC circuit containing pure resistance and pure inductance

▷ Discuss the properties of an inductive circuit

▷ Calculate values of inductive reactance

▷ Calculate values of impedance in circuit containing resistance and inductance

▷ Discuss true power, apparent power, and reactive power

▷ Discuss the power factor of an alternating-current circuit

Alternating-current circuits contain three basic types of loads: (1) resistive, (2) inductive, and (3) capacitive.

RESISTIVE CIRCUITS

The simplest of the AC loads is a circuit that contains only pure resistance, Figure 7–1. In a **pure-resistive circuit**, the voltage and current are **in phase** with each other, Figure 7–2. Voltage and current are in phase when they cross the zero line at the same point, and have their peak positive and negative values at the same time. A pure-resistive circuit is very similar to a direct-current circuit in the respect that true power or watts is equal to the voltage times the current. Examples of pure-resistive circuits are the heating elements of an electric range, an electric hot-water heater, and the resistive elements of an electric furnace.

► **Figure 7–1**
Pure-resistive circuit.

► **Figure 7–3**
A pure-inductive circuit.

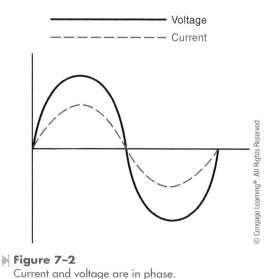

▬▬▬▬▬▬ Voltage
– – – – – – – Current

► **Figure 7–2**
Current and voltage are in phase.

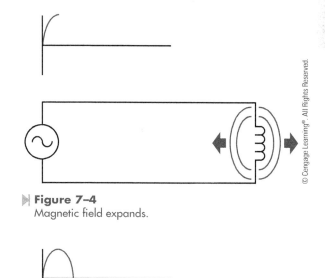

► **Figure 7–4**
Magnetic field expands.

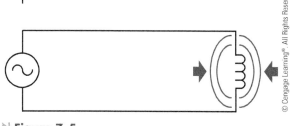

► **Figure 7–5**
Magnetic field collapses.

INDUCTIVE CIRCUITS

An inductive circuit contains an inductor or coil as the load instead of a resistor, Figure 7–3. The two most common types of inductive circuits are motors and transformers. Inductors are measured by a unit called the **henry**. The unit of inductance is named in honor of Joseph Henry, a physicist who studied electricity. The electrical symbol for inductance is *L*.

Inductors differ from resistors in several ways. One way is that the current of an inductor is not limited by the resistance of the coil. When an inductor is connected into an AC circuit and the voltage begins to rise from zero toward its peak value, a **magnetic field** is created around the coil, Figure 7–4. As the expanding magnetic field cuts through the wires of the coil, a voltage is induced in the coil. *The voltage induced in a coil is always opposed*

to the voltage that creates it. As the applied voltage begins to drop from its peak value back toward zero, the magnetic field around the inductor begins to collapse, Figure 7–5. Notice that the induced voltage is opposite in polarity to the applied voltage. An induced voltage is 180° out of phase with the applied voltage, Figure 7–6.

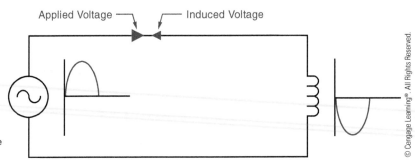

Applied Voltage ⎯⎯   ⎯⎯ Induced Voltage

Figure 7–6
The induced voltage is opposed to the
applied voltage.

Because the induced voltage is opposed to the applied voltage, it limits current flow through the circuit just as resistance does. Although the induced voltage of a coil limits the current flow "like" resistance, it is not resistance and cannot be treated as resistance. The unit of measure used to describe the current-limiting effect of an induced voltage is **reactance** and is given the electrical symbol X. Because this reactance is caused by an inductance, it is called **inductive reactance** and is given the electrical symbol X_L (pronounced "X sub L"). Inductive reactance is measured in ohms just as resistance is.

The inductive reactance of a coil is determined by two factors:

1. The inductance of the coil
2. The frequency of the applied voltage

If these factors are known, a formula can be used to find the inductive reactance of the coil. This formula is as follows:

$X_L = 2 \times \pi \times F \times L$
X_L = inductive reactance
π = the Greek letter pi, which has a value of 3.1416
F = the frequency of the AC voltage
L = the inductance of the coil in henrys

EXAMPLE

In the circuit shown in Figure 7–7, a coil has an inductance of 0.7 henrys, and is connected to a 120-volt, 60 Hz line. Find the current flow in the circuit.

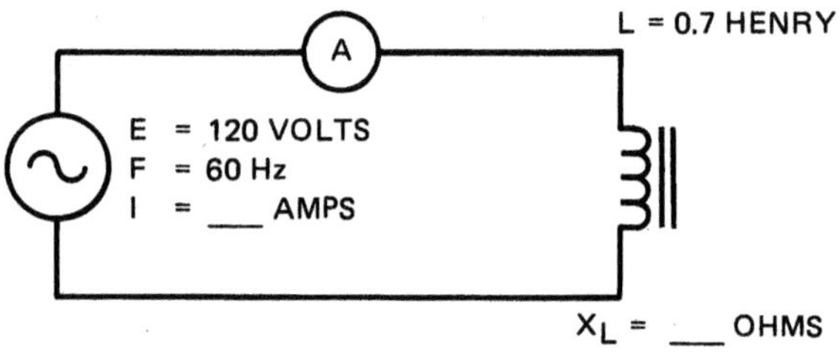

L = 0.7 HENRY

A

E = 120 VOLTS
F = 60 Hz
I = ___ AMPS

X_L = ___ OHMS

Figure 7–7
Current flow is limited by inductive reactance.

To solve the problem, the first step is to find the amount of inductive reactance in the circuit.

$X_L = 2 \times \pi \times F \times L$
$X_L = 2 \times 3.1416 \times 60 \times 0.7$
$X_L = 263.9$ ohms

Now that the inductive reactance of the coil is known, the current flow in the circuit can be calculated. If the value of inductive reactance is used like resistance, the formula to find current in a pure-inductive circuit is $I = E/X_L$.

$I = \dfrac{120}{263.9}$
$I = 0.455$ amp

VOLTAGE AND CURRENT RELATIONS

As stated previously, the voltage and current in a pure-resistive circuit are in phase with each other. In a **pure-inductive circuit**, however, the current lags behind the voltage by 90°, Figure 7–8.

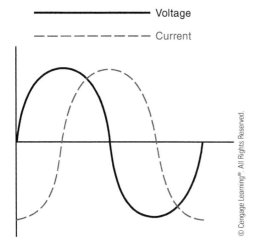

Figure 7–8
Current lags voltage by 90° in a pure inductive circuit.

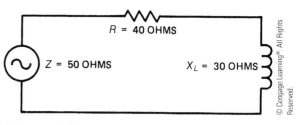

Figure 7–9
A series circuit containing resistance and inductance.

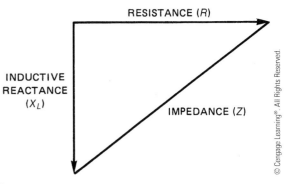

Figure 7–10
Impedance is the sum of resistance and inductive reactance.

In this type of circuit, there is no true power or watts. In a resistive circuit, the resistor limits the current flow by converting the energy of the moving electrons into heat. This conversion of one form of energy into another represents a true power loss. In an inductive circuit, the energy of the moving electrons is stored in the magnetic field created around the inductor. When the magnetic field collapses, this energy is given back into the circuit. Notice that the resistor used the electrical energy by converting it into heat, but the inductor stored the energy and then returned it to the circuit.

IMPEDANCE

In an alternating-current circuit that contains only resistance, the current is limited by the value of the resistor only. In this type of circuit the current flow can be calculated using the Ohm's law formula $I = E/R$.

In an AC circuit that contains only inductance, the current is limited only by the value of inductive reactance. In this type of circuit, the current flow can be calculated using the formula $I = E/X_L$.

In a circuit like the one shown in Figure 7–9, there are elements of both resistance and inductive reactance contained in the same circuit. In this type of circuit, it cannot be said that the current is

limited by resistance, because there is also inductive reactance. It can also not be said that the current is limited by inductive reactance because of the resistance. Alternating-current circuits use a different value to represent the total amount of opposition to current flow in the circuit regardless of what type of components are found in the circuit. This value is known as impedance and is given the symbol Z.

The resistor and inductor in Figure 7–9 are connected in series. Because these two components are connected in series, they will be added. They cannot be added in the normal way, however, because the inductance is not in phase with the resistance. Because inductive reactance is 90° out of phase with resistance, the total amount of opposition to current flow will be the value of the hypotenuse of the right triangle formed by the resistance and inductive reactance, Figure 7–10. To find the total value of impedance in this circuit, the formula $Z = \sqrt{R^2 + X_L^2}$ can be used. If the resistor has a value of 40 ohms and the inductor has an inductive reactance of 30 ohms, the impedance will be:

$$Z = \sqrt{R^2 + X_L^2}$$
$$Z = \sqrt{40^2 + 30^2}$$
$$Z = \sqrt{1600 + 900}$$
$$Z = \sqrt{2500}$$
$$Z = 50 \text{ ohms}$$

APPARENT POWER

In a direct-current circuit, the true power or watts is always equal to the voltage multiplied by the current, because the current and voltage are never out of phase with each other. This also is true for an AC circuit that contains only pure resistance, because the voltage and current are in phase.

In a circuit that contains pure inductance, however, there is no true power or watts. In this type of circuit, the voltage multiplied by the current equals a value known as **VARs**, which stands for **volt-amps reactive**. VARs is often referred to as wattless power.

The **apparent power** or **volt-amps** of an AC circuit is the applied voltage multiplied by the current flow in the circuit. The amount of apparent power as compared to the true power or VARs is determined by the elements of the circuit itself. In the circuit shown in Figure 7–11, the amount

of true power is 400 watts. The amount of reactive power is 300 VARs. The apparent power is 500 volt-amps. Notice that the apparent power is found by adding the watts and VARs together in the same manner that the resistance and inductive reactance were added to find the total value of impedance. Volt-amps can be calculated by the formula:

$$\text{Volt-amps} = \sqrt{W^2 + \text{VARs}^2}$$

POWER FACTOR

The **power factor** of an alternating-current circuit is a ratio of the apparent power compared to the true power. Power factor is important because utility companies charge industries large penalties for a poor power factor. The power factor of the circuit shown in Figure 7–11 can be found by

$$PF = \frac{W}{VA}$$
$$PF = \frac{400}{500}$$
$$PF = 0.8$$
$$PF = 80\%$$

Notice in this circuit, the power factor is 80%. This means 80% of the load is resistive and 20% is reactive. If the load is pure resistive, the power factor will be 100%, or **unity**.

Utility companies become very concerned about power factor because they must furnish the amount of current needed to produce the volt-amp value. The company, however, is charged by the amount of true power or watts used. In this instance, if the applied voltage is 120 volts, the utility company must supply 4.16 amps to operate the load (500 volt-amps/120 volts = 4.16 amps). The actual amount of current being used to operate the load, however, is 3.33 amps (400 watts/120 volts = 3.33 amps). Because the air-conditioning load is often the major part of the electrical power consumed by an industry or office building, power factor can become an important consideration to the service technician.

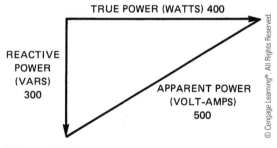

TRUE POWER (WATTS) 400
REACTIVE POWER (VARS) 300
APPARENT POWER (VOLT-AMPS) 500

▶ **Figure 7–11**
Volt-amps is the vector sum of watts and VARs.

SUMMARY

- ⊙ In a pure resistive circuit, the current and voltage are in phase with each other.
- ⊙ In a pure inductive circuit, the current lags the voltage by 90°.
- ⊙ Inductive reactance (X_L) is the current-limiting property of inductance.
- ⊙ Inductive reactance is actually a counter voltage, which opposes the applied or line voltage.
- ⊙ Inductive reactance is proportional to two factors:
    A. The inductance of the coil
    B. The frequency of the applied voltage
- ⊙ Inductive reactance is measured in ohms like resistance.
- ⊙ Impedance (Z) is a measurement of the total current-limiting effect in an alternating-current circuit.
- ⊙ Impedance is a combination of all current-limiting properties of an AC circuit and is measured in ohms.
- ⊙ True power is measured in watts.
- ⊙ Watts is a measurement of the amount of electrical energy converted to some other form such as heat or mechanical.
- ⊙ VARs is a measure of the amount of power in a pure inductive circuit, sometimes referred to as wattless power.
- ⊙ Apparent power or volt amperes (VA) is a combination of true power, or watts, and VARs.
- ⊙ In an AC circuit, apparent power is determined by multiplying the applied voltage by the circuit current.
- ⊙ Power factor (PF) is a comparison of the amount of true power with the apparent power in an alternating-current circuit.
- ⊙ Power factor is measured in a percent.

KEY TERMS

apparent power	magnetic field	reactance
henry	power factor	unity
inductive reactance	pure-inductive circuit	VARs (Volt-Amps Reactive)
in phase	pure-resistive circuit	volt-amps

REVIEW QUESTIONS

1. Name the three basic types of alternating-current loads.
2. What type of load always has its voltage and current in phase with each other?

3. In a pure-inductive circuit, how many degrees out of phase is the current with the voltage?

4. Does the current lead or lag the voltage in question 3?

5. What electrical value is used to measure inductance?

6. What is inductive reactance?

7. What electrical value is used to measure the total opposition to current flow in an AC circuit?

8. What is power factor?

UNIT 8 ▷

Capacitance

OBJECTIVES

After studying this unit, the student should be able to:

▷ Discuss the operating theory of a capacitor

▷ List the factors that determine the amount of capacitance a capacitor will have

▷ Discuss the voltage and current relationship in an AC circuit containing pure capacitance

▷ Discuss the voltage and current relationship in an AC circuit containing resistance and capacitance

▷ Compute values of capacitive reactance

▷ Compute values of impedance for circuits that contain both resistance and capacitive reactance

▷ Discuss power factor correction

▷ Compute the amount of capacitance needed to correct the power factor of a motor

▷ Discuss different types of capacitors and methods for testing

The third type of alternating-current load to be discussed is **capacitance**. A capacitor can be made by separating two metal plates with an insulating material, Figure 8–1. The insulating material used to isolate the plates from each other is called the **dielectric**. Three factors determine how much capacitance a capacitor will have:

1. The surface area of the plates
2. The distance between the plates
3. The type of dielectric material used between the plates

CHARGING A CAPACITOR

In Figure 8–2, the terminals of a capacitor have been connected to a battery. Electrons are negative particles. Therefore, the positive terminal of

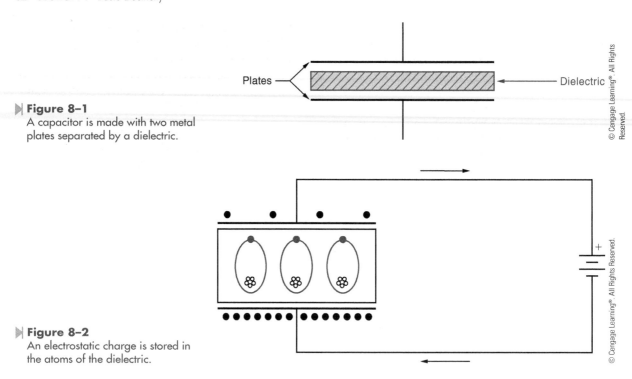

▶ **Figure 8–1**
A capacitor is made with two metal plates separated by a dielectric.

▶ **Figure 8–2**
An electrostatic charge is stored in the atoms of the dielectric.

the battery attracts electrons from one plate of the capacitor. The negative terminal of the battery causes electrons to flow to the other capacitor plate. This flow of current continues until the voltage across the capacitor plates is equal to the battery voltage. If the battery is disconnected, the capacitor will be left in a charged state. *CAUTION: It is the habit of some people to charge a capacitor to a high voltage and then hand the capacitor to another person. Although some people think this is comical, it is an extremely dangerous practice. Capacitors have the ability to supply an almost infinite amount of current. Under certain conditions, a capacitor can have enough power to cause a person's heart to go into fibrillation.*

ELECTROSTATIC CHARGE

Notice the illustration of the atoms in the dielectric material in Figure 8–2. When a capacitor has been charged, the negative electrons of the dielectric material are repelled from the negative plate of the capacitor and attracted to the positive plate. This causes the electron orbit of the atoms in the

dielectric to extend. This places the atoms of the dielectric material in tension. This is known as **dielectric stress**. Placing the atoms of the dielectric under stress has the same effect as drawing back a bow and arrow and holding it, Figure 8–3.

The amount of dielectric stress is determined by the voltage between the plates. The greater the volt-

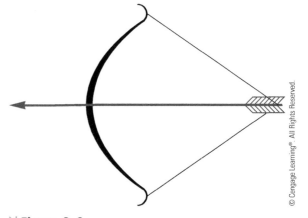

▶ **Figure 8–3**
Dielectric stress is similar to drawing back a bow and arrow, and holding it.

age, the greater the dielectric stress. If the voltage becomes too great, the dielectric will break down and destroy the capacitor. This is the reason capacitors have a voltage rating that must be followed.

The energy of a capacitor is stored in the dielectric and is known as an **electrostatic charge**. It is this electrostatic charge that permits the capacitor to produce extremely high currents under certain conditions. If the leads of a charged capacitor are shorted together, it has the same effect as releasing the drawn bow in Figure 8–3. The arrow will be propelled forward at great speed. The same is true for the electrons of the capacitor. When the electron orbits of the dielectric snap back, the electrons stored on the negative capacitor plate are propelled toward the positive plate at great speed.

CAPACITANCE RATINGS

Capacitance is rated in units called **farads**. The farad is actually such a large amount of capacitance it is not practical to use. For this reason, a unit called the **microfarad** is generally used. A microfarad is one millionth of a farad. The Greek lowercase letter mu is used to symbolize micro, μ. The term microfarad is indicated by combining mu and lowercase f, μf. Because the letter mu is not included on a standard typewriter, the term *microfarad* is sometimes shown as uf or mf. All of these terms mean the same thing.

Another term used is the **picofarad**. This term is used for extremely small capacitors found in electronics applications. A picofarad is 1 millionth of a microfarad and is generally shown as $\mu\mu$f or pf.

VOLTAGE RATINGS

When installing a capacitor in the circuit, you must stay within the capacitor's voltage rating. The voltage rating indicates the amount of voltage the dielectric can withstand without breaking down and permitting current to flow between the plates. Capacitor voltage ratings can be listed in different ways. If the nameplate lists the voltage as volts AC (VAC), the voltage rating is an RMS value, Figure 8–4. Capacitor intended for use in alternating current circuits may also list the voltage as volts DC (VDC) or working volts DC (WVDC); both mean the same thing. If the voltage is listed as VDC or WVDC, it is a peak, or maximum, voltage value. It will be necessary to convert the peak value to the RMS value to make certain that the capacitor can be safely connected in an AC circuit. To do this, multiply the VDC value by 0.707.

EXAMPLE 1: An oil-filled AC capacitor has a voltage rating of 600 WVDC. Can this capacitor be connected in a 480-volt AC circuit without damaging the capacitor?

SOLUTION: $600 \times 0.707 = 424.2$ volts. The capacitor cannot be connected in a 480-volt AC circuit without damaging the capacitor.

The importance of staying within the voltage rating cannot be overstressed. Many years ago the United States Military made a study of the effect to the dielectric at different voltages. The study determined that if the capacitor voltage rating is double the amount of voltage applied to it, the capacitor will have a life span approximately eight times longer than a capacitor with a voltage rating that is the same as the voltage applied to it. It is always permissible to connect a capacitor in a circuit that has a higher voltage rating than the voltage connected to it, but under no circumstances should a capacitor with a voltage rating less than the connected voltage be used.

CAPACITIVE REACTANCE

When AC voltage is applied to a capacitor, Figure 8–5, the plates of the capacitor are alternately charged and discharged each time the current changes direction of flow. When a capacitor is charged, the voltage across its plates becomes the same as this applied voltage. As the voltage across the plates of a capacitor increases, it offers resistance to the flow of current. The applied voltage must continually overcome the voltage of the capacitor to produce current flow. The current in a **pure-capacitive circuit** is limited by the voltage of the charged capacitor. Because current is limited by a counter voltage and not resistance, the counter voltage of the capacitor is referred to as reactance. Recall that the symbol for reactance is X. Because this reactance is caused by capacitance, it is called **capacitive reactance** and is symbolized by X_c (pronounced "X sub c").

2MDV6A

7.5μF ± 5%

370VAC(VCA) 60/50HZ

Rated to **85°C**Max Temp.

Protected **10,000**AFC

NON PCB

▶ **Figure 8–4**
If the capacitor nameplate lists the voltage as VAC, it is an RMS value.

▶ **Figure 8–5**
A pure capacitive circuit.

The amount of capacitive reactance in a circuit is determined by two factors:

1. Frequency of the AC voltage
2. The size of the capacitor

If the frequency of the line and the capacitance rating of the capacitor are known, the capacitive reactance can be found using the following formula:

$$X_C = \frac{1}{2 \times \pi \times F \times C}$$

The value of capacitive reactance is measured in ohms. In the formula to find capacitive reactance:

X_C = Capacitive reactance

π = The Greek letter pi, which has a value of 3.1416

F = Frequency in Hz

C = The value of capacitance in farads. Because most capacitors are rated in microfarads, be sure to write the capacitance value in farads. This can be done by dividing the microfarad rating by 1,000,000, or moving the decimal point six places to the left. Example: to change

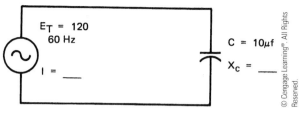

▶ **Figure 8–6**
Capacitive reactance limits current flow.

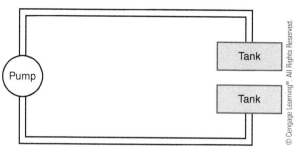

▶ **Figure 8–7**
Water flows in this system in a manner similar to the way current flows in a capacitive circuit.

a 50 µf capacitor to a value expressed in farads, move the decimal point after the 50 six places to the left. This capacitor has a value of 0.000050 farad.

EXAMPLE 2: Find the current flow in the circuit shown in Figure 8–6.

SOLUTION: To find the current flowing in this circuit, the amount of capacitive reactance of the capacitor must first be found.

$$X_c = \frac{1}{2 \times \pi \times F \times C}$$

$$X_c = \frac{1}{2 \times 3.1416 \times 60 \times 0.000010}$$

$$X_c = \frac{1}{0.0037699}$$

$$X_c = 265.2 \text{ ohms}$$

Now that the capacitive reactance of the circuit is known, the value of current can be found using the formula: $I = E/X_c$.

$$I = \frac{120}{265.2}$$

$$I = 0.452 \text{ amp}$$

CURRENT FLOW IN A CAPACITIVE CIRCUIT

Notice that a capacitor is constructed of two metal plates separated by an insulator. One of the metal plates is connected to one side of the circuit, and the other metal plate is connected to the other side of the circuit. Because there is an insulator separating the two plates, current cannot flow through a capacitor. When a capacitor is connected into a direct-current circuit, current will flow until the capacitor has been charged to the value of the applied voltage, and then stop. When a capacitor is connected into an alternating-current circuit, current will "appear" to flow through the capacitor. This is because the plates of the capacitor are alternately charged and discharged each time the current reverses direction. To understand this concept better, refer to the water circuit shown in Figure 8–7. In this illustration, a water pump is connected to two tanks. The pump is used to pump water back and forth between the two tanks. When one tank becomes full, the direction of the pump is reversed and water is pumped from the full tank back into the empty tank. Notice that there is no complete loop in this hydraulic circuit for water to flow from one side of the pump to the other, but water does flow because it is continuously pumped from one tank to the other.

CURRENT AND VOLTAGE RELATIONSHIPS

In a pure-capacitive circuit, the voltage and current are out of phase with each other. Figure 8–8 shows that the current in a pure-capacitive circuit leads the voltage by 90°. Because the voltage and current are 90° out of phase with each other, there is no true power or watts consumed in a pure-capacitive circuit. The capacitor stores the energy in an electrostatic field, and then returns it to the circuit at the end of each half cycle.

In the circuit shown in Figure 8–9, a resistor and capacitor are connected in series with each other. Because this circuit contains elements of both resistance and capacitive reactance, the current is

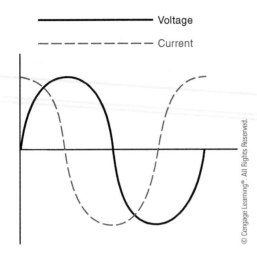

Figure 8–8
Current leads voltage by 90° in a pure capacitive circuit.

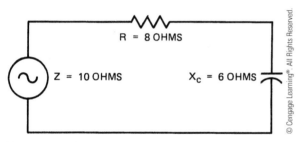

Figure 8–9
Impedance must be used to determine the current flow in a circuit that contains resistance and capacitive reactance.

limited by impedance. The impedance for a circuit of this type can be found by using the formula $Z = \sqrt{R^2 + X_C^2}$. Notice this is the same basic formula as the one used to find the impedance of a circuit that contains both resistance and inductive reactance. The impedance of the circuit shown in Figure 8–9 can be found by the following:

$$Z = \sqrt{R^2 + X_C^2}$$
$$Z = \sqrt{8^2 + 6^2}$$
$$Z = \sqrt{64 + 36}$$
$$Z = \sqrt{100}$$
$$Z = 10 \text{ ohms}$$

Figure 8–10 shows a vector diagram of the circuit in Figure 8–9.

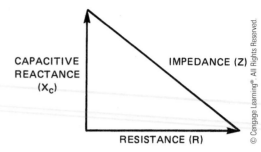

Figure 8–10
Impedance is the vector sum of R and X_C.

POWER FACTOR CORRECTION

Because the current flow in a capacitive circuit leads the voltage by 90° and the current in an inductive circuit lags the voltage by 90°, the current of a capacitive circuit is in direct opposition to the current of an inductive circuit. Figure 8–11 illustrates the currents of capacitive and inductive circuits as compared with each other. These two currents

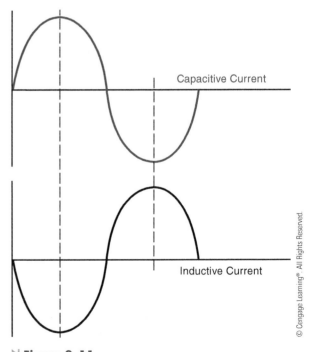

Figure 8–11
Capacitive and inductive current are 180° out of phase with each other.

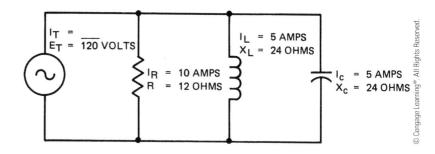

▶ **Figure 8–12**
A parallel circuit has resistance, inductance, and capacitance.

are 180° out of phase with each other. When the capacitive current is at its peak positive value, the inductive current is at its peak negative value. When the capacitive current is at its peak negative value, the inductive current is at its peak positive value. Because these two currents are in direct opposition, one can be used to cancel the other.

The circuit shown in Figure 8–12 shows a parallel circuit that contains a resistor, an inductor, and a capacitor. The applied voltage of the circuit is 120 volts at 60 Hz. Because this is a parallel circuit, the voltage applied to each component will be the same—120 volts. The resistor has a resistance of 12 ohms. This permits a current flow of 10 amps through the resistor (120/12 = 10). The inductor has an inductive reactance of 24 ohms. This permits a current flow through the inductor of 5 amps (120/24 = 5). The capacitor has a capacitive reactance of 24 ohms, which permits a current flow through the capacitor of 5 amps.

QUESTION

What is the total current flow in the circuit? In a parallel circuit, current is added. Therefore, it would appear that the current flow would be 20 amps (10 + 5 + 5 = 20). The currents of this circuit, however, are out of phase with each other. Figure 8–13 shows a vector diagram of this circuit. Notice that the 5 amps of capacitive current is 180° out of phase with the 5 amps of inductive current. These two currents will cancel each other. The AC alternator sees only the resistance in this circuit. The current is therefore the same as the current flow through the resistor, or 10 amps.

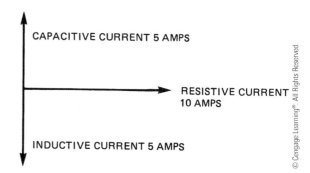

▶ **Figure 8–13**
Capacitive current and inductive current are 180° out of phase with each other.

POWER FACTOR CORRECTION OF A MOTOR

In the circuit shown in Figure 8–14, an AC induction motor is connected to a 120-volt line. A wattmeter is used to measure the amount of true power in the circuit. For this example, it is assumed that the wattmeter has a reading of 720 watts. An ammeter has also been inserted in the circuit. Assume the ammeter has a reading of 10 amps. The apparent power or volt-amp value for this circuit is 1200 VA (120 volts × 10 amps = 1200 VA). The power factor of this circuit can now be computed using the formula ($PF = W/VA$).

$$PF = \frac{W}{VA}$$

$$PF = \frac{720}{1200}$$

$$PF = 0.6 \text{ or } 60\%.$$

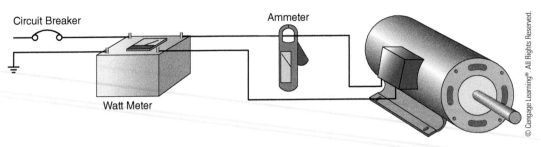

▶ **Figure 8–14**
Finding the power factor of a motor.

If the power factor of this motor is to be corrected, it must be determined how much of this circuit is comprised of true power and how much is composed of reactive power. Because the true power (watts) and the apparent power (volt-amps) is known, the reactive power (VARs) can be found using the following formula:

$$VARs = \sqrt{VA^2 - W^2}$$

$$VARs = \sqrt{1200^2 - 720^2}$$

$$VARs = \sqrt{1,440,000 - 518,400}$$

$$VARs = \sqrt{921,600}$$

$$VARs = 960$$

Because a motor is an inductive device, the reactive power in this circuit can be canceled by an equal amount of capacitive VARs. If a capacitor of the correct value is connected in parallel with the motor, the power factor will be corrected. To find the correct value of capacitance, determine the amount of capacitance needed to produce a VAR reading of 960. The amount of capacitive reactance can be found using the formula:

$$X_C = \frac{E^2}{VARs}$$

$$X_C = \frac{120^2}{960}$$

$$X_C = 15 \text{ ohms}$$

The amount of capacitance needed to produce 15 ohms of capacitive reactance at 60 Hz can be calculated using the following formula:

$$C = \frac{1}{2 \times \pi \times F \times X_C}$$

$$C = \frac{1}{2 \times 3.1416 \times 60 \times 15}$$

$$C = \frac{1}{5654.88}$$

$$C = 0.0001768 \text{ farad}$$

The answer for the value of C is in farads. To convert farads to microfarads, multiply the answer by 1,000,000, or move the decimal point 6 places to the right; 0.0001768 farad becomes 176.8 µf. If a capacitor of this value is connected in parallel with the motor as shown in Figure 8–15, the power factor will be corrected.

CAPACITOR TYPES

The most common types of capacitors used in the air-conditioning field fall into two categories. One kind is known as an oil-filled type. Figure 8–16 shows a photograph of this type of capacitor. The **oil-filled capacitor** is made with two metal foil plates separated by paper, Figure 8–16. The paper is soaked in a special dielectric oil. These capacitors are true AC capacitors and are generally used as the run capacitors on many single-phase air-conditioning compressors. They are also used as the starting capacitors on some units. The important ratings on these capacitors are the microfarad rating and the voltage rating. The voltage rating of a capacitor should never be exceeded. It is permissible to use a capacitor of

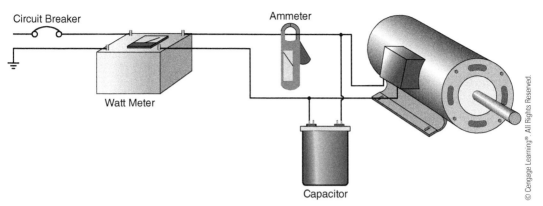

▶ **Figure 8–15**
A capacitor corrects the motor power factor.

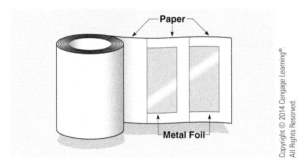

▶ **Figure 8–16**
An oil filled paper capacitor is made by separating two metal foil plates by paper soaked in a dielectric oil.

▶ **Figure 8–17**
AC electrolytic capacitor.

higher voltage rating, but never use a capacitor with less voltage rating.

The second type of capacitor frequently used in air-conditioning systems is the **AC electrolytic capacitor**, Figure 8–17. The AC electrolytic capacitor is used as the starting capacitor on many small single-phase motors. This type of capacitor is designed to be used for a short period of time only. If an AC

electrolytic capacitor were to be used in a continuous circuit such as the running capacitor of a compressor, it would fail in a short period of time. The advantage of the AC electrolytic capacitor is that a large amount of capacitance can be housed in a small case size. This makes the AC electrolytic capacitor a good choice for starting circuits, because the capacitor is in the circuit for only a few seconds when the motor is started.

TESTING A CAPACITOR

Capacitors can be tested for a short with an ohmmeter. If an ohmmeter is connected across the terminals of a capacitor as shown in Figure 8–18,

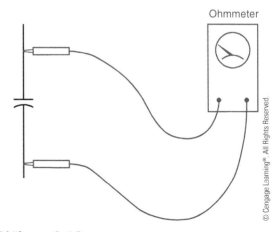

▶ **Figure 8–18**
Testing a capacitor with an ohmmeter.

the meter should show a deflection up scale and then return to infinity ohms. The deflection up scale indicates current flow to the capacitor when it is being charged by the ohmmeter battery. If the leads of the ohmmeter are reversed, the meter should deflect twice as far up scale and then return to infinity ohms.

The ohmmeter test basically indicates if the capacitor is shorted or not. A short indicates the dielectric has been punctured. This test will not indicate a broken plate, which would result in a lower capacitance value. Many digital meters contain a capacitance testing function, as shown in Figure 8–19. These meters actually measure the capacitance value, which can be compared to the rating marked on the capacitor.

Neither of these tests, however, can measure the dielectric strength. A capacitor may test okay with an ohmmeter or digital meter but break down when connected to line voltage. Ohmmeters and common digital meters do not supply enough voltage to test the dielectric at rated voltage. To test the dielectric strength, a dielectric test set should be used, as shown in Figure 8–20. The dielectric test set is sometimes referred to as a *hipot* because it provides a high potential or high voltage. The dielectric tester can provide rated voltage to the capacitor, and a microamperes meter measures any leakage current.

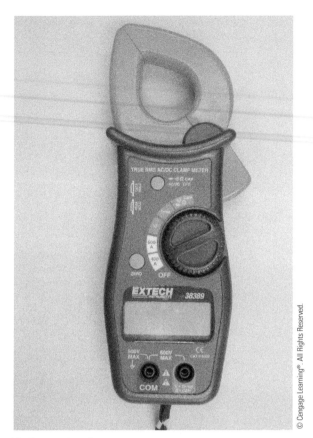

Figure 8–19
Digital meter capable of measuring capacitance.

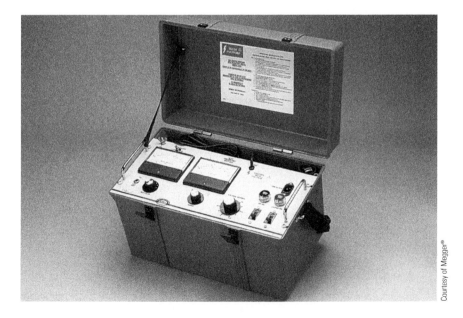

Figure 8–20
A dielectric test set.

SUMMARY

▶ A capacitor can be constructed by separating two metal plates with an insulating material.

▶ The insulating material is called the dielectric.

▶ Three factors that determine the amount of capacitance a capacitor will have are

A. The surface area of the plates.

B. The distance between the plates.

C. The type of dielectric material used.

▶ Most of the energy of a capacitor is stored in an electrostatic charge.

▶ Capacitors can produce extremely high current for a short period of time.

▶ The basic unit of capacitance is the farad.

▶ Capacitance values are generally rated in microfarads (µf), which are one-millionth of a farad.

▶ Capacitor voltage ratings given as VAC indicate the RMS value of voltage.

▶ Capacitor voltage ratings given as VDC or WVDC indicate the peak value of voltage.

▶ In an AC circuit containing pure capacitance, the current is limited by capacitive reactance.

▶ In a pure-capacitive circuit the current leads the voltage by 90 electrical degrees.

▶ Capacitors are often used to correct the power factor of a motor.

KEY TERMS

AC electrolytic capacitor
capacitance
capacitive reactance
dielectric

dielectric stress
electrostatic charge
farad
microfarad

oil-filled capacitor
picofarad
pure-capacitive circuit

REVIEW QUESTIONS

1. What three factors determine the capacitance of a capacitor?

2. What is the dielectric of a capacitor?

3. In what type of field is the energy of a capacitor stored?

4. In a pure-capacitive circuit, how many degrees are the current and voltage out of phase with each other?

5. Does a capacitive current lead the voltage or lag the voltage?

6. What limits the current in a pure capacitive circuit?

7. Name two common types of capacitors used in the air-conditioning field.

8. What type of capacitor is generally used as the running capacitor on many air-conditioning compressors?

9. What is the advantage of an AC electrolytic capacitor?

10. What is the disadvantage of an AC electrolytic capacitor?

11. A capacitor has a voltage rating of 150 VAC. Can this capacitor be safely connected to a 120 VAC circuit?

SECTION 2

Control Circuits

Schematics and Wiring Diagrams

Schematics and **wiring diagrams** are the written language of control circuits. It is impossible for a service technician to become proficient in troubleshooting electrical faults if he or she cannot read and interpret electrical diagrams. Learning to read electrical diagrams is not as difficult as many people first believe it to be. Once a few basic principles are understood, reading schematics and wiring diagrams will become no more difficult than reading a newspaper.

TWO-WIRE CIRCUITS

Control circuits are divided into two basic types, the two-wire and the three-wire. Figure 9–1 shows a simple **two-wire control circuit**. In this circuit, a simple switch is used to control the power applied to a small motor. If the switch is open, there is no complete path for current flow, and the motor will not operate. If the switch is closed, power is supplied to the motor, and it then operates.

THREE-WIRE CIRCUITS

Three-wire control circuits are used because they are more flexible than two-wire circuits. Three-wire circuits are characterized by the fact that they are operated by a magnetic relay or motor starter. These circuits are generally controlled by one or more pilot devices. Three-wire control circuits receive their name from the fact that three conductors or wires are required to make connection from a start–stop pushbutton station to a motor starter, Figure 9–2.

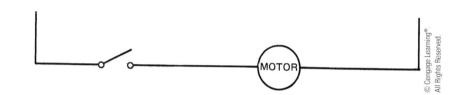

© Cengage Learning®. All Rights Reserved.

▶ **Figure 9–1**
Two-wire control circuit.

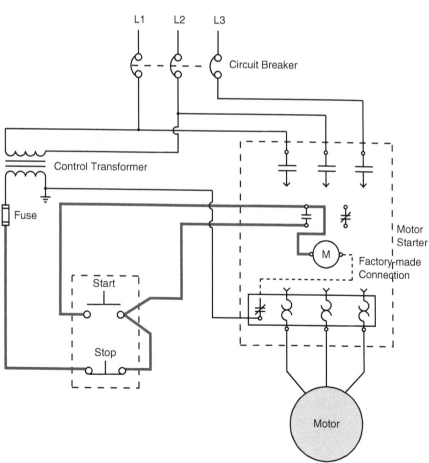

© Cengage Learning® All Rights Reserved.

▶ **Figure 9–2**
Basic three-wire control circuit.

ELECTRICAL SYMBOLS

When a person first learns to read, he or she learns a set of symbols that are used to represent different sounds. This set of symbols is generally referred to as the alphabet. When learning to read electrical diagrams, it is necessary to learn the symbols used to represent different devices and components. The symbols shown on the following page are commonly used on control schematics and wiring diagrams. These are not all the symbols used. Unfortunately, there is no set standard for the use of electrical symbols.

To better understand electrical symbols, it is helpful to understand some of the common terms used to describe these symbols. The terms **movable** and **stationary contacts**, for example, refer to electrical contacts located on different components. A simple switch contains both a movable and a stationary contact, Figure 9–3. Stationary contacts refer to contacts that cannot be moved or changed. Movable contacts can be moved from one position to another. The switch shown in Figure 9–3 can also be described as a **normally open** switch because

1. The movable contact is shown not touching the stationary contact.
2. The movable contact is drawn *below* the stationary contact.

Control components are drawn in the position they should be in when the circuit is de-energized or turned off. Normally open means that there is no electrical connection or complete circuit made between the movable and stationary contacts of the switch.

The switch in Figure 9–3 can be described as a *single-pole single-throw* (SPST) switch. Single pole indicates that the switch contains only one movable contact. Single throw means that the movable contact will complete a circuit when *thrown* or moved in only one direction.

A normally closed single-pole single-throw switch is shown in Figure 9–4. The switch is **normally closed** because

1. The movable contact is shown touching the stationary contact.
2. The movable contact is drawn above the stationary contact.

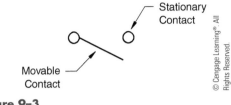

▶ **Figure 9–3**
Normally open, single-pole single-throw switch.

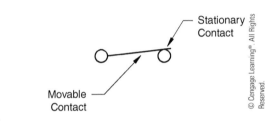

▶ **Figure 9–4**
Normally closed, single-pole single-throw switch.

When a component is normally closed, it indicates that a complete circuit exists through the component.

Although these basic rules apply to switches, they do not necessarily apply to all control components. When possible, components are generally drawn to indicate how they function. Pushbuttons, for example, are drawn differently than switches. A normally closed pushbutton is drawn with the movable contact below instead of above the stationary contacts, Figure 9–5. It is drawn in this manner to illustrate that when pressure is applied to the spring-loaded stationary contact. It causes the movable contact to move downward, breaking the connection between the two stationary contacts. A normally open pushbutton symbol is drawn with the movable contact above instead of below the stationary contacts. Pressure forces the movable contact downward, bridging the gap between the two stationary contacts and completing a circuit.

Pressure

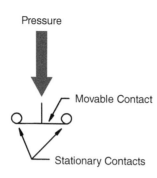

Movable Contact

Stationary Contacts

Normally Closed Pushbutton

Pressure

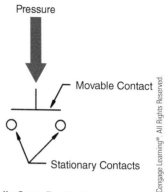

Movable Contact

Stationary Contacts

Normally Open Pushbutton

▶ **Figure 9–5**
Common pushbutton symbols.

Most of the following symbols are approved by the **National Electrical Manufacturers Association (NEMA)**:

1. Normally closed pushbutton. Generally used to represent a stop button.

2. Normally open pushbutton. Generally used to represent a start button.

3. Double-acting pushbutton. Contains both normally closed and normally open contacts on one pushbutton.

4. Double-acting pushbutton drawn differently but meaning the same as number 3.

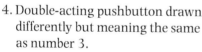

5. Double-acting pushbutton. The dashed line indicates mechanical connection. This means that when one button is pushed, the other moves at the same time.

6. Single-pole single-throw switch (SPST).

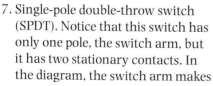

7. Single-pole double-throw switch (SPDT). Notice that this switch has only one pole, the switch arm, but it has two stationary contacts. In the diagram, the switch arm makes

contact with the upper stationary contact. When the switch is thrown, contact is made between the switch arm and the lower stationary contact. The switch arm, or pole, of the switch is generally referred to as the common because it can make contact with either of the two stationary contacts.

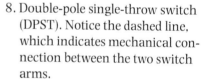

8. Double-pole single-throw switch (DPST). Notice the dashed line, which indicates mechanical connection between the two switch arms.

9. Double-pole double-throw switch (DPDT).

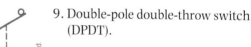

10. Off-Automatic-Manual control switch. This switch is basically a single-pole double-throw that has a center off position.

11. Normally open relay contact.

12. Normally closed relay contact.

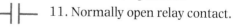

13. Fuse.

14. Fuse.

15. Transformer.

16. Coil.

17. Coil. (Generally used to represent the coil of a relay or motor starter in a control schematic.)

18. Pilot light or lamp.

19. Lamp or lightbulb.

20. Thermal heater element.

21. Thermal heater element. (Generally used to represent the overload heater element in a motor control circuit.)

22. Solenoid coil.

23. Fixed resistor.

24. Variable resistor.

25. Variable resistor.

26. Single-pole circuit breaker.

27. Double-pole circuit breaker.

28. Capacitor.

29. Normally closed float switch.

30. Normally open float switch.

31. Normally closed pressure switch.

32. Normally open pressure switch.

33. Normally closed temperature switch. (Normally closed thermostat.)

34. Normally open temperature switch.

35. Normally closed flow switch. This symbol is used to represent both liquid- and air-sensing flow switches.

36. Normally open flow switch.

37. Normally closed limit switch.

38. Normally open limit switch.

39. Normally closed ON-DELAY timer contact. Often shown on schematics as DOE, which stands for "delay on energize."

40. Normally open ON-DELAY timer contact.

41. Normally closed OFF-DELAY timer contact. Often shown on schematics as DODE, which stands for delay on de-energize.

42. Normally open OFF-DELAY timer contact.

43. Battery.

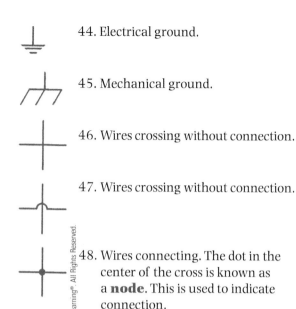

44. Electrical ground.

45. Mechanical ground.

46. Wires crossing without connection.

47. Wires crossing without connection.

48. Wires connecting. The dot in the center of the cross is known as a **node**. This is used to indicate connection.

49. Rotary switch.

▶ **Figure 9–6**
Normally closed pressure switch.

▶ **Figure 9–7**
Normally closed, held-open pressure switch.

▶ **Figure 9–8**
Normally open pressure switch.

▶ **Figure 9–9**
Normally open, held-closed pressure switch.

The contact symbols shown are standard and relatively simple to understand. There can be instances, however, in which symbols can be used to show something that is not apparent. For example, the symbol for a normally closed pressure switch is shown in Figure 9–6. Notice that this symbol not only shows the movable arm making contact with the stationary contact, but it also shows the movable arm drawn above the stationary contact. In Figure 9–7, the contact arm is shown not making connection with the stationary contact. This symbol, however, is not a normally open contact symbol because the contact arm is drawn above the stationary contact. This symbol indicates a normally closed, held-open pressure switch. This symbol is indicating that the switch is actually connected as a normally closed switch, but pressure is used to keep the contact open. If pressure decreases to a certain point, the switch contact will close.

Figure 9–8 shows a normally open pressure switch. Notice that the contact arm is drawn below the stationary contact. Figure 9–9 shows the same symbol except that the movable arm is making connection with the stationary contact. This symbol represents a normally open, held-closed pressure switch. This switch symbol indicates that the pressure switch is wired normally open, but pressure holds the contact closed. If the pressure decreases to a certain level, the switch will open and break connection to the rest of the circuit.

SCHEMATIC DIAGRAMS

Schematic diagrams *show components in their electrical sequence without regard for physical location.* Schematic diagrams are used to troubleshoot and install control circuits. Schematics are generally easier to read and understand than wiring diagrams.

WIRING DIAGRAMS

Wiring diagrams *show components mounted in their general location with connecting wires.* A wiring diagram is used to represent how the circuit generally appears. To help illustrate the differences between wiring diagrams and schematics, a basic control circuit will first be explained as a schematic and then shown as a wiring diagram.

READING SCHEMATIC DIAGRAMS

To read a schematic diagram, a few rules must first be learned. Commit the following rules to memory:

1. Reading a schematic diagram is similar to reading a book. It is read from left to right and from top to bottom.

2. Electrical symbols are always shown in their off or de-energized position.

3. Relay contact symbols are shown with the same numbers or letters that are used to designate the relay coil. All contact symbols that have the same number or letter as a coil are controlled by that coil, regardless of where in the circuit they are located.

4. When a relay is energized, or turned on, all of its contacts change position. If a contact is shown as normally open, it will close when the coil is energized. If the contact is shown as normally closed, it will open when the coil is turned on.

5. There must be a complete circuit before current can flow through a component.

6. Components used to provide a function of stop are generally wired normally closed and connected in series. Figure 9–10 illustrates this concept. Both switches A and B are normally closed and connected in series. If either switch is opened, connection to the lamp will be broken and current will stop flowing in the circuit.

7. Components used to provide the function of start are generally wired normally open and connected in parallel. In Figure 9–11, switches A and B are normally open and connected in parallel with each other. If either switch is closed, a current path will be provided for the lamp and it will turn on.

DASHED LINES

Often the service technician must be able to determine what dashed lines indicate in a schematic diagrams.

In the schematic diagram shown in Figure 9–12, dashed lines indicate several different conditions.

1. Mechanical connection between two components such as those shown in electrical symbols 5, 8, and 9. Each of these symbols show a dashed line connected between different components. The dashed line indicates that when one component is changed, the other one is changed at the same time. The double-acting pushbuttons illustrated in number 5 will both operate when one of them is pushed.

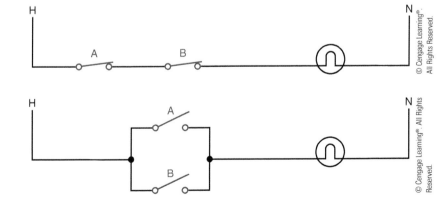

▶ **Figure 9–10**
Components used to perform the function of stop are normally closed and connected in series.

▶ **Figure 9–11**
Components used to perform the function of start are normally open and connected in parallel.

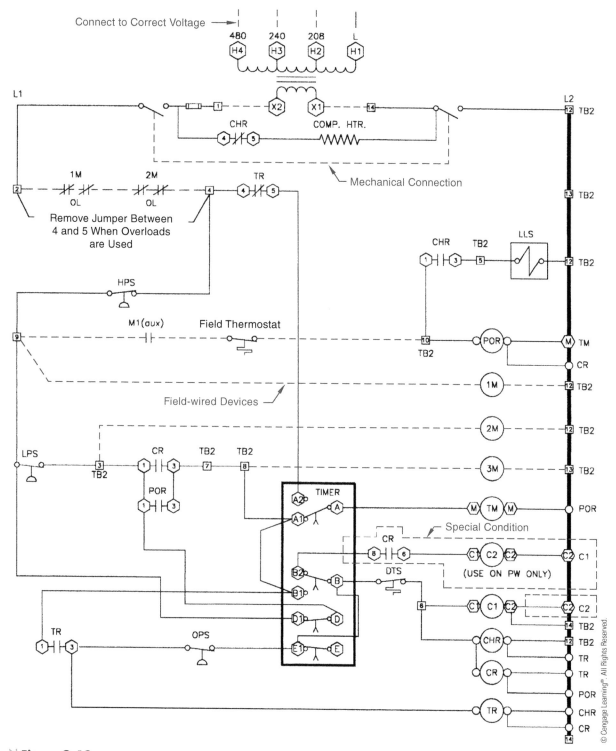

▶ **Figure 9–12**
Dashed lines can represent different conditions.

2. Field-wired or-installed components.

3. Components used only in special circumstances.

4. Factory-wired or-installed components.

At the top of the diagram, dashed lines indicate that only one set of the primary terminals is to be connected, depending on the amount of input voltage. An input voltage of 480 volts, for example, would have one line connected to the H1 terminal and the other to the H4 terminal. An input voltage of 208 volts would have one line connected to the H1 terminal and the other connected to the H2 terminal.

Another dashed line connected between two switches indicates that these two switches are mechanically connected. In reality, this would be a double-pole single-throw switch, Figure 9–13. The dashed line indicates that when one switch is opened or closed, the other one is opened or closed also.

Several components, such as 1M, 2M, and 3M are connected with dashed lines. The dashed lines in this instance indicate that these devices are field-wired and not part of the assembled unit. Wiring to these devices is connected during the installation of the equipment. It should be noted that some diagrams use dashed lines to indicate factory-installed wiring, and solid lines illustrate field-connected devices. There is no hard rule. It is generally necessary for the service person to determine the meaning of the lines on a particular schematic.

Another set of components shown in Figure 9–12 is surrounded by dashed lines. These dashed lines mark components that are used under special circumstances. Contactor coil, C2, and contact CR are installed only if the compressor motor utilizes part winding starting. If the motor does not employ part winding starting, these components will not be present.

EXAMPLE

The first circuit to be discussed is a basic control circuit used throughout industry. Figure 9–14 shows a start–stop pushbutton circuit. This schematic shows both the control circuit and the motor circuit. Schematic diagrams do not always show both control and motor connections. Many schematic diagrams show only the control circuit.

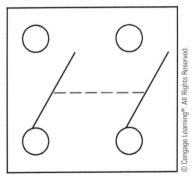

▶ **Figure 9–13**
Double-pole single-throw switch.

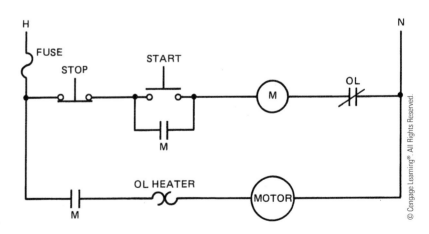

▶ **Figure 9–14**
Start–stop pushbutton control circuit.

Notice in this schematic that there is no complete circuit to M motor starter coil because of the open start pushbutton and open-M auxiliary contacts. There is also no connection to the motor because of the open-M load contacts. The open-M contacts connected in parallel with the start button are small contacts intended to be used as part of the control circuit. This set of contacts is generally referred to as the holding, sealing, or maintaining contacts. These contacts are used to provide a continued circuit to the M coil when the start button is released.

The second set of M contacts is connected in series with the overload heater element and the motor, and are known as **load** contacts. These contacts are large and designed to carry the current needed to operate the load. Notice that these contacts are normally open, and there is no current path to the motor.

When the start button is pushed, a path for current flow is provided to the M motor starter coil. When the M coil energizes, both M contacts close, Figure 9–15. The small auxiliary contact provides a continued current path to the motor starter coil when the start button is released and returns to its open position. The large M load contact closes and provides a complete circuit to the motor and the motor begins to run. The motor will continue to operate in this manner as long as the M coil remains energized.

If the stop button is pushed, Figure 9–16, the current path to the M coil is broken and the coil de-energizes. This causes both M contacts to return to their normally open position. When M holding contacts open, there is no longer a complete circuit provided to the coil when the stop button is returned to its normal position. The circuit remains in the off position until the start button is again pushed.

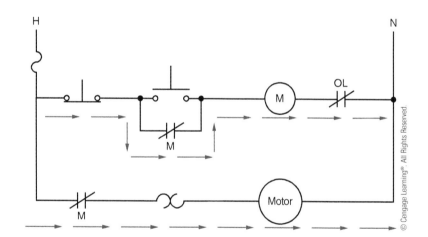

▶ **Figure 9–15**
Current path through the circuit.

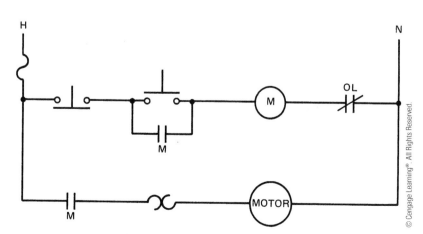

▶ **Figure 9–16**
The stop button breaks the circuit.

Notice that the overload contact is connected in series with the motor starter coil. If the overload contact should open, it has the same effect as pressing the stop button. The fuse is connected in series with both the control circuit and the motor. If the fuse should open, it has the effect of disconnecting power from the line.

A wiring diagram for the start–stop pushbutton circuit is shown in Figure 9–17. Although this diagram looks completely different, it is electrically the same as the schematic diagram. Notice the pushbutton symbols indicate double-acting pushbuttons. The stop button, however, uses only the normally closed section, and the start button uses only the normally open section. The motor starter shows three load contacts and two auxiliary contacts. One auxiliary contact is open and one is closed. Notice that only the open contact has been used.

The overload unit shows two different sections. One section contains the thermal heater element connected in series with the motor, and the normally closed contact is connected in series with the coil of the M motor starter.

EXAMPLE

The circuit shown in Figure 9–18 controls the operation of an oil-fired boiler. A high-pressure pump motor is used to inject fuel oil into a combustion chamber where it is burned. A blower motor is used to supply combustion air to the chamber. The circuit does not permit fuel oil to be injected into the chamber unless the blower motor is operating. The circuit also permits the blower motor to continue operation for a period of 1 minute after the thermostat is satisfied. This permits any residual smoke or fumes to be removed from the combustion chamber.

The first step in understanding the operation of the circuit is to examine the components and determine what they control. The thermostat is a normally closed, held-open switch. It is normally closed because the movable contact is drawn above the stationary contact. The movable contact is not making connection to the stationary contact, however. This indicates that the contact is being held open. The thermal symbol indicates that the contact is controlled by temperature. The thermal symbol represents a bimetal helix. An increase in temperature causes the helix to expand and push upward on the contact. A decrease in temperature causes the helix to contract. If the helix contracts enough, the movable contact will make connection with the stationary contact and close the switch. This thermostat symbol indicates that an increase of temperature will open the switch and a decrease of temperature will close the switch. This is the normal operation of a heating thermostat.

The high temperature switch is a thermally activated switch also. The switch is shown normally closed. If the temperature should increase high enough, the switch will open and break connection to the high pressure pump motor relay and time-delay relay.

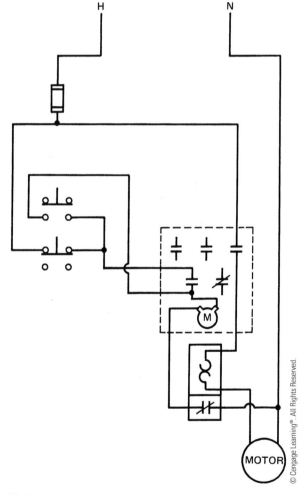

▶ **Figure 9–17**
Wiring diagram of start–stop pushbutton control circuit.

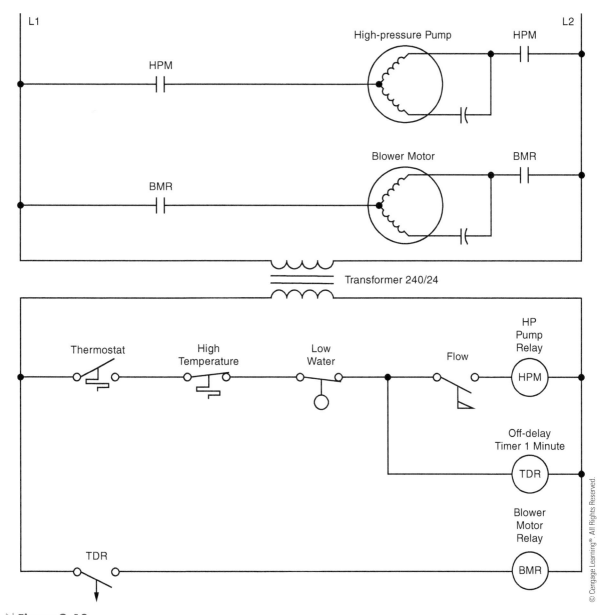

Figure 9–18
Boiler control circuit.

The low-water switch is a normally open, held-closed switch. The switch is normally open because the movable contact is drawn below the stationary contact. Because the movable contact is touching the station contact, it is being held closed. This switch is drawn to indicate that a drop in liquid level will cause the switch contacts to open and break the circuit to the high-pressure pump motor relay and time-delay relay. One of the most dangerous conditions for a boiler is a low water level. If the water level should drop below a preset point, the switch will open.

The flow switch is normally open. A flow of air causes the switch contacts to close. The flow switch is used to ensure that there is a flow of combustion air into the combustion chamber before fuel oil is injected into the chamber.

The time-delay contact is connected in series with the blower motor relay coil. The symbol indicates that the timer is an off-delay timer. The arrow always points in the direction the contacts will move after the delay period. The arrow indicates that the contacts will delay reopening after they have changed position.

CIRCUIT OPERATION

When the thermostat contact closes, a circuit is completed through the closed high-temperature switch, low-water switch, and coil of the time-delay relay. Because the timer is an off-delay timer, TDR contacts close immediately and energize the blower motor relay, Figure 9–19. This causes the BMR

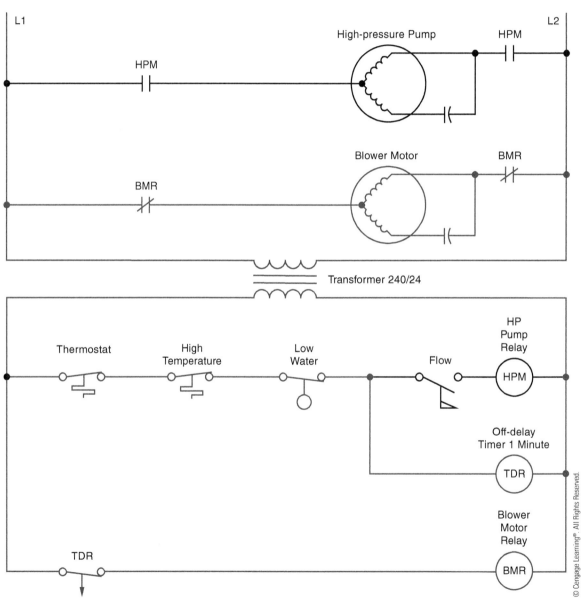

▶ **Figure 9–19**
The thermostat contacts close.

contacts to close and connect the blower motor to the power line.

The airflow produced by the blower motor causes the flow switch to close and energize the coil of the high-pressure pump motor relay, Figure 9–20. This causes the HPM contacts to close and connect the high-pressure pump to the line. The circuit is now in full operation and will continue to operate in this manner until the thermostat contacts reopen, Figure 9–21. When the thermostat contacts reopen, power is disconnected from the high-pressure pump motor relay and the time-delay relay. Because the

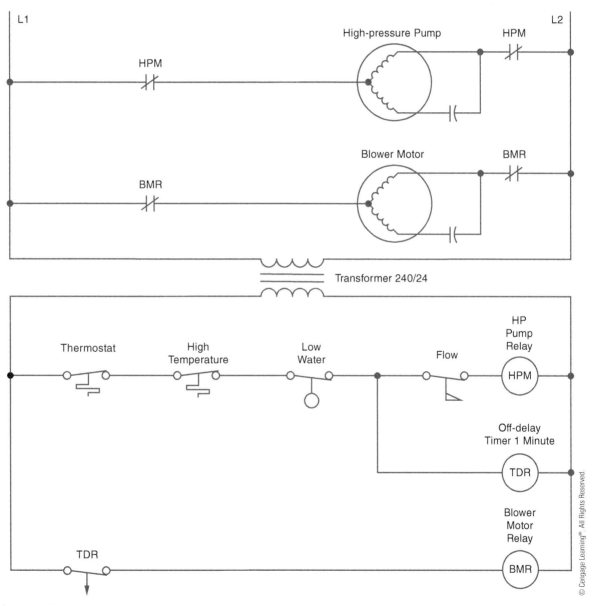

▶ **Figure 9–20**
The circuit is in full operation.

time-delay relay is an off-delay timer, it does not start timing until the coil is de-energized. The TDR contacts will remain closed for a period of 1 minute before reopening. This permits the blower motor to remove any smoke and fumes from the combustion chamber. When the TDR contacts open, the circuit is back in its original state, as shown in Figure 9–18. If the high-temperature switch or low-water switch should open, it would have the same effect as opening the thermostat.

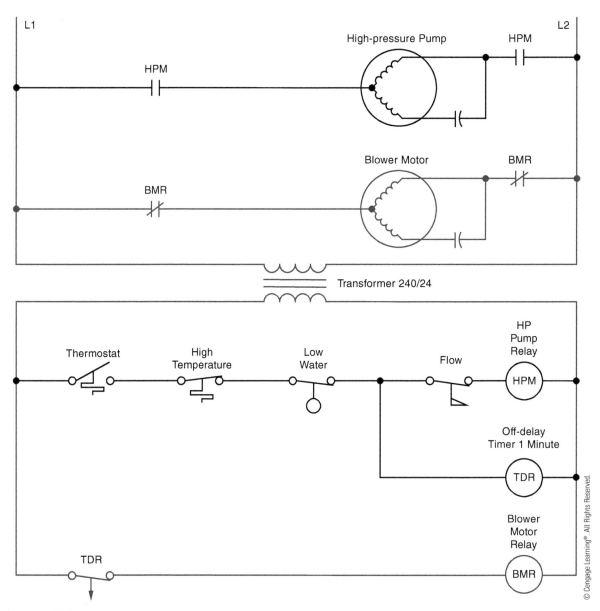

▶ **Figure 9–21**
The thermostat contacts open.

SUMMARY

- ▷ Schematics and wiring diagrams are the written language of control systems.
- ▷ Motor control symbols are generally drawn to pictorially represent their function.
- ▷ The way a motor control symbol is drawn indicates how it is to be connected in a circuit.
- ▷ Schematic diagrams show components in their electrical sequence without regard for physical location.
- ▷ Wiring diagrams show a pictorial representation of the circuit with connecting wires.
- ▷ Schematics and wiring diagrams always show the circuit in its de-energized, or off, position.

KEY TERMS

load

movable contact

National Electrical
   Manufacturers
   Association (NEMA)

node

normally closed

normally open

schematics

stationary contact

three-wire control circuits

two-wire control circuit

wiring diagrams

REVIEW QUESTIONS

1. What are the two basic types of motor controls?

2. Define a schematic diagram.

3. Define a wiring diagram.

4. Components used for the function of stop are generally wired _____ and connected in _____.

5. Components used for the function of start are generally wired _____ and connected in _____.

6. When reading a schematic diagram, are the components shown in their energized or de-energized position?

7. What does this symbol represent?

8. What does this symbol represent?

9. What does a dashed line drawn between components represent?

10. What is an auxiliary contact?

11. Make a schematic drawing of a cooling thermostat that turns on the coil of a contactor. Label the contactor "CC," which stands for "compressor contactor." Make sure that the circuit is drawn so that an increase in temperature will cause the contact to close.

12. Draw a schematic that controls the operation of a sump pump. A float switch is used to turn the pump on when the water rises to a high enough level. It is assumed that the float switch has contacts rated high enough to control the motor without the use of a relay.

13. Draw a schematic diagram for a low-water-level alarm. A float switch is used to detect the level of water in a tank. If it should fall below a predetermined level, a warning light will come on and an alarm will sound.

14. Draw a circuit for a low-pressure cutoff switch. If the pressure falls below a certain level, it will cause a contactor coil to be disconnected. Label the contactor "CC."

15. Add a thermal switch to the circuit in question 14. The switch is to be installed so that a rise in temperature will disconnect the contactor coil.

Developing Wiring Diagrams

OBJECTIVES

After studying this unit, the student should be able to:

- ▷ Discuss the operation of an electric circuit by interpreting a schematic diagram
- ▷ Place wire numbers on a schematic diagram and develop a wiring diagram from the schematic

In this unit, two schematic diagrams are discussed, including their operation and development into wiring diagrams. Developing a wiring diagram from a schematic is the same basic procedure that is followed when installing a control system. Understanding this process is also a great advantage when troubleshooting existing circuits.

DEVELOPING CIRCUIT 1

The first circuit discussed is shown in Figure 10–1. In this circuit, a fan motor is controlled by relay **FR (fan relay)**. The circuit is designed so that a switch can be used to turn the circuit completely off, operate the fan manually, or permit the fan to be operated by a thermostat. If the control switch is moved to the "MAN" position, as shown in Figure 10–2, a complete circuit is provided to the

111

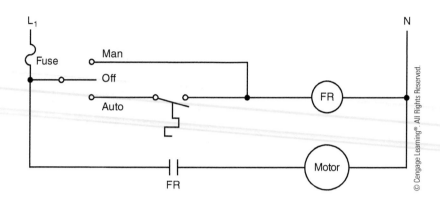

▶ **Figure 10–1**
Fan control circuit.

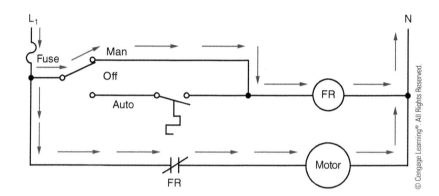

▶ **Figure 10–2**
Fan relay coil is energized by
control switch.

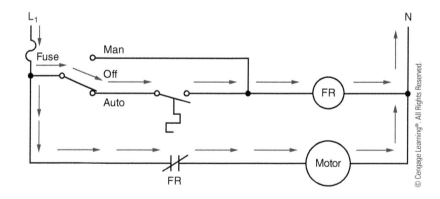

▶ **Figure 10–3**
Fan relay is controlled by the
thermostat.

fan relay coil. Then the relay energizes, and the FR contact closes and connects the motor to the line. This setting permits the fan to be operated at any time, regardless of the condition of the thermostat.

If the control switch is moved to the "AUTO" position, as shown in Figure 10–3, the fan will be controlled by the action of the thermostat. When the temperature increases to a predetermined level, the thermostat contact will close. This completes a circuit to FR coil. When FR coil energizes, the FR contact closes and connects the fan motor to the line. When the temperature decreases sufficiently, the thermostat contact opens and breaks the circuit to FR coil. When FR coil de-energizes, FR contact opens and disconnects the motor from the line.

This schematic will now be developed into a wiring diagram. To aid in the connection of this circuit, a simple numbering system will be used. To use

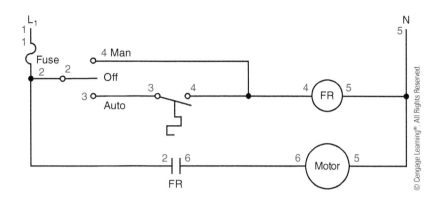

▶ Figure 10–4
Schematic is numbered to aid in connection of the circuit.

this numbering system, the following rules will be followed:

1. All components connected to the same line will receive the same number.
2. Any time you go through a component has, the number will change.
3. A set of numbers can be used only once.

Figure 10–4 shows the numbers placed on the schematic. Notice that a 1 is placed at the incoming power line and a 1 is also placed at one side of the fuse. Because the fuse is a component, the number must change on the other side of it. Therefore, the fuse has a 2 on the other side. There is also a 2 placed beside the common terminal of the OFF-AUTOMATIC-MANUAL switch, and a 2 placed beside one side of FR contact. Notice that all of these components have the same number because there is no break between them.

The AUTO side of the switch has been numbered 3, and one side of the thermostat has also been numbered 3. The other side of the thermostat is numbered 4; the MAN side of the switch is number 4; and one side of FR coil is numbered 4. The other side of the coil has been numbered 5, the neutral line is numbered 5, and one side of the motor is numbered 5. The other side of the motor is numbered 6, and the other side of FR contact is numbered 6.

Notice that all the points that are electrically connected together have the same number. Notice also that no set of numbers was used more than once.

The components of the system are shown in Figure 10–5. Notice that numbers have been placed beside certain components. These numbers

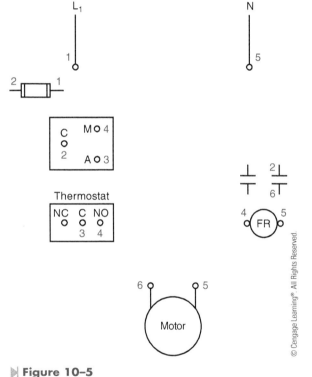

▶ Figure 10–5
Circuit components are numbered with the same numbers that appear on the schematic.

correspond to the numbers in the schematic. For example, the fuse in the schematic is shown with a 1 on one side and a 2 on the other side. The fuse in the wiring diagram is shown with a 1 on one side and a 2 on the other side. Notice the OFF-AUTOMATIC-MANUAL switch shown on the schematic. The common terminal is numbered 2, the

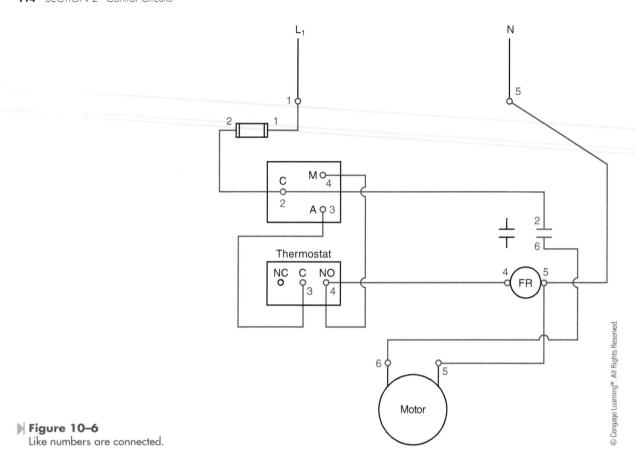

▶ **Figure 10–6**
Like numbers are connected.

MAN terminal is numbered 4, and the AUTO terminal is numbered 3. Now notice the same switch on the wiring diagram. The common terminal is numbered 2; the MAN terminal is numbered 4; and the AUTO terminal is numbered 3. The thermostat in the schematic has been numbered 3 on one terminal and 4 on the other terminal. The thermostat shown in the wiring diagram has three terminals. One terminal is common; one terminal is marked NC; and the other terminal is marked NO. This is a common arrangement for many control components. This shows that the thermostat is a single-pole double-throw switch. Because the thermostat shown in the schematic is normally open, the 3 will be placed beside the common terminal, and the 4 will be placed beside the NO contact. Notice that one of the contacts on FR relay is numbered 2 on one side and 6 on the other side. FR coil is numbered

4 on one side and 5 on the other. One motor terminal is numbered 5 and the other is numbered 6.

Now that the component parts have been numbered with the same numbers as those used on the schematic, connection can be made easily and quickly. To connect the circuit, connect all the like numbers. For example, all the number 1s will connect together; all the number 2s will connect together; and so forth. The connected circuit is shown in Figure 10–6.

DEVELOPMENT OF CIRCUIT 2

The schematic diagram for circuit 2 is shown in Figure 10–7. Notice that this schematic shows both the control circuit and the motor connection. This circuit is designed to turn off a compressor if the pressure in the system reaches a predetermined

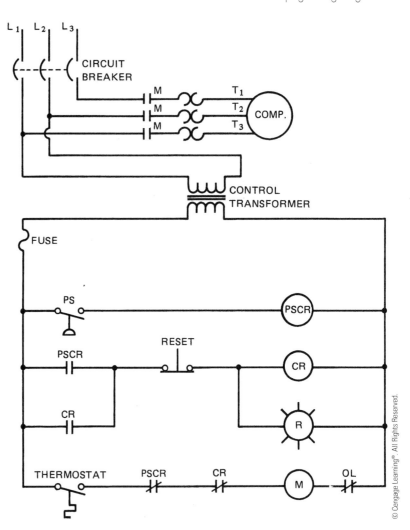

▌ **Figure 10–7**
High pressure locks
compressor off.

level. If the pressure becomes high enough to cause the pressure switch contacts to close, the compressor motor will not only be disconnected from the power line but a warning light will also be turned on. Once the warning light has been turned on, the system must be manually reset by the service technician before the compressor can be restarted by the thermostat. The operation of the circuit is as follows.

When the thermostat contact closes, a circuit is completed to M motor starter coil. When M coil energizes, M contacts close and connect the compressor to the three-phase power line. When the temperature decreases, the thermostat contacts open and de-energize M coil. When M coil de-energizes, M contacts open and disconnect the compressor from the power line. Notice that in the normal action of this circuit, the compressor is controlled by the thermostat.

Now assume that the thermostat contacts are closed and that the compressor is connected to the power line. Also assume that the pressure in the system becomes too great and that the contacts of the pressure switch close. When the pressure switch contacts close, **PSCR (pressure switch control relay)** coil energizes. This causes both

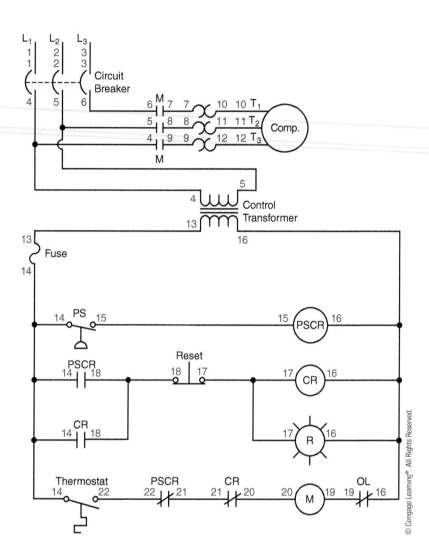

▶ **Figure 10–8**
Schematic is numbered to aid in
circuit connection.

PSCR contacts to change position. When the normally closed PSCR contact opens, the circuit to M coil is broken. This causes the compressor to be disconnected from the line. When the normally open PSCR contact closes, a circuit is completed to CR (control relay) coil. When CR coil energizes, both CR contacts change position. The normally open CR contact closes to maintain a circuit to CR coil in the event that the pressure in the system decreases and opens the pressure switch contacts. This would cause PSCR relay to de-energize and return both PSCR contacts to their normal position. The normally closed CR contact will open. This prevents M coil from being energized by the

thermostat if PSCR contact should reclose. Notice that the warning light is connected in parallel with the coil of CR relay. The warning light will be turned on as long as CR relay coil is turned on. As long as CR relay is energized, the compressor cannot be restarted by the thermostat.

Now assume that the pressure in the system has returned to normal and the problem that caused the excessive pressure has been corrected. When the pressure switch contact reopened, PSCR coil de-energized and reset both PSCR contacts to their normal position. When the service technician presses the reset button, CR coil de-energizes and both CR contacts return to their normal positions. The circuit

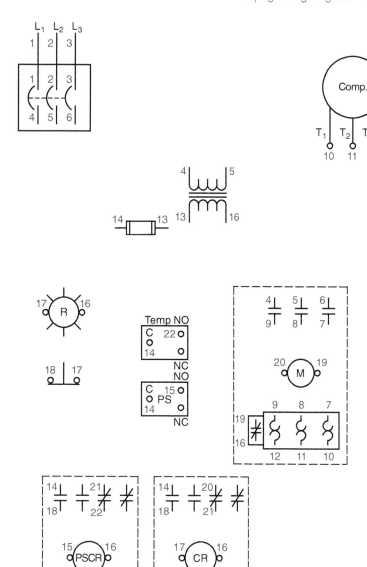

▶ **Figure 10–9**
Components are numbered the same in
the schematic.

is now back in its original position and ready for normal operation.

This schematic diagram is now being developed
into a wiring diagram. As before, the schematic is
numbered in the same manner as the first example.
The numbered schematic is shown in Figure 10–8.
Notice that all components that are electrically tied
together have the same number. Also notice that no
number set has been used more than once.

The control components are shown in
Figure 10–9. Notice that the numbers on the
components correspond with like numbers on
the schematic diagram. For example, on the
schematic diagram the primary of the control
transformer is numbered 4 and 5. The secondary
leads are numbered 13 and 16. Notice the same
is true on the wiring diagram. Note the number
of each component on the schematic and then

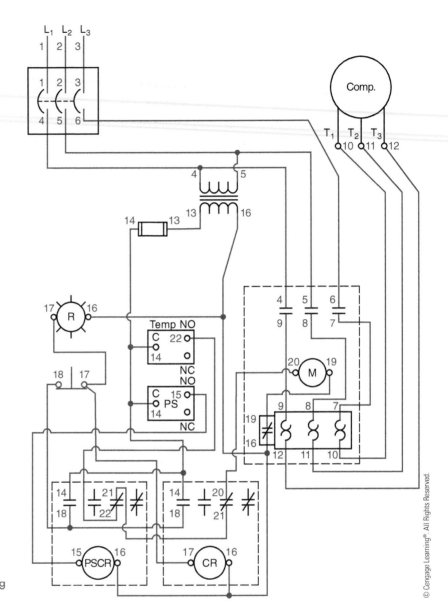

▶ **Figure 10–10**
Connection is made by connecting
like numbers.

find the corresponding number beside the proper component used on the wiring diagram.

Once the components of the wiring diagram have been numbered with the same numbers as those on the schematic, the circuit can be connected by connecting like numbers. The circuit connection is shown in Figure 10–10. Again, notice that the wiring diagram appears to be completely different from the schematic, but both are the same electrically.

SUMMARY

- Wiring diagrams are generally developed from a schematic diagram.
- All components in a schematic diagram are shown in their de-energized, or off, position.
- Wire numbers are often placed on schematic diagrams to aid in connecting a wiring diagram.
- Three basic rules for placing wire numbers on a schematic diagram are as follows:
    A. All components connected to the same line will receive the same number.
    B. Any time you go through a component has, the number will change.
    C. A set of numbers can be used only once.

KEY TERMS

FR (fan relay)
PSCR (pressure switch control relay)

REVIEW QUESTIONS

Refer to Figure 10–1 for the following questions.

1. Explain the action of the circuit if the thermostat should fail to operate.

2. Explain the action of the circuit if FR contacts should become shorted together.

Refer to Figure 10–7 for the following questions.

3. Explain the action of the circuit if the overload (OL) contact should open.

4. Explain the action of the circuit if the pressure switch contacts should become shorted.

5. Explain the action of the circuit if the CR coil should open.

SECTION 3

Motors

UNIT 11 ▷ Split-Phase Motors

OBJECTIVES

After studying this unit, the student should be able to:

≫ List the basic types of split-phase motors

≫ Discuss two-phase power

≫ Discuss the operation of a resistance-start induction-run motor

≫ Discuss the operation of a capacitor-start induction-run motor

≫ Discuss the operation of a permanent split-capacitor motor

≫ Reverse the direction of rotation of a split-phase motor

≫ Connect dual voltage motors for 120- or 240-volt operation

≫ Identify the terminal markings of an oil-filled capacitor, and discuss the proper connection to a permanent split-capacitor motor

Because three-phase power is not available to small business and residential locations, the air-conditioning equipment for these areas is powered by single-phase electric motors. The single-phase motors used in air-conditioning systems are generally one of two types: the split-phase and the shaded-pole induction motor.

SPLIT-PHASE MOTORS

Split-phase motors fall into three general classifications.

1. The **resistance-start induction-run motor**,

2. The **capacitor-start induction-run motor**,

3. The **permanent split-capacitor motor (PSC)**. It should be noted that although all permanent split-capacitor motors contain a run capacitor, some are equipped with a separate starting capacitor to improve starting torque. The motors with the separate starting capacitor are often referred to as **capacitor-start capacitor-run motors**.

Although all of these motors have different operating characteristics, they are similar in construction. Split-phase motors get their name from the manner in which they operate. These motors operate on the principle of a **rotating magnetic field**. A rotating magnetic field, however, cannot be produced with only one phase. Split-phase motors literally split single-phase power in order to imitate a two-phase power system. A rotating magnetic field can be produced with two separate phases.

THE TWO-PHASE SYSTEM

In some parts of the world, **two-phase power** is produced. A two-phase system is produced by having an alternator with two sets of coils wound 90° out of phase with each other, Figure 11–1. The voltages of a two-phase system are therefore 90° out of phase with each other, Figure 11–2. The two out-of-phase voltages can be used to produce a rotating magnetic field. Because there have to

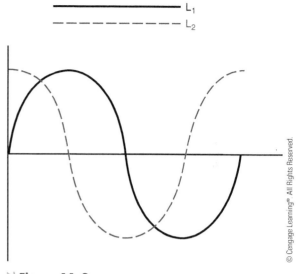

Figure 11–2
Two-phase voltages are 90° out of phase with each other.

be two voltages or currents out of phase with each other to produce a rotating magnetic field, single-phase motors use two separate windings and create a phase difference between the currents in each of these windings. These motors literally "split" one phase and produce a second phase, hence the name split-phase motor.

RESISTANCE-START INDUCTION-RUN AND CAPACITOR-START INDUCTION-RUN MOTORS

Resistance-start induction-run and capacitor-start induction-run motors are very similar in construction. The stator winding of both motors contains both a **start winding** and a **run winding**. The start winding is made of smaller wire and placed higher in the metal core material than the run winding, as shown in Figure 11–3. Because the start winding is made with smaller wire than the run winding, it exhibits a higher resistance than the run winding. Placing the run winding deeper in the metal core material causes it to exhibit a greater amount of inductance than the start winding. Electrically, the winding appears similar to the circuit shown in Figure 11–4. The stator is constructed in this

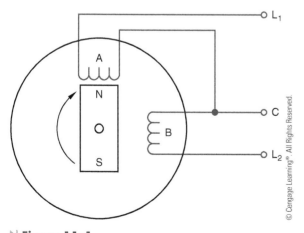

Figure 11–1
Two-phase alternator.

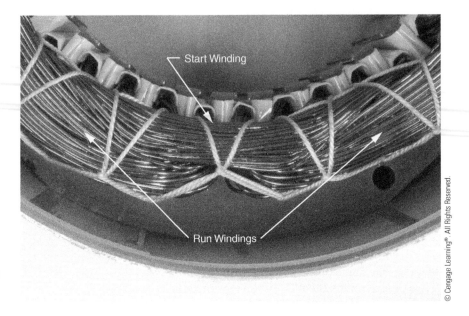

Figure 11–3
Stator winding of a resistance-start induction-run or capacitor-start induction-run motor.

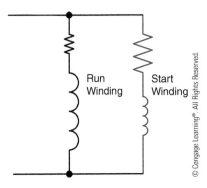

Figure 11–4
The run winding has more inductance and less resistance than the start winding. The start winding has more resistance and less inductance than the run winding.

manner to produce a phase shift between the current flowing through the run winding and the current flowing through the start winding. Both the resistance-start and capacitor-start induction-run motors start rotation by producing a rotating magnetic field in the stator winding. Recall that a rotating magnetic field cannot be produced with a single phase.

Resistance-Start Induction-Run Motor

The rotating magnetic field of the resistance-start induction-run motor is produced by the out-of-phase currents in the run and start windings. Because the run winding appears more inductive and less resistive than the start winding, the current flow in the run winding will be close to 90° out of phase with the applied voltage. The start winding appears more resistive and less inductive than the run winding, causing the start winding's current to be less out of phase with the applied voltage, as shown in Figure 11–5. The phase-angle difference between current in the run winding and current in the start winding of a resistance-start induction-run motor is generally 35° to 40°. This is enough phase-angle difference to produce a weak rotating field, and consequently a weak torque, to start the motor. Once the motor reaches about 75% of its rated speed, the start winding is disconnected from the circuit and the motor continues to operate on the run winding. In nonhermetically sealed motors, the start winding is generally disconnected with a **centrifugal switch.** A centrifugal switch is shown in Figure 11–6. The contacts of the centrifugal switch are connected in series with the start

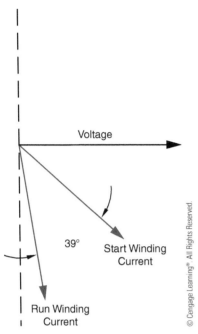

Figure 11–5
The run winding current and start winding current of a resistance-start induction-run motor will generally be between 35° and 40° out of phase with each other.

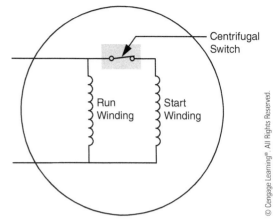

Figure 11–7
The centrifugal switch contacts are connected in series with the start winding.

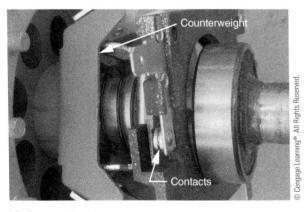

Figure 11–6
A centrifugal switch is used to disconnect the start windings when the motor reaches about 75% of rated speed.

Figure 11–8
Squirrel-cage rotor.

winding, as shown in Figure 11–7. When the motor is at rest or not running, the contacts of the centrifugal switch are closed and provide a circuit to the start winding. When the motor is started and reaches about 75% of its rated speed, a counterweight on the centrifugal switch moves outward because of centrifugal force, causing the contacts to open and disconnect the start winding from power. The motor continues to operate on the run winding.

When the start winding is disconnected from the circuit, a rotating magnetic field is no longer produced in the stator. This type of motor continues to operate because of current induced in the squirrel-cage windings in the rotor. Squirrel-cage rotors are so named because they contain bars inside the rotor that would resemble a squirrel cage if the laminations were removed, as shown in Figure 11–8.

A squirrel cage is a device that is often placed inside the cage of small pets such as hamsters to permit them to exercise by running inside the squirrel cage. A squirrel-cage rotor that has been cut in half clearly shows the bars and motor shaft, as shown in Figure 11–9. The bars of the turning squirrel-cage rotor winding cut through lines of magnetic flux, causing an induced voltage in the rotor. Because the rotor bars are shorted together at each end, current flow through the rotor bars produces a magnetic field in the rotor. Alternate magnetic fields are produced in the rotor, causing the motor to continue operating, as shown in Figure 11–10. This is the same principle that permits a three-phase motor to continue operating if one phase is lost and the motor is connected to single-phase power. The main difference is that the split-phase motor is designed to operate in this condition and the three-phase motor is not. Resistance-start and capacitor-start induction-run motors are rugged and will provide years of service with little maintenance. Their operating characteristics, however, are not as desirable as those of other types of single-phase motors. Due to the way they operate, they have a low power factor. They draw almost as much current when the motor is running at no load as they do when the motor is running at full load. Typically, if the motor has a full-load current draw of 8 amperes, the no-load current may be 6.5 to 7 amperes.

Capacitor-Start Induction-Run Motors

Capacitor-start induction-run motors are very similar to resistance-start induction-run motors. The design of the stator winding is basically the same. The main difference is that a capacitor is connected in series with the start winding, as shown in Figure 11–11. Inductive loads cause the current to lag the applied voltage. Capacitors, however, cause the current to lead the applied voltage. If the starting capacitor is sized correctly, the start winding current will lead the applied voltage by an amount that will result in a 90° phase shift between the run winding current and the start winding current, producing an increase in the amount of starting torque, as shown in Figure 11–12. If the capacitance of the start capacitor is too great, it will cause the start winding current to shift more than 90° out of phase with the run winding current, and starting torque will be reduced. When replacing the start capacitor for this type

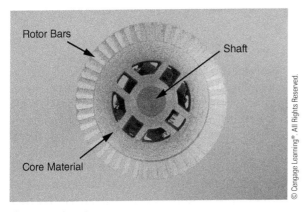

▶ **Figure 11–9**
Bars and shaft of a squirrel-cage rotor.

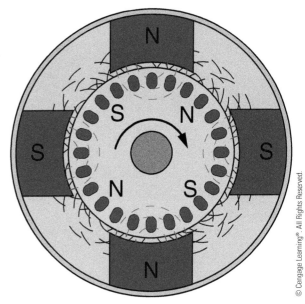

▶ **Figure 11–10**
The rotor continues to turn because of magnetic fields produced by the current induced in the rotor of the motor.

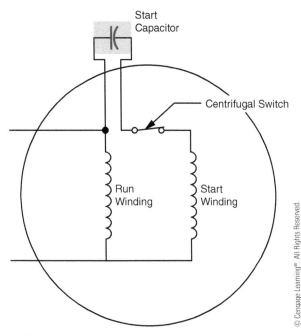

▶ **Figure 11–11**
A capacitor is connected in series with the start winding.

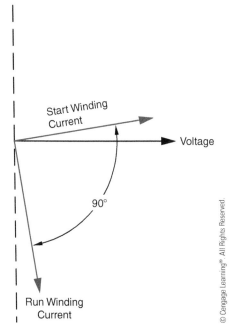

▶ **Figure 11–12**
The capacitor causes a 90° phase shift between run winding current and start winding current.

▶ **Figure 11–13**
A capacitor-start induction-run motor.

▶ **Figure 11–14**
Typical starting capacitor.

of motor, the microfarad rating recommended by the manufacturer should be followed. It is permissible to use a capacitor with a higher voltage rating, but never install a capacitor with a lower voltage rating. A capacitor-start induction-run motor is shown in Figure 11–13. A typical starting capacitor is shown in Figure 11–14.

TESTING THE STATOR WINDING

The stator winding of a single-phase motor is generally tested with an ohmmeter. The ohmmeter test can be used to determine whether a winding is open or grounded. Many single-phase motors have one lead of the run and start windings connected as shown in Figure 11–15. To test the windings for an open, connect one ohmmeter lead to the common motor terminal, and the other meter lead to the run winding. The ohmmeter should indicate continuity through the winding. The resistance of the run winding of a single-phase motor can vary greatly from one motor to another. The winding resistance of a single-speed motor may be only 1 or 2 ohms, whereas the resistance of a multispeed fan motor may be 10 to 15 ohms.

To test the start winding for an open, connect the ohmmeter leads to the common terminal and the S terminal. The start winding should indicate continuity and should have a higher resistance than the run winding. This difference of resistance may not be great, but the start winding should have a higher resistance than the run winding.

To test the stator winding for a ground, connect one of the ohmmeter leads to the case of the motor, Figure 11–16. Alternately check each motor terminal with the other ohmmeter lead. The ohmmeter should indicate no continuity between either winding and the case of the motor.

A shorted start winding can sometimes be detected by the fact that the motor will not start but will run if the shaft is turned by hand. The motor will produce a humming sound but will not turn when power is first applied to it. The shaft can be turned in either direction by hand, and the motor will continue to run in that direction.

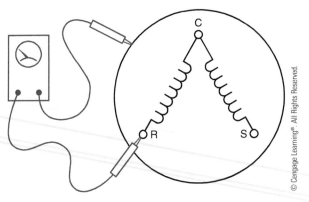

▶ **Figure 11–16**
Testing a split-phase motor for a grounded winding.

REVERSING DIRECTION OF ROTATION

The direction of rotation of a split-phase motor can be reversed by changing the start winding leads or the run winding leads, but not both. The rotation is generally reversed by changing the start winding leads with respect to the run winding. Some motor manufacturers bring both start winding leads to the outside of the motor. This permits the service technician to decide the direction of rotation the motor is to turn when it is installed on the unit.

DUAL-VOLTAGE MOTORS

Single-phase motors can also be constructed to operate on two separate voltages. These motors are designed to be connected to 120 or 240 volts. A common connection for this type of motor contains two run windings and one start winding, Figure 11–17. The run windings are labeled T1–T2, and T3–T4. The start winding is labeled T5 and T6. In the circuit shown in Figure 11–17, the windings have been connected for operation on a 240-volt line. Each winding is rated at 120 volts. The two run windings are connected in series, which causes each to have a voltage drop of 120 volts when connected to 240 volts. Notice that the start winding has been connected in parallel with one of the run windings. This causes the start winding to have an applied

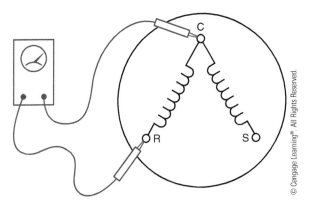

▶ **Figure 11–15**
Testing the split-phase motor for an open winding.

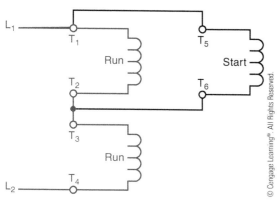

Figure 11–17
The run windings are connected in series for a high-voltage connection.

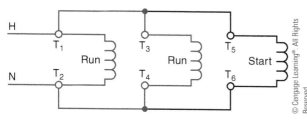

Figure 11–18
Because these windings are connected in parallel, each will have 120 volts applied to it.

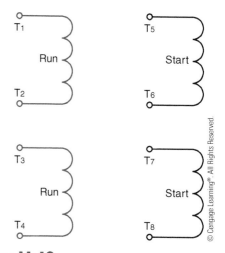

Figure 11–19
Some dual-voltage single-phase motors contain two start windings as well as two run windings.

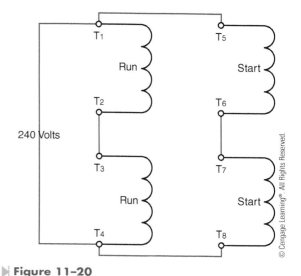

Figure 11–20
The motor is connected for operation on 240 volts.

voltage of 120 volts also. Notice that each of the windings has 120 volts connected to it, which is the rating of the windings.

If the motor is to be operated on a 120-volt line, the windings are connected in parallel, as shown in Figure 11–18. Because these windings are connected in parallel, each will have 120 volts applied to it.

Some dual-voltage motors contain two start windings as well as two run windings, as shown in Figure 11–19. The run windings are labeled T_1 through T_4, the same as dual-voltage motors that contain only one start winding. One of the start windings is labeled T_5 and T_6. The second start winding is labeled T_7 and T_8. If the motor is to be connected for operation on 240 volts, the run windings are connected in series by connecting T_2 and T_3 together, and the start windings are connected in series by connecting T_6 and T_7 together. The start windings are then connected in parallel with the run windings, as shown in Figure 11–20. The direction of rotation can be changed by reversing T_5 and T_8.

If the motor is operated on 120 volts, the run and start windings are connected in parallel, as shown in Figure 11–21. If the motor is to be reversed, leads T_5 and T_7 are changed with leads T_6 and T_8. Some

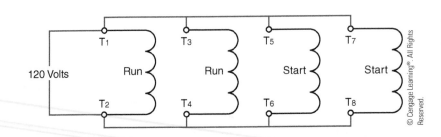

▶ **Figure 11–21**
The motor is connected for operation on 120 volts.

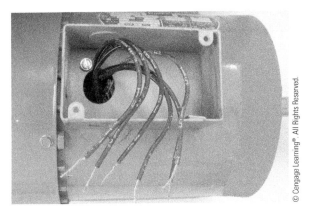

▶ **Figure 11–22**
Connection leads of a dual-voltage single-phase motor.

manufacturers label the start winding leads T_5 and T_8 even if they contain only one start winding. The connection leads of a dual-voltage single-phase motor are shown in Figure 11–22.

MOTOR POWER CONSUMPTION

It should be noted that the motor does not use less energy when connected to 240 volts than it does when connected to 120 volts. Power is measured in watts, and the watts will be the same regardless of the connection. When the motor is connected to operate on 240 volts, it will have half the current draw as it does on a 120-volt connection. Therefore, the amount of power used is the same. For example, assume the motor has a current draw of 5 amps when connected to 240 volts and 10 amps when connected to 120 volts. Watts can be computed from multiplying volts by amps. When the motor is connected to 240 volts, the amount of power used is $240 \times 5 = 1200$ watts. When the motor is connected to 120 volts, the power used is $120 \times 10 = 1200$ watts.

The 240-volt connection is generally preferred, however, because the lower current draw causes less voltage drop on the line supplying power to the motor. If the motor is located a long distance from the panel, voltage drop of the wire can become very important to the operation of the unit.

PERMANENT SPLIT-CAPACITOR MOTOR

The permanent split-capacitor motor has greatly increased in popularity for use in the air-conditioning field over the past years. This type of split-phase motor does not disconnect the start windings from the circuit when it is running. This eliminates the need for a centrifugal switch or starting relay to disconnect the start windings from the circuit when the motor reaches about 75% of its full speed, Figure 11–23. This motor has good starting torque

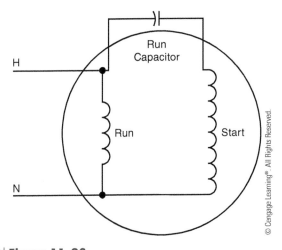

▶ **Figure 11–23**
A schematic for a permanent split-capacitor motor.

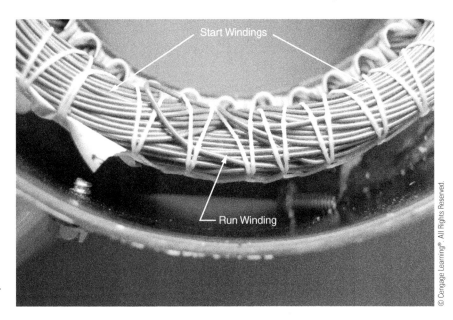

Start Windings

Run Winding

▶ **Figure 11–24**
The run and start windings
of a permanent split-capacitor
motor.

and good running torque. Because the capacitor remains in the circuit during operation, it helps correct power factor for the motor. The stator winding of the permanent split-capacitor (PSC) motor is different from the stator windings of the resistance-start induction-run or capacitor-start induction-run motors. The PSC motor stator winding still contains a run and start winding, but the start winding will generally have the same size wire and just as many turns as the run winding, as shown in Figure 11–24. The run winding is placed lower in the core material, which helps increase the inductance.

The capacitor used in this type of motor is generally the AC oil-filled type. Condenser and ventilating fan motors are generally of this type, Figure 11–25. Because this capacitor remains connected in the circuit, an AC electrolytic capacitor cannot be used to replace the run capacitor for this type of motor.

The permanent split-capacitor motor sometimes uses an extra capacitor to aid in starting. When this is done, the start capacitor is connected in parallel with the run capacitor. During the time of starting, both of these capacitors are connected in the circuit, Figure 11–26. When the motor has

▶ **Figure 11–25**
Condenser and ventilating fans generally contain a permanent split-capacitor motor.

accelerated to about 75% of full speed, the start capacitor is disconnected from the circuit. If the motor is an open type, the start capacitor will be disconnected by a centrifugal switch. If the motor is sealed, such as a hermetically sealed compressor, the start capacitor will be disconnected by a starting relay.

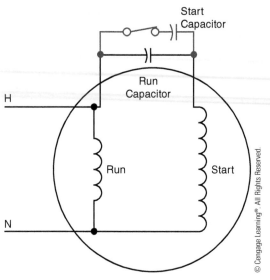

▶ **Figure 11–26**
An extra starting capacitor used with a permanent split-capacitor motor.

IDENTIFYING CAPACITOR TERMINALS

Most run capacitors and some starting capacitors are of the oil-filled type, Figure 8–16. This is especially true for high-current motors such as those used to operate compressors. Many manufacturers

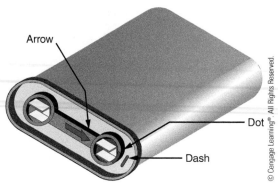

▶ **Figure 11–27**
The markings indicate the terminal that connects to the plate located closest to the case of the capacitor.

of oil-filled capacitors identify one terminal with an arrow, a painted dot, or by stamping a dash in the capacitor can, Figure 11–27. This identified terminal marks the connection to the plate that is located nearer to the metal container or can. It has long been known that when a capacitor's dielectric breaks down and permits a short circuit to ground, it is most often the plate nearer to the outside case that becomes grounded. For this reason, it is desirable to connect the identified capacitor terminal to the line side instead of to the motor start winding.

In Figure 11–28, the run capacitor has been connected in such a manner that the identified terminal is connected to the start winding of a compressor

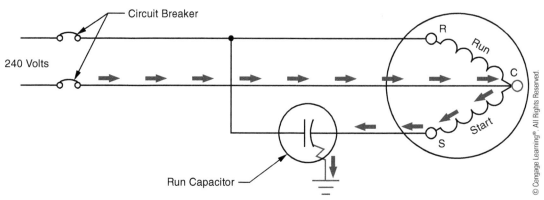

▶ **Figure 11–28**
Identified capacitor terminal connected to motor start winding.

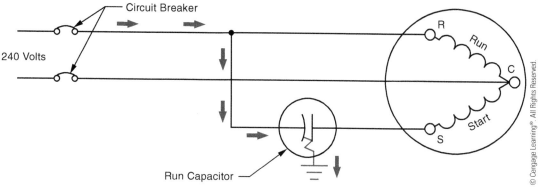

▷ **Figure 11–29**
Identified capacitor terminal connected to the line.

motor. If the capacitor shorts to ground, a current path will exist through the motor start winding. The start winding is an inductive-type load, and inductive reactance will limit the value of current flow to ground. Because the flow of current is limited, it will take the circuit breaker or fuse time to open the circuit and disconnect the motor from the power line. This time delay can permit the start winding to overheat and become damaged.

In Figure 11–29, the run capacitor has been connected in such a manner that the identified terminal is connected to the line side. If the capacitor shorts to ground, a current path will exist directly to ground, bypassing the motor start winding. When the capacitor is connected in this manner, the start winding does not limit current flow and allows the fuse or circuit breaker to open almost immediately.

▷ SUMMARY

◉ Split-phase motors fall into three general classifications:

A. The resistance-start induction run.

B. The capacitor-start induction run.

C. The permanent split-capacitor motor.

◉ The voltages of a two-phase system are 90° out of phase with each other.

◉ Split-phase motors receive their name from the fact that they split the current flow through two windings to produce an out-of-phase condition that produces a rotating magnetic field.

◉ The resistance-start induction-run and capacitor-start induction-run motors disconnect their start winding when they have accelerated to about 75% of their rated speed.

◉ Split-phase motors that are not hermetically sealed generally use a centrifugal switch to disconnect the start winding when the motor has reached about 75% of its rated speed.

◉ The direction of rotation of a split-phase motor can be reversed by changing the connections of the run winding or the start winding, but not both.

◉ Dual-voltage, split-phase motors can be connected to operate on 120 or 240 volts.

⊙ The capacitor-start induction-run motor develops a higher starting torque than the resistance-start induction-run motor.

⊙ The capacitor-start motor develops a higher starting torque by using the capacitor to produce a 90° phase shift between the current flow in the run winding and the current flow in the start winding.

⊙ Maximum starting torque for the split-phase motor is produced when the run winding current and start winding current are 90° out of phase with each other.

⊙ Permanent split-capacitor motors do not disconnect the start winding when the motor is in operation.

⊙ Some permanent split-capacitor motors use an extra capacitor during the starting period.

⊙ The identifying mark on an oil-filled capacitor should be connected to the line side of the circuit.

KEY TERMS

capacitor-start
  capacitor-run motor
capacitor-start
  induction-run motor
centrifugal switch

permanent split-
  capacitor motor (PSC)
resistance-start
  induction-run motor
rotating magnetic field

run winding
split-phase motor
start winding
torque
two-phase power

REVIEW QUESTIONS

1. What is a split-phase motor?
2. What are the three basic types of split-phase motors?
3. Explain the difference in construction between run windings and start windings.
4. How many degrees out of phase should the current in the start winding be with the current in the run winding to develop maximum starting torque?
5. What type of capacitor is generally used with a capacitor-start induction-run motor?
6. Can the microfarad value of this capacitor be increased to improve starting torque?
7. What type of capacitor is used with a permanent split-capacitor motor?
8. Does the capacitor of a capacitor-start induction-run motor help correct power factor?
9. If necessary, can an AC electrolytic capacitor of higher voltage rating be used as the starting capacitor?
10. What is a centrifugal switch used for?

The Shaded-Pole Induction Motor

OBJECTIVES

After studying this unit, the student should be able to:

▷ Discuss the operation of a shaded-pole induction motor

▷ Define a shading coil

▷ List common uses for the shaded-pole induction motor

The **shaded-pole induction motor** is another type of AC single-phase motor used to a large extent in the air-conditioning field. This motor is popular because of its simplicity and long life. The shaded-pole motor contains no start winding or centrifugal switch. The rotating magnetic field is created by a **shading coil** wound around one side of each pole piece.

THE SHADING COIL

The shading coil is wound around one end of the pole piece, Figure 12–1. The shading coil is actually a large loop of copper wire or a copper band. Both ends of the loop are connected together to form a complete circuit. The shading coil acts in the same manner as a transformer with a shorted secondary winding. When the voltage of the AC

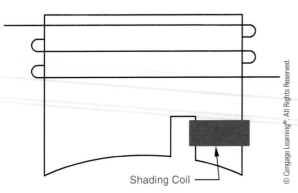

▶ **Figure 12–1**
A shaded pole.

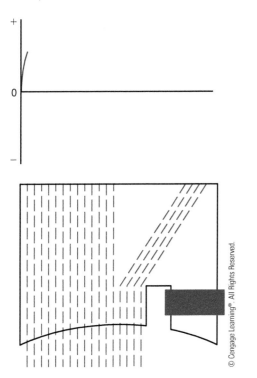

▶ **Figure 12–2**
The shading coil opposes a change of magnetic flux as voltage increases.

waveform increases from zero toward its positive peak, a magnetic field is created in the pole piece. As magnetic lines of flux cut through the shading coil, a voltage is induced in the coil. Because the coil is a low-resistance short circuit, a large amount of current flows in the loop. This current flow causes an **opposition** to the change of **magnetic flux**, Figure 12–2. As long as voltage is induced into

the shading coil, there will be an opposition to the change of magnetic flux.

When the AC voltage reaches its peak value, it is no longer changing, and there is no voltage being induced into the shaded coil. Because there is no current flow in the shading coil, there is no opposition to the magnetic flux. The magnetic flux of the pole piece is now uniform across the pole face, Figure 12–3.

When the AC voltage begins to decrease from its peak value back toward zero, the magnetic field of the pole piece begins to collapse. A current is again induced into the shading coil. The induced current opposes the change of magnetic flux, Figure 12–4. This causes the magnetic flux to be concentrated in the shaded section of the pole piece.

When the AC voltage passes through zero and begins to increase in the negative direction, the same set of events happen, except that the polarity of the magnetic field is reversed. If these events were to be

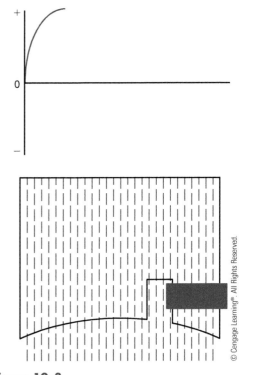

▶ **Figure 12–3**
There is no opposition to magnetic flux when the voltage is not changing.

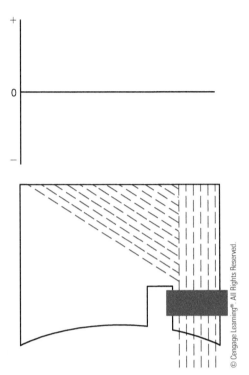

▶| **Figure 12–4**
The shading coil opposes a change of flux when the voltage decreases.

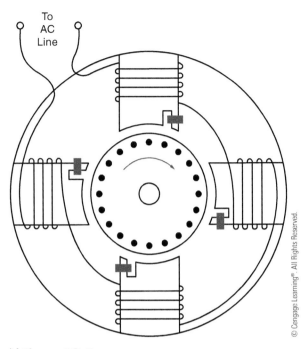

▶| **Figure 12–5**
A four-pole shaded-pole induction motor.

seen in rapid order, it could be seen that the magnetic field rotates across the face of the pole piece.

SPEED

The speed of the shaded-pole induction motor is determined by the same factors that determine the synchronous speed of other induction motors: frequency and number of stator poles. Shaded-pole motors are commonly wound as four- or six-pole motors. Figure 12–5 shows a drawing of a four-pole motor.

REVERSING DIRECTION OF ROTATION

The direction the magnetic field moves across the face of the pole piece is determined by the side of the pole piece that has the shaded coil. The rotor will turn in the direction of the shaded pole, as shown by the arrow in Figure 12–5. If the direction of rotation must be changed, it can be done by removing the stator winding and turning it around. This is not a common practice, however. As a general rule, the shaded-pole induction motor is considered to be nonreversible.

GENERAL OPERATING CHARACTERISTICS

The shaded-pole motor contains a standard squirrel-cage rotor. The amount of torque produced is determined by the strength of the magnetic field of the stator, the strength of the magnetic field of the rotor, and the phase-angle difference between rotor current and stator current. The shaded-pole motor has a low starting torque and running torque. This motor is generally used in applications that do not require a large amount of starting torque, such as fans and blowers. Figure 12–6 shows a photograph of a shaded-pole induction motor.

▶ **Figure 12–6**
Stator winding and rotor of a shaded-pole induction
motor.

SUMMARY

⊙ The shaded-pole induction motor is popular because of its simplicity and long life.

⊙ Shaded-pole induction motors do not contain a start winding or centrifugal switch.

⊙ Shaded-pole induction motors operate on the principle of a rotating magnetic field.

⊙ A shading coil or loop is used to produce an out-of-phase flux across the face of the pole piece, thus producing a rotating magnetic field.

⊙ The speed of a shaded-pole induction motor is determined by the number of stator poles and the frequency of the applied voltage.

⊙ Shaded-pole induction motors are generally considered to be nonreversible.

KEY TERMS

magnetic flux   shaded-pole induction motor   shading coil
opposition

REVIEW QUESTIONS

1. What is a shading coil?

2. What determines the synchronous speed of a shaded-pole motor?

3. In general, how is the direction of a shaded-pole induction motor reversed?

4. What type of rotor does the shaded-pole motor contain?

5. Name two advantages of the shaded-pole motor over the split-phase induction motor.

UNIT 13 ▷

Multispeed Motors

OBJECTIVES

After studying this unit, the student should be able to:

▷ Discuss the operation of a consequent pole motor

▷ List the factors that determine the synchronous field speed of an AC motor

▷ Discuss the operation of multispeed fan motors

▷ Connect a multispeed fan motor for operation at different speeds

Multispeed AC motors have been used to a great extent in the air-conditioning field for many years. There are two basic types of multispeed motors used. One type is known as the **consequent pole motor**. The other type is generally a permanent split-capacitor motor.

THE CONSEQUENT POLE MOTOR

The speed of the rotating magnetic field of an AC induction motor can be changed in either of two ways:

1. Change the frequency of the AC voltage.
2. Change the number of stator poles.

The consequent pole motor changes the motor speed by changing the number of its stator poles. The run winding in Figure 13–1 has been **tapped**

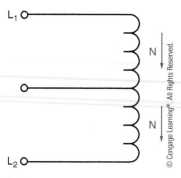

▶ **Figure 13–1**
Center-tapped run winding.

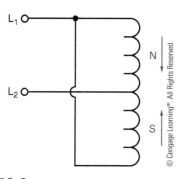

▶ **Figure 13–2**
Two magnetic poles are produced.

in the center. If the AC line is connected to each end of the winding as shown, current flows through the winding in only one direction. Therefore, only one magnetic **polarity** is produced in the winding. If the winding is connected as shown in Figure 13–2, current flows in opposite directions in each half of the winding. Because current flows through each half of the winding in opposite directions, the polarity of the magnetic field is different in each half of the winding. The run winding now has two polarities instead of one. There are now two magnetic poles instead of one. If the windings of a two-pole motor were to be tapped in this manner, the motor could become a four-pole motor. The synchronous speed of a two-pole motor is 3600 RPM, and the synchronous speed of a four-pole motor is 1800 RPM.

The consequent pole motor has the disadvantage of having a wide variation in speed. When the speed is changed, it changes from a synchronous speed of 3600 RPM to 1800 RPM. The speed cannot be

changed by a small amount. This wide variation in speed makes the consequent pole motor unsuitable for some loads, such as fans and blowers.

The consequent pole motor, however, does have some advantages over the other type of multispeed motor. When the speed of the consequent pole motor is reduced, its torque increases. For this reason, the consequent pole motor can be used to operate heavy loads, such as **two-speed compressors**.

MULTISPEED FAN MOTORS

Multispeed fan motors have been used in the air-conditioning industry for many years. These motors are generally wound for two to five steps of speed, and are used to operate fans and squirrel-cage blowers. A schematic drawing of a three-speed motor is shown in Figure 13–3. Notice that the run winding has been tapped to produce low, medium, and high speed. The start winding is connected in parallel with the run winding section. The other end of the start lead is connected to an external oil-filled run capacitor. This motor obtains a change in speed by inserting inductance in series with the run winding. The actual run winding for this motor is between the terminals marked "High" and "C". The windings shown between High and Medium are connected in series with the main run winding. When the rotary switch is connected to the medium-speed position, the inductive reactance of this coil limits the amount

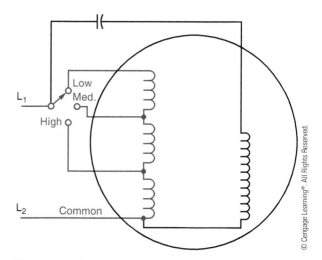

▶ **Figure 13–3**
Three-speed fan motor.

of current flow through the run winding. When the current of the run winding is reduced, the strength of the magnetic field of the run winding is reduced and the motor produces less torque. This causes the motor speed to decrease.

If the rotary switch is changed to the low position, more inductance is connected in series with the run winding. This causes less current to flow through the winding and another reduction in torque. When the torque is reduced, the motor speed decreases again.

Common speeds for a four-pole motor of this type are 1625, 1500, and 1350 RPM. Notice that this motor does not have the wide range between speeds as the consequent pole motor does. Most induction motors would overheat and damage the

motor windings if the speed were to be reduced to this extent. This motor, however, has much higher impedance in its windings than most motors. The run windings of most split-phase motors have a wire resistance of 1 to 4 ohms. This motor generally has a resistance of 10 to 15 ohms in its run winding. It is the high impedance of the windings that permits the motor to be operated in this manner without damage.

Because this motor is designed to slow down when load is added, it is not used to operate high-torque loads. This type of motor is generally used to operate only low-torque loads, such as fans and blowers. The schematic in Figure 13–4 shows a multispeed fan motor and switch.

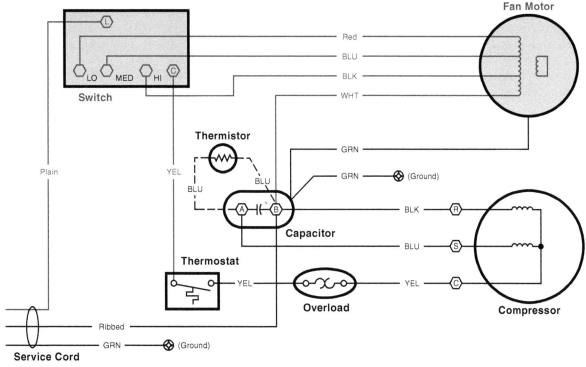

Switch Position	Contacts Made
LO	L to C, L to LO
MED	L to C, L to MED
HI	L to C, L to HI
Off	None

▶ **Figure 13–4**
A multispeed fan motor and switch.

VARIABLE FREQUENCY DRIVES

One of the factors that determines the speed of the rotating magnetic field of an AC induction motor is the frequency of the applied voltage. If the frequency is changed, the speed of the rotating magnetic field changes also. A four-pole stator has a synchronous speed (speed of the rotating magnetic field) of 1800 RPM when connected to a 60 Hz line. When the frequency is lowered to 30 Hz, the synchronous speed decreases to 900 RPM.

When the frequency is lowered, care must be taken not to damage the stator windings. The current flow through the winding is limited to a great extent by inductive reactance. When the frequency is lowered, inductive reactance is lowered also ($X_L = 2\pi FL$). For this reason, **variable frequency** drives must employ some method of lowering the applied voltage to the stator as frequency is reduced.

In the air-conditioning field, variable frequency drive is often used to control the speed of blower motors. This method of controlling airflow can be more efficient than inserting dampers into the duct system. Variable frequency drives are very popular in zone controlled systems.

Most variable frequency drives operate by first changing the AC voltage into DC and then changing it back to AC at the desired frequency. A variable frequency drive is shown in Figure 13–5. There are several methods used to change the DC voltage back into AC. The method employed is generally determined by the manufacturer, age of the equipment, and the size of the motor the drive must control. Variable frequency drives intended to control the speed of motors up to 500 horsepower generally use transistors. In the circuit shown in Figure 13–6, a three-phase bridge changes the three-phase alternating-current into direct current. The bridge rectifier uses **SCRs (silicon-controlled rectifiers)** instead of diodes. The SCRs permit the output voltage of the rectifier to be controlled. As the frequency decreases, the SCRs fire later in the cycle and lower the output voltage to the transistors. A choke coil and capacitor bank are used to filter the output voltage before transistors Q1 through Q6 change the DC voltage back into AC. An electronic control unit is connected to the bases of transistor Q1 through Q6. The control

Courtesy of Toshiba International Corp.

▶ **Figure 13–5**
Variable frequency drive to control speed of motors rated from 300 to 600 hp[1].

unit converts the DC voltage back into three-phase alternating current by turning transistors on or off at the proper time and in the proper sequence. Assume, for example, that transistors Q1 and Q4 are switched on at the same time. This permits stator winding T_1 to be connected to a positive voltage and T_2 to be connected to a negative voltage. Current can flow through Q4 to T_2, through the motor stator winding and through T_1 to Q1.

Now assume that transistors Q1 and Q4 are switched off and transistors Q3 and Q6 are switched on. Current will now flow through Q6 to stator winding T_3, through the motor to T_2, and through Q3 to the positive of the power supply.

Because the transistors are turned completely on or completely off, the waveform produced is a square wave instead of a sine wave, Figure 13–7. Induction motors operate on a square wave without

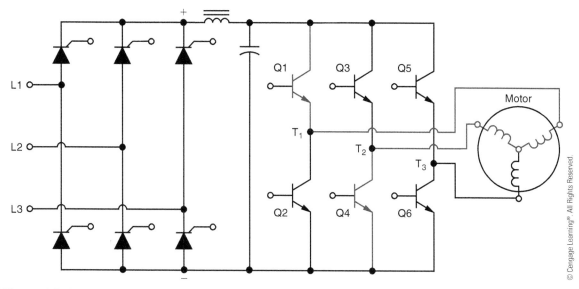

▶ **Figure 13–6**
Variable frequency drive using bipolar transistor to change the direct current back into alternating current.

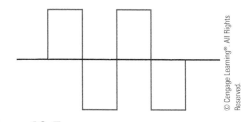

▶ **Figure 13–7**
Square wave.

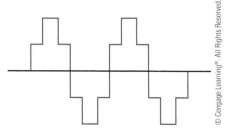

▶ **Figure 13–8**
Stepped wave.

Some Related Problems

The circuit illustrated in Figure 13–6 employs the use of SCRs in the power supply and junction transistors in the output stage. SCR power supplies control the output voltage by chopping the incoming waveform. This can create **harmonics** on the line leading to overheating of transformers and motors, and can cause fuses to blow and circuit breakers to trip. When bipolar junction transistors are employed as switches, they are generally driven into saturation by supplying them with an excessive amount of base-emitter current. Saturating the transistor causes the collector-emitter voltage to drop to between 0.04 and 0.03 volt. This small voltage drop allows the transistor to control large amounts of current without being destroyed. When a transistor is driven into saturation, however, it cannot recover or turn off as quickly as normal. This greatly limits the frequency response of the transistor.

IGBTs

Many transistor-controlled variable drives now employ a special type of transistor called an **insulated gate bipolar transistor (IGBT)**. IGBTs

much of a problem. Some manufacturers design units that produce a stepped waveform, as shown in Figure 13–8. The stepped waveform is used because it closely approximates a sine wave.

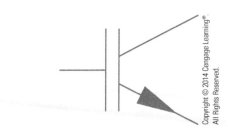

▶ **Figure 13–9**
Schematic symbol for an insulated gate biolar transitor.

have an insulated gate very similar to some types of field effect transistors (FETs). Because the gate is insulated, it has a very high impedance. The IGBT is a voltage controlled device, not a current controlled device. This gives it the ability to turn off very quickly. IGBTs can be driven into saturation to provide a very low voltage drop between emitter and collector, but they do not suffer from the slow recovery time of common junction transistors. The schematic symbol for an IGBT is shown in Figure 13–9.

Drives using IGBTs generally use diodes to rectify the AC voltage into DC, not SCR, Figure 13–10. The three-phase rectifier supplies a constant DC voltage to the transistors. The output voltage to the motor is controlled by **pulse width modulation (PWM)**. PWM is accomplished by turning the transistor on and off several times during each half cycle, Figure 13–11. The output voltage is an average of the peak, or maximum, voltage and the amount of time the transistor is turned on or off.

Assume that 480-volt, three-phase AC is rectified to DC and filtered. The DC voltage applied to the IGBTs is approximately 630 volts. The output voltage to the motor is controlled by the switching of the transistors. Assume that the transistor is on for 10 microseconds and off for 20 microseconds. In this example, the transistor is on for one-third of the time and off for two-thirds of the time. The voltage applied to the motor would be 210 volts (630/3).

Advantages and Disadvantages of IGBT Drives

A great advantage of drives using IGBTs is the fact that SCRs are generally not used in the power supply, and this greatly reduces problems with line harmonics. The greatest disadvantage is that the fast switching rate of the transistors can cause voltage spikes in the range of 1600 volts to be applied to the motor. These voltage spikes can destroy some motors. Line length from the drive to the motor is of great concern with drives using IGBTs. The shorter the line length, the better.

Inverter Rated Motors

Because of the problem of excessive voltage spikes caused by IGBT drives, some manufacturers produce a motor that is "inverter rated." These motors are specifically designed to be operated by variable

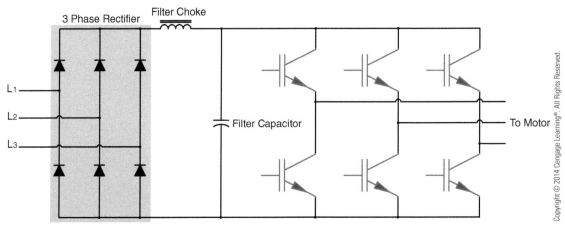

3 Phase Rectifier   Filter Choke

L1
L2   Filter Capacitor   To Motor
L3

▶ **Figure 13–10**
Variable frequency drives using IGBTs generally use diodes in the rectifier instead of SCRs.

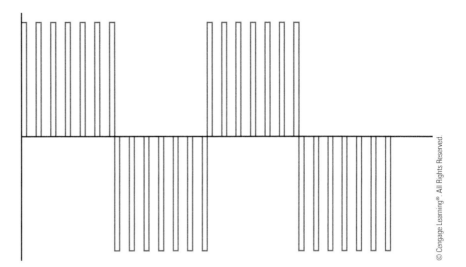

Figure 13–11
Pulse width modulation is accomplished by turning the voltage on and off several times during each half cycle.

frequency drives. They differ from standard motors in several ways:

1. Many inverter rated motors contain a separate blow to provide continuous cooling for the motor, regardless of the speed. Many motors use a fan connected to the motor shaft to help draw air though the motor. When the motor speed is reduced, the fan cannot maintain sufficient airflow to cool the motor.

2. Inverter rated motors generally have insulating paper between the windings and the stator core, Figure 13–12. The high voltage spikes produce high currents that produce a high magnetic field. This increased magnetic field causes the motor windings to move. This movement can eventually cause the insulation to wear off the wire and produce a grounded motor winding.

3. Inverter rated motors generally have phase paper added to the terminal leads. Phase paper is insulating paper added to the terminal leads that exit the motor. The high voltage spikes affect the beginning lead of a coil much more than the wire inside the coil. The coil is an inductor that naturally opposes a change of current. Most of the insulation stress caused by high voltage spikes occurs at the beginning of a winding.

4. The magnet wire used in the construction of the motor windings has a higher rated insulation than other motors.

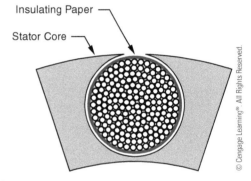

Insulating Paper
Stator Core

Figure 13–12
Insulating paper is between the windings and the stator frame.

5. The case size is larger than most three phase motors. The case size is larger because of added insulating paper between the windings and the stator core. Also, a larger case size helps cool the motor by providing a larger surface area for the dissipation of heat.

Variable Frequency Drives Using SCRs and GTOs

Variable frequency drives intended to control motors over 500 horsepower generally use SCRs or GTOs (gate turn-off devices). GTOs are similar to SCRs except that conduction through the GTO can be stopped by applying a negative voltage—negative with respect to the cathode—to the gate. SCRs and

GTOs are thyristors and have the ability to handle a greater amount of current than transistors. An example of a single-phase circuit used to convert DC voltage to AC voltage with SCRs is shown in Figure 13–13. In this circuit, the SCRs are connected to a control unit that controls the sequence and rate at which the SCRs are gated on. The circuit is constructed so that SCRs A and A' are gated on at the same time and SCRs B and B' are gated on at the same time. Inductors L_1 and L_2 are used for filtering and wave shaping. Diodes D_1 through D_4 are clamping diodes and are used to prevent the output voltage from becoming excessive. Capacitor C_1 is used to turn one set of SCRs off when the other set is gated on. This capacitor must be a true AC capacitor because it will be charged to the alternate polarity each half cycle. In a converter intended to handle large amounts of power, capacitor C_1 will be a bank

of capacitors. To understand the operation of the circuit, assume that SCRs A and A' are gated on at the same time. Current will flow through the circuit as shown in Figure 13–14. Notice the direction of current flow through the load, and also that capacitor C_1 has been charged to the polarity shown. When an SCR is gated on, it can only be turned off by permitting the current flow through the anode-cathode section to drop below a certain level called the holding current level. As long as the current continues to flow through the anode-cathode, the SCR will not turn off.

Now assume that SCRs B and B' are turned on. Because SCRs A and A' are still turned on, two current paths now exist through the circuit. The positive charge on capacitor C_1, however, causes the negative electrons to see an easier path. The current will rush to charge the capacitor to the opposite

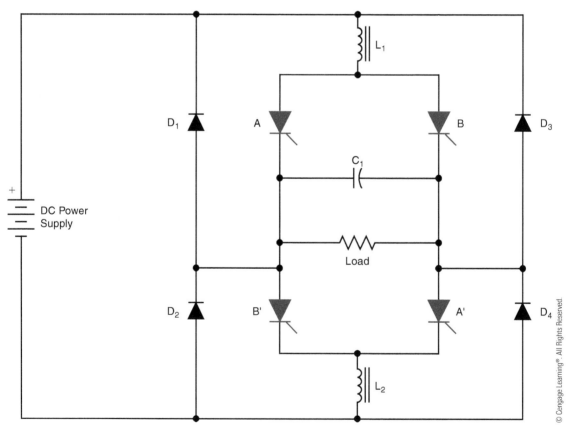

▶ **Figure 13–13**
Changing DC to AC using SCRs.

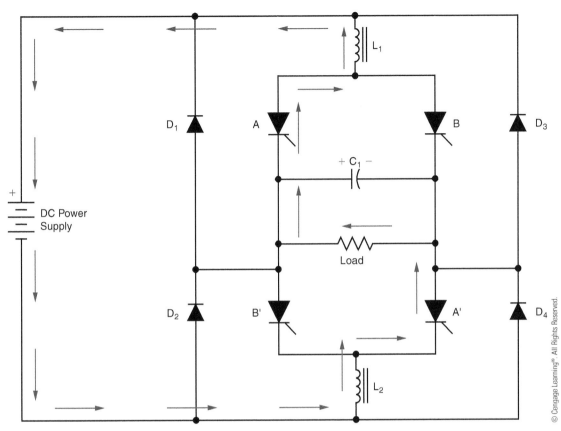

▶ **Figure 13–14**
Current flows through SCRs A and A'.

polarity, stopping the current flowing through SCRs A and A', thus permitting them to turn off. The current now flows through SCRs B and B' and charges the capacitor to the opposite polarity, Figure 13–15. Notice that the current now flows through the load in the opposite direction, which produces alternating current across the load.

To produce the next half cycle of AC current, SCRs A and A' are gated on again. The positively charged side of the capacitor now causes the current to stop flowing through SCRs B and B' permitting them to turn off. The current again flows through the load in the direction indicated in Figure 13–15. The frequency of the circuit is determined by the rate at which the SCRs are gated on.

FEATURES OF VARIABLE FREQUENCY CONTROL: Although the primary purpose of a variable frequency drive

is to provide speed control for an AC motor, most drives provide functions that other types of controls do not. Many variable frequency drives can provide the low speed torque characteristic that is so desirable in DC motors. This feature permits AC squirrel-cage motors to replace DC motors for many applications.

Many variable frequency drives also provide current limit and automatic speed regulation for the motor. Current limit is generally accomplished by connecting current transformers to the input of the drive and sensing the increase in current as load is added. Speed regulation is accomplished by sensing the speed of the motor and feeding this information back to the drive.

Another feature of variable frequency drives is acceleration and deceleration control, sometimes

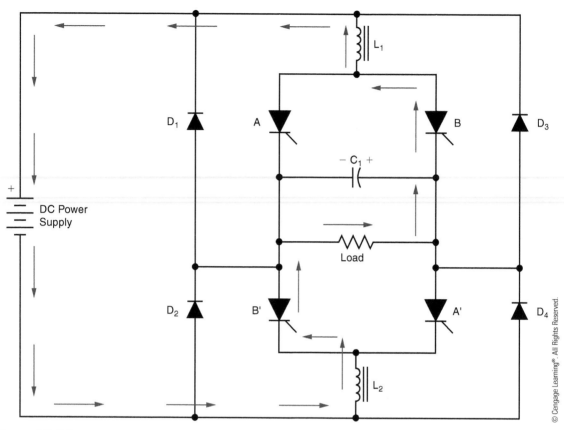

▶ **Figure 13–15**
Current flows through SCRs B and B'.

called ramping. Ramping is used to accelerate or decelerate a motor over some period of time. Ramping permits the motor to bring the load up to speed slowly as opposed to simply connecting the motor directly to the line. Even if the speed control is set in the maximum position when the start button is pressed, ramping permits the motor to accelerate the load from zero to its maximum RPM over several seconds. This feature is a real advantage for some types of loads, especially gear drive loads. In some units, the amount of acceleration and deceleration time can be adjusted by setting potentiometers on the main control board. Other units are completely digitally controlled with the acceleration and deceleration times programmed into the computer memory.

Some other adjustments that can usually be set by changing potentiometers or programming the unit are as follows:

Current Limit: These controls set the maximum amount of current the drive is permitted to deliver to the motor.

Volts per Hertz: This sets the ratio by which the voltage increases as frequency increases, or decreases as frequency decreases.

Maximum Hertz: These controls set the maximum speed of the motor.

Minimum Hertz: This sets the minimum speed the motor is permitted to run.

SUMMARY

- ⊙ Consequent pole motors change speed by changing the number of stator poles.
- ⊙ The disadvantage of consequent pole motors is that they have a wide range between speeds.
- ⊙ The advantage of the consequent pole motor is that it maintains a high torque.
- ⊙ Consequent pole motors are generally used to operate two-speed compressors because of their ability to maintain high torque.
- ⊙ Multispeed fan motors insert inductance in series with the main run winding to produce a change of speed.
- ⊙ The run windings of multispeed fan motors have a high resistance so they will not overheat when the motor slows down.
- ⊙ Multispeed fan motors cannot be used to operate high torque loads.
- ⊙ Variable frequency drives change the speed of the motor by changing the frequency of the applied voltage.
- ⊙ The synchronous speed of an induction motor is the speed of the rotating magnetic field.
- ⊙ Synchronous speed is determined by two factors: number of stator poles per phase and frequency of the applied voltage.
- ⊙ When the frequency to the motor is reduced, the voltage must be reduced also.
- ⊙ Variable frequency drives up to 500 horsepower generally use transistors to change the DC voltage back into AC voltage.
- ⊙ Variable frequency drives above 500 horsepower generally use SCRs or GTOs to change the DC voltage back into AC voltage.
- ⊙ Insulated gate bipolar transistors are used in many variable frequency drives because they can be switched on or off at a faster rate.
- ⊙ Units employing IGBTs can produce voltage spikes on the motor as high as 1600 volts.
- ⊙ Inverter rated motors are designed to operate with variable frequency drives.
- ⊙ Some variable frequency drives use potentiometers to change settings, and others are digital and must have the setting programmed into the drive.

KEY TERMS

consequent pole motor
harmonics
insulated gate bipolar
   transistors (IGBTs)

multispeed AC motors
polarity
pulse width
   modulation (PWM)

silicon-controlled
   rectifiers (SCRs)
tapped
two-speed compressors
variable frequency

▶ REVIEW QUESTIONS

1. Name two ways of changing the speed of a rotating magnetic field.
2. How does the consequent pole motor change speed?
3. Name a disadvantage of the consequent pole motor.
4. Name an advantage of a consequent pole motor.
5. How many steps of speed are common to a multispeed fan motor?
6. Refer to Figure 13–3. Explain what would happen to motor operation if the winding between low and medium should become open.
7. What is an advantage of the multispeed fan motor over the consequent pole motor?
8. What is a disadvantage of the multispeed fan motor when compared with the consequent pole motor?
9. How much wire resistance is common for the run winding of most split-phase motors?
10. How much wire resistance is common for the multispeed fan motor?
11. What is synchronous speed?
12. What is the disadvantage of a variable frequency drive unit that uses SCRs to convert the AC voltage into DC voltage?
13. What is the disadvantage of driving a common bipolar junction transistor into saturation?
14. What is the main advantage of the insulated grate bipolar transistor over the common bipolar junction transistor?
15. What is an inverter rated motor?

▶ TROUBLESHOOTING QUESTIONS

Refer to the schematic shown in Figure 13–4 to answer the following questions.

1. If the switch is set in the HI position, the thermostat will control the operation of:
   A. The compressor only.
   B. The fan motor only.
   C. The speed of the fan motor.
   D. Both the compressor and the speed of the fan motor.
2. When the switch is set in the HI position, both the fan motor and compressor operate. If the switch is changed to the MED or LOW position, the compressor continues to operate, but the fan motor stops. Which of the following could cause this problem?
   A. The fan motor start winding is open.
   B. The section of run winding between the red and blue wires is open in the fan motor.
   C. The section of run winding between the blue and black wires is open in the fan motor.
   D. The section of run winding between the black and white wires is open in the fan motor.

3. The fan motor will operate in any of its three speeds, but the compressor motor will not start. Which of the following could cause this problem?

A. The switch is not making connection between L and C.

B. The switch is not making connection between L and LO.

C. The switch is not making connection between L and MED.

D. The switch is not making connection between L and HI.

4. If it is assumed that this unit operates on 120 volts AC, how will the neutral conductor be identified on the schematic shown in Figure 13–4?

A. The wire color will be green.

B. The wire color will be white.

C. The conductor will be ribbed.

D. The conductor will be plain.

5. If the unit is in operation and the overload protector should open:

A. Both the fan motor and compressor will stop operating.

B. Only the compressor will stop operating.

C. Only the fan motor will stop operating.

D. Both the fan motor and compressor will continue to operate.

Three-Phase Motor Principles

OBJECTIVES

After studying this unit, the student should be able to:

» List the three major types of three-phase motors

» Discuss the operating principle of a three-phase motor

» List the factors that determine synchronous speed

» Discuss the operation of dual-voltage motors

» Connect dual-voltage three-phase motors for operation on low voltage or high voltage

There are three basic types of **three-phase motors**:

1. The squirrel-cage induction motor
2. The wound rotor induction motor
3. The synchronous motor

The type of three-phase motor is determined by the rotor, or rotating member. The stator windings for any of these motors is the same. In this unit, the basic principles of operation for three-phase motors are discussed.

The principle of operation for all three-phase motors is the **rotating magnetic field**. There are three factors that cause the magnetic field to rotate:

1. The voltages of a three-phase system are 120° out of phase with each other.

2. The three voltages change polarity at regular intervals.

3. The stator windings are arranged around the inside of the motor.

Figure 14–1A shows three AC voltages 120° out of phase with each other, and the stator winding of a three-phase motor. The stator illustrates a two-pole, three-phase motor. Two-pole means that there are two poles per phase. AC motors do not generally have actual pole pieces as shown in Figure 14–1A, but they are used here to aid in understanding how the rotating magnetic field is created in a three-phase motor. Notice that pole pieces 1A and 1B are located opposite each other. The same is true for poles 2A and 2B, and 3A and 3B. The pole pieces 1A and 1B are wound with wire that is connected to phase 1 of the three-phase system. Notice also that the pole pieces are wound in such a manner that they will always have opposite magnetic polarities. If pole piece 1A has a north magnetic polarity, pole piece 1B will have a south magnetic polarity at the same time.

The windings of pole pieces 2A and 2B are connected to line 2 of the three-phase system. The windings of pole pieces 3A and 3B are connected to line 3 of the three-phase system. These pole pieces are also wound in such a manner as to have the opposite polarity of magnetism.

To understand how the magnetic field rotates around the inside of the motor, refer to Figure 14–1B. Notice a line, labeled A, has been drawn through the three voltages of the system. This line is used to illustrate the condition of the three voltages at this point in time. The arrow drawn inside the motor indicates the greatest strength of the magnetic field at the same point in time. It is to be assumed that the arrow is pointing in the direction of the north magnetic field. Notice in Figure 14–1B, that phase 1 is at its maximum positive peak and that phases 2 and 3 are less than maximum. The magnetic field is therefore strongest between pole pieces 1A and 1B.

In Figure 14–1C, line B indicates that the voltage of line 3 is zero. The voltage of line 1 is less than maximum positive; and line 2 is less than maximum negative. The magnetic field at this point is concentrated between the pole pieces of phase 1 and phase 2.

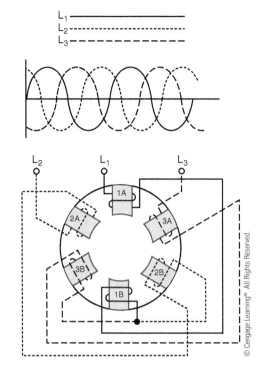

▶ **Figure 14–1A**
Basic stator winding.

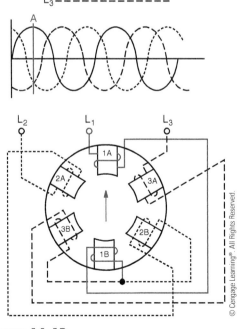

▶ **Figure 14–1B**
The magnetic field is concentrated between the poles of phase 1.

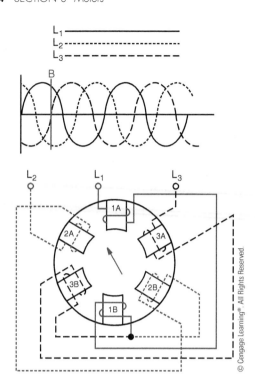

Figure 14–1C
The magnetic field is concentrated between phases 1 and 2.

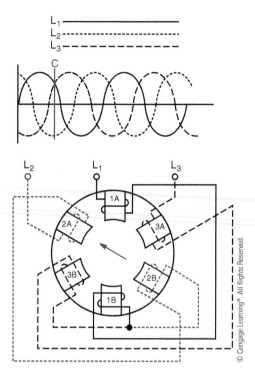

Figure 14–1D
The magnetic field is concentrated between the poles of phase 1.

In Figure 14–1D, line C indicates that line 2 is at its maximum negative peak and that lines 1 and 3 are less than maximum positive. The magnetic field at this point is concentrated between pole pieces 2A and 2B.

In Figure 14–1E, line D indicates that line 1 is zero. Lines 2 and 3 are less than maximum and in opposite directions. At this point, the magnetic field is concentrated between the pole pieces of phase 2 and phase 3.

In Figure 14–1F, line E indicates that phase 3 is at its maximum positive peak, and lines 1 and 2 are less than maximum and in the opposite direction. The magnetic field at this point is concentrated between pole pieces 3A and 3B.

In Figure 14–1G, line F indicates that phase 2 is zero. Line 3 is less than maximum positive, and line 1 is less than maximum negative. The magnetic field at this time is concentrated between the pole pieces of phase 1 and phase 3.

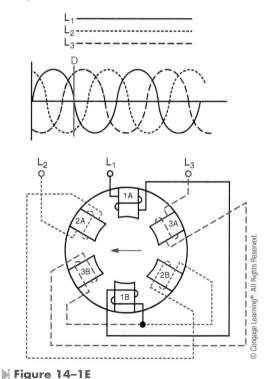

Figure 14–1E
The magnetic field is concentrated between phases 2 and 3.

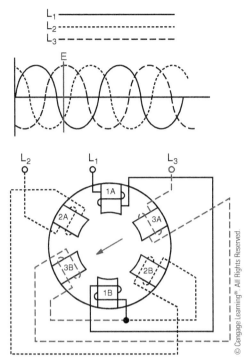

> **Figure 14–1F**
The magnetic field is concentrated between the poles of phase 3.

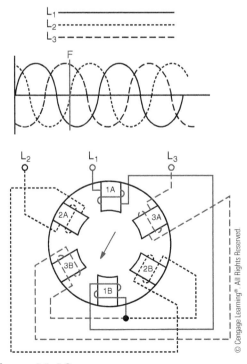

> **Figure 14–1G**
The magnetic field is concentrated between phases 1 and 3.

In Figure 14–1H, line G indicates that phase 1 is at its maximum negative peak, and phases 2 and 3 are less than maximum and in the opposite direction. Notice that the magnetic field is again concentrated between pole pieces 1A and 1B. This time, however, the magnetic polarity is reversed because the current has reversed in the stator winding.

In Figure 14–1I, line H indicates phase 2 is at its maximum positive peak, and phases 1 and 3 are less than maximum and in the negative direction. The magnetic field is concentrated between pole pieces 2A and 2B.

In Figure 14–1J, line 1 indicates that phase 3 is maximum negative, and phases 1 and 2 are less than maximum in the positive direction. The magnetic field at this point is concentrated between pole pieces 3A and 3B.

In Figure 14–1K, line J indicates that phase 1 is at its positive peak, and phases 2 and 3 are less than maximum and in the opposite direction. The

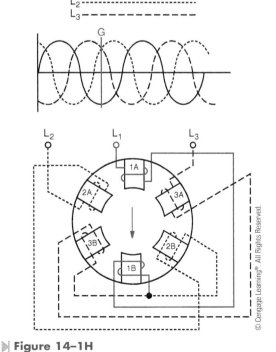

> **Figure 14–1H**
The magnetic field is concentrated between the poles of phase 1.

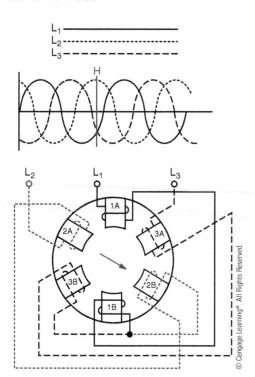

Figure 14-1I
The magnetic field is concentrated between the poles of phase 2.

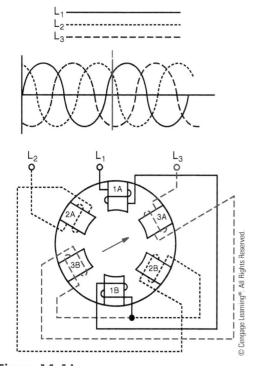

Figure 14-1J
The magnetic field is concentrated between the poles of phase 3.

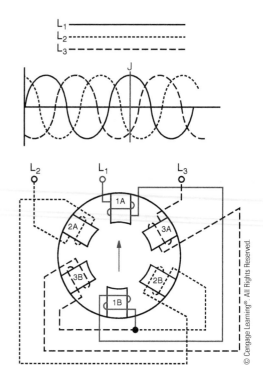

Figure 14-1K
The magnetic field has rotated 360°.

magnetic field is again concentrated between pole pieces 1A and 1B. Notice that in one complete cycle of three-phase voltage, the magnetic field has rotated 360° around the inside of the stator winding.

If any two of the stator leads are connected to a different line, the relationship of the voltages will change and the magnetic field will rotate in the opposite direction. The direction of rotation of a three-phase motor can be reversed by changing any two stator leads.

SYNCHRONOUS SPEED

The speed at which the magnetic field rotates is known as the **synchronous speed**. The synchronous speed of a three-phase motor is determined by two factors:

1. The number of stator poles
2. The frequency of the AC line

Because 60 Hz is a standard frequency throughout the United States and Canada, the following gives the synchronous speeds for motors with different numbers of poles.

2 Poles	3600 RPM
4 Poles	1800 RPM
6 Poles	1200 RPM
8 Poles	900 RPM

STATOR WINDINGS

The stator windings of three-phase motors are connected in either a wye or a delta. Some stators are designed in such a manner that they can be connected in either wye or delta, depending on the operation of the motor. Some motors, for example, are started as a wye-connected stator to help reduce starting current, and then changed to a delta connection for running.

Many three-phase motors have dual-voltage stators. These stators are designed to be connected to 240 volts or 480 volts. The leads of a dual-voltage stator use a standard numbering system. Figure 14–2 shows a dual-voltage wye-connected stator. Notice the stator leads have been numbered in a spiral. This diagram shows that numbers 1 and 4 are opposite ends of the same coil. Lead number 7 begins another coil, and this coil is to be connected to the same phase as leads 1 and 4. Leads 2 and 5 are opposite ends of the same coil. Coil number 8 must be connected with the same phase as leads 2 and 5. Leads 3 and 6 are opposite ends of the same coil and must be connected with lead number 9. Keep in mind that Figure 14–2

is a schematic diagram, and that when connecting a three-phase motor for operation at the proper voltage, the leads will look more like Figure 14–3. This figure illustrates the leads coming out of the terminal connection box on the motor. Some leads are numbered with metal or plastic bands on the wires, and some leads have numbers printed on the insulation of the wire.

Figure 14–4 shows the stator connection for operation on a 480-volt line. Figure 14–5 shows

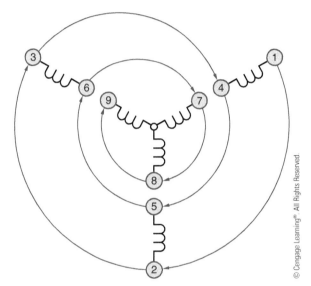

▶ **Figure 14–2**
Numbering a dual-voltage stator.

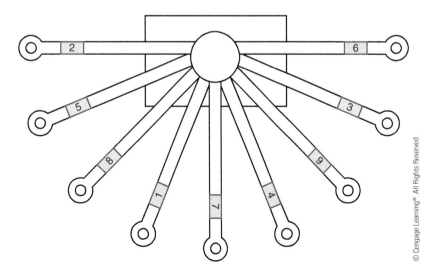

▶ **Figure 14–3**
Leads of a dual-voltage motor.

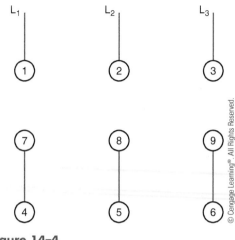

Figure 14–4
High-voltage connection.

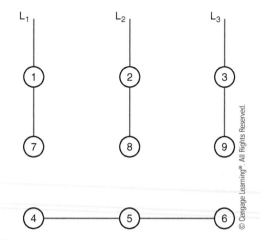

Figure 14–6
Low-voltage connection.

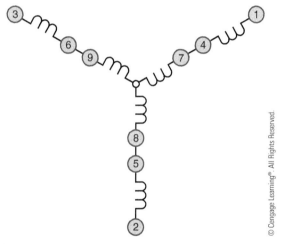

Figure 14–5
Windings connected in series.

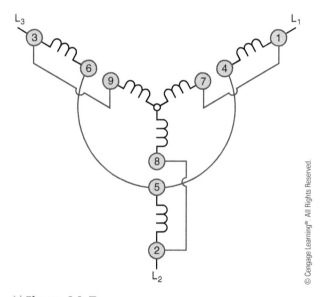

Figure 14–7
Windings connected in parallel.

the schematic equivalent of this connection. Notice that the windings have been connected in series. Figure 14–6 shows the stator connection for operation on a 240-volt line. Figure 14–7 shows the schematic equivalent of this connection. When the motor is to be operated on 240 volts, the stator windings are connected in parallel. Notice that leads 4, 5, and 6 are connected together to form another center point. This center point is electrically the same as the point where leads 7, 8, and 9 join together. Figure 14–8 shows the equivalent circuit.

When a motor is operated on a 240-volt line, the current draw of the motor is double the current draw of a 480-volt connection. For example, if a motor draws 10 amps of current when connected to 240 volts, it will draw 5 amps when connected to 480 volts. The reason for this is the difference of impedance in the windings between a 240-volt connection and a 480-volt connection. For instance, assume the stator windings of a motor have an

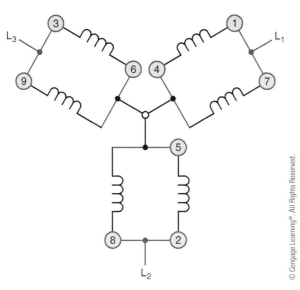

▶ Figure 14–8
Equivalent parallel circuit.

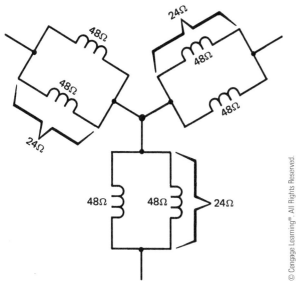

▶ Figure 14–9
Total impedance of a parallel connection.

impedance of 48 ohms. If the stator windings are connected in parallel, as shown in Figure 14–9, the total impedance of the windings is 24 ohms.

$$Rt = \frac{R1 \times R2}{R1 + R2}$$

$$Rt = \frac{48 \times 48}{48 + 48}$$

$$Rt = \frac{2304}{96}$$

$$Rt = 24 \text{ ohms}$$

If 240 volts is applied to this connection, 10 amps of current will flow.

$$I = \frac{E}{R}$$

$$I = \frac{240}{24}$$

$$I = 10 \text{ amps}$$

If the windings are connected in series for operation on a 480-volt line, as shown in Figure 14–10, the total impedance of the winding is 96 ohms.

$$Rt = R1 + R2$$

$$Rt = 48 + 48$$

$$Rt = 96 \text{ ohms}$$

If 480 volts is applied to this winding, 5 amps of current will flow.

$$I = \frac{E}{R}$$

$$I = \frac{480}{96}$$

$$I = 5 \text{ amps}$$

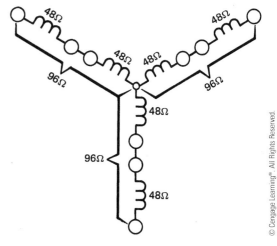

▶ Figure 14–10
Impedance adds in series.

DELTA CONNECTIONS

Three-phase motors are also connected in delta. The same standard numbering system is used for delta-connected motors. If a dual-voltage motor is to be connected in delta, there must be 12 leads instead of 9 leads brought out at the terminal box.

Figure 14–11 shows the schematic diagram of a motor connected for operation as a high-voltage delta. Notice that the stator windings for each phase have been connected in series for operation on high voltage. If the motor is to be connected for operation on a low voltage, the windings will be connected in parallel, as shown in Figure 14–12. Figure 14–13 shows an equivalent parallel connection.

SPECIAL CONNECTIONS

Some three-phase motors designed for operation on voltages higher than 600 volts may have more than 9 or 12 leads brought out at the terminal box. A motor with 15 or 18 leads can be found in high-voltage installations. A 15-lead motor has three coils per phase instead of two. Figure 14–14 shows the proper number sequence for a 15-lead motor. Notice the leads are numbered in the same spiral as a 9-lead motor.

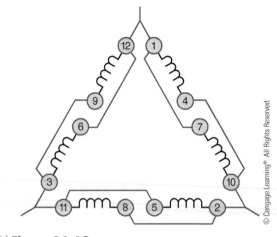

▶ **Figure 14–12**
Low-voltage delta connection.

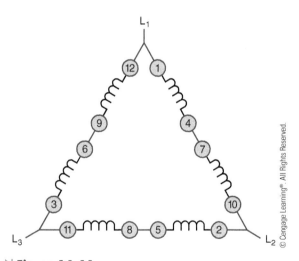

▶ **Figure 14–11**
High-voltage delta connection.

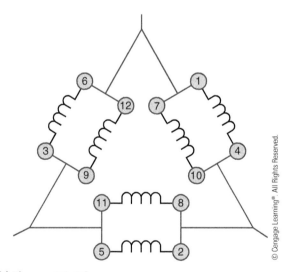

▶ **Figure 14–13**
Equivalent parallel delta connection.

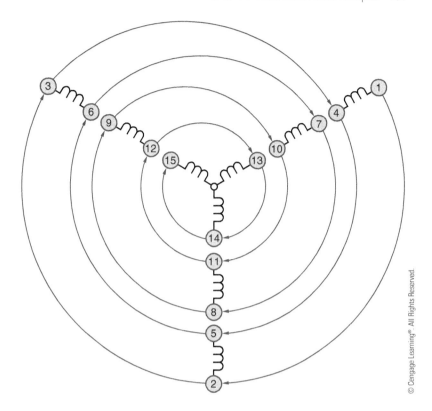

> **Figure 14–14**
Fifteen-lead motor.

SUMMARY

⊙ The three basic types of three-phase motors are:
  A. Squirrel-cage induction.
  B. Wound rotor induction.
  C. Synchronous.
⊙ All three-phase motors operate on the principle of a rotating magnetic field.
⊙ Three factors that cause a magnetic field to rotate are:
  A. The voltages of a three-phase system are 120° out of phase with each other.
  B. The three voltages change polarity at regular intervals.
  C. The arrangement of the stator winding around the inside of the motor.
⊙ The speed of the rotating magnetic field is determined by two factors:
  A. The number of stator poles.
  B. The frequency of the applied voltage.
⊙ The ends of the stator windings are numbered in a standard manner.
⊙ The stator windings of some three-phase motors are connected in a wye configuration, and some are connected in a delta configuration.

⊙ When a dual-voltage motor is to be connected for operation at low voltage, the stator windings will be connected in parallel.

⊙ When a dual-voltage motor is to be connected for operation on high voltage, the stator windings will be connected in series.

KEY TERMS

rotating magnetic field
synchronous speed
three-phase motors

REVIEW QUESTIONS

1. What are the three basic types of three-phase motors?

2. Name three factors that produce a rotating magnetic field.

3. What is synchronous speed?

4. What two factors determine the synchronous speed of a three-phase motor?

5. How is the direction of rotation of a three-phase motor changed?

6. What is the synchronous speed of a four-pole motor when connected to a 60-Hz line?

7. A dual-voltage three-phase motor has a current draw of 50 amps when connected to a 240-volt line. How much current will flow if the motor is connected for operation on 480 volts?

8. If the stator windings of a three-phase motor are connected for operation on high voltage, will the windings be connected in series or parallel?

9. If a dual-voltage motor is connected for operation on low voltage, and the motor is then connected to high voltage, will the motor operate at a faster speed?

10. Why does a dual-voltage motor draw more current when connected to low voltage than it does when connected to high voltage?

UNIT 15 ▷

The Squirrel-Cage Induction Motor

OBJECTIVES

After studying this unit, the student should be able to:

▷ Discuss the principle of operation of a squirrel-cage three-phase motor

▷ List the factors that determine the amount of torque developed by a squirrel-cage motor

▷ Discuss code letters and their meaning

▷ Perform an ohmmeter test on a three-phase squirrel-cage motor

The **squirrel-cage induction motor** receives its name from the fact that the rotor contains a set of bars that resemble a squirrel cage. If the soft-iron laminations were to be removed from the rotor, it would be seen that the rotor contains a set of metal bars joined together at each end by a metal ring, Figure 15–1. Figure 15–2 shows a complete squirrel-cage rotor and stator winding.

PRINCIPLE OF OPERATION

The squirrel-cage motor is an induction motor. This means that the current flow in the rotor is produced by induced voltage from the rotating magnetic field of the stator. In Figure 15–3, a squirrel-cage rotor is shown inside the stator winding of a three-phase motor. It will be assumed that the motor shown in Figure 15–3 contains four poles per phase, which

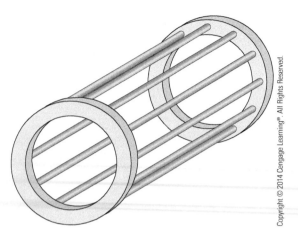

▶ **Figure 15–1**
Squirrel-cage bars.

▶ **Figure 15–2**
Rotor and stator of a three-phase, squirrel-cage motor.

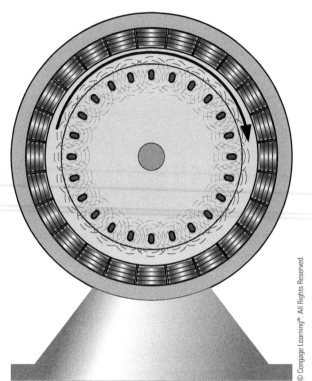

▶ **Figure 15–3**
Voltage is induced into the rotor by the rotating
magnetic field.

produces a synchronous speed of 1800 RPM when
the stator is connected to a 60 Hz line. When power
is first connected to the stator, the rotor is not turning. The magnetic field of the stator cuts the **rotor
bars** at a rate of 1800 RPM. Three factors that
determine the amount of induced voltage follow:

1. The strength of the magnetic field.
2. The number of turns of wire cut by the magnetic field. (This is sometimes stated as length
   of conductor.)
3. Speed of the cutting action.

Because the rotor is stationary at this time, maximum voltage is induced into the rotor. The induced
voltage causes current to flow through the rotor
bars. As current flows through the rotor, a magnetic
field is produced around each rotor bar. The magnetic field of the rotor is attracted to the magnetic
field of the stator, and the rotor begins to turn in the
same direction as the rotating magnetic field.

As the speed of the rotor increases, the rotating
magnetic field cuts the rotor bars at a slower rate.
For example, assume the rotor has accelerated to
a speed of 600 RPM. The synchronous speed of
the rotating magnetic field is 1800 RPM. Therefore, the rotor is being cut at a rate of 1200 RPM.
(1800 RPM – 600 RPM = 1200 RPM), Because
the rotor is being cut at a slower rate, less voltage
is induced into the rotor. This produces less current
flow through the rotor. When the current flow in
the rotor is reduced, the current flow in the stator is
reduced also.

As the rotor continues to accelerate, the rotating magnetic field cuts the rotor bars at a slower rate. This causes less voltage to be induced into the rotor and therefore less current flow in the rotor. Notice that the maximum amount of induced voltage and current occurs when the rotor is not turning at the instant of start. This is the reason that AC induction motors require more current to start than to run.

TORQUE

Torque is the amount of turning or twisting force developed by a motor. It is generally measured in pound-inches or pound-feet depending on the application. Imagine a bar 1 foot in length attached to the shaft of a motor. A torque of 1 pound-foot would be the force exerted by applying a pressure of 1 pound on the end of the bar.

The amount of torque produced by an AC induction motor is determined by three factors:

1. The strength of the magnetic field of the stator
2. The strength of the magnetic field of the rotor
3. The phase angle difference between stator and rotor flux

Notice that one of the factors that determines the amount of torque produced by an induction motor is the strength of the magnetic field of the rotor. *An induction motor cannot run at synchronous speed.* If the rotor were to accelerate to the speed of the rotating magnetic field, there would be no cutting action of the squirrel-cage bars and therefore no current flow in the rotor. If there were no current flow in the rotor, there could be no rotor magnetic field and therefore no torque.

When an induction motor is operating with no load connected to it, it will run close to the synchronous speed. For example, a four-pole motor that has a synchronous speed of 1800 RPM could run at 1795 RPM at no load. The speed of an AC induction motor is determined by the amount of torque needed. When the motor is operating at no load, it produces only the amount of torque needed to overcome its own friction and windage losses. This low torque requirement permits the motor to operate at a speed close to that of the rotating magnetic field.

If a load is connected to the motor, it must furnish more torque to operate the load. This causes the motor to slow down. When the motor speed decreases, the rotating magnetic field cuts the rotor bars at a faster rate, causing more voltage to be induced in the rotor and therefore more current. The increased current flow produces a stronger magnetic field in the rotor, which causes more torque to be produced by the motor. As the current flow increases in the rotor, it causes more current flow to be produced in the stator. This is why motor current increases as load is added to the motor.

Another factor that determines the amount of torque produced by an induction motor is the phase angle difference between rotor and stator flux. Motor torque is basically the attracting force of two magnetic fields. Imagine two bar magnets representing the magnetic fields of the stator and rotor, Figure 15–4. If the north end of one magnet is placed close to the south end of the other, they will be attracted to each other, Figure 15–5. Torque can be compared to the amount of force necessary to separate the two magnets. When the magnets are in line with each other, as shown in Figure 15–5, the attraction is strongest, and the amount of force necessary to separate them is the greatest. This compares to the stator flux and rotor flux being in phase with each other.

Now assume that the two magnets are placed at an angle of 10° to each other, Figure 15–6. The 10° angle produces a greater amount of separation between the two magnetic fields. The magnets can now be separated with less force. Now assume that the magnets are placed at a 30° angle to each other, Figure 15–7. The amount of force necessary

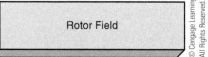

▶ **Figure 15–4**
Stator and rotor magnetic fields are compared to two bar magnets.

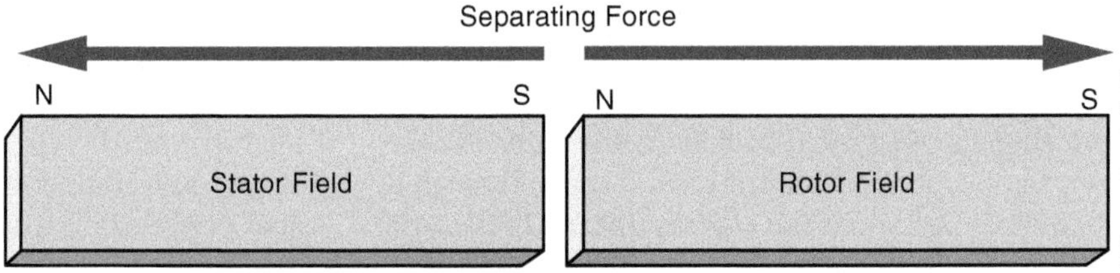

Separating Force

N   Stator Field   S   N   Rotor Field   S

Figure 15–5
Torque is the force attracting the two magnets.

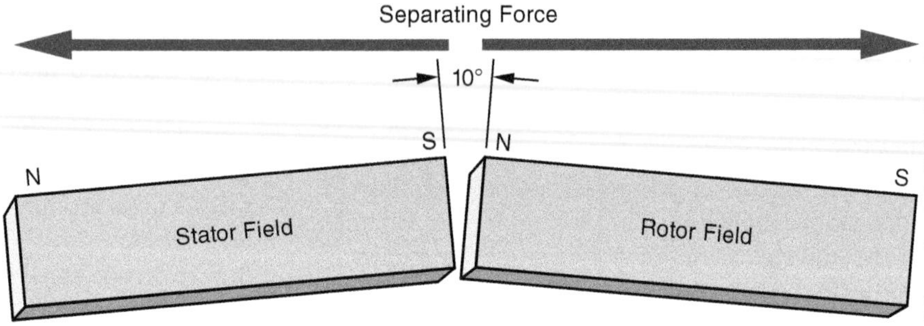

Separating Force

10°

S | N

N   Stator Field   S   Rotor Field   S

Figure 15–6
Magnets are separated by an angle of 10°.

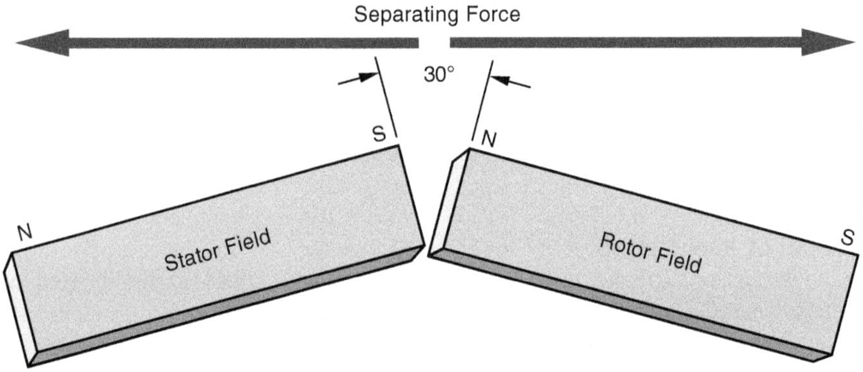

Separating Force

30°

S   N

N   Stator Field   Rotor Field   S

Figure 15–7
Magnets are separated by an angle of 30°.

to separate the two magnets is less than it was at a 10° angle. The greater the angle between the two magnets, the less force is required to separate them. This corresponds to the phase angle difference between rotor and stator flux. The greater the phase angle between rotor and stator flux, the less torque is developed by the motor.

CODE LETTERS

Squirrel-cage rotors are not all the same. Rotors are made with different types of bars. The type of rotor bars used in the construction of the rotor determines the operating characteristics of the motor. AC motors are given a **code letter** on their nameplate. The code letter indicates the type of bars used in the rotor. Figure 15–8 shows a rotor with type "A" bars. A type "A" rotor has the highest resistance of any squirrel-cage rotor. This means that the starting torque is high because the rotor current is closer to being 90° out of phase with the stator current than the other types of rotors. The high resistance of the rotor bars limits the amount of current flow in the rotor when starting. This produces a low starting current for the motor. A rotor with type "A" bars has very poor running characteristics, however. Because the bars are resistive, a large amount of voltage must be induced into the rotor to produce an

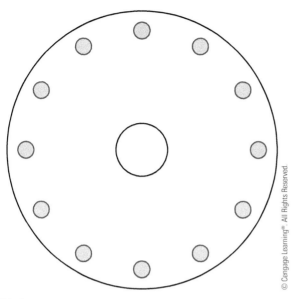

▶ **Figure 15–8**
Type "A" rotor.

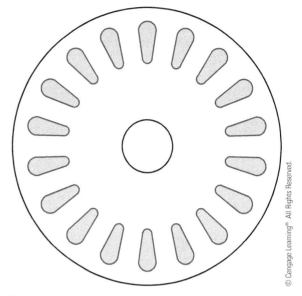

▶ **Figure 15–9**
Type "B–E" rotor.

increase in rotor current and therefore an increase in the rotor magnetic field. This means that when load is added to the motor, the rotor must slow down a great amount to produce enough current in the rotor to increase the torque.

Figure 15–9 shows a rotor with bars similar to those found in rotors with code letters B through E. These rotor bars have lower resistance than the type "A" rotor. This rotor has fair starting torque, low starting current, and fair speed regulation.

Figure 15–10 shows a rotor with bars similar to those found in rotors with code letters F through V. This rotor has low starting torque, high starting current, and good running torque. This type of rotor also has good speed regulation.

The code letter found on the nameplate is also used to determine the amount of **locked rotor current** for the motor. Locked rotor current is the amount of current the motor will draw at the moment of starting. Figure 15–11 shows *Table 430.7(B)* of the *National Electrical Code*. This table is used to determine the locked rotor current for a squirrel-cage rotor. To use the table, the horsepower, code letter, and voltage of the motor must be known. For this example, assume the motor is 10 horsepower, has a code letter J, and is operated on

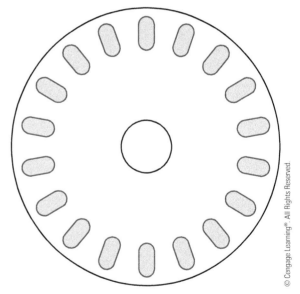

▶ **Figure 15–10**
Type "F–V" rotor.

a 480-volt line. The table lists the locked rotor currents in kilovolt-amperes per horsepower. The table shows that code letter J is 7.1 to 7.99. For this calculation, a midvalue of 7.5 will be used. Because the

Table 430.7(B) Locked-Rotor Indicating Code Letters

Code Letter	Kilovolt-Amperes per Horsepower with Locked Rotor
A	0–3.14
B	3.15–3.54
C	3.55–3.99
D	4.0–4.49
E	4.5–4.99
F	5.0–5.59
G	5.6–6.29
H	6.3–7.09
J	7.1–7.99
K	8.0–8.99
L	9.0–9.99
M	10.0–11.19
N	11.2–12.49
P	12.5–13.99
R	14.0–15.99
S	16.0–17.99
T	18.0–19.99
U	20.0–22.39
V	22.4 and up

▶ **Figure 15–11**
NFPA 70®, National Electrical Code and NEC® are registered trademarks of the National Fire Protection Association, Quincy, MA.

MOTOR NAMEPLATE	
HP	Phase
10	3
Volts	Amps
240/480	28/14
Hz	FL Speed
60	1745 RPM
Code	SF
J	1.25
Frame	Model No.
XXXX	XXXX

▶ **Figure 15–12**
Motor nameplate.

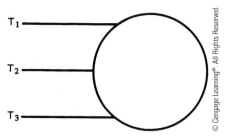

▶ **Figure 15–13**
Schematic symbol of a three-phase squirrel-cage induction motor.

values are listed in kilovolt-amperes, 7.5 is actually 7500 volt-amperes. To find the locked rotor current, multiply the kVA rating by the horsepower, and then divide by the voltage.

$$I = 7500 \text{ VA} \times 10 \text{ hp}$$

$$I = \frac{75{,}000}{480}$$

$$I = 156.25 \text{ amps}$$

THE NAMEPLATE

Electric motors have **nameplates** that give a great deal of information about the motor. Figure 15–12 illustrates the nameplate of a three-phase induction motor. The nameplate shows that the motor is 10 horsepower, is a three-phase motor, and operates on 240 or 480 volts. The full-load running current of the motor is 28 amps when operated on 240 volts, or 14 amps when operated on 480 volts. The motor is designed to be operated on a 60 Hz AC voltage, and has a full-load speed of 1745 RPM. The speed indicates that this motor has four poles per phase. Because the full-load speed is 1745 RPM, the synchronous speed would be 1800 RPM. The motor contains a type J squirrel-cage rotor, and has a service factor of 1.25. The service factor is used to determine the amperage rating of the overload protection for the motor. The frame indicates the type of mounting the motor has. Figure 15–13 shows the schematic symbol for a three-phase squirrel-cage induction motor.

Ohmmeter

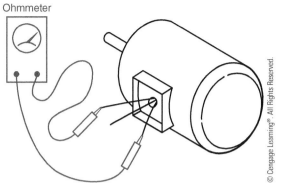

▶ **Figure 15–14**
Testing the stator winding for opens.

Ohmmeter

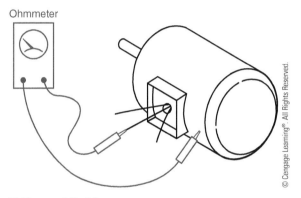

▶ **Figure 15–15**
Testing the stator winding for grounds.

TESTING THE MOTOR

Most service technicians test a three-phase motor with an ohmmeter. The ohmmeter can be used to check the stator winding for an open condition or a grounded condition. To test the stator winding for an open condition, check the continuity of each winding by measuring the resistance between each of the three windings, as shown in Figure 15–14. The resistance of each pair of windings should be the same. To test for a grounded motor, connect one ohmmeter lead to the case of the motor, and the other lead to one of the motor leads. There should be no continuity between any of the leads and the case of the motor, Figure 15–15. The ohmmeter, however, will not generally detect a shorted winding. The resistance of the stator windings of most large horsepower motors is so low that they appear under

normal conditions to be a short circuit to the ohmmeter. To test for a shorted winding, some method must be used to measure the reactance of the windings instead of their resistance. If the motor will run, an ammeter can be used to measure the current draw of the motor. The current of each line should be equal and within the full-load current rating of the motor. If one line has a higher current reading than the others, it is an indication of a shorted stator winding. If it is not possible to operate the motor, an instrument that measures the actual inductance of the winding can be used. If one winding has a lower inductance than the others, it is shorted.

STARTING METHODS FOR SQUIRREL-CAGE MOTORS

There are several methods that can be employed to start three-phase squirrel-cage motors, including these:

1. Across-the-line starting
2. Resistor/reactor starting
3. Autotransformer starting
4. Wye-delta starting
5. Part winding starting

Across-the-Line Starting

The simplest of these methods is across-the-line starting. Across-the-line starting is accomplished by connecting the motor directly to the power line, Figure 15–16. This method of starting is used when the inrush current, called locked rotor current, is not so great as to adversely affect the power line. Across-the-line starting is employed for all small horsepower motors. Motors of several hundred horsepower may be started across-the-line if the local power base is sufficient to provide the starting current.

Resistor/Reactor Starting

Often it is necessary to reduce the inrush current during the starting period of large motors. One method of reducing the starting current is to connect resistors or reactors (choke coils or inductors) in series with the motor during the starting period, Figure 15–17. This permits the resistance of the resistors or the inductive

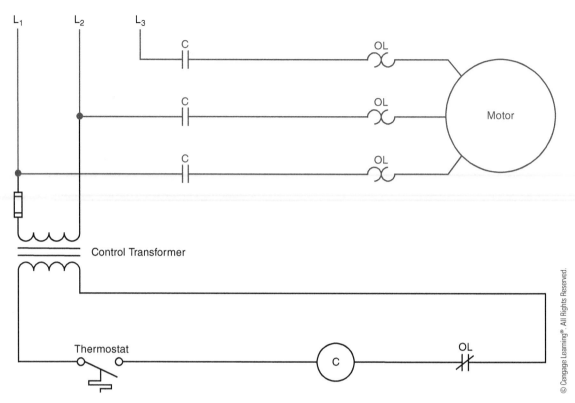

▶ **Figure 15–16**
Across-the-line starting is accomplished by connecting the motor directly to the power line.

reactance of the reactors to limit the inrush current when the motor is first started. After the motor has accelerated to near full load speed, the resistors or reactors are shunted out of the circuit. Several methods may be employed for shunting the resistors or reactors out of the line. The circuit shown in Figure 15–17 uses a time delay relay to energize R contactor after some period of time. The circuit operates as follows:

- When the thermostat contact closes, the coils of the compressor contactor (C) and the on-delay timer (TR) energize.
- When the compressor contactor energizes, C contacts close and connect the motor and reactors to the power line.
- After some period of time, timed contact TR closes and energizes the R contactor, causing the R contacts connected in parallel with the current-limiting inductors to close and shunt the inductors out of the line. The motor is now connected to full voltage.

Autotransformer Starting

Autotransformer starting differs from resistor or reactor starting by decreasing the voltage applied to the motor during the starting period instead of inserting resistance or inductive reactance in series with the motor. It should be noted that when the voltage to the motor is reduced, the torque is reduced also. A 50% reduction of voltage produces a 50% reduction of current also, but the torque is reduced to 25% of the value produced by full voltage starting.

Several methods can be employed when using autotransformer starting. Some starters used three transformers, but most use two transformers connected in an open delta, Figure 15–18. In this circuit, notice the addition of the normally closed R contact connected in series with S coil and the normally closed S contact connected in series with the R coil. This is referred to as **interlocking**.

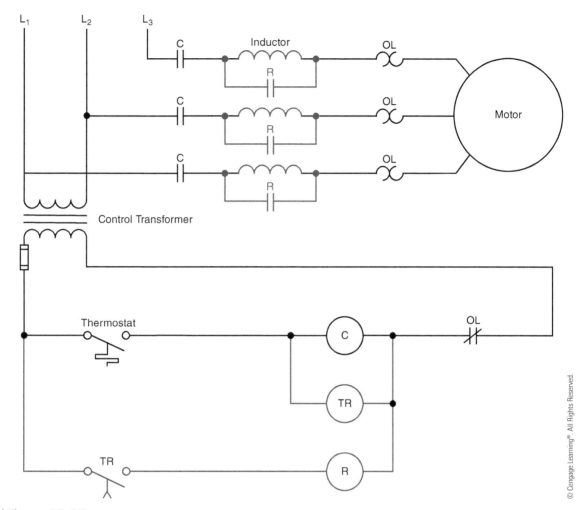

▶| **Figure 15–17**
Reactor starter for a three-phase squirrel-cage motor.

Interlocking is used to prevent one contactor being energized while some other contactor is energized. In this circuit, coils R and S can never be energized at the same time. The circuit operates as follows (it will be assumed that timer TR is set for a delay of 3 seconds):

• When the thermostat contact closes, coils S and TR energize.

• All S contacts change position. The normally open S contacts close and connect the autotransformer to the line. The motor is now connected to reduced voltage. The normally closed S contact connected

in series with the coil of the R contactor opens and prevents the possibility of the R coil being energized.

• After a delay of 3 seconds, the TR timed contacts change position.

• The normally closed TR contact connected in series with the S coil opens and disconnects it from the line.

• When the S coil de-energizes, all S contacts return to their normal position. The autotransformer is now disconnected from the line, and the contact connected in series with R coil is now closed.

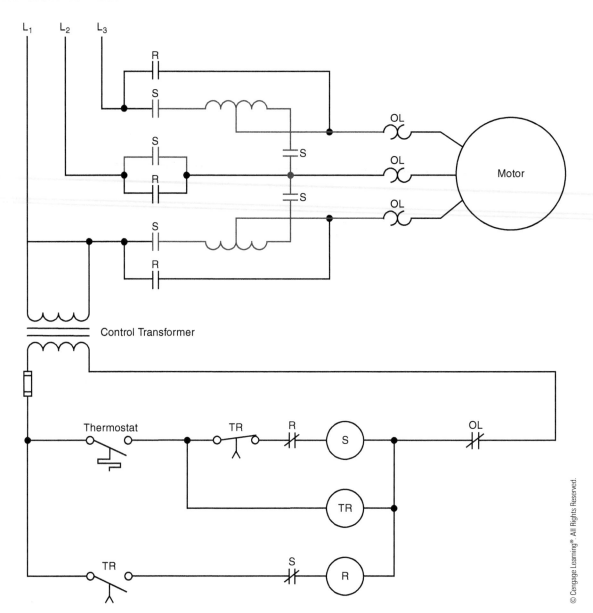

▶ **Figure 15–18**
Autotransformer starter using an open delta connection.

- When the normally open TR contact closes, the R coil energizes and all R contacts change position.
- The normally open R contacts close and connect the motor directly to the power line. The normally closed R contact opens and prevents the possibility of S coil being energized at the same time.

Wye-Delta Starting

Wye-delta starting is also known as star-delta starting. Wye-delta starting is used to reduce the inrush current during the starting period by connecting the stator windings of the motor in a wye configuration during the starting period and then reconnecting them in delta during the run period. If the stator windings are connected in wye, the inrush current will be one-third the value it would be if they were connected in delta. Assume that the motor stator windings have an impedance of 2.5 ohms each and that the windings are connected in delta, Figure 15–19. It will also be assumed that the motor is connected to a line voltage of 480 volts. In a delta connection, the phase voltage is the same as the line voltage. Therefore, when the C contacts close, 480 volts will be applied across 2.5 ohms. This produces a current of 192 amperes in each phase.

$$I_{PHASE} = \frac{480}{2.5}$$

In a delta connection, the line current is greater than the phase current by a factor of the square root of 3 (1.732). Therefore, the line current supplied to the motor is 332.5 amperes (192 × 1.732).

Now assume that the same stator windings are connected in wye, Figure 15–20. In a wye connection, the phase voltage is less than the line voltage by a factor of the square root of 3 (1.732). Therefore, 277 volts is applied across each stator winding instead of 480 volts (480/1.732). This produces a phase current of 110.8 amperes. Because the stator windings are now connected in wye instead of delta, the line current is the same as the phase current. The motor inrush current has been reduced from 332.5 amperes to 110.8 amperes by reconnecting the stator windings from delta to wye.

There are two conditions that must be met when wye-delta starting is to be used:

1. The motor must be designed to operate with its stator windings connected in delta during the normal run period.
2. All stator winding leads must be accessible at the terminal connection box. This would be leads T1 through T6 for a single-voltage motor and T1 through T12 for a dual-voltage motor.

Another consideration when using wye-delta starting is that the overload heaters should be sized for the phase current value and not the line current value. The basic stator connection for wye-delta starting is shown in Figure 15–21. Notice that the overload heaters are connected in the phase windings, not in the line leads.

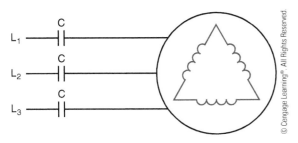

Figure 15–19
The stator windings are connected in delta.

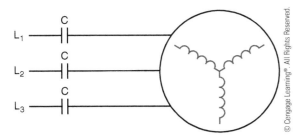

Figure 15–20
The stator windings are connected in wye.

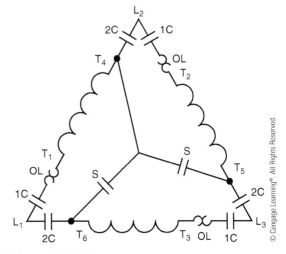

Figure 15–21
Stator connection for a single-voltage wye-delta starter.

A basic control schematic for a wye-delta starter is shown in Figure 15–22. To understand the operation of this circuit assume that timer TR has been set for a delay of 3 seconds.

- When the thermostat contact closes, a current path exists through coils 1C, TR, and S, Figure 15–23.
- The 1C load contacts close to supply power to the motor T leads.
- The S contacts change position also. The two S load contacts close and connect the stator winding in a wye configuration by shorting T4, T5, and T6 together. The normally open S contact connected in series with coil 2C opens to provide interlocking protection.
- After a delay of 3 seconds, both TR contacts change position, Figure 15–24.
- The normally closed TR contact connected in series with S coil opens and de-energizes S contactor,

causing all S contacts to return to their normal position. The normally closed S contact connected in series with coil 2C recloses to provide a current path to coil 2C.

- When the normally open TR contact connected in series with coil 2C closes, coil 2C energizes, causing all 2C contacts to change position. The 2C load contacts close and reconnect the motor stator windings in a delta configuration. The normally closed 2C contact connected in series with S coil opens to provide interlocking protection.

Part Winding Starters

Another method for reducing the starting current of large three-phase squirrel-cage motors is part winding starting. Motors intended for use with part winding starting contains two separate stator windings, Figure 15–25. Each winding is

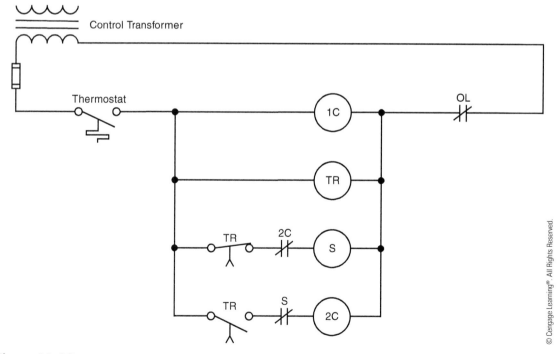

▶ **Figure 15–22**
Basic control schematic for wye-delta starting.

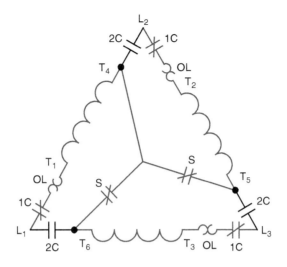

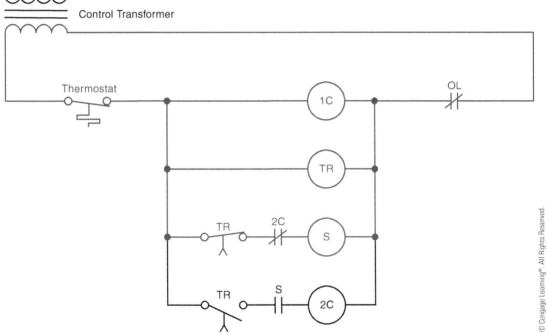

▶ **Figure 15–23**
The stator windings are connected in wye during starting.

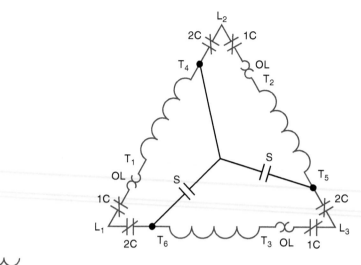

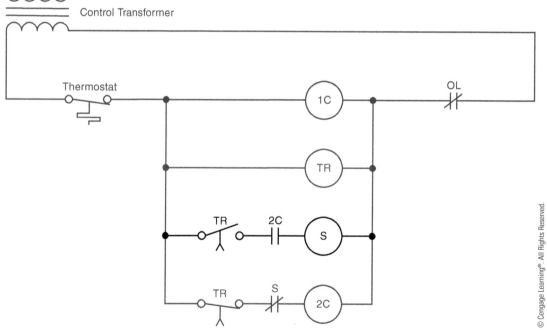

▶ **Figure 15–24**
The stator windings are connected in delta during the run period.

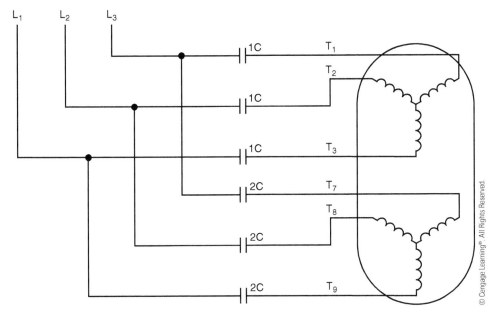

▶ **Figure 15–25**
A motor intended for part windings starting contains two stator windings.

rated for the intended line voltage. The impedance of a single stator winding is double that of the two windings connected in parallel. Dual-voltage motors can also be used for part winding starting provided the line voltage corresponds to the low-voltage rating of the motor. A 480/240-volt motor could employ part winding starting provided the motor is connected to 240 volts and not 480 volts. In this situation T4, T5, and T6 are connected, and power is connected to T1, T2, and T3 during the starting period. After some period of time, power is also connected to T7, T8, and T9. This has the same effect as connecting the stator windings in parallel.

A basic circuit for part winding starting is shown in Figure 15–26. To understand the operation of the circuit, assume that timer TR has been set for a delay of 3 seconds.

• When the thermostat contact closes, power is provided to coils C1 and TR, Figure 15–27. All 1C contacts close and connect T1, T2, and T3 to the power line.

• After a delay of 3 seconds, the normally open TR contact connected in series with 2C closes and energizes coil 2C, Figure 15–28.

• When the 2C load contacts close, power is connected to T7, T8, and T9. This has the same effect as connecting the two windings in parallel.

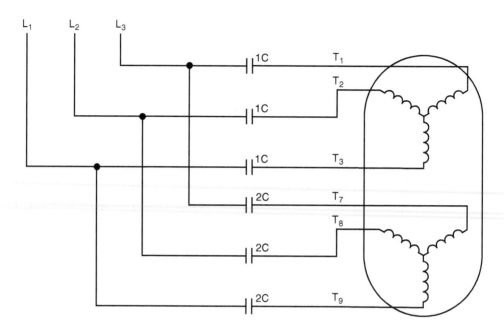

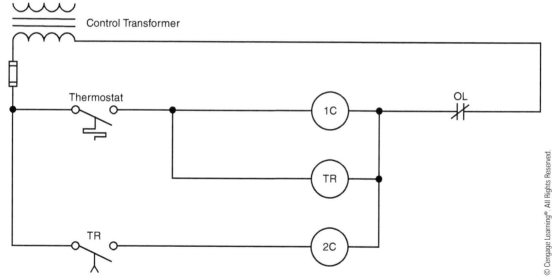

▶ **Figure 15–26**
Basic part winding starting circuit.

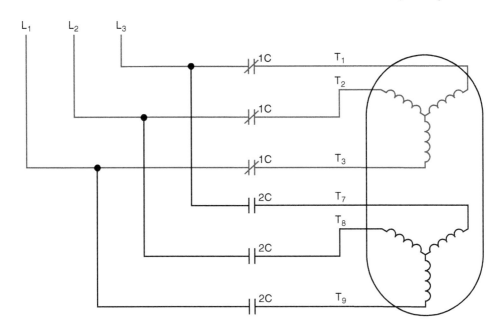

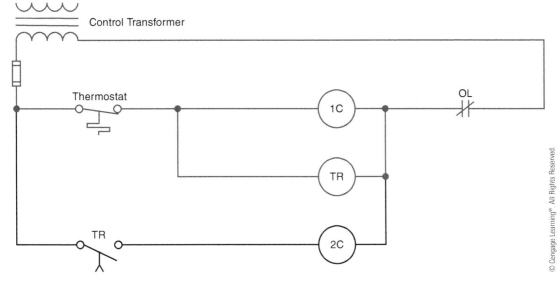

Figure 15–27
During the starting period, only one stator winding is energized.

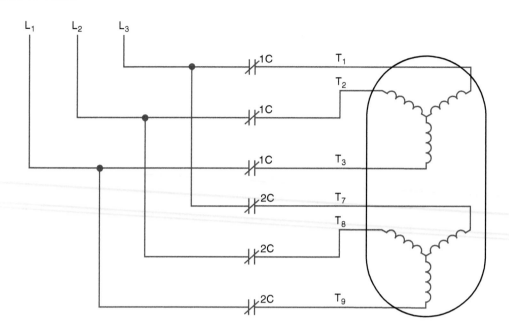

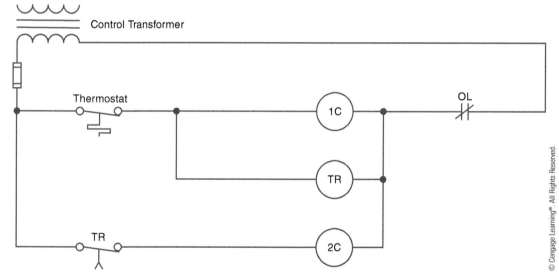

▶ **Figure 15–28**
During the run period, both stator windings are connected.

> ## SUMMARY

- ⊙ The squirrel-cage motor receives its name from the fact that the rotor contains a set of bars connected together at each end, and the entire assembly resembles a squirrel cage.
- ⊙ Three factors that determine the amount of voltage induced into the rotor are:
  - A. The strength of the magnetic field of the stator.
  - B. The number of stator bars contained in the rotor.
  - C. The difference in speed between the speed of the rotating magnetic field and the speed of the rotor.
- ⊙ The greatest amount of current draw for a squirrel-cage motor is during the starting period.
- ⊙ Three factors that determine the amount of torque produced by a squirrel-cage motor are:
  - A. The strength of the magnetic field of the stator.
  - B. The strength of the magnetic field of the rotor.
  - C. The phase angle difference between stator current and rotor current.
- ⊙ The direction of rotation of a squirrel-cage motor can be changed by reversing any two stator leads.
- ⊙ The code letter found on the motor nameplate indicates the type of bars used in the construction of the rotor.
- ⊙ The simplest method of starting a squirrel-cage motor is across-the-line starting.
- ⊙ Resistor/reactor starting is accomplished by connecting resistors or inductors in series with the motor during the starting period.
- ⊙ Autotransformer starting reduces starting current by lowering the applied voltage to the motor during the starting period.
- ⊙ If the applied voltage is reduced by 50% of normal during the starting period, the starting torque is reduced to 25% of normal.
- ⊙ Wye-delta starting is accomplished by connecting the stator windings in wye during the starting period and changing them to a delta connection during the normal run time.
- ⊙ A motor will draw one-third as much current during the starting period with its windings connected in wye as it will if they are connected in delta.
- ⊙ Two requirements for motors intended for wye-delta starting are:
  - A. All stator winding leads must be brought out at the terminal connection box.
  - B. The motor must be designed to run with its stator windings connected in delta.
- ⊙ Motors intended to be used for part winding starting have two separate stator windings.
- ⊙ Dual-voltage motors can be used for part winding starting provided the motor is connected for low-voltage operation.

KEY TERMS

code letter   nameplates   squirrel-cage induction
interlocking   rotor bars   motor
locked rotor current

REVIEW QUESTIONS

1. What three factors determine the amount of torque produced by an AC induction motor?

2. When does an AC induction motor draw more current when starting than it does when running?

3. Why does the current flow to the motor increase when load is added to the motor?

4. What does the code letter found on the nameplate of the motor indicate?

5. At what degree angle between the stator current and the rotor current is the maximum torque developed?

6. What type of squirrel-cage rotor has the highest starting torque?

7. What type of squirrel-cage rotor has the best speed regulation?

8. Why can an induction motor never operate at synchronous speed?

9. What does the locked rotor current of a motor indicate?

10. The nameplate of a squirrel-cage motor indicates that the motor has a full-load speed of 875 RPM. How many poles per phase does the motor have?

11. What is the simplest of all starting methods for a squirrel-cage motor?

12. Explain interlocking.

13. How does autotransformer starting differ from resistor/reactor starting?

14. A three-phase squirrel-cage induction motor has its stator windings connected in a delta connection. During the starting period, the motor has a current draw of 360 amperes. If the stator windings were reconnected to form a wye connection during the starting period, how much starting current would the motor draw?

15. What two conditions must be met before a motor can be used with a wye-delta starter?

The Wound Rotor Induction Motor

OBJECTIVES

After studying this unit, the student should be able to:

▷ Describe the construction of a wound rotor induction motor

▷ Discuss the difference in operation between wound rotor and squirrel-cage induction motors

▷ Discuss the starting and running characteristics of a wound rotor induction motor

▷ Connect a wound rotor induction motor for operation

▷ Draw the standard schematic symbol for a wound rotor induction motor

▷ Perform an ohmmeter test on a wound rotor induction motor

Another type of three-phase induction motor used for operating large air-conditioning units is the **wound rotor induction motor**. The stator winding of this motor is the same as the stator of a squirrel-cage induction motor. The rotor of the wound rotor motor, however, does not contain squirrel-cage bars. The rotor of this motor contains wound coils of wire, as illustrated in Figure 16–1. The rotor contains as many poles as there are stator poles. The motor shown in Figure 16–1 is for a two-pole stator. Notice that there are three separate windings on the rotor. The finish end of all the windings are connected together, forming a wye connection for the rotor winding. The start end of each winding is connected to a separate **slip ring** on the rotor shaft.

The slip rings permit the connection of external resistance to the rotor windings. Figure 16–2 shows

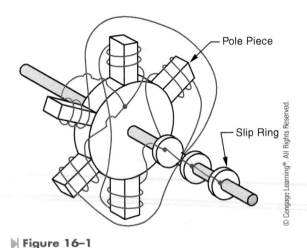

▶ **Figure 16–1**
A wound rotor.

a schematic diagram of the stator connection and rotor connection of a wound rotor motor. Notice that the wye-connected stator winding is connected directly to the incoming power. The wye-connected rotor is connected to three variable resistors. The dashed line drawn between the resistors indicates that they are mechanically connected together. If the resistance of one is changed, the resistance of the other two changes also.

Resistance is connected to the slip rings by means of **carbon brushes**, as shown in Figure 16–3. Because the resistance connection to the rotor is external, the amount of resistance used in the circuit can be controlled, thus permitting the amount of current flow in the rotor to be controlled. If the current flow in the rotor is limited by the amount of resistance connected in the circuit, the stator current is limited also. A great advantage of the

wound rotor motor is that it limits the amount of inrush current when the motor is first started, which eliminates the need for reduced voltage starters or wye-delta starting.

Another advantage of the wound rotor motor is its high starting torque. Because resistors are used to limit current flow in the rotor, the phase angle between the stator current and the rotor current is close to 90°. The schematic symbol for a wound rotor motor is shown in Figure 16–4.

MOTOR OPERATION

When power is applied to the stator winding, a rotating magnetic field is created in the motor. This magnetic field cuts through the windings of the rotor and induces a voltage into them. The amount of voltage induced in the rotor windings is determined by the same three factors that determined the amount of voltage induced in the squirrel-cage rotor. The amount of current flow in the rotor is determined by the amount of induced voltage and the amount of resistance connected to the rotor ($I = E/R$). When current flows through the rotor, a magnetic field is produced. This magnetic field is attracted to the rotating magnetic field of the stator.

As the rotor speed increases, the induced voltage decreases because of less cutting action between the rotor windings and rotating magnetic field. This produces less current flow in the rotor and therefore less torque. If resistance is reduced, more current can flow, which will increase motor torque, and the rotor will increase in speed. This action continues until the rotor is operating at maximum speed and all resistance has been shorted out of the rotor circuit. When all of the resistance has been

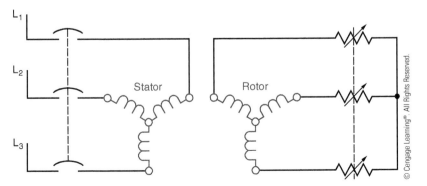

▶ **Figure 16–2**
A wye-connected stator and wye-connected rotor.

▶ **Figure 16–3**
External resistance is connected to the rotor circuit
with brushes and slip rings.

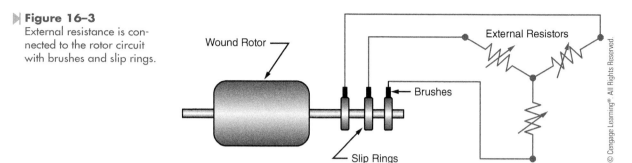

Wound Rotor

External Resistors

Brushes

Slip Rings

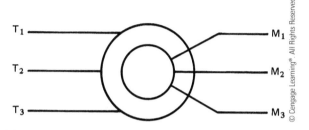

▶ **Figure 16–4**
Schematic symbol of a wound rotor induction motor.

shorted out, the motor operates like a squirrel-cage induction motor.

STARTING

Most large wound rotor motors use a method of **step starting** as opposed to actual variable resistors. Step starting is similar to shifting the gears in the transmission of an automobile. The transmission is placed in first gear when the car is first started. As the car gains speed, the transmission is shifted to second gear, then third gear, and so on, until the car is operating in its highest gear. When a wound rotor motor is step started, it begins with maximum resistance connected in the rotor circuit. As the motor speed increases, resistance is shorted out of the circuit until the windings of the rotor are shorted together. The number of steps can vary from one motor to another, depending on the size of the motor and how smooth a starting action is desired.

There are different control methods used to short out the steps of resistance when starting a wound rotor motor. Some controllers sense the amount of

current flow to the stator. This method is known as current limit control. Another method, known as slip frequency control, detects the speed of the rotor. One of the most common methods uses time relays to control when resistance is shorted out of the circuit. This method is known as **definite time control**. Figure 16–5 shows a schematic diagram of a time-controlled starter for a wound rotor motor. In this schematic, the motor circuit is shown at the top of the diagram. A control transformer is used to step the line voltage down to the value of voltage used in the control circuit. The operation of the circuit is as follows:

1. When the start button is pressed, a circuit is completed through M motor starter coil, TR_1 coil, and the overload contact. When M coil energizes, all M contacts close. The three large load contacts located at the top of the diagram close and connect the stator winding to the line. The M contact located beneath the start button is known as the **holding, sealing, or maintaining contact**. Its job is to provide a continued circuit to the M coil when the start button is released. The motor now begins to run at its lowest speed. Maximum resistance is connected in the rotor circuit.

2. TR_1 relay is a timer. For this example, it shall be assumed that all timers are set for a delay of 3 seconds. When TR_1 coil energizes, it begins a time operation. After 3 seconds, TR_1 contact closes. This completes a circuit to S_1 coil and TR_2 coil.

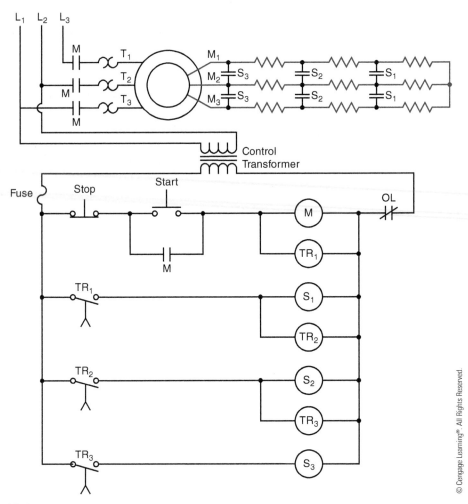

▶ **Figure 16–5**
Definite time starting for a wound rotor motor.

3. When S_1 coil energizes, both S_1 contacts close and short out the last three resistors in the rotor circuit, causing the motor to accelerate to the next higher speed. When TR_2 coil energizes, it begins timing.

4. At the end of a 3-second time period, TR_2 contact closes and completes a circuit to coil S_2 and TR_3.

5. When S_2 coil energizes, both S_2 contacts close and short out the next set of resistors, thus permitting the motor to accelerate to a higher speed. When TR_3 coil energizes, it begins its timing sequence.

6. After a 3-second time period, contact TR_3 closes and provides a complete circuit for coil S_3. This causes both S_3 contacts to close and short out the last set of resistors. The motor now accelerates to its highest speed.

7. When the stop button is pressed, the circuit to coil M and coil TR_1 is broken. When coil M de-energizes, all M contacts open. This disconnects the stator winding from the line. When TR_1 coil de-energizes, contact TR_1 opens immediately, de-energizing coil S_1 and coil TR_2. When coil S_1 de-energizes, both S_1 contacts return to their open position.

When coil TR$_1$ de-energizes, contact TR$_2$ opens immediately. When contact TR$_2$ opens, it breaks the circuit to coil S$_2$ and coil TR$_3$. When coil S$_2$ de-energizes, both S$_2$ contacts reopen. Contact TR$_3$ opens immediately when coil TR$_3$ de-energizes. This causes coil S$_3$ to de-energize and open both S$_3$ contacts.

8. If the fuse should blow or the overload contact open, it has the same effect as pressing the stop button.

TESTING A WOUND ROTOR MOTOR

Because the stator winding of the wound rotor motor is the same as the squirrel-cage motor, the same test procedure can be followed. Testing the rotor of a wound rotor motor is very similar to testing the stator. The rotor can be tested for an open winding with an ohmmeter by checking the continuity between each of the slip rings, Figure 16–6. The resistance readings should be the same between each pair of slip rings. To test the rotor for a ground, connect one ohmmeter lead to the shaft, and connect the other lead to each one of the slip rings, Figure 16–7. The ohmmeter should show no continuity between the rotor windings and ground. Like the stator winding, the rotor is difficult to test for a shorted winding. To test the rotor for a shorted winding, it is generally necessary to use equipment that measures the inductance of the winding instead of its resistance.

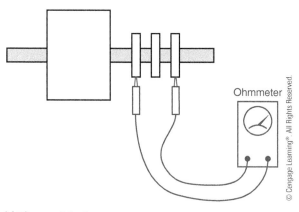

Figure 16–6
Testing a rotor for an open winding.

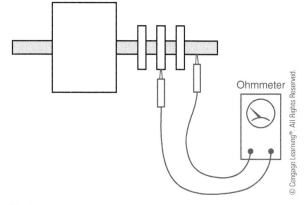

Figure 16–7
Testing a rotor for a grounded winding.

SUMMARY

- The stator winding of a wound rotor induction motor is the same as the stator winding of a squirrel-cage induction motor.
- The rotor of a wound rotor induction motor contains windings instead of squirrel-cage bars.
- The rotor of a wound rotor induction motor contains as many poles per phase as the stator winding.

⊙ The finish ends of each winding of a wound rotor are connected together to form a wye connection, and the other ends of each winding are connected to a slip ring on the shaft.

⊙ Wound rotor induction motors are sometimes called slip ring motors because they contain three slip rings on their shaft.

⊙ Wound rotor induction motors have a higher starting torque per amp of starting current than any other type of three-phase motor.

⊙ The speed of a wound rotor induction motor can be controlled by permitting resistance to remain in the rotor circuit during operation.

⊙ The brushes of a wound rotor induction motor are used to provide connection of the rotor windings to external resistors.

⊙ The stator winding leads of a wound rotor induction motor are labeled T_1, T_2, and T_3.

⊙ The rotor leads of a wound rotor induction motor are labeled M_1, M_2, and M_3.

▷ KEY TERMS

carbon brushes
definite time control
holding, sealing, or
    maintaining contact

slip ring
step starting

wound rotor induction
    motor

▷ REVIEW QUESTIONS

1. How many slip rings are located on the shaft of the rotor of a wound rotor induction motor?

2. What is the purpose of the slip rings?

3. Name two advantages of the wound rotor motor over the squirrel-cage motor.

4. What two factors determine the amount of current flow in the rotor of a wound rotor motor?

5. What does the dashed line drawn between the three resistors shown in Figure 16–2 indicate?

6. Why is the starting torque of a wound rotor induction motor higher than the starting torque of a squirrel-cage induction motor?

7. The stator of a wound rotor motor has a synchronous speed of 1200 RPM when connected to a 60 Hz line. How many poles per phase are there in the rotor?

8. Refer to Figure 16–5. Describe what would happen in this circuit if coil S_1 were open when the motor started.

9. Refer to Figure 16–5. Describe what would happen in this circuit if coil TR_2 were open when the motor was started.

10. Refer to Figure 16–5. Describe what would happen in this circuit if the two halves of holding contact M should become stuck together when the motor is started and not open.

TROUBLESHOOTING QUESTIONS

Refer to the schematic shown in Figure 16–5 to answer the following questions. It is to be assumed that all timers are set for a delay of 3 seconds.

1. When the start button is pressed, the motor starts in its lowest speed. After a delay of 6 seconds, the motor accelerates to third speed and 3 seconds later accelerates to the fourth, or highest, speed. Which of the following could cause this problem?

A. Coil TR_1 is open.

B. Timed contact TR_1 did not close.

C. Coil S_1 is open.

D. Coil TR_2 is open.

2. When the start button is pressed, the motor starts and accelerates through all speeds normally. When the stop button is pressed, the motor continues to operate normally. Which of the following could cause this condition?

A. The start button is shorted. (Shorted means contacts welded together.)

B. The stop button is shorted.

C. M auxiliary contact is shorted.

D. Any of the above has occurred.

3. When the start button is pressed, the motor accelerates through the first three steps of speed normally. When the motor tries to accelerate to the fourth speed, however, the motor stops. It is found that the control circuit fuse has blown. Which of the following conditions could cause this problem?

A. The overload (OL) contact has shorted.

B. Coil S_2 is shorted.

C. Coil TR_3 is shorted.

D. Coil S_3 is shorted.

4. When the start button is pressed, the motor immediately starts operating in third speed. Three seconds later, the motor accelerates to fourth speed. Which of the following could cause this condition?

A. TR_2 timed contact is shorted.

B. TR_3 timed contact is shorted.

C. S_2 load contacts are shorted.

D. None of the above could cause this.

5. When the start button is pressed, the motor will not start. Which of the following could cause this condition?

A. The control transformer is defective.

B. The control circuit fuse is blown.

C. The overload contact is open.

D. All of the above could be the cause.

UNIT 17

The Synchronous Motor

OBJECTIVES

After studying this unit, the student should be able to:

- Describe the construction of a synchronous motor
- Compare the operating characteristics of a synchronous motor with those of a squirrel-cage induction motor and a wound rotor induction motor
- Discuss the starting and running characteristics of a synchronous motor
- Describe the function of the field discharge resistor
- Discuss power factor correction, using the synchronous motor
- Perform an ohmmeter test of a synchronous motor

The third type of three-phase motor to be discussed is the **synchronous motor**. This motor has several characteristics that no other type of motor has, some of which are listed here:

1. The synchronous motor is not an induction motor. This means that it does not depend on induced voltage from the stator to produce a magnetic field in the rotor.

2. The synchronous motor will run at a constant speed from no load to full load.

3. The synchronous motor has the ability to not only correct its own power factor but also correct the power factor of other motors connected to the same line.

The synchronous motor has the same type of stator windings as the other two 3-phase motors. The rotor of a synchronous motor has

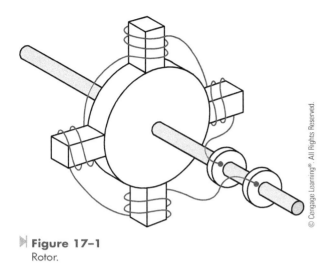

Figure 17–1
Rotor.

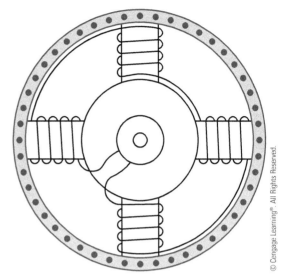

Figure 17–2
Amortisseur winding.

windings similar to the wound rotor induction motor, Figure 17–1. Notice that the winding in the rotor of a synchronous motor is different, however. The winding of a synchronous motor is one continuous set of coils instead of three different sets as is the case with the wound rotor motor. Notice also that the synchronous motor has only two slip rings on its shaft as opposed to three on the shaft of a wound rotor motor.

STARTING A SYNCHRONOUS MOTOR

The rotor of a synchronous motor also contains a set of type "A" squirrel-cage bars. This set of squirrel-cage bars is used to start the motor and is known as the **amortisseur** winding, Figure 17–2. When power is first connected to the stator, the rotating magnetic field cuts through the type "A" squirrel-cage bars. The cutting action of the field induces a current into the squirrel-cage bars. The current flow through the amortisseur winding produces a rotor magnetic field that is attracted to the rotating magnetic field of the stator. This causes the rotor to begin turning in the direction of rotation of the stator field. When the rotor has accelerated to a speed that is close to the synchronous speed of the field, DC is connected to the rotor through the

slip rings on the rotor shaft, Figure 17–3. When DC is applied to the rotor, the windings on the rotor become electromagnets. The **electromagnetic field** of the rotor locks in step with the rotating magnetic field of the stator. The rotor will now turn at the same speed as the rotating magnetic field. When the rotor begins to turn at the synchronous speed of the field, there is no more cutting action between the field and the amortisseur winding. This causes the current flow in the amortisseur winding to cease.

Notice that the synchronous motor starts as a squirrel-cage induction motor. Because the rotor bars used are type "A," they have a relatively high resistance, which gives the motor good starting torque and low starting current. A synchronous motor must never be started with DC connected to the rotor. If DC is applied to the rotor, the field poles of the rotor become electromagnets. When the stator is energized, the rotating magnetic field begins turning at synchronous speed. The electromagnets of the rotor are attracted to the rotating magnetic field of the stator and are alternately attracted and repelled 60 times a second. As a result, the rotor does not turn.

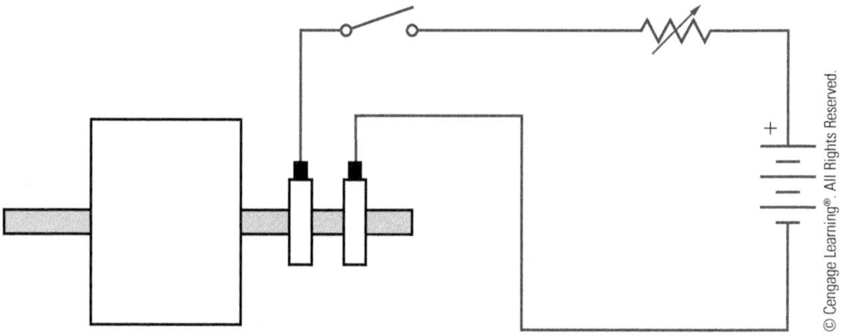

Figure 17-3
Direct current is applied to the
rotor through the slip rings.

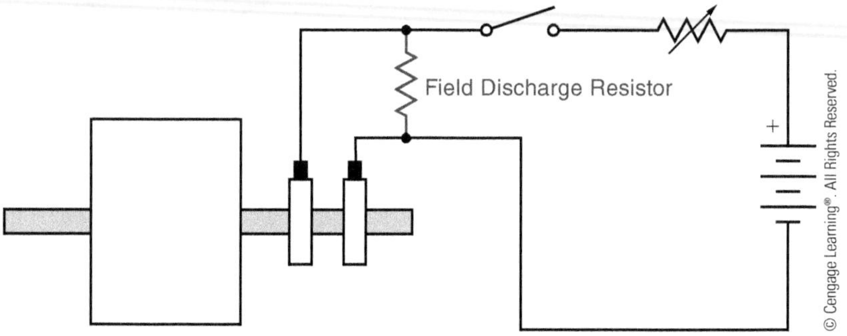

Figure 17-4
A field discharge resistor protects
the rotor circuit.

Field Discharge Resistor

THE FIELD DISCHARGE RESISTOR

When the stator winding is first energized, the rotating magnetic field cuts through the rotor winding at a fast rate of speed. This causes a large amount of voltage to be induced into the winding of the rotor. To prevent this voltage from becoming excessive, a resistor is connected across the winding. This resistor is known as the **field discharge resistor**, Figure 17-4. It also helps to reduce the voltage induced into the rotor by the collapsing magnetic field when the DC is disconnected from the rotor.

CONSTANT SPEED OPERATION

Although the synchronous motor starts as an induction motor, it does not operate as one. After the amortisseur winding has been used to accelerate the rotor to about 95% of the speed of the rotating magnetic field, direct current is connected to the rotor and the electromagnets lock in step with the rotating field. Notice that the synchronous motor does not depend on induced voltage from the stator field to produce a magnetic field in the rotor. The magnetic field of the rotor is produced by external DC applied to the rotor. This is the reason that the synchronous motor has the ability to operate at the speed of the rotating magnetic field. As load is added to the motor, the magnetic field of the rotor remains locked with the rotating magnetic field and the rotor continues to turn at the same speed.

POWER FACTOR CORRECTION

The power factor of the synchronous motor can be changed by adjusting the **DC excitation current** to the rotor. When the DC is adjusted to the point that the motor current is in phase with the voltage, the motor has a power factor of 100%. This is considered to be normal excitation for the motor. For this example, assume this current to be 10 amps. If the DC power supply is adjusted to a point that the excitation current is less than 10 amps, the rotor is under excited. This causes the motor to have a lagging power factor like an induction motor. If the excitation current is adjusted above 10 amps, the rotor is overexcited. This causes the motor to have a leading power factor like a capacitor. When a synchronous motor is operated at no load and used for power factor correction, it is generally referred to as a **synchronous condenser**. Utility companies generally charge industries extra for poor power factor in the plant. For this reason, synchronous motors are often used when a large horsepower motor must be used. Commercial and industrial air-conditioning systems are often the largest single load in a plant or building. It is not uncommon to find synchronous motors being used to operate the compressors of large air-conditioning systems.

THE POWER SUPPLY

DC power supply of a synchronous motor can be provided by several methods. The most common of these methods is either a small DC generator mounted to the shaft of the motor or an electronic power supply that converts the AC line voltage to DC voltage.

TESTING THE SYNCHRONOUS MOTOR

The procedure for testing the stator winding of a synchronous motor is the same as that described for testing the stator of a squirrel-cage induction motor. The rotor can be tested with an ohmmeter for an open winding or a grounded winding. To test the rotor for an open winding, connect one of the ohmmeter leads to each of the slip rings on the rotor shaft, Figure 17–5. Because the rotor winding of a synchronous motor is intended for DC current, the resistance of the wire will be high as compared with the wire resistance of a wound rotor motor. Owing to the fact that alternating current flows in the rotor of a wound rotor motor, the current is limited by the inductance of the coil and not its resistance.

To test the rotor for a grounded winding, connect one ohmmeter lead to the shaft of the motor and the other lead to one of the slip rings. There should be no continuity between the winding and the motor shaft, Figure 17–6.

Because the resistance of the rotor is relatively high, a shorted winding can often be found with

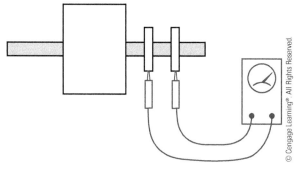

▶ **Figure 17–5**
Testing the rotor for an open winding.

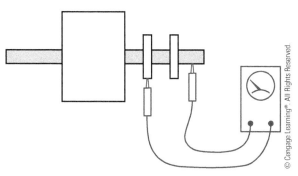

▶ **Figure 17–6**
Testing the rotor for a grounded winding.

the ohmmeter. In coils designed for DC current, the resistance of the wire is used to limit the flow of current. For example, assume the specifications of a synchronous motor indicate that the DC excitation voltage should be 125 volts and that maximum rotor current should be 10 amps. The resistance of the rotor can now be calculated by using Ohm's law:

$$R = \frac{E}{I}$$

$$R = \frac{125}{10}$$

$$R = 12.5 \text{ ohms}$$

If the ohmmeter measures a rotor resistance close to 12.5 ohms, the rotor is good. If the ohmmeter measures a much lower resistance, however, the rotor is shorted.

THE BRUSHLESS EXCITER

Many large synchronous motors use a brushless exciter instead of slip rings and brushes. The **brushless exciter** is constructed by incorporating a small three-phase alternator winding on the shaft of the motor, Figure 17–7. The three-phase winding is placed inside the field of electromagnets, Figure 17–8. A variable source of direct current is used to control the strength of the electromagnets. The amount of voltage induced into the armature winding is determined by three factors:

1. The number of turns of wire in the armature winding

2. The speed of the armature

3. The strength of the electromagnets

Because the amount of DC current that flows through the windings of the electromagnets determines their strength, the output voltage of the three-phase armature can be controlled by the DC excitation current.

The output voltage of the alternator winding is then rectified to direct current with a three-phase bridge rectifier, Figure 17–9. The bridge rectifier and all protective devices such as fuses are mounted on the motor shaft with the armature winding. The output of the bridge rectifier is connected to the

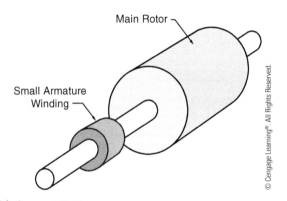

▶ **Figure 17–7**
A brushless exciter contains a small three-phase winding on the motor shaft.

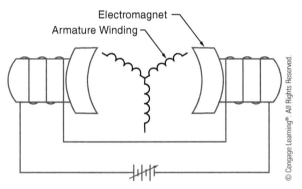

▶ **Figure 17–8**
The armature winding is between two electromagnets.

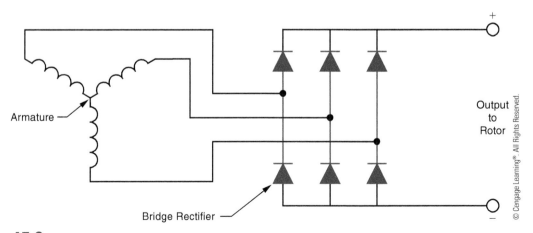

▶ **Figure 17–9**
The output of the armature winding is connected to a three-phase bridge rectifier.

Amortisseur Winding ⎯⎯ Brushless Exciter Winding ⎯⎯

Courtesy of GE Energy, Fort Wayne, Indiana

▶ **Figure 17–10**
Rotor of a large synchro-
nous motor.

winding of the main rotor. The amount of DC excitation in the rotor of the synchronous motor is now
controlled by the amount of DC excitation current
supplied to the windings of the electromagnets.

The rotor of a large synchronous motor is shown
in Figure 17–10. The amortisseur winding and the
brushless exciter winding can be seen in the photograph.

▶ SUMMARY

- ⊙ The synchronous motor is not an induction motor.
- ⊙ The synchronous motor must have an external source of direct current supplied to the rotor during normal operation.
- ⊙ The DC current supplied to the rotor is known as the excitation current.
- ⊙ Synchronous motors operate at a constant speed from no load to full load.
- ⊙ Synchronous motors run at the speed of the rotating magnetic field.
- ⊙ The two factors that determine the speed of a synchronous motor are:
  A. Number of stator poles per phase.
  B. Frequency of the applied voltage.
- ⊙ Synchronous motors use a special squirrel-cage winding called the amortisseur winding for starting.
- ⊙ A synchronous motor can be made to have a leading power factor by over excitation of the rotor current.

▷ KEY TERMS

amortisseur   electromagnetic field   synchronous motor
brushless exciter   field discharge resistor
DC excitation current   synchronous condenser

▷ REVIEW QUESTIONS

1. Name three characteristics of a synchronous motor that the squirrel-cage induction motor and the wound rotor motor do not have.

2. What is an amortisseur winding?

3. How many slip rings are located on the shaft of a synchronous motor?

4. How many slip rings are located on the shaft of a wound rotor induction motor?

5. Is a synchronous motor started with DC excitation voltage applied to the rotor?

6. What is the field discharge resistor used for?

7. A synchronous motor has an eight-pole stator. What will be the speed of the rotor when it is under full load?

8. How is it possible to know when a synchronous motor has normal excitation applied to its rotor?

9. How can a synchronous motor be made to have a leading power factor?

10. What is a synchronous condenser?

UNIT 18 ▷ Brushless DC Motors

OBJECTIVES

After studying this unit, the student should be able to:

▷ Describe the operation of a brushless DC motor

▷ Discuss applications for brushless DC motors

▷ List differences in construction between brushless DC motors and other types of motors

▷ Discuss the operation and advantages of variable-speed air handlers

Brushless direct current motors have become very popular for operating variable-speed air handlers found in many residential central heating and cooling systems. The typical air handler is powered by a multispeed motor that can generally operate at three speeds. When the fan motor is started, however, it operates at its set speed until it is turned off. Basically, the blower motor is either full on or completely off. **Variable-speed** air handlers permit the motor to operate at different speeds in accord with the demands of the system. Many variable-speed air handlers are operated at a low speed continuously to maintain air circulation throughout the dwelling at all times. When the heating or cooling system turns on, the fan speed can be increased gradually and in steps instead of all at once.

Variable-speed air handlers exhibit several advantages over air handlers that operate at a single speed. Some of these advantages are listed here:

Dust collection. Single-speed air handlers pull dust through the filter at a high rate of speed. This lessens the filter's ability to trap dust particles. Variable-speed air handlers operate at a very low speed most of the time, causing more dust particles to be collected by the filter.

Less temperature variation. Typically, the inside temperature varies from 3° to 5° before the heating or cooling system turns on. The variable-speed air handler can be operated at different speeds between the cycling of the heating or cooling system to provide more or less airflow as needed. This greatly reduces the temperature variation and permits the temperature to remain closer to the comfort zone setting.

Better humidity control during cooling. When the air-conditioning compressor starts, the speed of the air handler is increased gradually instead of all at once. This permits the warm air to flow across the evaporator coil at a slower rate, permitting the coil to rapidly cool down. This **rapid cooldown** results in increased moisture removal.

THE BRUSHLESS DC MOTOR

The brushless DC motor operates by converting direct current into three-phase alternating current at different frequencies. The motor contains a **permanent magnet rotor**, a stator, and the electronics necessary to change direct current into three-phase alternating current, Figure 18–1. The speed of the motor is determined by the **frequency** of the three-phase current. The motor operates on the principle of a **rotating magnetic field** very similar to that of a three-phase induction motor. The brushless DC motor, however, is *not* an induction motor. It does not depend on current being induced into the rotor to produce a rotor magnetic field. Because permanent magnets supply the rotor magnetic field, the brushless DC motor does not exhibit the high inrush current during starting that is a characteristic of induction-type motors. The

▶ **Figure 18–1**
Permanent magnet rotor and stator of a brushless DC motor.

starting current and the running current of the motor are basically the same. This feature greatly reduces the current-handling requirement of the electronic components needed to change direct current into three-phase alternating current.

The factors that determine the amount of torque produced by the brushless DC motor are the same as those for any other electric motor:

1. Strength of the magnetic field of the stator
2. Strength of the magnetic field of the rotor
3. Phase angle difference between rotor and stator flux

Because the rotor magnetic field is supplied by permanent magnets, the flux of the rotor and stator are always in phase with each other. This produces a strong torque for this type of motor.

The electronic components necessary to change single-phase alternating current into direct current and then into three-phase alternating current are located inside the motor housing, Figure 18–2. Most variable frequency type controls first change alternating current into direct current because it is a simpler process to produce multiphase alternating current from direct current than to change the frequency and number of phases of an existing alternating current source.

Brushless DC motors are operated in one of two modes, the thermostat mode or the variable-speed mode. When used in the thermostat mode, the motor is controlled by a 24 VAC signal from the thermostat. When used in the variable-speed mode, the motor is controlled by a pulse width modulating signal.

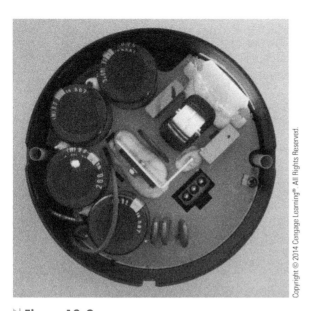

Copyright © 2014 Cengage Learning®. All Rights Reserved.

Figure 18–2
Electronic control unit for a brushless DC motor.

The motor is provided with two terminal connections. One, 16-pin connector called the **control connector**, receives the information necessary to control the operation of the motor. The other connection is a 5-pin connector called the **power connector**, which provides the power to operate the motor. A pin connection diagram is shown in Figure 18–3.

Brushless DC motors can generally be obtained in 1/3, 1/2, 3/4, or 1 horsepower ratings. They can exhibit efficiency as high as 82% and can operate on 120/240 and 277 VAC at 50 or 60 Hz. The speed range is from 300 to 1200 RPM. Although the motor can be operated at 1200 RPM, it is generally recommended not to operate the motor above 1050 RPM when used for continuous operation.

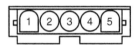

CONTROL CONNECTOR	
PIN	DESCRIPTION
1	C1
2	W/W1
3	C2
4	DELAY
5	COOL
6	Y1
7	ADJUST
8	OUT -
9	O
10	BK/PWM
11	HEAT
12	R
13	EM/W2
14	Y/Y2
15	G
16	OUT +

POWER CONNECTOR	
PIN	DESCRIPTION
1	JUMPER PIN 1 TO PIN 2 FOR 120 VAC INPUT ONLY
2	
3	CHASSIS GRD.
4	AC LINE
5	AC LINE

WARNING—APPLYING 240 VAC LINE INPUT WITH PINS 1 AND 2 JUMPERED WILL PERMANENTLY DAMAGE THE UNIT.

Figure 18–3
Pin connection diagram for a brushless DC motor.

© Cengage Learning®. All Rights Reserved.

Electronic Control

The brushless DC motor is controlled by a circuit board located in the air handler. The board supplies direct current to both the motor and the signals that determine motor speed. Some boards are preset at the factory for the specific type of air handler unit. Other boards contain DIP (dual inline package) switches that permit the air handler to be set for different operating conditions. Typical DIP switch configurations for some systems are as follows:

Switches 1 and 2: Set for system tonnage
Switches 3 and 4: Set for 300, 400, or 450 cfm/ton
Switches 5 and 6: Set for time delay or mode
Switches 7 and 8: Set for desired heating flow

SUMMARY

- ⊙ Brushless DC motors are generally used to power the blower in variable-speed air handlers.
- ⊙ The circuitry necessary to change single-phase alternating current into three-phase alternating current is located inside the motor housing.
- ⊙ Brushless DC motors operate on the principle of a rotating magnetic field.
- ⊙ Brushless DC motors are not induction motors and do not depend on an inducted current to produce a magnetic field in the rotor.
- ⊙ Brushless DC motors are operated in one of two modes.
- ⊙ Brushless DC motors can have efficiencies as high as 82%.
- ⊙ Because brushless DC motors are not induction motors, they do not exhibit the high starting current associated with induction-type motors.

KEY TERMS

brushless
control connector
frequency

permanent magnet rotor
power connector
rapid cooldown

rotating magnetic field
variable-speed

REVIEW QUESTIONS

1. What is the operating voltage of a brushless DC motor?
2. What is the recommended maximum operating speed for a brushless DC motor when used for continuous operation?
3. If the motor is to be operated on 120 VAC, what must be done to permit the motor to operate on this voltage?
4. Referring to the power connector, which pins are used to connect AC line voltage to the motor?
5. When the brushless DC motor is used in the thermostat mode, what controls the operation of the motor?

6. Name three factors that determine the amount of torque produced by a brushless DC motor.

7. Referring to the control connector, to which pin would Y/Y2 be connected?

8. Name three advantages of a variable-speed air handler over a single-speed air handler.

SECTION 4
Transformers

UNIT 19 ▷

Isolation Transformers

OBJECTIVES

After studying this unit, the student should be able to:

▷ Discuss the different types of transformers

▷ Calculate values of voltage, current, and turns for single-phase transformers, using formulas

▷ Calculate values of voltage, current, and turns for a single-phase transformer, using the turns ratio

▷ Connect a transformer and test the voltage output of different windings

Transformers are one of the most common devices found in the HVAC field. They range in size from being so small as to occupy a space of less than 1 cubic inch to being large enough that they requirie rail cars to move them after they have been broken into sections. Their ratings can range from mVA (millivolt amps) to GVA (gigavolt amps).

A transformer is a magnetically operated machine that can change values of voltage, current, and impedance without a change of frequency.

Transformers are the most efficient machines known. Their efficiencies commonly range from 90% to 99% at full load. Transformers can be divided into several classifications such as these:

A. Isolation

B. Auto

C. Current

A basic law concerning transformers is that *all values of a transformer are proportional to its turns ratio*. This does not mean that the exact number of turns of wire on each winding must be known to determine different values of voltage and current for a transformer. What must be known is the *ratio* of turns. For example, assume a transformer has two windings. One winding, the primary, has 1000 turns of wire and the other, the secondary, has 250 turns of wire, Figure 19–1. The turns ratio of this transformer is 4 to 1 or 4:1 (1000/250 = 4). This indicates there are four turns of wire on the primary for every one turn of wire on the secondary.

TRANSFORMER FORMULAS

There are different formulas that can be used to find the values of voltage and current for a transformer. The following is a list of standard formulas:

Where:

N_P = Number of turns in the primary

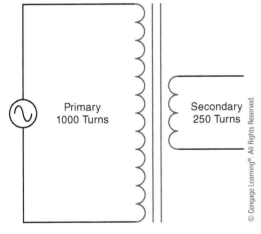

Figure 19–1
All values of a transformer are proportional to its turns ratio.

N_S = Number of turns in the secondary
E_P = Voltage of the primary
E_S = Voltage of the secondary
I_P = Current in the primary
I_S = Current in the secondary

$$\frac{E_P}{E_S} = \frac{N_P}{N_S} \quad \frac{E_P}{E_S} = \frac{I_S}{I_P} \quad \frac{N_P}{N_S} = \frac{I_S}{I_P}$$

or

$$E_P \times N_S = E_S \times N_P$$
$$E_P \times I_P = E_S \times I_S$$
$$N_P \times I_P = N_S \times I_S$$

The primary winding of a transformer is the power input winding, which is the winding that is connected to the incoming power supply. The secondary winding is the load winding, or output winding. It is the side of the transformer that is connected to the driven load, Figure 19–2. Any winding of a transformer can be used as a primary or secondary winding provided its voltage or current rating is not exceeded. Transformers can also be operated at a lower voltage than their rating indicates, but they cannot be connected to a higher voltage. Assume the transformer shown in Figure 19–2, for example, has a primary voltage rating of 480 volts and a secondary voltage rating of 240 volts. Now assume that the primary winding is connected to a 120-volt source. No damage would occur to the transformer, but the secondary winding would produce only 60 volts.

ISOLATION TRANSFORMERS

The transformers shown in Figure 19–1 and Figure 19–2 are **isolation transformers**. This means that the secondary winding is physically and electrically isolated from the primary winding. There

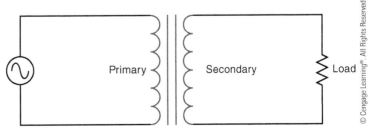

Figure 19–2
Isolation transformer.

is no electrical connection between the primary and secondary winding. This transformer is magnetically coupled, not electrically coupled. This "line isolation" is often a very desirable characteristic. Because there is no electrical connection between the load and power supply, the transformer becomes a filter between the two. The isolation transformer greatly reduces any voltage spikes that originate on the supply side before they are transferred to the load side. Some isolation transformers are built with a turns ratio of 1:1. A transformer of this type has the same input and output voltage and is used for the purpose of isolation only.

The reason that the transformer can greatly reduce any voltage spikes before they reach the secondary is because of the rise time of current through an inductor. The current in an inductor rises at an exponential rate, Figure 19–3. As the current increases in value, the expanding magnetic field cuts through the conductors of the coil and induces a voltage that is opposed to the applied voltage. The amount of induced voltage is proportional to the rate of change of current. This simply means

that the faster the current attempts to increase, the greater the opposition to that increase will be. Spike voltages and currents are generally of very short duration, which means that they increase in value very rapidly, Figure 19–4. This rapid change of value causes the opposition to the change to increase just as rapidly. By the time the spike has been transferred to the secondary winding of the transformer, it has been eliminated or greatly reduced, Figure 19–5.

Another purpose of isolation transformers is to remove some piece of electrical equipment from ground. It is sometimes desirable that a piece of electrical equipment not be connected directly to ground. This is often done as a safety precaution to eliminate the hazard of an accidental contact between a person at ground potential and the ungrounded conductor. If the case of the equipment should come in contact with the ungrounded conductor, the isolation transformer would prevent a circuit being completed to ground through someone touching the case of the equipment. Many alternating current circuits have one side connected to ground. A familiar example of this is the common

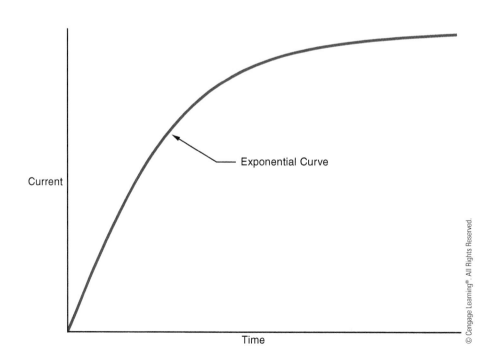

▶ Figure 19–3
The current through an inductor rises at an exponential rate.

Current

Exponential Curve

Time

120-volt circuit with a grounded neutral conductor, Figure 19–6. An isolation transformer can be used to remove a piece of equipment from circuit ground.

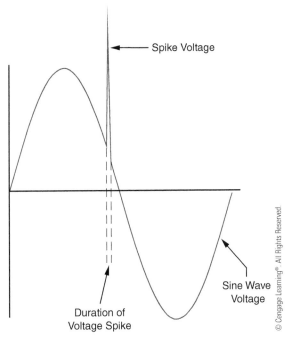

▶ **Figure 19–4**
Voltage spikes are generally of short duration.

TRANSFORMER CONSTRUCTION

The basic construction of an isolation transformer is shown in Figure 19–7. A metal core is used to provide good magnetic coupling between the two windings. The core is generally made of laminations stacked together. Laminating the core helps reduce power losses due to eddy current induction. Figure 19–7 shows the basic design of electrically separated winding.

TRANSFORMER CORE TYPES

There are several different types of cores used in the construction of transformers. Most cores are made from thin steel punchings laminated together to form a solid metal core. Laminated cores are preferred because a thin layer of oxide forms on the surface of each lamination, which acts as an insulator to reduce the formation of eddy currents inside the core material. The amount of core material needed for a particular transformer is determined by the power rating of the transformer. The amount of core material must be sufficient to prevent saturation at full load. The type and shape of the core generally determines the amount of magnetic coupling between the windings, and to some extent the efficiency of the transformer.

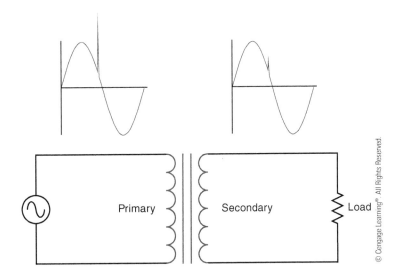

▶ **Figure 19–5**
The isolation transformer greatly reduces the voltage spike.

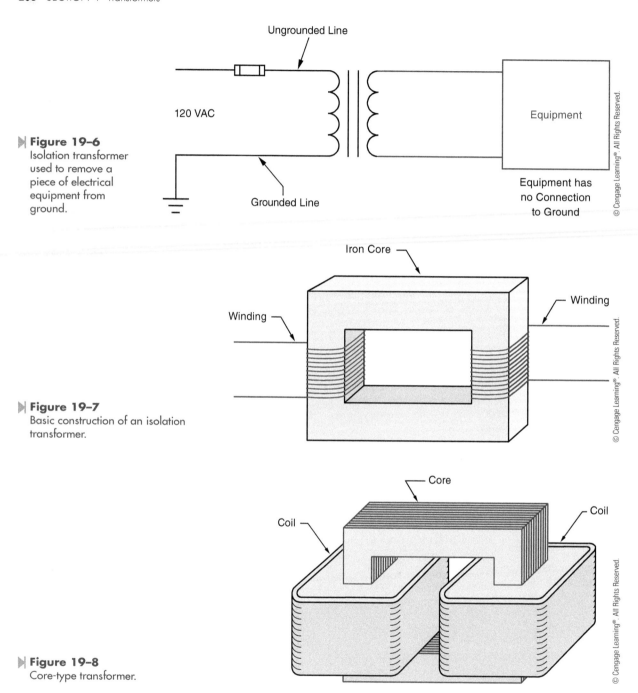

Figure 19–6
Isolation transformer used to remove a piece of electrical equipment from ground.

Figure 19–7
Basic construction of an isolation transformer.

Figure 19–8
Core-type transformer.

The transformer illustrated in Figure 19–8 is known as a **core transformer**. The windings are placed around each end of the core material. The metal core provides a good magnetic path between the two windings.

The **shell transformer** is constructed in a similar manner as the core type, except that the shell type has a metal core piece through the middle of the window, Figure 19–9. The primary and secondary windings are wound around the center core piece, with the low voltage winding being closest to the metal core. This arrangement permits the transformer to be surrounded by the core, which

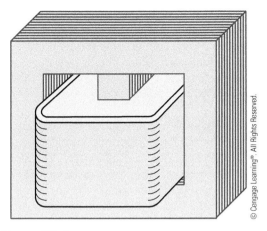

▶ **Figure 19–9**
Shell-type transformer.

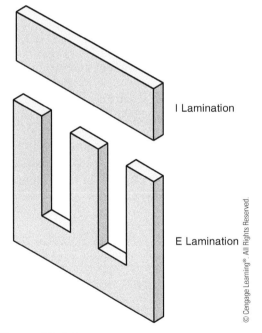

I Lamination

E Lamination

▶ **Figure 19–10**
Shell-type cores are made of E and I laminations.

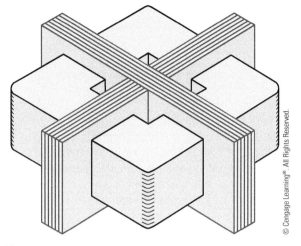

▶ **Figure 19–11**
Transformer with H-type core.

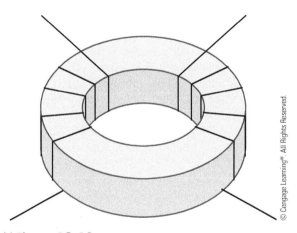

▶ **Figure 19–12**
Toroid transformer.

provides excellent magnetic coupling. When the transformer is in operation, all the magnetic flux must pass through the center core piece. It then divides through the two outer core pieces. Shell-type cores are sometimes referred to as **E-I cores** because the steel punchings used to construct the core are in the shape of an E and an I, Figure 19–10.

The H-type core shown in Figure 19–11 is similar to the shell-type core in that it has an iron core through its center, around which the primary and secondary windings are wound. The H core, however, surrounds the windings on four sides instead of two. This extra metal helps reduce stray leakage flux and improve the efficiency of the transformer. The H-type core is often found on high-voltage distribution transformers.

The **tape wound** core, or **toroid core**, Figure 19–12, is constructed by tightly winding one long continuous silicon steel tape into a spiral. The tape may or may not be housed in a plastic container, depending on the application. This type core does not require steel punchings that are then laminated together. Because the core is one continuous length of metal, flux leakage is kept to a minimum. The tape wound core is one of the most efficient core designs available.

BASIC OPERATING PRINCIPLES

In Figure 19–13, one winding of the transformer has been connected to an alternating current supply, and the other winding has been connected to a load. As current increases from zero to its peak positive point, a magnetic field expands outward around the coil. When the current decreases from its peak positive point toward zero, the magnetic field collapses. When the current increases toward its negative peak, the magnetic field again expands, but with an opposite polarity of that previously. The field again collapses when the current decreases from its negative peak toward zero. This continually expanding and collapsing magnetic field cuts the windings of the primary and induces a voltage into it. This induced voltage opposes the applied voltage and limits the current flow of the primary. When a coil induces a voltage into itself, it is known as **self-induction**. It is this induced voltage, inductive reactance, that limits the flow of current in the primary winding. If the resistance of the primary winding is measured with an ohmmeter, it will indicate only the resistance of the wire used to construct the winding and will not give an indication of the actual current-limiting effect of the winding. Most transformers with a large kVA rating appear to be almost a short circuit when measured with an ohmmeter. When connected to power, however, the actual no load current is generally relatively small.

EXCITATION CURRENT

There will always be some amount of current flow in the primary of a transformer, even if there is no load connected to the secondary. This is called the **excitation current** of the transformer. The excitation current is the amount of current required to magnetize the core of the transformer. The excitation current remains constant from no load to full load. As a general rule, the excitation current is such a small part of the full-load current, it is often omitted when making calculations.

MUTUAL INDUCTION

Because the secondary windings are wound on the same core as the primary, the magnetic field produced by the primary winding cuts the windings of the secondary also, Figure 19–14. This continually changing magnetic field induces a voltage into the secondary winding. The ability of one coil to induce a voltage into another coil is called **mutual induction**. The amount of voltage induced in the secondary is determined by the number of turns of wire in the secondary as compared with

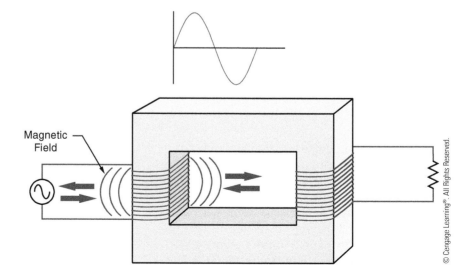

▶ **Figure 19–13**
Magnetic field produced by alternating current.

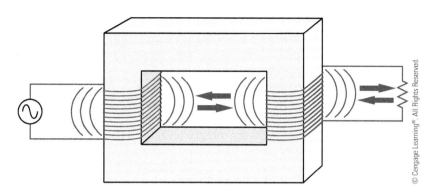

Figure 19–14
The magnetic field of the primary induces a voltage into the secondary.

the primary. For example, assume the primary has 240 turns of wire and is connected to 120 volts AC. This gives the transformer a volts-per-turn ratio of 0.5 (120 volts per 240 turns = 0.5 volt per turn). Now assume the secondary winding contains 100 turns of wire. Because the transformer has a volts-per-turn ratio of 0.5, the secondary voltage will be 50 volts (100 × 0.5 × 50).

TRANSFORMER CALCULATIONS

In the following examples, values of voltage, current, and turns for different transformers are computed.

EXAMPLE

Assume the isolation transformer shown in Figure 19–2 has 240 turns of wire on the primary and 60 turns of wire on the secondary. This is a ratio of 4:1 (240/60 = 4). Now assume that 120 volts is connected to the primary winding. What is the voltage of the secondary winding?

$$\frac{E_P}{E_S} = \frac{N_P}{N_S}$$

$$\frac{120}{E_S} = \frac{240}{60}$$

$$E_S = 30 \text{ volts}$$

The transformer in this example is known as a **step-down transformer** because it has a lower secondary voltage than primary voltage.

Now assume that the load connected to the secondary winding has an impedance of 5 Ω. The next problem is to calculate the current flow in the secondary and primary windings. The current flow of the secondary can be computed using Ohm's law because the voltage and impedance are known.

$$I = \frac{E}{Z}$$

$$I = \frac{30}{5}$$

$$I = 6 \text{ amps}$$

Now that the amount of current flow in the secondary is known, the primary current can be computed using the formula:

$$\frac{E_P}{E_S} = \frac{I_S}{I_P}$$

$$\frac{120}{30} = \frac{6}{I_P}$$

$$120 \, I_P = 180$$

$$I_P = 1.5 \text{ amps}$$

Notice that the primary voltage is higher than the secondary voltage, but the primary current is much less than the secondary current. A good rule for transformers is that *power in must equal power out.* If the primary voltage and current are multiplied together, the resulting number should equal the product of the voltage and current of the secondary:

Primary

120 × 1.5 = 180 volt amps

Secondary

30 × 6 = 180 volt amps

EXAMPLE

In the next example, assume that the primary winding contains 240 turns of wire and the secondary contains 1200 turns of wire. This is a turns ratio of 1:5 (1200/240 = 5). Now assume that 120 volts is connected to the primary winding. Compute the voltage output of the secondary winding.

$$\frac{E_p}{E_s} = \frac{N_p}{N_s}$$

$$\frac{120}{E_s} = \frac{240}{1200}$$

$$240\ E_s = 144,000$$

$$E_s = 600 \text{ volts}$$

Notice that the secondary voltage of this transformer is higher than the primary voltage. This type of transformer is known as a **step-up transformer**.

Now assume that the load connected to the secondary has an impedance of 2400 Ω. Find the amount of current flow in the primary and secondary windings. The current flow in the secondary winding can be computed using Ohm's law.

$$I = \frac{E}{Z}$$

$$I = \frac{600}{2400}$$

$$I = 0.25 \text{ amp}$$

Now that the amount of current flow in the secondary is known, the primary current can be computed using the formula:

$$\frac{E_p}{E_s} = \frac{I_s}{I_p}$$

$$\frac{120}{600} = \frac{0.25}{I_p}$$

$$120\ I_p = 150$$

$$I_p = 1.25 \text{ amps}$$

Notice that the amount of power input equals the amount of power output.

Primary

$$120 \times 1.25 = 150 \text{ volt amps}$$

Secondary

$$600 \times 0.25 = 150 \text{ volt amps}$$

CALCULATING TRANSFORMER VALUES USING THE TURNS RATIO

As illustrated in the previous examples, transformer values of voltage, current, and turns can be computed using formulas. It is also possible to compute these same values using the turns ratio. There are several ways in which turns ratios can be expressed. One method is to use a whole number value such as 13:5 or 6:21. The first ratio indicates that one winding has 13 turns of wire for every 5 turns of wire in the other winding. The second ratio indicates that there are 6 turns of wire in one winding for every 21 turns in the other.

A second method is to use the number 1 as a base. When using this method, the number 1 is always assigned to the winding with the lowest voltage rating. The ratio is found by dividing the higher voltage by the lower voltage. The number on the left side of the ratio represents the primary winding, and the number on the right of the ratio represents the secondary winding. For example, assume a transformer has a primary rated at 240 volts and a secondary rated at 96 volts, Figure 19–15. The

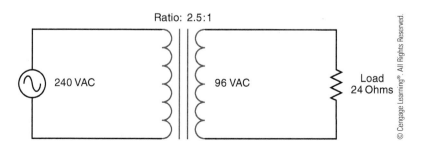

▶ **Figure 19–15**
Computing transformer values using the turns ratio.

turns ratio can be computed by dividing the higher voltage by the lower voltage.

$$\text{Ratio} = \frac{240}{96}$$

$$\text{Ratio} = 2.5:1$$

Notice in this example that the primary winding has the higher voltage rating and the secondary has the lower. Therefore, the 2.5 is placed on the left, and the base unit, 1, is placed on the right. This ratio indicates that there are 2.5 turns of wire in the primary winding for every 1 turn of wire in the secondary.

Now assume that there is a resistance of 24 Ω connected to the secondary winding. The amount of secondary current can be found using Ohm's law.

$$I_S = \frac{96}{24}$$

$$I_S = 4 \text{ amps}$$

The primary current can be found using the turns ratio. Recall that the volt amps of the primary must equal the volt amps of the secondary. Because the primary voltage is greater, the primary current must be less than the secondary current. Therefore, the secondary current will be divided by the turns ratio.

$$I_P = \frac{I_S}{\text{Turns Ratio}}$$

$$I_P = \frac{4}{25}$$

$$I_P = 1.6 \text{ amps}$$

To check the answer, find the volt amps of the primary and secondary.

Primary

$$240 \times 1.6 = 384$$

Secondary

$$96 \times 4 = 384$$

Now assume that the secondary winding contains 150 turns of wire. The primary turns can also be found by using the turns ratio. Because the primary voltage is higher than the secondary voltage, the primary must have more turns of wire. Because the primary must contain more turns of wire, the secondary turns will be multiplied by the turns ratio.

$$N_P = N_S \times \text{Turns Ratio}$$

$$N_P = 150 \times 2.5$$

$$N_P = 375 \text{ turns}$$

In the next example, assume a transformer has a primary voltage of 120 volts and a secondary voltage of 500 volts. The secondary has a load impedance of 1,200 Ω. The secondary contains 800 turns of wire, Figure 19–16. The turns ratio can be found by dividing the higher voltage by the lower voltage.

$$\text{Ratio} = \frac{500}{120}$$

$$\text{Ratio} = 1:4.17$$

The secondary current can be found using Ohm's law.

$$I_S = \frac{500}{1200}$$

$$I_S = 0.417 \text{ amp}$$

In this example, the primary voltage is lower than the secondary voltage. Therefore, the primary current must be higher. To find the primary current, multiply the secondary current by the turns ratio.

$$I_P = I_S \times \text{Turns Ratio}$$

$$I_P = 0.417 \times 4.17$$

$$I_P = 1.74 \text{ amps}$$

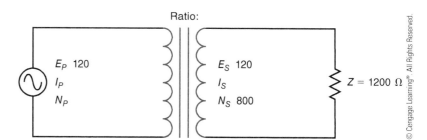

Ratio:

E_P 120
I_P
N_P

E_S 120
I_S
N_S 800

$Z = 1200 \ \Omega$

▶ **Figure 19–16**
Calculating transformer values.

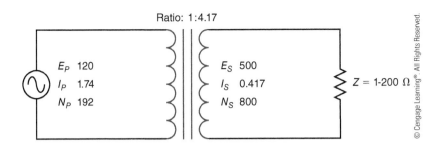

Ratio: 1:4.17

E_P 120
I_P 1.74
N_P 192

E_S 500
I_S 0.417
N_S 800

Z = 1-200 Ω

Figure 19–17
Transformer with completed values.

To check this answer, compute the volt amps of both windings.

Primary

$120 \times 1.74 = 208.8$

Secondary

$500 \times 0.417 = 208.5$

The slight difference in answers is caused by rounding off the values.

Because the primary voltage is less than the secondary voltage, there will be fewer turns of wire in the primary also. The primary turns are found by dividing the turns of wire in the secondary by the turns ratio.

$$N_p = \frac{N_s}{\text{Turns Ratio}}$$

$$N_p = \frac{800}{4.17}$$

$$N_p = 192 \text{ turns}$$

Figure 19–17 shows the transformer with all completed values.

MULTIPLE TAPPED WINDINGS

It is not uncommon for transformers to be designed with windings that have more than one set of lead wires connected to the primary or secondary. The transformer shown in Figure 19–18 contains a secondary winding rated at 24 volts. The primary winding contains several taps, however. One of the primary lead wires is labeled C and is the common for the other leads. The other leads are labeled 120, 208, and 240, respectively. This transformer is designed in such a manner that it can be connected to different primary voltages without changing the value of the secondary voltage. In this example, it is assumed that the secondary winding has a total of 120 turns of wire. To maintain the proper turns ratio, the primary would have 600 turns of wire between C and 120; 1040 turns between C and 208; and 1200 turns between C and 240.

The transformer shown in Figure 19–19 contains a single primary winding. The secondary winding, however, has been tapped at several points. One of the secondary lead wires is labeled C and is common to the other lead wires. When rated voltage is applied to the primary, voltages of 12, 24, and 48 volts can be obtained at the secondary. It should also be noted that this arrangement of taps permits the transformer to be used as a center-tapped transformer for two of the voltages. If a load is placed across the lead wires labeled C and 24, the lead wire labeled 12 becomes a center tap. If a load is placed across the C and 48 lead wires, the 24-lead wire becomes a center tap.

In this example, it is assumed the primary winding has 300 turns of wire. To produce the proper turns ratio would require 30 turns of wire between C and 12, 60 turns of wire between C and 24, and 120 turns of wire between C and 48.

The transformer shown in Figure 19–20 is similar to the transformer in Figure 19–19. The transformer in Figure 19–20, however, has multiple secondary windings instead of a single secondary winding with multiple taps. The advantage of the transformer in Figure 19–20 is that the secondary windings are electrically isolated from each other. These secondary windings can be either step-up or step-down depending on the application of the transformer.

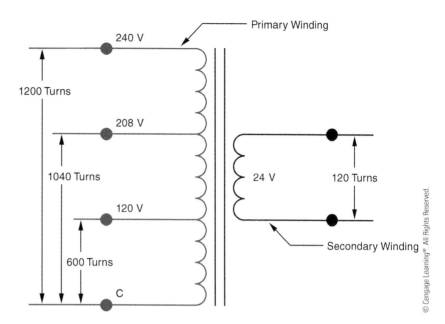

▶ **Figure 19–18**
Transformer with multiple
tap primary winding.

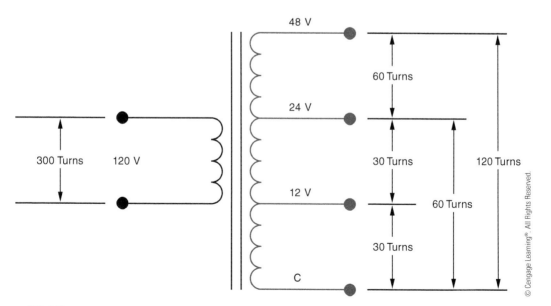

▶ **Figure 19–19**
Transformer secondary with multiple taps.

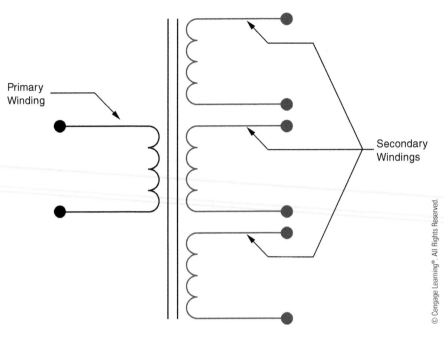

Primary
Winding

Secondary
Windings

▶| **Figure 19–20**
Transformer with
multiple secondary
windings.

COMPUTING VALUES FOR TRANSFORMERS WITH MULTIPLE SECONDARIES

EXAMPLE: When computing the values of a transformer with multiple secondary windings, each secondary must be treated as a different transformer. For example, the transformer in Figure 19–21 contains one primary winding and three secondary windings. The primary is connected to 120 volts AC and contains 300 turns of wire. One secondary has an output voltage of 560 volts and a load impedance of 1000 Ω. The second secondary has an output voltage of 208 volts and a load impedance of 400 Ω, and the third secondary has an output voltage of 24 volts and a load impedance of 6 Ω. The current, turns of wire, and ratio for each secondary and the current of the primary are found next.

SOLUTION: The first step is to compute the turns ratio of the first secondary. The turns ratio can be found by dividing the smaller voltage into the larger.

$$\text{Ratio} = \frac{E_{S1}}{E_P}$$

$$\text{Ratio} = \frac{560}{120}$$

$$\text{Ratio} = 1{:}4.67$$

The current flow in the first secondary can be computed using Ohm's law.

$$I_{S1} = \frac{560}{1000}$$

$$I_{S1} = 0.56 \text{ amp}$$

The number of turns of wire in the first secondary winding is found using the turns ratio. Because this secondary has a higher voltage than the primary, it must have more turns of wire. The number of primary turns is then multiplied by the turns ratio.

$$N_{S1} = N_p \times \text{Turns Ratio}$$

$$N_{S1} = 300 \times 4.67$$

$$N_{S1} = 1{,}401 \text{ turns}$$

The amount of primary current needed to supply this secondary winding can be found using the turns ratio also. Because the primary has less voltage, it requires more current. The primary current can be determined by multiplying the secondary current by the turns ratio.

$$I_{P(FIRST\ SECONDARY)} = I_{S1} \times \text{Turns Ratio}$$

$$I_{P(FIRST\ SECONDARY)} = 0.56 \times 4.67$$

$$I_{P(FIRST\ SECONDARY)} = 2.61 \text{ amps}$$

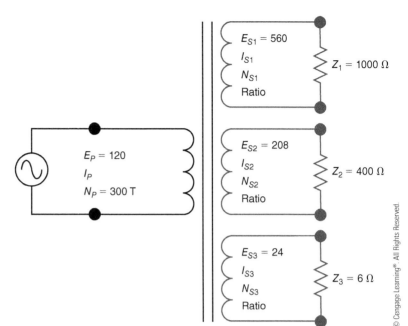

$E_{S1} = 560$

I_{S1}

N_{S1} Ratio

$Z_1 = 1000 \, \Omega$

$E_P = 120$

I_P

$N_P = 300 \, T$

$E_{S2} = 208$

I_{S2}

N_{S2} Ratio

$Z_2 = 400 \, \Omega$

$E_{S3} = 24$

I_{S3}

N_{S3} Ratio

$Z_3 = 6 \, \Omega$

▶ **Figure 19–21**
Computing values for a transformer with multiple secondary windings.

The turns ratio of the second secondary winding will be found by dividing the higher voltage by the lower voltage.

$$\text{Ratio} = \frac{208}{120}$$

$$\text{Ratio} = 1{:}1.73$$

The amount of current flow in this secondary can be determined using Ohm's law.

$$I_{S2} = \frac{208}{400}$$

$$I_{S2} = 0.52 \text{ amp}$$

Because the voltage of this secondary is greater than the primary, it will have more turns of wire than the primary. The turns of this secondary is found by multiplying the turns of the primary by the turns ratio.

$$N_{S2} = N_P \times \text{Turns Ratio}$$

$$N_{S2} = 300 \times 1.73$$

$$N_{S2} = 519 \text{ turns}$$

The voltage of the primary is less than this secondary. The primary will, therefore, require a greater amount of current. The amount of primary current required to operate this secondary is computed by multiplying the secondary current by the turns ratio.

$$I_{P(SECOND\ SECONDARY)} = I_{S2} \times \text{Turns Ratio}$$

$$I_{P(SECOND\ SECONDARY)} = 0.52 \times 1.732$$

$$I_{P(SECOND\ SECONDARY)} = 0.9 \text{ amp}$$

The turns ratio of the third secondary winding is computed in the same way as the other two. The larger voltage is divided by the smaller.

$$\text{Ratio} = \frac{120}{24}$$

$$\text{Ratio} = 5{:}1$$

The primary current is found using Ohm's law.

$$I_{S3} = \frac{24}{6}$$

$$I_{S3} = 4 \text{ amps}$$

Because the output voltage of the third secondary is less than that of the primary, the number of turns of wire for this secondary will be fewer than the primary turns. To find the number of secondary turns, divide the primary turns by the turns ratio.

$$N_{S3} = \frac{N_P}{\text{Turns Ratio}}$$

$$N_{S3} = \frac{300}{5}$$

$$N_{S3} = 60 \text{ turns}$$

The primary has a higher voltage than this secondary. The primary current will therefore be less than the secondary current by the amount of the turns ratio.

$$I_{P(THIRD\ SECONDARY)} = \frac{I_{S3}}{Turns\ Ratio}$$

$$I_{P(THIRD\ SECONDARY)} = \frac{4}{5}$$

$$I_{P(THIRD\ SECONDARY)} = 0.8 \text{ amp}$$

The primary must supply current to each of the three secondary windings. Therefore, the total amount of primary current is the sum of the currents required to supply each secondary.

$$I_{P(TOTAL)} = I_{P1} + I_{P2} + I_{P3}$$

$$I_{P(TOTAL)} = 2.61 + 0.9 + 0.8$$

$$I_{P(TOTAL)} = 4.31 \text{ amps}$$

The transformer with all computed values is shown in Figure 19–22.

DISTRIBUTION TRANSFORMERS

A very common type of isolation transformer is the distribution transformer, Figure 19–23. This transformer is used to supply power to most homes and many businesses. In this example, it is assumed that the primary is connected to a 7200 volt line. The secondary is 240 volts with a center tap. The center tap is grounded and becomes the neutral conductor. If voltage is measured across the entire secondary, a voltage of 240 volts will be seen. If voltage is measured from either line to the center tap, half of the secondary voltage, or 120 volts, will be seen, Figure 19–24. Loads that are intended to operate on 240 volts, such as water heaters, electric resistance heating units, and central air conditioners, are connected directly across the lines of the secondary. Loads intended to operate on 120 volts connect from the center tap or neutral to one of the secondary lines. The function of the neutral is to carry the difference in current between the two secondary lines and maintain a balanced voltage. In the example shown in Figure 19–25, it is assumed that one of the secondary lines has a current flow of 30 amperes and the other has a current flow of 24 amperes. The neutral will conduct the sum of the unbalanced load.

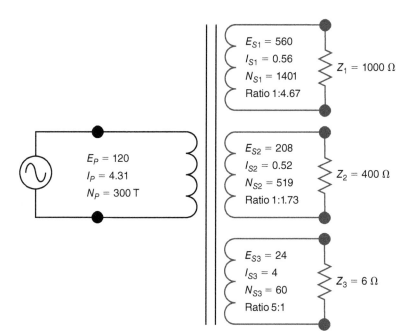

▶ **Figure 19–22**
The transformer with all computed values.

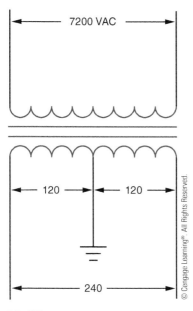

Figure 19–23
Distribution transformer.

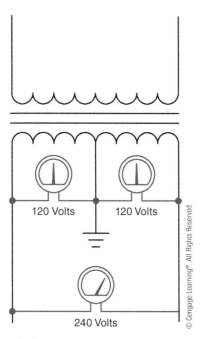

Figure 19–24
The voltage from either line to neutral is 120 volts. The voltage across the entire secondary winding is 240 volts.

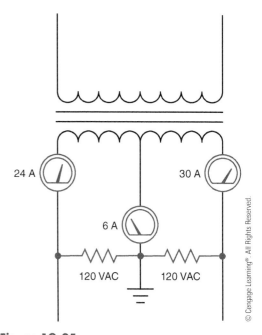

Figure 19–25
The neutral carries the sum of the unbalanced current.

In this example, the neutral current will be 6 amperes (30 – 24 = 6).

TESTING THE TRANSFORMER

There are several tests that can be made to determine the condition of the transformer. A simple test for grounds, shorts, or opens can be made with an ohmmeter, Figure 19–26. Ohmmeter A is connected to one lead of the primary and one lead of the secondary. This test checks for shorted windings between the primary and secondary. The ohmmeter should indicate infinity. If there is more than one primary or secondary winding, all isolated windings should be tested for shorts. Ohmmeter B illustrates testing the windings for grounds. One lead of the ohmmeter is connected to the case of the transformer and the other is connected to the winding. All windings should be tested for grounds, and the ohmmeter should indicate infinity for each winding. Ohmmeter C illustrates testing the windings for continuity. The wire resistance of the winding should be indicated by the ohmmeter. Each winding should be tested for continuity. If the transformer appears to be in good condition after the ohmmeter

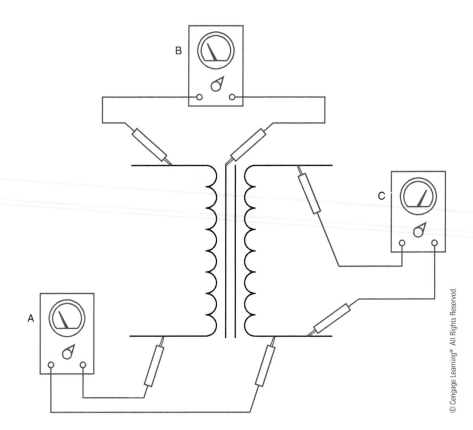

▶ **Figure 19–26**
Testing a transformer with
an ohmmeter.

test, it should then be tested for shorts and grounds with a megohmmeter, or "megger." A megger will reveal problems of insulation breakdown that an ohmmeter will not.

TRANSFORMER RATINGS

Most transformers contain a nameplate that lists information concerning the transformer. The information listed is generally determined by the size, type, and manufacturer. Almost all nameplates list the primary voltage, secondary voltage, and kVA rating. Transformers are rated in kilovolt amps, not kilowatts, because the true power is determined by the power factor of the load. Other information that may or may not be listed is frequency, temperature rise in °C, % impedance (%Z), type of insulating oil, gallon of insulating oil, serial number, type number, model number, and whether the transformer is single phase or three phase.

DETERMINING MAXIMUM CURRENT

Notice that the nameplate does not list the current rating of the windings. Because power input must equal power output, the current rating for a winding can be determined by dividing the kVA rating by the winding voltage. For example, assume a transformer has a kVA rating of 0.5 kVA, a primary voltage of 480 volts, and a secondary voltage of 120 volts. To determine the maximum current that can be supplied by the secondary, divide the kVA rating by the secondary voltage.

$$I_s = \frac{kVA}{E_s}$$

$$I_s = \frac{500}{120}$$

$$I_s = 4.16 \text{ amps}$$

The primary current can be computed in the same way.

$$I_p = \frac{kVA}{E_p}$$

$$I_p = \frac{500}{480}$$

$$I_p = 1.04 \text{ amps}$$

Transformers with multiple secondary windings generally have the current rating listed with the voltage rating.

TRANSFORMER LOSSES

Although transformers are probably the most efficient machines known, they are not perfect. A transformer operating at 90% efficiency has a power loss of 10%. Some of these losses are I^2R losses, eddy current losses, hysteresis losses, and magnetic flux leakage. Most of these losses result in heat production. Recall that I^2R is one of the formulas for finding power or watts. In the case of a transformer, it describes the power loss associated with heat due to the resistance of the wire in both primary and secondary windings.

Eddy currents are currents that are induced into the metal core material by the changing magnetic field as alternating current produces a changing flux. Eddy currents are so named because they circulate around inside the metal in a similar manner as the swirling eddies in a river, Figure 19–27. These swirling currents produce heat, which is a power loss. Transformers are constructed with laminated cores to help reduce eddy currents. The surface of each lamination forms a layer of iron oxide, which acts as an insulator to help prevent the formation of eddy currents.

Hysteresis losses are losses due to molecular friction. As discussed previously, the reversal of the direction of current flow causes the molecules of iron in the core to realign themselves each time the current changes direction. The molecules of iron are continually rubbing against each other as they realign magnetically. The friction of the molecules rubbing together causes heat, which is a power loss. Hysteresis loss is proportional to frequency. The higher the frequency, the greater the loss. A special steel called **silicon steel** is often used in transformer cores to help reduce hysteresis loss. The power loss due to hysteresis and eddy currents is often called core loss.

Magnetic flux leakage does not produce heat, but does constitute a power loss. Flux leakage is caused by magnetic lines of flux radiating away from the transformer and not cutting the secondary windings. Flux leakage can be reduced by better core designs.

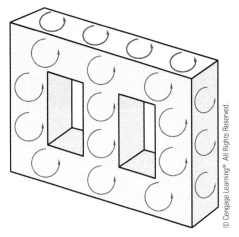

Figure 19–27
Eddy currents circulate inside the core material.

SUMMARY

- ◉ All values of voltage, current, and impedance in a transformer are proportional to the turns ratio.
- ◉ Transformers can change values of voltage, current, and impedance, but cannot change the frequency.
- ◉ The primary winding of a transformer is connected to the power line.
- ◉ The secondary winding is connected to the load.
- ◉ A transformer that has a lower secondary voltage than primary voltage is a step-down transformer.
- ◉ A transformer that has a higher secondary voltage than primary voltage is a step-up transformer.
- ◉ An isolation transformer has its primary and secondary windings electrically and mechanically separated from each other.
- ◉ When a coil induces a voltage into itself, it is known as self-induction.
- ◉ When a coil induces a voltage into another coil, it is known as mutual induction.
- ◉ Either winding of a transformer can be used as the primary or secondary as long as its voltage or current ratings are not exceeded.
- ◉ Isolation transformers help filter voltage and current spikes between the primary and secondary side.

KEY TERMS

core transformer	mutual induction	step-down transformer
E-I core	self-induction	step-up transformer
excitation current	shell transformer	tape wound or toroid
isolation transformer	silicon steel	core

REVIEW QUESTIONS

1. What is a transformer?
2. What are common efficiencies for transformers?
3. What is an isolation transformer?
4. All values of a transformer are proportional to its _____ _____.
5. A transformer has a primary voltage of 480 volts and a secondary voltage of 20 volts. What is the turns ratio of the transformer?
6. If the secondary of the transformer in question 5 supplies a current of 9.6 amperes to a load, what is the primary current? (Disregard excitation current.)
7. Explain the difference between a step-up and a step-down transformer.

8. A transformer has a primary voltage of 240 volts and a secondary voltage of 48 volts. What is the turns ratio of this transformer?

9. A transformer has an output of 750 volt amps. The primary voltage is 120 volts. What is the primary current?

10. A transformer has a turns ratio of 1:6. The primary current is 18 amperes. What is the secondary current?

▶ PRACTICE PROBLEMS

Refer to Figure 19–28 to answer the following questions. Find all missing values.

Ratio: _____

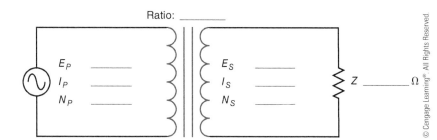

E_P _____
I_P _____
N_P _____

E_S _____
I_S _____
N_S _____

Z _____ Ω

▶ **Figure 19–28**
Practice problems 1 through 6.

1.

$E_P = 120$	$E_S = 24$
$I_P =$	$I_S =$
$N_P = 300$	$N_S =$
RATIO:	$Z = 3\ \Omega$

2.

$E_P = 240$	$E_S = 320$
$I_P =$	$I_S =$
$N_P =$	$N_S = 280$
RATIO:	$Z = 500\ \Omega$

3.

$E_P = $ _____	$E_S = 160$
$I_P = $ _____	$I_S = $ _____
$N_P = $ _____	$N_S = 80$
RATIO: 1:2.5	$Z = 12\ \Omega$

4.

$E_P = 48$	$E_S = 240$
$I_P = $ _____	$I_S = $ _____
$N_P = 220$	$N_S = $ _____
Ratio: _____	$Z = 360\ \Omega$

5.

$E_P = $ _____	$E_S = $ _____
$I_P = 16.5$	$I_S = 3.25$
$N_P = $ _____	$N_S = 450$
Ratio: _____	$Z = 56\ \Omega$

6.

$E_P = 480$	$E_S = $ _____
$I_P = $ _____	$I_S = $ _____
$N_P = 275$	$N_S = 525$
Ratio:	$Z = 1.2\ k\Omega$

Refer to Figure 19–29 to answer the following questions. Find all missing values.

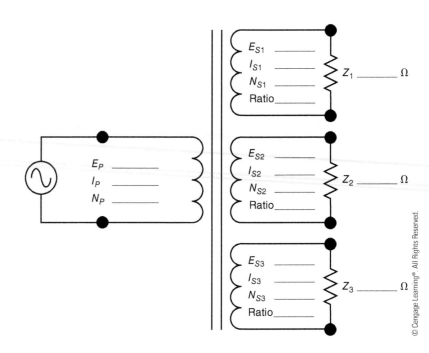

▶ **Figure 19–29**
Practice problems 7 and 8.

7.

$E_P = 208$	$E_{S1} = 320$	$E_{S2} = 120$	$E_{S3} = 24$
$I_P = \underline{\quad}$	$I_{S1} = \underline{\quad}$	$I_{S2} = \underline{\quad}$	$I_{S3} = \underline{\quad}$
$N_P = 800$	$N_{S1} = \underline{\quad}$	$N_{S2} = \underline{\quad}$	$N_{S3} = \underline{\quad}$
	Ratio$_1$ $\underline{\quad}$	Ratio$_2$ $\underline{\quad}$	Ratio$_3$ $\underline{\quad}$
	$R_1 = 12 \text{ k}\Omega$	$R_2 = 6 \ \Omega$	$R_3 = 8 \ \Omega$

8.

$E_P = 277$	$E_{S1} = 480$	$E_{S2} = 208$	$E_{S3} = 120$
$I_P = \underline{\quad}$	$I_{S1} = \underline{\quad}$	$I_{S2} = \underline{\quad}$	$I_{S3} = \underline{\quad}$
$N_P = 350$	$N_{S1} = \underline{\quad}$	$N_{S2} = \underline{\quad}$	$N_{S3} = \underline{\quad}$
	Ratio$_1$ $\underline{\quad}$	Ratio$_2$ $\underline{\quad}$	Ratio$_3$ $\underline{\quad}$
	$R_1 = 200 \ \Omega$	$R_2 = 60 \ \Omega$	$R_3 = 24 \ \Omega$

UNIT 20

Autotransformers

OBJECTIVES

After reading this unit, the student should be able to:

≫ Discuss the operation of an autotransformer

≫ List differences between isolation transformers and autotransformers

≫ Compute values of voltage, current, and turns ratios for autotransformers

≫ Connect an autotransformer for operation

The word *auto* means self. An autotransformer is literally a **self-transformer**. It uses the same winding as both the primary and secondary. Recall that the definition of a **primary winding** is a winding that is connected to the source of power, and the definition of a **secondary winding** is a winding that is connected to a load. Autotransformers have very high efficiencies, most in the range of 95% to 98%.

In Figure 20–1, the entire winding is connected to the power source, and part of the winding is connected to the load. In this illustration, all the turns of wire form the primary, and part of the turns form the secondary. Because the secondary part of the winding contains fewer turns than the primary section, the secondary produces less voltage. This autotransformer is a step-down transformer.

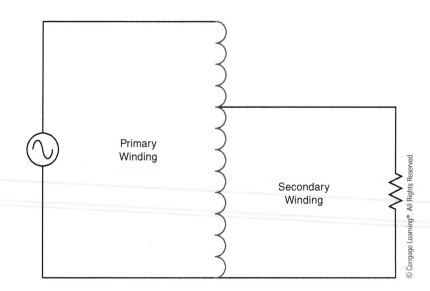

Figure 20–1
Autotransformer used as a
step-down transformer.

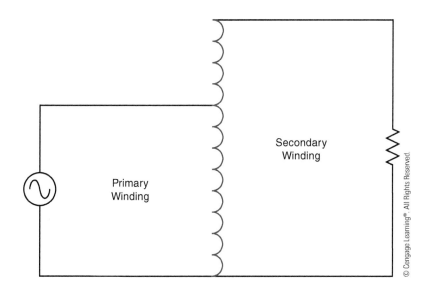

Figure 20–2
Autotransformer used as a
step-up transformer.

In Figure 20–2, the primary section is connected across part of a winding, and the secondary is connected across the entire winding. In this illustration the secondary section contains more windings than the primary. This autotransformer is a step-up transformer. Notice that autotransformers, like isolation transformers, can be used as step-up or step-down transformers.

DETERMINING VOLTAGE VALUES

Autotransformers are not limited to a single secondary winding. Many autotransformers have **multiple taps** to provide different voltages, as shown in Figure 20–3. In this example, there are 40 turns of wire between taps A and B, 80 turns of wire between taps B and C, 100 turns of wire between taps C and D,

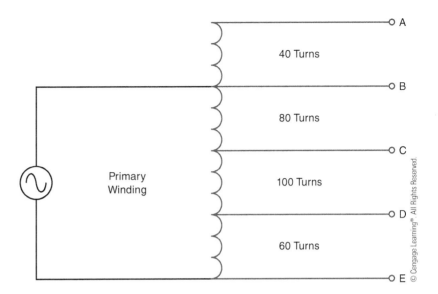

○ A

40 Turns

○ B

80 Turns

○ C

100 Turns

○ D

60 Turns

Primary
Winding

○ E

▶ Figure 20–3
Autotransformer with multiple taps.

and 60 turns of wire between taps D and E. The primary section of the windings is connected between taps B and E. It is assumed that the primary is connected to a source of 120 volts. The voltage across each set of taps will be determined.

There is generally more than one method that can be employed to determine values of a transformer. Because the number of turns between each tap is known, the **volts-per-turn** method will be used in this example. *The volts-per-turn for any transformer is determined by the primary winding.* In this illustration, the primary winding is connected across taps B and E. The primary turns are therefore the sum of the turns between taps B and E (80 + 100 + 60 = 240 turns). Because 120 volts is connected across 240 turns, this transformer will have a volts-per-turn ratio of 0.5 (240 turns/120 volts = 0.5 volts per turn). To determine the amount of voltage between each set of taps, it becomes a simple matter of multiplying the number of turns by the volts per turn.

A–B (40 turns × 0.5 = 20 volts)
A–C (120 turns × 0.5 = 60 volts)
A–D (220 turns × 0.5 = 110 volts)
A–E (280 turns × 0.5 = 140 volts)
B–C (80 turns × 0.5 = 40 volts)
B–D (180 turns × 0.5 = 90 volts)
B–E (240 turns × 0.5 = 120 volts)

C–D (100 turns × 0.5 = 50 volts)
C–E (160 turns × 0.5 = 80 volts)
D–E (60 turns × 0.5 = 30 volts)

USING TRANSFORMER FORMULAS

The values of voltage and current for autotransformers can also be determined by using standard transformer formulas. The primary winding of the transformer shown in Figure 20–4 is between points B and N, and has a voltage of 120 volts applied to it. If the turns of wire are counted between points B and N, it can be seen that there are 120 turns of wire. Now assume that the selector switch is set to point D. The load is now connected between points D and N. The secondary of this transformer contains 40 turns of wire. To compute the amount of voltage to be applied to the load, the following formula can be used:

$$\frac{E_P}{E_S} = \frac{N_P}{N_S}$$

$$\frac{120}{E_S} = \frac{120}{40}$$

$$120\, E_S = 4800$$

$$E_S = 40 \text{ volts}$$

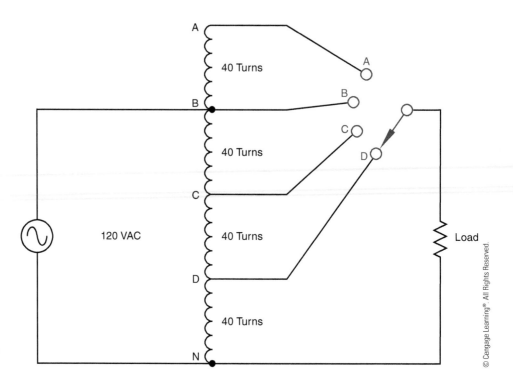

▶ **Figure 20–4**
Determining voltage
and current values.

Assume that the load connected to the secondary
has an impedance of 10 Ω. The amount of current
flow in the secondary circuit can be computed using
the formula:

$$I = \frac{E}{Z}$$

$$I = \frac{40}{10}$$

$$I = 4 \text{ amps}$$

The primary current can be computed by using the
same formula that was used to compute primary
current for an isolation type of transformer.

$$\frac{E_P}{E_S} = \frac{I_S}{I_P}$$

$$120\, I_P = 160$$

$$\frac{120}{40} = \frac{4}{I_P}$$

$$I_P = 1.333 \text{ amps}$$

The amount of power input and output for the autotransformer must also be the same.

Primary
$120 \times 1.333 = 160$ volt-amps

Secondary
$40 \times 4 = 160$ volt-amps

Now assume that the rotary switch is connected to
point A. The load is now connected to 160 turns
of wire. The voltage applied to the load can be
computed by:

$$\frac{E_P}{E_S} = \frac{N_P}{N_S}$$

$$\frac{120}{E_S} = \frac{120}{60}$$

$$120\, E_S = 19{,}200$$

$$E_S = 160 \text{ volts}$$

The amount of secondary current can be computed
using the formula:

$$I = \frac{E}{Z}$$

$$I = \frac{160}{10}$$

$$I = 16 \text{ amps}$$

The primary current can be computed using the formula:

$$\frac{E_P}{E_S} = \frac{I_S}{I_P}$$

$$\frac{120}{160} = \frac{16}{I_P}$$

$$120\,I_P = 2560$$

$$I_P = 21.333 \text{ amps}$$

The answers can be checked by determining whether the power in and power out are the same.

Primary
$120 \times 21.333 = 2,560$ volt-amps

Secondary
$160 \times 16 = 2,560$ volt-amps

CURRENT RELATIONSHIPS

An autotransformer with a 2:1 turns ratio is shown in Figure 20–5. It is assumed that a voltage of 480 volts is connected across the entire winding. Because the transformer has a turns ratio of 2:1, a voltage of 240 volts will be supplied to the load.

Ammeters connected in series with each winding indicate the current flow in the circuit. It is assumed that the load produces a current flow of 4 amperes on the secondary. Note that a current flow of 2 amperes is supplied to the primary.

$$I_{PRIMARY} = \frac{I_{SECONDARY}}{Ratio}$$

$$I_P = \frac{4}{2}$$

$$I_P = 2 \text{ amperes}$$

If the rotary switch shown in Figure 20–4 were removed and replaced with a sliding tap that made contact directly to the transformer winding, the turns ratio could be adjusted continuously. This type of transformer is commonly referred to as a Variac or Powerstat, depending on the manufacturer, Figure 20–6. The windings are wrapped around a tape-wound toroid core inside a plastic case. The tops of the windings have been milled flat, similar to a commutator. A carbon brush makes contact with the windings. When the brush is moved across the windings, the turns ratio changes, which changes the output voltage. This type of autotransformer provides a very efficient means of controlling AC voltage.

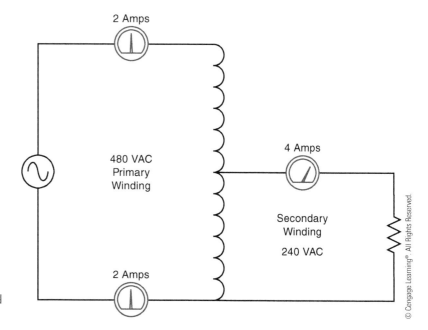

2 Amps

480 VAC
Primary
Winding

4 Amps

Secondary
Winding

240 VAC

2 Amps

▷ **Figure 20–5**
Current divides between primary and secondary.

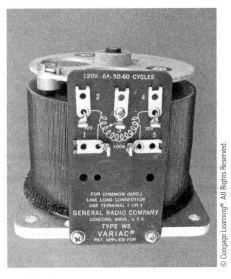

▶ **Figure 20–6**
Variable autotransformer.

Autotransformers are often used by power companies to provide a small increase or decrease to the line voltage. They help provide voltage regulation to large power lines.

The autotransformer does have one disadvantage. Because the load is connected to one side of the power line, there is no **line isolation** between the incoming power and the load. This can cause problems with certain types of equipment and must be a consideration when designing a power system.

◆ SUMMARY

- ⊙ The autotransformer has only one winding that is used as both the primary and secondary.
- ⊙ Autotransformers have efficiencies that range from about 95% to 98%.
- ⊙ Values of voltage, current, and turns can be computed in the same manner as an isolation transformer.
- ⊙ Autotransformers can be step-up or step-down transformers.
- ⊙ Autotransformers can be made to provide a variable output voltage by connecting a sliding tap to the windings.
- ⊙ Autotransformers have the disadvantage of no line isolation between primary and secondary.
- ⊙ One of the simplest ways of computing values of voltage for an autotransformer when the turns are known is to use the volts-per-turn method.

◆ KEY TERMS

line isolation	primary winding	self-transformer
multiple taps	secondary winding	volts per turn

▷ REVIEW QUESTIONS

1. An AC power source is connected across 325 turns of an autotransformer, and the load is connected across 260 turns. What is the turns ratio of this transformer?

2. Is the transformer in question 1 a step-up or a step-down transformer?

3. An autotransformer has a turns ratio of 3.2:1. A voltage of 208 volts is connected across the primary. What is the voltage of the secondary?

4. A load impedance of 52 Ω is connected to the secondary winding of the transformer in question 3. How much current will flow in the secondary?

5. How much current will flow in the primary of the transformer in question 4?

6. The autotransformer shown in Figure 20–3 has the following number of turns between windings: A–B (120 turns), B–C (180 turns), C–D (250 turns), and D–E (300 turns). A voltage of 240 volts is connected across B and E. Find the voltages between each of the following pairs of points:

A–B _____   A–C _____   A–D _____   A–E _____

B–C _____   B–D _____   B–E _____   C–D _____

C–E _____   D–E _____

Current Transformers

OBJECTIVES

After studying this unit, the student should be able to:

- Discuss the operation of a current transformer
- Describe how current transformers differ from voltage transformers
- Discuss safety precautions that should be observed when using current transformers
- Connect a current transformer in a circuit

Current transformers differ from voltage transformers in that the primary winding is generally part of the power line. The primary winding of a current transformer must be connected in series with the load, Figure 21–1. Current transformers are used to change the full-scale range of AC ammeters. Most **in-line ammeters** (ammeters that must be connected directly into the line) that have multiple range values use a current transformer to provide the different ranges, Figure 21–2. The full-scale value of the ammeter is changed by changing the turns ratio. Assume that the ammeter illustrated in Figure 21–2 is to provide range values of **5 amperes**, 2.5 amperes, 1 ampere, and 0.5 ampere. Also assume that the meter movement requires a current flow of 100 mA (0.100) to deflect the meter full scale and that the primary of the current transformer contains 5 turns of wire. Transformer formulas can

Current Transformer

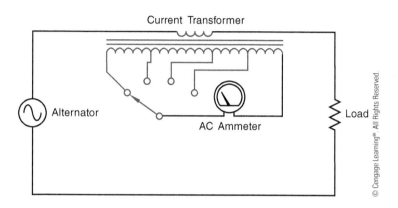

▶ Figure 21–1
The primary winding of a current
transformer is connected in series
with a load.

Current Transformer

Alternator

AC Ammeter

Load

▶ Figure 21–2
A current transformer is used
to change the range of an
AC ammeter.

be used to determine the number of secondary turns needed to produce the desired ranges. Turns needed for a full-scale range of 5 amperes.

$$\frac{N_P}{N_S} = \frac{I_S}{I_P}$$

$$\frac{5}{N_S} = \frac{0.1}{5}$$

$$0.1 N_S = 25$$

$$N_S = 250 \text{ turns}$$

Turns needed for a full-scale range of 2.5 amperes.

$$\frac{5}{N_S} = \frac{0.1}{2.5}$$

$$0.1 N_S = 12.5$$

$$N_S = 125 \text{ turns}$$

Turns needed for a full-scale range of 1 ampere.

$$\frac{5}{N_S} = \frac{0.1}{1}$$

$$0.1 N_S = 5$$

$$N_S = 50 \text{ turns}$$

Turns needed for a full-scale range of 0.5 ampere.

$$\frac{5}{N_S} = \frac{0.1}{0.5}$$

$$0.1 N_S = 2.5$$

$$N_S = 25 \text{ turns}$$

When a large amount of AC current must be measured, a different type of current transformer is connected in the power line. These transformers have ratios that start at 100:5 and can have ratios of several thousand to five. These **current transformers**, generally referred to in industry as **CTs**, have a standard secondary current rating of 5 amps AC. They are designed to be operated with a 5-amp AC ammeter connected directly to their secondary winding, which produces a short circuit. CTs are designed to operate with the secondary winding shorted. *The secondary winding of a CT should never be opened when there is power applied to the primary. This will cause the transformer to produce a step-up in voltage that could be high enough to kill anyone who comes in contact with it.*

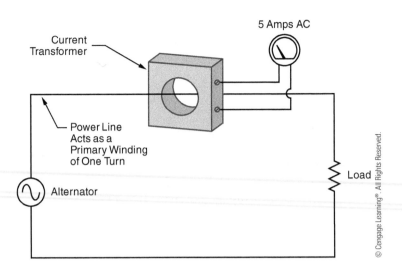

▶**Figure 21–3**
Current transformer used to
change the scale factor of an
AC ammeter.

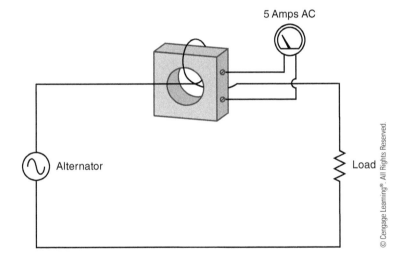

▶**Figure 21–4**
The primary conductor loops through
the CT to produce a second turn,
changing the turns ratio.

A current transformer of this type is basically
a toroid transformer. A toroid transformer is constructed with a hollow core, similar to a donut
in that it has a hole in the middle. When current transformers are used, the main power line is
inserted through the opening in the transformer,
Figure 21–3. The power line acts as the primary of
the transformer and is considered to be one turn.

The turns ratio of the transformer can be changed
by looping the power wire through the opening in
the transformer to produce a primary winding of
more than one turn. For example, assume a current
transformer has a ratio of 600:5. If the primary
power wire is inserted through the opening, it will
require a current of 600 amps to deflect the meter
full scale. If the primary power conductor is looped
around and inserted through the window a second time, the primary now contains two turns of
wire instead of one, Figure 21–4. It now requires
300 amps of current flow in the primary to deflect
the meter full scale. If the primary conductor is
looped through the opening a third time, it would
require only 200 amps of current flow to deflect the
meter full scale.

Utility companies generally use current transformers to meter the current entering a commercial or industrial location. The output of the current transformer is connected to the meter that measures the amount of power usage. Current transformers of this type are shown in Figure 21–5. Another location that current transformers are commonly employed is on the starter of large horsepower motors, Figure 21–6. Large horsepower motors can have current draws of several hundred amperes. The current transformers are used to supply power to the heaters of overload relays, Figure 21–7. Overload relay heaters do have a high enough current rating to be connected in series with a motor that has a current draw of several hundred amperes. The current transformer reduces the current to some level between 0 and 5 amperes.

The current transformers shown in Figure 21–6 have a ratio of 300:5. Assume that the motor in this circuit has a full load running current of 256 amperes. A ratio of the running current as compared to the transformer ratio can be used to determine the overload heater size needed for this motor.

$$\frac{256}{300} = \frac{X}{5}$$

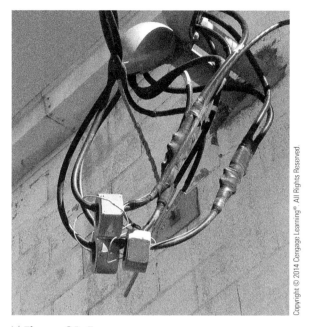

Figure 21–5
Current transformers used to meter the incoming power.

Cross multiply: $(X \times 300 = 300 X)(256 \times 5 = 1280)$

$$300X = 1280$$
$$X = \frac{1280}{300}$$
$$X = 4.27$$

Figure 21–6
Current transformers are often used on motor starters that control large horsepower motor.

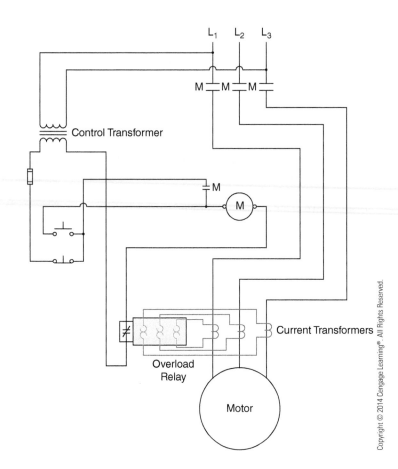

▶ **Figure 21–7**
Current transformers supply
power to the overload relay
heaters.

The overload heaters will be sized for a motor
with a full load current of 4.27 amperes.

CLAMP-ON AMMETERS

Many service technicians use the clamp-on type of
AC ammeter. To use this type of meter, the jaw of
the meter is clamped around one of the conductors supplying power to the load, Figure 21–8. The
meter is clamped around only one of the lines. If
the meter is clamped around more than one line, the
magnetic fields of the wires cancel each other and
the meter indicates zero.

This type of meter uses a current transformer to
operate the meter. The jaw of the meter is part of the
core material of the transformer. When the meter
is connected around the current-carrying wire, the
changing magnetic field produced by the AC current

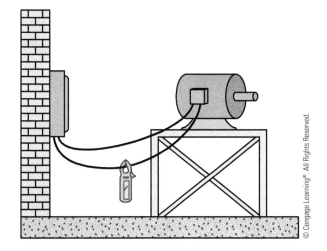

▶ **Figure 21–8**
The clamp-on ammeter connects around
one conductor.

induces a voltage into the current transformer. The strength of the magnetic field and its frequency determines the amount of voltage induced in the current transformer. Because 60 Hz is a standard frequency throughout the United States and Canada, the amount of induced voltage is proportional to the strength of the magnetic field.

The clamp-on ammeter can have different range settings by changing the turns ratio of the secondary of the transformer just as the in-line ammeter does. The primary of the transformer is the conductor the movable jaw is connected around. If the ammeter is connected around one wire, the primary has one turn of wire as compared to the turns of the secondary. The turn's ratio can be changed in the same manner as changing the ratio of the CT. If two turns of wire are wrapped around the jaw of the ammeter, Figure 21–9, the primary winding now contains two turns instead of one, and the turns ratio of the transformer is changed. The ammeter will now indicate double the amount of current in the circuit. The reading on the scale of the meter would have to be divided by two to get the correct reading. The ability to change the turns ratio of a clamp-on ammeter can be very useful for measuring low currents.

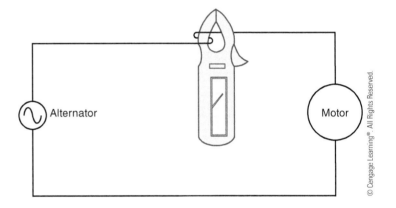

▶| Figure 21–9
Looping the conductor around the jaw of the ammeter changes the ratio.

▶ SUMMARY

- ⊙ Current transformers have their primary winding connected in series with a load.
- ⊙ Current transformers are often used to provide multiple scale values for in-line AC ammeters.
- ⊙ Current transformers are often referred to as CTs.
- ⊙ CTs are used to measure large amounts of AC current.
- ⊙ CTs have a standard secondary current value of 5 amperes.
- ⊙ CTs are designed to be operated with their secondary winding shorted.
- ⊙ The short circuit connected across the secondary of the CT should never be removed when power is connected to the circuit because the secondary voltage can become very high.
- ⊙ Many clamp-on AC ammeters operate on the principle of a current transformer.
- ⊙ The movable jaw of the clamp-on ammeter is the core of the transformer.
- ⊙ The secondary current value of a current transformer can be changed by changing the turns of wire of the primary.

KEY TERMS

5 amperes   current transformers
CTs   in-line ammeters

REVIEW QUESTIONS

1. Explain the difference in connection between the primary winding of a voltage trans-former and the primary winding of a current transformer.

2. What is the standard current rating for the secondary winding of a CT?

3. Why should the secondary winding of a CT never be disconnected from its load when there is current flow in the primary?

4. A current transformer has a ratio of 600:5. If three loops of wire are wound through the transformer core, how much primary current is required to produce 5 amperes of current in the secondary winding?

5. Assume that a primary current of 75 amperes flows through the windings of the transformer in question 4. How much current will flow in the secondary winding?

6. What type of core is generally used in the construction of a CT?

7. A current transformer has four turns of wire in its primary winding. How many turns of wire are needed in the secondary winding to produce a current of 2 amperes when a current of 60 amperes flows in the primary winding?

8. A 1500:5 CT develops a voltage of 3 volts across the primary winding. If the sec-ondary should be disconnected from its load, how much voltage would be developed across the secondary terminals?

9. A CT has a current flow of 80 amperes in its primary winding and a current of 2 amperes in its secondary winding. What is the ratio of the CT?

10. What is the most common use for a CT?

SECTION 5

Control Components

Overloads

OBJECTIVES

After studying this unit, the student should be able to:

- Explain why motors should be protected from an overload condition
- List the different types of overload protectors
- Describe the operation of a solder-melting and bimetal type overload
- Connect an overload relay into a motor circuit
- Perform an ohmmeter test on an overload relay

Overload relays are designed to protect the motor circuit from damage due to **overloads**. Most overload relays are operated by heat. Because the overload unit must be sensitive to motor current, the heater of the overload relay is connected in series with the motor. In this manner, the amount of current that flows through the motor winding also flows through the overload heater. There are two basic types of overload units used in the air-conditioning field, the **solder-melting type** and the **bimetal type**.

SOLDER-MELTING TYPE OVERLOAD RELAY

The solder-melting type of overload unit is used to a large extent on commercial and industrial air-conditioning units. This type of overload unit

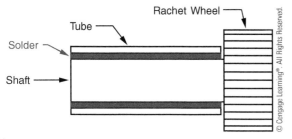

Figure 22-1
Shaft is held stationary by solder.

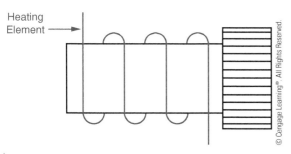

Figure 22-2
An electric heating element is wound around the tube.

Figure 22-3
The overload heater is connected in series with the motor.

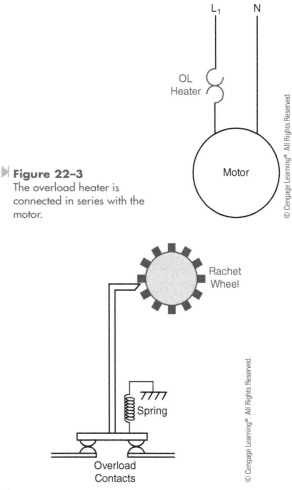

Figure 22-4
The ratchet wheel holds the contacts closed.

contains a ratchet wheel that is held stationary by solder. Figure 22–1 illustrates the principle of operation. A serrated wheel is attached to a shaft. The shaft is inserted in a hollow tube. The shaft would be free to rotate inside the tube except for the solder that bonds the two units together.

An electric **heating element** is wound around the tube, as shown in Figure 22–2. The heating element is connected in series with the motor, Figure 22–3. The current that flows through the motor windings also flows through the heating element. The heating element is calibrated to produce a certain amount of heat when a predetermined amount of current flows through it. As long as the current flowing through it does not exceed a certain amount, there is not enough heat produced to melt the solder. If the motor should become overloaded, an excessive amount of current will flow through the heater and the solder will melt. When the solder melts, the shaft is free to turn.

The ratchet wheel is used to mechanically hold a set of spring-loaded contacts closed, as shown in Figure 22–4. When the solder melts, the ratchet wheel is free to turn and the spring causes the contacts to open. The normally closed contacts are connected in series with the coil of the motor starter used to control the motor the overload relay is protecting, Figure 22–5. When the overload contacts open, the motor starter coil de-energizes and disconnects the motor from the line.

Notice that this overload relay has two separate sections, the **heater section** that is connected in series with the motor, and the **contact section**,

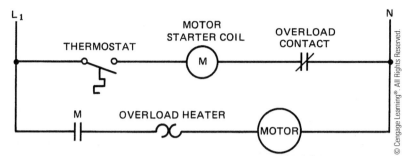

▶ **Figure 22–5**
When the overload contact opens, the
motor is disconnected from the line.

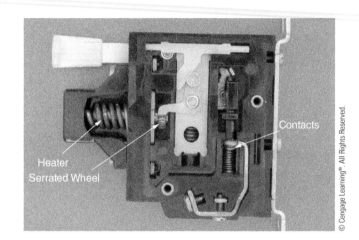

▶ **Figure 22–6**
Solder-melting type of overload relay.

which is connected in series with the coil of the
motor starter. Notice also that the overload contacts
are not used to disconnect the motor from the line.
They are used to disconnect the motor starter coil
from the line. This type of overload relay has a set
of small auxiliary contacts that are intended to
interrupt the current flow in the control circuit only.
After the overload has tripped, it must be allowed to
cool down enough for the solder to re-harden before
it can be reset. This is generally true of any type of
thermal overload. Figure 22–6 shows a photograph
of this type of overload.

BIMETAL TYPE OF OVERLOAD

The bimetal type of overload operates very similarly
to the solder-melting type, except a **bimetal strip**
is used to cause the contacts to open, Figure 22–7.
In this unit, the bimetal strip is used to mechanically

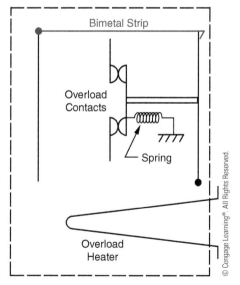

▶ **Figure 22–7**
Bimetal type of overload.

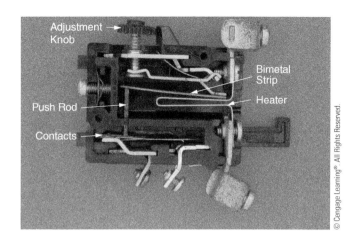

▶ **Figure 22-8**
Bimetal-type overload relay.

hold the spring-loaded contacts closed. If the current flow through the heater becomes excessive, the bimetal strip will warp and permit the spring to open the contacts. After the overload unit has tripped, the bimetal strip must be allowed some time to cool before it can be reset. This type of unit has an advantage over the solder-melting type in that it can be adjusted for manual reset or automatic reset, Figure 22-8. The solder-melting type of overload unit must be manually reset.

PROTECTING THREE-PHASE MOTORS

Each phase of a three-phase motor should be protected by an overload relay. There are two methods employed to provide this protection. One method is to connect a single overload relay like those shown in Figure 22-6 and Figure 22-8 into each phase; see Figure 22-9. When this is done, the three sets of overload contacts are connected in series with each other, as shown in Figure 22-10. Because the overload contacts are connected in series, if any one set of contacts should open, the starter coil will be disconnected, causing the load contacts to open and disconnect the motor from the line.

The second method used to protect three-phase motors is with a three-phase overload relay. This overload relay contains three separate heaters, but only one set of overload contacts, Figure 22-11. If any one of the heaters should trip, it will open the set of contacts and disconnect the starter coil from the line, Figure 22-12.

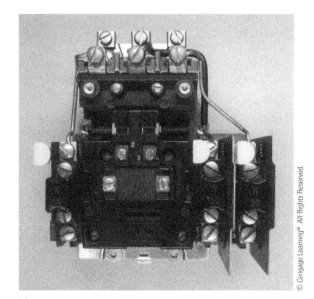

▶ **Figure 22-9**
Three single overload relays are used to protect a three-phase motor.

PROTECTING SINGLE-PHASE MOTORS

Single-phase motors are generally protected with a small **automatic reset overload**, Figure 22-13. These units are constructed in one of two ways. One unit has a small heater connected in series with the motor current, Figure 22-14. In this unit, the bimetal strip is constructed of a spring metal that provides a snap action when it warps. Notice that the

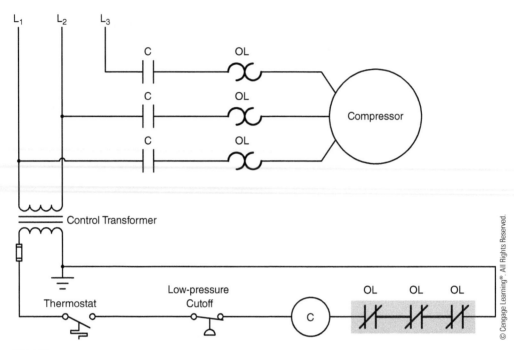

Figure 22–10
When three single overload relays are employed to protect a three-phase motor, all normally closed overload contacts are connected in series.

Figure 22–11
Three-phase overload relay.

contacts are connected directly to the bimetal strip. This means that the motor current flows not only through the heating element but also through the bimetal strip. If the motor current becomes excessive, the heater causes the bimetal strip to snap the

contacts open and disconnect the motor from the line. Notice that the contacts of this unit are used to interrupt the motor current. When the bimetal strip has cooled enough, it snaps back to its original position and recloses the contacts.

The second type of small overload unit does not contain a heating element, Figure 22–15. In this type of unit, the bimetal strip is used as the heating element. As current flows through the bimetal strip, it begins to heat. If motor current does not become excessive, the bimetal strip does not become heated enough to cause the contacts to open. If the current does become excessive, however, the contacts snap open and disconnect the motor from the line.

These overload units can be tested with an ohmmeter for a complete circuit. If the ohmmeter indicates no continuity through them when they are cool, they are defective and must be replaced. Care must be taken to replace these units with the correct size. Overload units are designed to open their contacts when the motor current reaches 115% to 125% of full-load current. The exact rating is

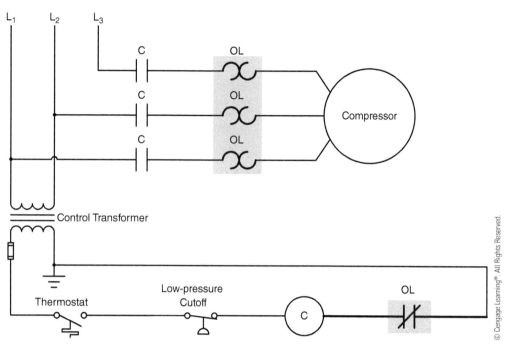

▷ Figure 22–12
A three-phase overload relay contains three heaters and one set of contacts.

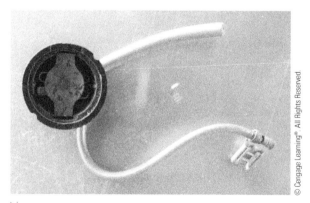

▷ Figure 22–13
Bimetal overload often used on fractional horsepower single-phase motors.

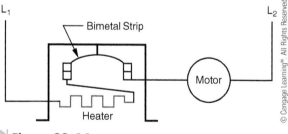

▷ Figure 22–14
Small overload unit with heater.

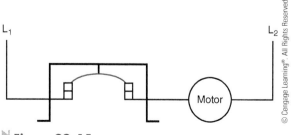

▷ Figure 22–15
Small overload unit without heater.

determined by the national electrical code. If an overload unit of too small a rating is installed, it will trip when there is no overload on the motor. If an overload unit of too high a value is used, the motor may be destroyed before the overload contacts open and disconnect the motor.

Notice that all of the overload units discussed are operated by sensing heat. For this reason, a heavy motor overload will cause more heat production and the unit will trip faster than it will under a light overload. Another factor that can affect these units is **ambient air temperature**. Overload relays trip faster in hot weather than they do in cool weather. In certain parts of the country, it is often necessary to replace the heater elements of industrial-type overload units to match the season. In winter it may be desirable to use a slightly smaller heater element than normal, and in summer it may be necessary to use a slightly larger heater.

PROTECTING LARGE MOTORS

Large horsepower motors often have current draws that are much greater than the rating of any standard overload heating element. When this is the case, current transformers are used to supply current to the overload heaters, Figure 22–16. Assume that the motor operating a large compressor has a nameplate current of 234 amperes. Now assume that three current transformers with a ratio of 300:5 are to be used to reduce the current to the overload heaters. The current rating of the overload heater can be calculated using the ratio of the current transformer.

$$\frac{300}{5} = \frac{234}{X}$$

$$300\,X = 1170$$

$$X = 3.9 \text{ amps}$$

If the overload heaters are sized for a motor with a running current of 3.9 amperes the motor will be protected. Two NEMA size 5 starters are shown in Figure 22–17. These starters contain current transformers used to reduce the current to the overload heaters.

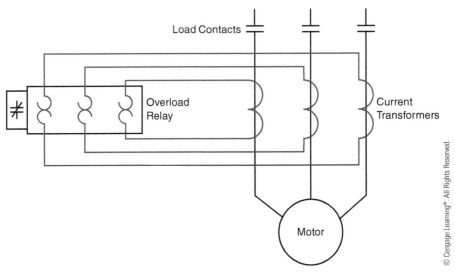

Figure 22–16
Current transformers reduce the motor current.

▷ **Figure 22–17**
NEMA size 5
starters with current
transformers.

SUMMARY

- Overload relays are designed to protect the motor against an overload condition.
- Most overload relays sense motor current by connecting an electric heater in series with the motor.
- Two basic types of overload relays are:
  A. Solder-melting type.
  B. Bimetal strip type.
- Solder-melting type overload relays permit the electric heater to melt solder at a predetermined temperature and open a set of contacts.
- The bimetal strip type of overload relays open a set of contacts when the electric heating element causes a bimetal strip to warp a certain amount.
- Overload relays contain two separate sections, the heater section and the contact section.
- The contacts of an overload relay are connected in series with a motor starter coil.
- Because overload relays operate by sensing heat, they are affected by ambient temperature.

KEY TERMS

ambient air
  temperature
automatic reset
  overload

bimetal strip
bimetal type
contact section
heater section

heating element
overloads
solder-melting type

REVIEW QUESTIONS

1. What are the two basic types of industrial overload units?

2. What is the advantage of the bimetal type of industrial overload unit?

3. Industrial overload units are divided into two sections. What are they?

4. At what percentage of full-load motor current are overload units generally set to trip?

5. When using an industrial type of overload unit, what are the contacts connected in series with?

6. What is the difference between the two types of small overload units?

7. In the small overload unit that does not contain a heater, what is used to sense the current flow through the motor?

Relays, Contactors, and Motor Starters

OBJECTIVES

After studying this unit, the student should be able to:

⫸ Describe the construction of a relay

⫸ Discuss the principle of operation of relays and contactors

⫸ Discuss different types of relays and contactors

⫸ Describe the difference between a contactor and a motor starter

The **relay** is a magnetically operated switch. This switch, however, can have multiple sets of contacts, and the contacts can be open or closed. The advantage of the relay is control. A single pilot device can be used to control the input or coil of the relay, and the output or contacts can control several different devices. An example of this is shown in the circuit of Figure 23–1. A flow switch is used to control the coil of a magnetic relay. When the flow switch closes, the coil of relay **FSCR (flow switch control relay)** is connected to the line. When current flows through the coil, the relay is energized, and all FSCR contacts change position. Notice that one FSCR contact is connected in series with the compressor motor. This contact does not actually start the motor, but it permits the thermostat to control the motor. This particular type of control is known as interlocking. Interlocking is used to prevent some

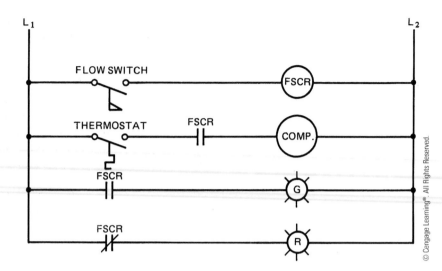

Figure 23–1
One relay controls several devices.

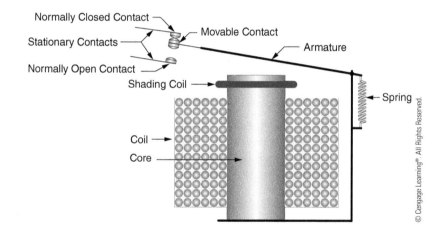

Figure 23–2
Simple relay.

function from happening until some other function has occurred. In this case, the thermostat cannot start the compressor until there is airflow in the system.

The second FSCR contact is normally open. When the FSCR coil energizes, this contact closes and turns on a green pilot light to indicate there is airflow in the system. The third FSCR contact is normally closed. It is used to turn off a red pilot light, which indicates there is no airflow in the system.

PRINCIPLE OF OPERATION

The relay operates on the **solenoid** principle. A solenoid is an electrical device that converts electrical energy to linear motion. This principle is illustrated in Figure 23–2. A coil of wire is wound around an iron core. When current flows through the coil, a magnetic field is developed in the iron core. The magnetic field of the iron core attracts the movable arm, known as the **armature**, and overcomes the strength of the spring holding the arm away from the iron core. Notice that a movable contact is connected to the armature. In its present position, the movable contact makes connection with a stationary contact. This contact set is normally closed. When the armature is attracted to the iron core, the movable contact breaks connection with one stationary contact and makes connection with another. This relay has both a normally open and a normally closed set of contacts. Notice that the movable contact is common to both of the

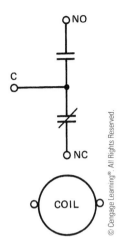

Figure 23–3
Schematic of a simple relay.

stationary contacts. The movable contact would be the common and the stationary contacts would be labeled normally open and normally closed. A schematic of this type of relay is shown in Figure 23–3. This illustration shows a relay with only one set of contacts. In practice, it is common to find this type of relay with several sets of contacts.

Notice in Figure 23–2 that a shading coil has been added to the iron core. The shading coil is used with AC relays to prevent contact chatter and hum. The shading coil operates in the same way it does in the shaded-pole motor. It opposes a change of magnetic flux. The shading coil is used to provide a continuous magnetic flow to the armature when the voltage of the AC waveform is zero. DC-operated relays do not contain a shading coil because the magnetic flux is constant.

Another type of relay is shown in Figure 23–4. This relay uses a **plunger-type** solenoid. Notice that the coil is surrounded by the iron core. There is an opening in the iron core through which the shaft of the armature can pass. When the coil is energized, the armature is attracted to both ends of the core. This creates a stronger magnetic field than the relay discussed in Figure 23–2. Notice the shading coils around both ends of the core. Notice also that the core and armature are constructed of laminated sheets. The core and armature are laminated to help prevent the induction of **eddy currents** into the core. Eddy currents are currents induced in the core material by the magnetic field of the coil. Eddy currents are generally unwanted because they heat the core and cause a power loss.

The plunger-type solenoid is generally used with relays that use double-break contacts. A **double-break contact** is one that breaks connection at two points, as shown in Figure 23–5. Notice there are two stationary contacts and one movable contact. The movable contact is used to bridge the gap between the two stationary contacts. This type of contact arrangement is preferred for relays that must control high voltage and current.

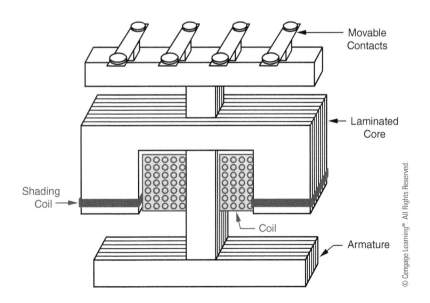

Figure 23–4
Plunger type of solenoid.

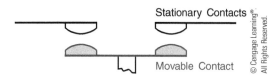

Figure 23–5
A set of double-break contacts.

Figure 23–6
Contactor.

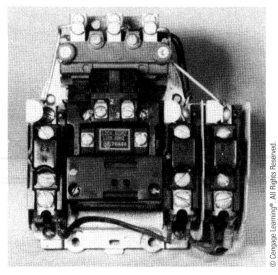

Figure 23–7
Motor starter with overload relays.

Notice that the surface of the contact is curved. This curved surface provides a wiping action when the contacts make connection. The wiping action helps to keep contact surfaces clean. Contact surfaces should never be filed flat. This would permit oil and dirt to collect on the surface of the contact and cause poor connection.

CONTACTORS AND MOTOR STARTERS

The term *relay* is often used to describe any type of magnetically operated switch. A relay is actually a control device that contains small auxiliary contacts designed to operate only low-current loads.

A **contactor** is very similar to a relay except that a contactor contains large-load contacts designed to control large amounts of current. In the heating and air-conditioning field, contactors are often used to connect power to resistance heater banks. A photograph of a contactor is shown in Figure 23–6. Contactors may contain auxiliary contacts as well as load contacts.

Motor starters are basically contactors with the addition of overload relays. Motor starters generally contain auxiliary contacts as well as load contacts. The auxiliary contacts are used as part of the control circuit, and the load contacts are used to connect the motor to the line. A photograph of a motor starter is shown in Figure 23–7.

SUMMARY

- ⊙ A relay is a magnetically operated switch.
- ⊙ A relay is a single-input, multi-output device.
- ⊙ Relays can have contacts that are normally open or normally closed.
- ⊙ On a schematic diagram, the contacts of a relay are always shown in the de-energized condition.
- ⊙ Interlocking is used to prevent some function from happening until some other function has occurred.
- ⊙ Most relays are basically electric solenoids with a set of contacts attached.
- ⊙ A solenoid is a device that converts electrical energy into linear motion.
- ⊙ AC relays use shading coils on the iron core to prevent contact chatter.
- ⊙ Contact surfaces are generally curved to provide a wiping action.
- ⊙ A relay is actually a control device that has only small auxiliary contacts that are used as part of the control circuit or to operate low-current devices.
- ⊙ Contactors contain large load contacts intended to connect the load to the line.
- ⊙ Motor starters are contactors with the addition of overload relays.

KEY TERMS

armature
contactor
double-break contact
eddy currents

FSCR (flow switch
   control relay)
motor starters
plunger-type

relay
solenoid

REVIEW QUESTIONS

1. What is a solenoid?
2. What type of relays contains a shading coil?
3. What purpose does the shading coil serve?
4. What is the movable part of a relay called?
5. Why is the core material of a relay laminated?
6. What are eddy currents?
7. What effect do eddy currents have on a relay?
8. Why are contact surfaces curved?
9. What is the difference between a relay and a contactor?
10. What is the difference between a contactor and a motor starter?

UNIT 24 ▷

The Solid-State Relay

OBJECTIVES

After studying this unit, the student should be able to:

▷ Describe the construction of a solid-state relay

▷ Discuss the principle of opto-isolation

▷ Describe the internal devices used for a relay intended to control a DC and an AC load

▷ Describe zero switching

▷ Connect a solid-state relay in a circuit

The **solid-state relay** is a device that has become increasingly popular for switching applications. The solid-state relay has no moving parts, is resistant to shock and vibration, and is sealed against dirt and moisture. The greatest advantage of the solid-state relay, however, is the fact that the control input voltage is isolated from the line device the relay is intended to control. Refer to Figure 24–1.

Solid-state relays can be used to control either a DC load or an AC load. If the relay is designed to control a DC load, a **power transistor** is used to connect the load to the line, as shown in Figure 24–2. The relay shown in Figure 24–2 has a light-emitting diode (LED) connected to the input or control voltage. When the input voltage turns the LED on, a photo detector connected to the base of the transistor turns the transistor on and connects

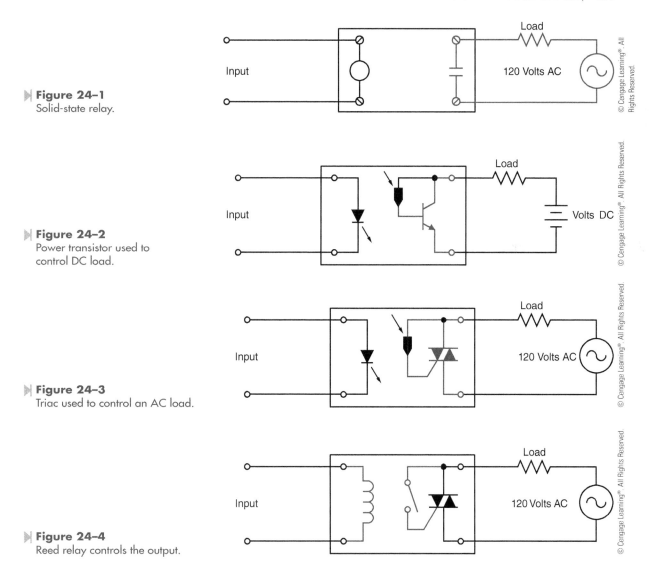

Figure 24-1
Solid-state relay.

Figure 24-2
Power transistor used to control DC load.

Figure 24-3
Triac used to control an AC load.

Figure 24-4
Reed relay controls the output.

the load to the line. This optical coupling is a very common method used with solid-state relays. The relays that use this method of coupling are referred to as being **opto-isolated**, which means the load side of the relay is optically isolated from the control side of the relay. Because a light beam is used as the control medium, no voltage spikes or electrical noise produced on the load side of the relay can be transmitted to the control side of the relay.

Solid-state relays intended for use as AC controllers have a **triac** connected to the load circuit in place of a power transistor. Refer to Figure 24–3. In

this example, an LED is used as the control device just as it was in the previous example. When the photo detector "sees" the LED, it triggers the gate of the triac and connects the load to the line.

Although opto-isolation is probably the most common method used for the control of a solid-state relay, it is not the only method used. Some relays use a small **reed relay** to control the output. Refer to Figure 24–4. A small set of reed contacts are connected to the gate of the triac. The control circuit is connected to the coil of the reed relay. When the control voltage causes a current to flow through the coil, a magnetic field is produced around the coil of the relay. This magnetic field closes the reed contacts, which causes the triac to turn on. In this type of solid-state relay, a magnetic field is used to isolate the control circuit from the load circuit instead of a light beam.

The control voltage for most solid-state relays ranges from about 3 to 32 volts and can be DC or AC. If a triac is used as the control device, load voltage ratings of 120 to 240 VAC are common and current ratings can range from 5 to 25 amps. Many solid-state relays have a feature known as **zero switching**. Zero switching means that if the relay is told to turn off when the AC voltage is in the middle of a cycle, it will continue to conduct until the AC voltage drops to a zero level and then turn off. For example, assume the AC voltage is at its positive peak value when the gate tells the triac to turn off. The triac will continue to conduct until the AC voltage drops to a zero level before actually turning off. Zero switching can be a great advantage when used with some inductive loads such as transformers. The core material of a transformer can be left saturated on one end of the flux swing if power is removed from the primary winding when the AC voltage is at its positive or negative peak. This can cause inrush currents of up to 600% of the normal operating current when power is restored to the primary.

Solid-state relays are available in different case styles and power ratings. Figure 24–5 shows a typical solid-state relay. Some solid-state relays are designed to be used as time-delay relays. One of the most common uses for the solid-state relay is the I/O (eye-oh) track of a programmable controller, which is covered in a later unit.

▌**Figure 24–5**
Solid-state relays.

▶ SUMMARY

- ⊙ Solid-state relays have no moving parts.
- ⊙ Solid-state relays are sealed against dirt and moisture and are resistant to mechanical shock and vibration.
- ⊙ The control side of a solid-state relay is totally isolated from the load side.
- ⊙ Solid-state relays used to control AC loads use a triac connected in series with the load.
- ⊙ Solid-state relays used to control DC loads use a power transistor connected in series with the load.
- ⊙ Many solid-state relays use a light-emitting diode (LED) as the control device.
- ⊙ Opto-isolation is used to separate the load side of the circuit from the control side. This prevents any electrical noise from being transferred from the load side to the control side of the circuit.
- ⊙ Solid-state relays are generally used in the I/O track of programmable controllers.

KEY TERMS

opto-isolated

power transistor

reed relay

solid-state relay

triac

zero switching

REVIEW QUESTIONS

1. What electronic component is used to control the output of a solid-state relay used to control a DC voltage?

2. What electronic component is used to control the output of a solid-state relay used to control an AC voltage?

3. Explain opto-isolation.

4. Explain magnetic isolation.

5. What is meant by zero switching?

The Control Transformer

OBJECTIVES

After studying this unit, the student should be able to:

- ⟫ Discuss the principle of operation of a transformer
- ⟫ Define mutual induction
- ⟫ Discuss the voltage and current relationships in a transformer
- ⟫ Find values of voltage, current, and turns of wire, using transformer formulas and Ohm's law
- ⟫ Connect an industrial control transformer for operation on low and high voltage
- ⟫ Perform an ohmmeter test on a control transformer

The **transformer** is a device that has the ability to change the value of AC voltage and current without a change of frequency. Most of the transformers used in the air-conditioning and refrigeration field are known as isolation transformers. This means that the primary and secondary windings are magnetically coupled but electrically isolated from each other. Figure 25–1 illustrates the basic principle of operation of a transformer. This transformer contains two separate windings, the primary and the secondary. The primary is the winding that is connected to the power source and brings power to the transformer. The secondary winding is used to supply power to the load. Notice that there is no electrical connection between the two windings. If one lead of an ohmmeter is connected to one of the primary leads and the other ohmmeter lead is connected to one of the secondary leads, the ohmmeter

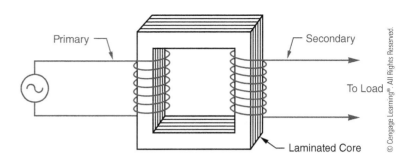

> **Figure 25-1**
> A basic transformer.

should indicate no continuity between the two windings.

PRINCIPLE OF OPERATION

The transformer operates by magnetic induction. When current flows through the primary winding, a magnetic field is created in the winding. Because the secondary winding is wound on the same core as the primary, the magnetic field of the primary induces a voltage into the secondary. This action is known as mutual induction. The amount of voltage induced into the secondary is determined by the ratio of the number of turns of wire in the primary as compared with the number of turns of wire in the secondary. For example, assume that the primary winding shown in Figure 25–1 contains 120 turns of wire and is connected to 120 volts AC. This means that each turn of the primary has a voltage drop of 1 volt. If the secondary winding also has 120 turns of wire, and 1 volt is induced into each turn, then the output voltage of the secondary is 120 volts also. This transformer has a turns ratio of 1:1, which is to say that the primary contains 1 turn of wire for each turn of wire in the secondary.

Now assume that the number of turns of wire in the secondary has been changed to 60. If the number of turns in the primary has not been changed, there is still 1 volt for each turn of wire. This will produce a secondary voltage of 60 volts ($60 \times 1 = 60$).

If the number of turns of wire on the secondary is changed to 240, the output voltage of the secondary will be 240 volts ($240 \times 1 = 240$). Notice that the transformer has the ability to increase or decrease the amount of the secondary voltage. If the voltage of the secondary is less than the primary voltage, the transformer is known as a step-down transformer. If the secondary voltage is greater than the primary voltage, it is known as a step-up transformer.

VOLTAGE AND CURRENT RELATIONSHIPS

It would first appear that the transformer has the ability to give more than it receives. This is not the case, however. Transformers are extremely efficient devices; they generally operate at 95% to 98% efficiency. For this reason, when working with transformers it is generally assumed that the power out of the transformer is equal to the power being put into the transformer.

Figure 25–2 shows the schematic symbol for a transformer. The primary has been connected to 120 volts. The secondary has a voltage of 480 volts and is connected to a load resistor of 960 ohms.

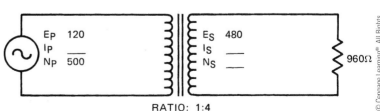

$$E_P \quad 120$$
$$I_P \quad \underline{}$$
$$N_P \quad 500$$

$$E_S \quad 480$$
$$I_S \quad \underline{}$$
$$N_S \quad \underline{}$$

960Ω

RATIO: 1:4

> **Figure 25-2**
> Schematic for transformer.

This transformer has a turns ratio of 1:4, which is to say that there is 1 turn of wire in the primary for every 4 turns of wire in the secondary. The amount of current in the secondary (I_S) can be computed by using Ohm's law.

$$I = \frac{E}{R}$$

$$I = \frac{480}{960}$$

$$I = 0.5 \text{ amp}$$

If the secondary of this transformer has a current flow of 0.5 amp, how much current is required in the primary? There are actually several methods that can be used to solve this problem. The most accepted method is to use the formulas shown in Figure 25–3. Because both the primary and secondary voltages are known, the formula that contains voltage and current will be used to solve the problem.

$$\frac{E_P}{E_S} = \frac{I_S}{I_P}$$

$$\frac{120}{480} = \frac{0.5}{I_P}$$

$$120 \, I_P = 240$$

$$I_P = 2$$

Notice that the transformer must have a current draw of 2 amps on the primary to supply a current of 0.5 amp at the secondary. If the power (volts × amps) is computed for both the primary and the secondary, it will be seen that they are equal.

$\dfrac{E_P}{E_S} = \dfrac{N_P}{N_S}$	$\dfrac{E_P}{E_S} = \dfrac{I_P}{I_S}$	$\dfrac{N_P}{N_S} = \dfrac{I_S}{I_P}$

E_P — Voltage of the primary

E_S — Voltage of the secondary

N_P — Number of turns of wire in the primary

N_S — Number of turns of wire in the secondary

I_P — Current flow in the primary

I_S — Current flow in the secondary

▶ **Figure 25–3**
Transformer formulas.

Primary

$$120 \times 2 = 240$$

Secondary

$$480 \times 0.5 = 240$$

The number of turns of wire can now be computed.

$$\frac{E_P}{E_S} = \frac{N_P}{N_S}$$

$$\frac{120}{480} = \frac{500}{N_S}$$

$$120 \, N_S = 240,000$$

$$N_S = 2000$$

Notice that the secondary has 2000 turns of wire compared to 500 turns in the primary. This is consistent with the turns ratio, which states there is 1 turn of wire in the primary for every 4 turns in the secondary.

The transformer shown in Figure 25–4 is a step-down transformer that has a primary voltage of 120 volts and a secondary voltage of 24 volts. The secondary is connected to a load resistance of 6 ohms. The current flow in the secondary winding is:

$$I = \frac{E}{R}$$

$$I = \frac{24}{6}$$

$$I = 4 \text{ amps}$$

Now that the secondary current is known, the amount of primary current can be computed.

$$\frac{E_P}{E_S} = \frac{I_S}{I_P}$$

$$\frac{120}{24} = \frac{4}{I_P}$$

$$120 \, I_P = 96$$

$$I_P = 0.8 \text{ amp}$$

If the amount of power for the primary and that for the secondary are computed, it will be seen that they are the same.

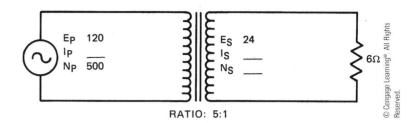

Figure 25–4
Step-down transformer.

RATIO: 5:1

Primary

$120 \times 0.8 = 96$

Secondary

$24 \times 4 = 96$

The number of turns of wire in the secondary can now be computed.

$$\frac{E_p}{E_S} = \frac{N_p}{N_s}$$

$$\frac{120}{24} = \frac{500}{N_S}$$

$$120\, N_S = 12000$$

$$N_S = 100$$

Notice that the number of turns of wire in the secondary, 100, as compared with the turns of wire in the primary, 500, is consistent with the turns ratio of 5:1.

RESIDENTIAL CONTROL TRANSFORMERS

Control transformers are used to change the value of line voltage to the value needed for the control circuit. Most residential air-conditioning systems operate on a control voltage of 24 volts AC. The amount of current needed varies from one system to another, but it is generally less than 1 amp. The primary voltage for residential control transformers is 120 or 240 volts. A photograph of a control transformer used in residential applications is shown in Figure 25–5. The primary lead wires for most of these transformers are black in color. The color of the secondary leads varies from one manufacturer to another.

Figure 25–5
A 24-volt transformer.

INDUSTRIAL CONTROL TRANSFORMERS

Most industrial and commercial air-conditioning systems operate on 240 or 480 volts. The control voltage for most of these units is 120 or 24 volts AC. Most industrial control transformers contain two primary windings and one or two secondary windings. In the following explanation, it is assumed that the transformer has two primary windings and one secondary winding. In this type of transformer, each primary winding has a voltage rating of 240 volts, and the secondary winding has a voltage rating of 120 volts. There is a turns ratio of 2:1 between each of the primary windings and the secondary winding.

There is a standard for marking the terminals of control transformers. One of the primary windings

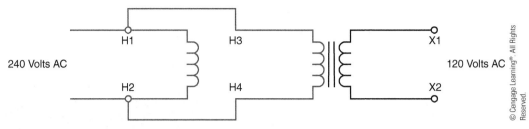

Figure 25–6
Primaries connected in parallel for 240-volt operation.

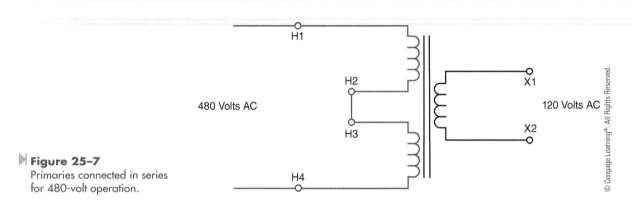

Figure 25–7
Primaries connected in series
for 480-volt operation.

will be identified with terminal markings of H1 and H2. The other primary winding will be identified with terminal markings of H3 and H4. The secondary winding will be identified with terminal markings of X1 and X2.

If the transformer is to be used to change a primary voltage of 240 volts into 120 volts, the two primary windings will be connected in parallel as shown in Figure 25–6. Because the two primary windings are connected in parallel, each will receive the same voltage. This produces a turns ratio of 2:1 between the primary windings and the secondary windings. If 240 volts is connected to the primary of a 2:1 ratio transformer, the secondary voltage will be 120 volts.

If the transformer is to be used to change 480 volts to 120 volts, the primary windings will be connected in series, as shown in Figure 25–7. In this connection, H2 of one primary winding is connected to H3 of the other primary winding. This series connection of the two windings produces a turns ratio of 4:1. When 480 volts is connected to the primary, 120 volts will be produced in the secondary.

The primary windings of most control transformers have leads H2 and H3 crossed, as shown in Figure 25–8. This is done to aid in the connection of the primary. For example, if it is desired to operate the transformer with the primary windings connected in parallel, a metal link is used to connect leads H1 and H3 together. Another metal link is used to connect leads H2 and H4 together,

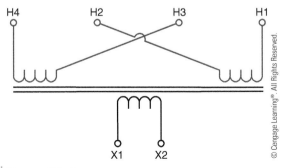

Figure 25–8
Primary leads are crossed.

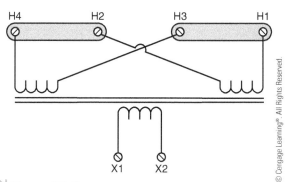

Figure 25–9
Metal links used to make a 240-volt connection.

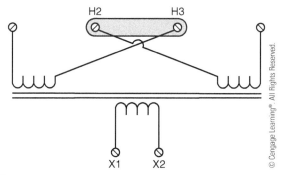

Figure 25–10
Metal link used to make a 480-volt connection.

Figure 25–11
Industrial control transformer.

Figure 25–9. Compare this lead connection with the schematic shown in Figure 25–6.

If it is desired to connect the primary windings for operation on 480 volts, terminals H2 and H3 are joined together with a metal link, Figure 25–10. Compare this connection with the schematic shown in Figure 25–7. A photograph of an industrial control transformer is shown in Figure 25–11.

TESTING THE TRANSFORMER

An ohmmeter is generally used to test the windings of the transformer. To test the transformer, check for continuity through each set of windings. For example, there should be continuity between leads H1 and H2; H3 and H4; and X1 and X2. There should be no continuity between any of the windings such as H1 and H3, or H1 and X1. Also check

for a grounded winding by testing to be sure there is no continuity between any of the windings and the case of the transformer.

The output voltage of the transformer should be tested with an AC voltmeter. If the output voltage is not close to the rated voltage, the transformer is probably defective. If the transformer is tested without a load connected to the secondary, it is normal for the secondary voltage to be slightly higher than the rated voltage. For example, a 24-volt transformer may have an output voltage as high as 28 volts without load connected to it. The voltage rating of the transformer assumes it is supplying full-rated current to the load. If the voltage is tested when the transformer is under full load, the rated voltage should be seen. Notice the use of the control transformer in the schematic shown in Figure 25–12.

Legend	
CMT	Condenser Motor Thermostat
CAP	Capacitor
CC	Compressor Contactor
CCH	Crankcase Heater
CFM	Condenser Fan Motor
Comp	Compressor
EFC	Evaporator Fan Contactor
F	Fuse
RR	Reset Relay
T	Transformer
TB	Terminal Block
TOP	Thermal Overload Protector
TS	Terminal Strip
EFM	Evaporator Fan Motor
LPC	Low-pressure Control
HPC	High-pressure Control
OL	Overload Protector

Warning

Disconnect Electrical Power
Source to Prevent Injury or
Death from Electrical Shock

Caution

Use Copper Conductors Only
to Prevent Equipment Damage

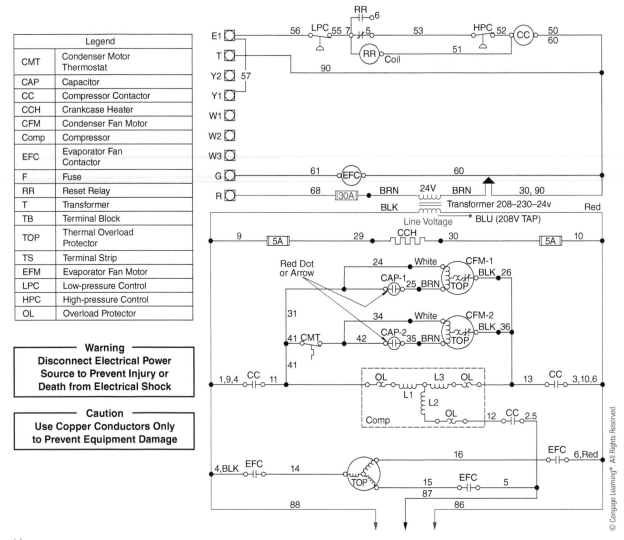

▶ **Figure 25–12**
Control transformer used to provide low voltage for the control circuit.

▷ **SUMMARY**

⊙ A transformer is a device that can change values of AC voltage and current without a change of frequency.

⊙ Isolation transformers have their primary and secondary sides physically and electrically separated from each other.

⊙ Isolation transformers operate on the principle of mutual induction.

⊙ The primary is the winding of the transformer that is connected to the incoming power line.

⊙ The secondary is the winding that is connected to the load.

⊙ All values of voltage and current in a transformer are proportional to the turns ratio.

⊙ A step-up transformer has a higher secondary voltage than primary voltage.

⊙ A step-down transformer has a lower secondary voltage than primary voltage.

KEY TERMS

control transformers
transformer

REVIEW QUESTIONS

1. What is an isolation transformer?
2. Define a step-up transformer.
3. Define a step-down transformer.
4. The primary of a transformer is connected to 120 volts AC. The secondary has a voltage of 30 volts and is connected to a resistance of 5 ohms. How much current will flow in the primary of the transformer?
5. What is the amount of control voltage used in most residential air-conditioning systems?
6. What is the amount of control voltage used in most industrial air-conditioning systems?
7. What is the color of the primary leads of most control transformers used for residential service?
8. How many primary windings are generally contained in an industrial control transformer?
9. What is the turns ratio of each of these primary windings as compared with the secondary winding?
10. When an industrial control transformer is to be operated on 480 volts, are the primary windings connected in parallel or in series?

UNIT 26 ▷ Starting Relays

OBJECTIVES

After studying this unit, the student should be able to:

- ▷ List the common types of starting relays
- ▷ Describe the operation of a hot-wire relay
- ▷ Connect a hot-wire relay in a circuit
- ▷ Describe the operation of a current relay
- ▷ Connect a current relay in a circuit
- ▷ Describe the operation of a potential relay
- ▷ Connect a potential relay in a circuit
- ▷ Describe the operation of a solid-state starting relay
- ▷ Connect a solid-state starting relay in a circuit
- ▷ Describe the operation, construction, and connection of a solid-state hard starting kit

When a split-phase motor is started, it is often necessary to disconnect the start windings when the motor reaches about 75% of full speed. In an open case motor, this job is generally done by the centrifugal switch. Some single-phase motors are hermetically sealed, however, and a centrifugal switch cannot be used. When this is the case, a **starting relay** must be used. A starting relay is located away from the motor and is used to disconnect the start windings when the motor has reached about 75% of its full speed. There are four basic types of starting relays in general use:

1. The **hot-wire relay**
2. The **current relay**
3. The **potential relay**
4. The **solid-state relay**

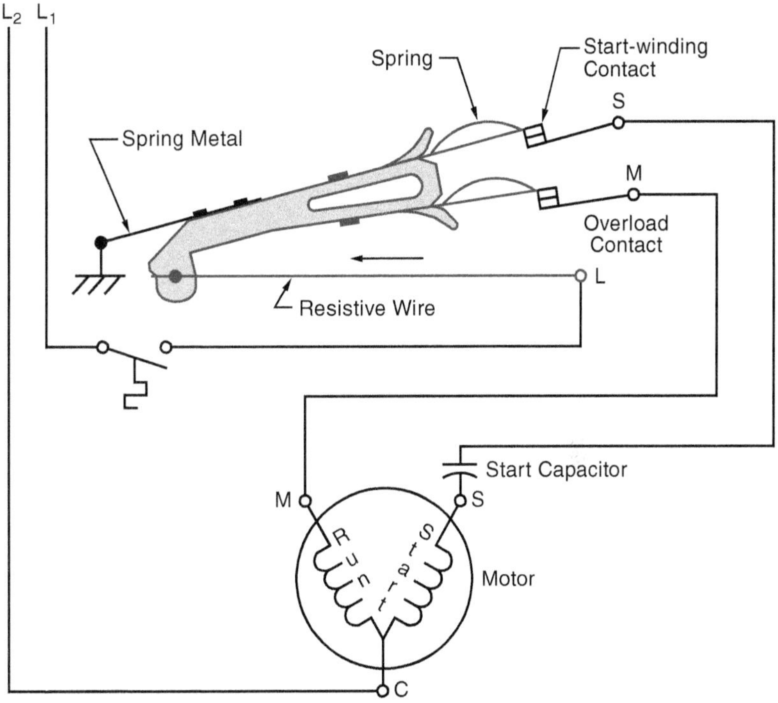

L₂ L₁

Spring — Start-winding Contact

S

Spring Metal

M

Overload Contact

L

Resistive Wire

Start Capacitor

M

S

R u n

S t a r t

Motor

C

▶ **Figure 26–1**
Hot-wire relay connection.

THE HOT-WIRE RELAY

The **hot-wire relay** is so named because it uses a length of resistive wire connected in series with the motor to sense motor current. A diagram of this type of relay is shown in Figure 26–1. When the thermostat contact closes, current can flow from line 1 to terminal L of the relay. Current then flows through the resistive wire, the movable arm, and the normally closed contacts to the run and start windings. When current flows through the resistive wire, its temperature increases. This increase of temperature causes the wire to expand in length. When the length of the resistive wire increases, the movable arm is forced to move down. As the arm moves down, tension is applied to the springs of both contacts. The relay is so designed that the start contact will snap open first. When the start winding is disconnected, the current flow to the motor will decrease. If the motor current is not excessive, the resistive wire will not expand enough to cause the run contact to open. If the current flow is excessive, however, the wire will continue to expand, and the contact connected in series with the run winding will open.

▶ **Figure 26–2**
Hot-wire type of starting relay.

Notice that this type of relay is used as both a starting relay and an overload relay. One disadvantage of the hot-wire relay is that it must be permitted to cool after each operation. A motor using this type of starting relay cannot be started in rapid succession. Figure 26–2 shows a photograph of the hot-wire type of starting relay.

Testing this relay is difficult. An ohmmeter can be used to check for continuity between the L terminal and the start and main winding terminals. To properly test this relay, an ammeter should be used to make certain that the start contact opens and disconnects the start winding. If the relay is opening on overload, the ammeter can be used to check the current draw of the motor. This will determine whether the motor is actually overloaded, or whether the relay is opening when it should not. A good rule to follow concerning starting relays is to always test them if the motor has been damaged. It makes poor business sense to damage a new motor because of not checking the starting relay.

When replacing this relay, it is necessary to use the correct replacement. Because the relay is operated by motor current, it has been designed to open its contacts when a specific amount of current flows through the circuit. The relay must therefore be matched to the characteristics of the motor it is intended to control.

THE CURRENT RELAY

The **current relay** also operates by sensing the amount of current flow in the circuit. This type of relay operates on the principle of a magnetic field instead of expanding metal. The current relay contains a coil of a few turns of large wire, and a set of normally open contacts, Figure 26–3. The coil of the relay is connected in series with the run winding of the motor, as shown in Figure 26–4. The contacts are connected in series with the start winding. When the thermostat contact closes and connects

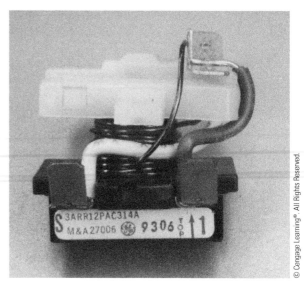

▶ **Figure 26–3**
Current type of starting relay.

power to the motor, the starting contacts of the relay are open. Because no power is applied to the start winding, the motor cannot start. This causes a current of about three times the normal full-load current to flow in the run winding. The high current flow through the coil of the start relay produces a strong magnetic field. The magnetic field is strong enough to cause the solenoid to close the starting contacts. When the starting contacts close, power is applied to the start winding and the motor begins to turn. As the motor accelerates, the current flow through the run winding decreases rapidly. When the current flow through the relay coil decreases,

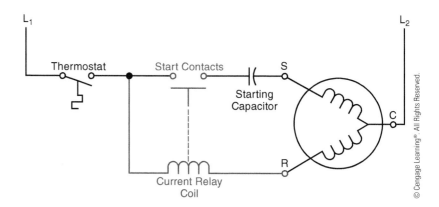

▶ **Figure 26–4**
Current relay connection.

the strength of the magnetic field becomes weaker. When the motor has reached about 75% of full speed, the magnetic field is weak enough to permit the solenoid to reopen the starting contacts. This disconnects the start winding from the circuit, and the motor continues to operate normally.

Notice that the current relay is used to disconnect the start windings only and does not provide overload protection. A motor using this type of starting relay must be provided with separate overload protection.

If it is necessary to replace this type of relay, the correct size must be used. The current relay is matched to the characteristics of the motor it is designed to be used with. This type of relay is also sensitive to the position it is mounted in. The current relay generally uses the force of gravity to open the starting contacts. When installing a new relay, it must be mounted in the correct position. If it is installed upside down, the starting contacts will be closed instead of open.

When testing this type of relay, an ohmmeter can be used to check the continuity of the contacts. When the relay is held in the correct position, the ohmmeter should show an open circuit across the contacts. If it does not, the contacts are shorted. If the relay is held upside down, the contacts should indicate continuity. The coil of the relay is generally exposed, and a visual inspection will reveal shorted windings. The best method of testing the relay is with an ammeter. If the ammeter is used to measure the current flow to the start winding, it can be seen if the motor starts and the relay contacts disconnect the start winding.

THE POTENTIAL RELAY

The **potential** (voltage) **relay** operates by sensing an increase in the voltage developed in the start winding when the motor is operating. A potential relay is shown in Figure 26–5. A schematic diagram for a potential starting relay circuit is shown in Figure 26–6. In this circuit, the potential relay is used to disconnect the starting capacitor from the circuit when the motor reaches about 75% of its full speed. SR (starting relay) coil is connected in

▶ **Figure 26–5**
Potential starting relay.

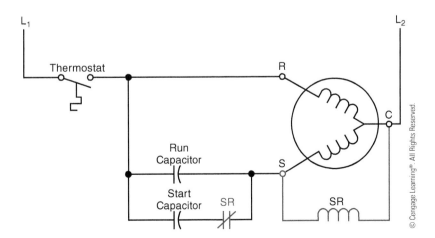

▶ **Figure 26–6**
Potential relay connection.

parallel with the start winding of the motor. A normally closed SR contact is connected in series with the starting capacitor. When the thermostat contact closes, power is applied to both the run and start windings of the motor. Notice that both the run capacitor and the start capacitor are in the circuit at this time.

The rotating magnetic field of the stator cuts through the bars of the squirrel-cage rotor and induces a current into them. The current flow in the rotor produces a magnetic field around the rotor. As the rotor begins to turn, its magnetic field cuts through the start winding and induces a voltage into it. The induced voltage causes the total voltage across the start winding to be higher than the voltage applied by the line. When the motor has accelerated to about 75% of full speed, the induced voltage in the start winding is high enough to energize SR coil. When SR coil energizes, SR contact opens and disconnects the start capacitor from the circuit, Figure 26–7.

Notice that this type of relay depends on the induced voltage created in the start winding by the magnetic field of the rotor and the run winding. The start winding acts like the secondary winding of a transformer and produces a step-up in voltage. For this reason, the amount of voltage necessary to energize the coil of a potential relay is greater than line voltage. Once the relay has been energized it can be held in by less voltage than that required to energize it. The potential relay is primarily used to disconnect the starting capacitor of a permanent split-capacitor motor, as shown in Figure 26–6. Although the potential relay is still in use, it is being replaced by the solid-state relay.

The potential type of starting relay is often used with compressors that use the permanent split-capacitor start motor. The coil of the relay can be tested for an open circuit with an ohmmeter. When the ohmmeter is connected across the coil, it should indicate continuity. The actual amount of resistance can vary from one type of relay to another. The best method for testing the starting relay is with an ammeter. If an ammeter is connected to the start capacitor, it can be seen whether the capacitor is energized when the motor is started and whether the relay disconnects it from the circuit.

AUTHOR'S NOTE: A couple of things need to be pointed out concerning the schematic in Figure 26–7. One is that there is a mistake in the terminal numbers of the potential relay. Terminal numbers 1 and 2 should be switched. The terminal shown as 2 is actually terminal 1, and the terminal shown as 1 is actually terminal 2. The second issue is not a mistake, but due to how the schematic was drawn, it could cause confusion. Notice that the low-pressure switch has been drawn upside down. The switch appears to be a normally closed switch. In reality, it is a normally open, held-closed switch. If the switch were drawn properly, it would be apparent that the switch is actually a normally open, held-closed switch. Although this schematic does contain a mistake and a component not drawn properly, I chose to retain it in the text. This is an excellent example of problems that occur in the field. The service technician must be aware that mistakes like this do happen.

SOLID-STATE STARTING RELAYS

Another type of starting relay is known as the **solid-state starting relay**, Figure 26–8. This relay is intended to replace the current-type starting relay and has several advantages over the current relay. Two of these advantages are listed here:

1. The solid-state relay contains no moving parts and no contacts, which can become burned or pitted.
2. The solid-state relay can be used to replace almost any current relay. This interchangeability makes it possible for the service technician to stock only a few solid-state relays instead of a large number of current relays.

The solid-state starting relay is actually an electronic component known as a **thermistor**. A thermistor is a device that exhibits a change of resistance with a change of temperature. This particular thermistor has a positive coefficient of resistance, which means that the resistance of the device increases with an increase of temperature.

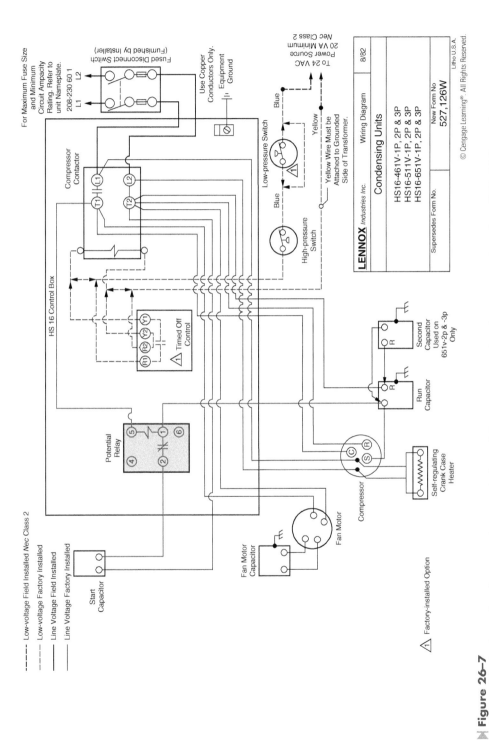

▲ Figure 26-7
A potential relay is used to disconnect the start windings of the compressor when the motor reaches about 75% of full speed.

The schematic diagram in Figure 26–9 illustrates the connection for a solid-state starting relay. Notice that this is the same basic connection used for the connection of a current starting relay, Figure 26–4. The solid-state relay, however, does not contain a coil or contacts. When the solid-state relay is used, a current path exists between the line connection terminal and the terminal marked M for MAIN winding. The thermistor is connected between the line connection and the terminal marked S for START winding.

When power is first applied to the circuit, the thermistor has a relatively low resistance. This permits current to flow through both the start and run windings of the motor. The temperature of the thermistor increases because of the current flowing through it. The increase of temperature causes the

resistance to change from a very low value of 3 or 4 ohms to several thousand ohms. This increase of resistance is very sudden and has the effect of opening a set of contacts connected in series with the start winding. Although the start winding is never completely disconnected from the power line, the amount of current flow through it is very small, typically 0.03 to 0.05 amp, and does not affect the operation of the motor. This small amount of **leakage current** maintains the temperature of the thermistor and prevents it from returning to a low resistance. After power has been disconnected from the motor, a cooldown period of about 2 minutes should be allowed before restarting. This cooldown period is needed for the thermistor to return to a low value of resistance.

TESTING THE SOLID-STATE STARTING RELAY

A continuity test can be made with an ohmmeter. If the probes of an ohmmeter are connected to the M and S terminals of the relay, a low value of resistance, typically 2 to 5 ohms, should be seen. The most accurate test is made by connecting the relay in the motor circuit. A clamp-on ammeter set to its lowest scale can be used to measure the current in the start winding. After the motor has been started, the ammeter should give an indication very close to zero amps.

SOLID-STATE HARD STARTING KIT

Another device that uses a solid-state relay, the **solid-state hard starting kit,** is shown in Figure 26–10. This device is intended to increase

Figure 26–8
Solid-state starting relay.

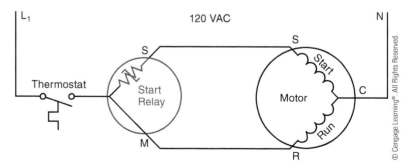

Figure 26–9
Solid-state starting relay connection.

Figure 26–10
Hard starting kit increases starting torque of permanent split-capacitor motors.

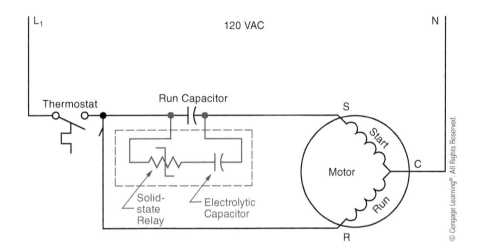

Figure 26–11
Hard starting kit.

the starting torque of a permanent split-capacitor motor. The kit contains a solid-state relay and an AC electrolytic capacitor similar to those used as the starting capacitor for a capacitor-start induction-run motor. The kit connects directly across the terminals of the existing run capacitor, as shown in Figure 26–11. When the thermostat contact closes and connects power to the motor circuit, the resistance of the solid-state relay is very low. A current path exists through the run winding, run capacitor, solid-state relay, electrolytic capacitor, and start winding. Because the run capacitor and electrolytic capacitor are connected in parallel, their values of capacitance add, providing extra capacitance to

the motor during the starting period. The current flowing through the solid-state relay and electrolytic capacitor causes the temperature of the relay to increase, resulting in an increase in resistance. The increased resistance reduces the current flow through the electrolytic capacitor to a very low value. This has the effect of disconnecting the electrolytic capacitor from the circuit. The leakage current through the relay and electrolytic capacitor prevents the relay from returning to a low value of resistance. After power has been disconnected from the motor circuit, a cooldown period of 2 to 3 minutes should be given to permit the solid-state relay to return to a low value of resistance.

SUMMARY

- ⊙ Starting relays are used to disconnect the start winding of a hermetically sealed split-phase motor.
- ⊙ The four basic types of starting relays are:
    - A. The hot-wire relay
    - B. The current relay
    - C. The potential relay
    - D. The solid-state relay
- ⊙ The hot-wire relay senses motor current by connecting a piece of resistive wire in series with the motor. The motor current heats the wire, causing it to expand and open a set of contacts.
- ⊙ Hot-wire relays can also provide overload protection for the motor.
- ⊙ Current relays use a coil connected in series with the motor run winding. The magnetic field of the coil is used to close a set of contacts connected in series with the start winding.
- ⊙ The potential relay is used primarily with permanent split-capacitor motors.
- ⊙ The coil of the potential relay is connected in parallel with the motor start winding.
- ⊙ The contact of the potential relay is connected in series with the starting capacitor.
- ⊙ Solid-state starting relays are generally used to replace current starting relays.
- ⊙ Solid-state starting relays use a thermistor, which rapidly changes its resistance when heated.
- ⊙ Solid-state hard starting kits generally consist of an AC electrolytic capacitor and a solid-state starting relay that connect in parallel with the existing run capacitor.

KEY TERMS

current relay
hot-wire relay
leakage current

potential relay
solid-state hard starting kit
solid-state starting relay

starting relay
thermistor

REVIEW QUESTIONS

1. What are the four types of starting relays?
2. On what type of motor is it necessary to use a starting relay?
3. What principle is used to operate the hot-wire relay?
4. What principle is used to operate the current relay?
5. What type of starting relay does not sense motor current to operate?
6. What type of starting relay can be used for overload protection for the motor?
7. What type of motor can the potential relay be used with?

8. Is the start contact of a hot-wire relay open or closed when power is first applied to the motor?

9. Is the start contact of a current relay open or closed when power is first applied to the motor?

10. Refer to the circuit shown in Figure 26–4. What would happen if the coil of the current relay were open when the thermostat connected power to the motor circuit?

▷ TROUBLESHOOTING QUESTIONS

Refer to the schematic shown in Figure 26–7 to answer the following questions.
NOTE: *The "TIMED OFF CONTROL" shown in the schematic is a short-cycle timer, which will be discussed in Unit 36.*

1. What voltage is used to operate the coil of the compressor contactor?

A. 208 VAC

B. 230 VAC

C. 60 VAC

D. 24 VAC

2. This schematic does not show the thermostat connection, which is relatively common with air-conditioning schematics. In which wire would the thermostat contact normally be connected?

A. L1

B. L2

C. The blue wire

D. The yellow wire

3. Which of the following components is not controlled by the operation of the compressor contactor?

A. The self-regulating crankcase heater

B. The fan motor

C. The compressor

D. All circuit components are controlled by the compressor contactor.

4. To which line is the common of the compressor connected?

A. The blue wire of the 24-volt circuit

B. The yellow wire of the 24-volt circuit

C. Line 1 of the main power

D. Line 2 of the main power

5. Assume that this unit is a model 651V-3P. How many capacitors are connected to the compressor start winding during the initial starting period?

A. 1

B. 2

C. 3

D. None

UNIT 27 ▷

Variable-Speed Motor Control

OBJECTIVES

After studying this unit, the student should be able to:

▷ Describe different types of variable-speed motors

▷ Discuss autotransformer control

▷ Connect an autotransformer speed controller in a circuit

▷ Discuss the use of a triac to control motor speed

▷ Connect a triac speed controller in a circuit

▷ Discuss the operation of a series impedance speed control

▷ Connect a series impedance speed controller in a circuit

The use of small **variable-speed motors** has increased greatly in the last few years. These motors are commonly used to operate light loads such as ceiling fans and blower motors. There are two types of motors used for these applications, the shaded-pole and the permanent split-capacitor motor. These motors are used because they operate without having to disconnect a set of start windings with a centrifugal switch or starting relay. Motors intended to be used in this manner are wound with high-impedance stator windings. The high impedance of the stator limits the current flow through the motor when the speed of the rotor is decreased. Speed control for these motors is accomplished by controlling the amount of voltage applied to the motor or by inserting impedance in series with the stator winding.

VARIABLE-VOLTAGE CONTROL

The amount of voltage applied to the motor can be controlled by several methods. One method is to use an **autotransformer** with several taps, Figure 27–1. This type of controller has several steps of speed control. Notice that the applied voltage, 120 volts in this illustration, is connected across the entire transformer winding. When the rotary switch is moved to the first tap, 30 volts is applied to the motor. This produces the lowest motor speed for this controller. When the rotary switch is moved to the second tap, 60 volts is applied to the motor. This provides an increase in motor speed. When the switch has been moved to the last position, the full 120 volts is applied to the motor, operating it at the highest speed.

Another type of variable-voltage control uses a triac to control the amount of voltage applied to the motor, Figure 27–2. This type of speed control provides a more linear control because the voltage can be adjusted from zero to the full applied voltage. At first appearance, many people assume this controller to be a **variable resistor** connected in series with the motor. A variable resistor large enough to control even a small motor would produce several hundred watts of heat and could never be mounted in a switch box. The variable resistor in this circuit is used to control the amount of phase shift for the triac. The triac controls the amount of voltage applied to the motor by turning on at different times during the AC cycle.

A triac speed control is very similar to a triac light dimmer used in many homes. A light dimmer, however, should never be used as a motor speed controller. Triac light dimmers are intended to be used with resistive loads such as incandescent lamps. Light dimmer circuits sometimes permit one half of the triac to start conducting before the other half. The wave form shown in Figure 27–3 illustrates this condition. Notice that only part of the positive half of the waveform is being conducted to the load. Because only positive voltage is being applied to the load, it is DC. Operating a resistive load, such as an incandescent lamp with DC, will do no damage. Operating an inductive load such as the winding of a motor can do a great deal of damage, however. When direct current is applied to a motor winding, there is no inductive reactance to limit the current. The actual wire resistance of the stator is the only current-limiting factor. The motor winding or

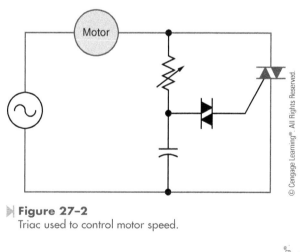

Figure 27-2
Triac used to control motor speed.

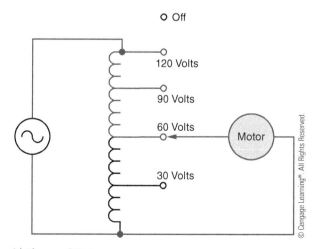

Figure 27-1
Autotransformer controls motor voltage.

Figure 27-3
Triac conducts only the positive half of the waveform.

the controller can easily be destroyed if direct current is applied to the motor. For this reason, only triac controllers designed for use with inductive loads should be used for motor control. A photograph of a triac speed controller is shown in Figure 27–4.

SERIES IMPEDANCE CONTROL

Another common method of controlling the speed of small AC motors is to connect impedance in series with the stator winding. This is the same basic method of control used with multispeed fan motors.

The circuit in Figure 27–5 shows a tapped inductor connected in series with the motor. When the motor is first started, it is connected directly to the full voltage of the circuit. As the rotary switch is moved from one position to another, steps of inductance are connected in series with the motor. As more inductance is connected in series with the stator, the amount of current flow decreases. This produces a weaker magnetic field in the stator. **Rotor slip** increases because of the weaker magnetic field and causes the motor speed to decrease. A photograph of this type of controller is shown in Figure 27–6.

▶ **Figure 27–4**
Triac speed control.

▶ **Figure 27–6**
Fan speed control using a tapped inductor connected in series with the motor.

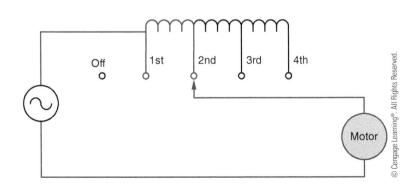

▶ **Figure 27–5**
Series inductor changes impedance of circuit.

SUMMARY

▶ The two main types of single-phase AC motors used for variable-speed control are the permanent split-capacitor motor and the shaded-pole motor.

▶ Permanent split-capacitor motors and shaded-pole motors are generally used for variable-speed control because they do not have to disconnect the start windings when the motor reaches a certain speed.

▶ Variable-speed control for small motors is accomplished by controlling the amount of voltage applied to the motor or by inserting impedance in series with the motor.

▶ Two common methods used to control the voltage applied to the motor are autotransformer control and triac control.

▶ Only triac controllers designed for use as motor speed controllers should be used for motor speed control.

▶ Series impedance control is accomplished by connecting a tapped inductor in series with the motor.

KEY TERMS

autotransformer
rotor slip

variable resistor
variable-speed motors

REVIEW QUESTIONS

1. What two types of small AC motors are used with variable-voltage speed control?

2. Why are these two types of motors used?

3. Name two methods of variable-voltage control for small AC motors.

4. What solid-state device is used to control the voltage applied to the motor?

5. Why is it necessary to use only controllers designed for use with inductive loads?

6. Name a method other than variable voltage used to control the speed of small AC motors.

UNIT 28

The Defrost Timer

OBJECTIVES

After studying this unit, the student should be able to:

▷ Describe the construction of a defrost timer

▷ Discuss the operation of a continuous run and cumulative compressor run timer

▷ Connect a defrost timer for continuous run

▷ Connect a defrost timer for cumulative compressor run

▷ Discuss the operation of commercial defrost timers

▷ Perform an ohmmeter test on a defrost timer

Many of the refrigeration appliances used in the home are "frost-free." The frost-free appliance could more accurately be termed "automatic defrost." The brain of the frost-free appliance is the **defrost timer**. The job of this timer is to disconnect the compressor circuit and connect a **resistive heating element** located near the **evaporator** at regular time intervals. The defrost heater is thermostatically controlled and is used to melt any frost formation on the evaporator. The defrost heater is permitted to operate for some length of time before the timer disconnects it from the circuit and permits the compressor to operate again.

TIMER CONSTRUCTION

The defrost timer is operated by a single-phase synchronous motor like those used to operate electric wall clocks, Figure 28–1. The contacts are operated

▶ **Figure 28–1**
Defrost timer.

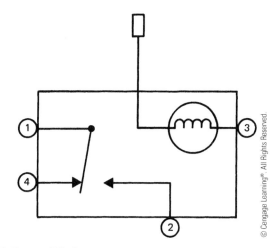

▶ **Figure 28–2**
Schematic of a defrost timer.

by a **cam** that is gear driven by the clock motor. A schematic drawing of the timer is shown in Figure 28–2. Notice that terminal 1 is connected to the common of a single-pole double-throw switch. Terminals 2 and 4 are connected to stationary contacts of the switch. In the normal operating mode, the switch makes connection between contacts 1 and 4. When the defrost cycle is activated, the contact changes position and makes connection between terminals 1 and 2. Terminal 3 is connected to one lead of the motor. The other motor lead is

brought outside the case. This permits the timer to be connected in one of two ways:

1. The **continuous run timer**
2. The **cumulative compressor run timer**

It should be noted that the schematic drawing can be a little misleading. In the schematic shown, the timer contact can only make connection between terminals 1 and 4, or terminals 1 and 2. In actual practice, a common problem with this timer is that the movable contact becomes stuck between terminals 4 and 2. This causes the compressor and defrost heater to operate at the same time.

THE CONTINUOUS RUN TIMER

The schematic for the continuous run timer is shown in Figure 28–3. Notice in this circuit that the pigtail lead of the motor has been connected to terminal 1, and that terminal 1 is connected directly to the power source. Terminal 3 is connected directly to the neutral. This places the timer motor directly across the power source, which permits the motor to operate on a continuous basis.

Figure 28–4 shows the operation of the timer in the compressor run cycle. Notice there is a current path through the timer motor and a path through the timer contact to the thermostat. This permits power to be applied to the compressor and evaporator motor when the thermostat closes.

Figure 28–5 shows the operation of the circuit when the timer changes the contact and activates the defrost cycle. Notice there is still a complete circuit through the timer motor. When the timer contact changes position, the circuit to the thermostat is open and the circuit to the defrost heater is closed. The heater can now melt any frost accumulation on the evaporator. At the end of the defrost cycle, the timer contact returns to its normal position and permits the compressor to be operated by the thermostat.

THE CUMULATIVE COMPRESSOR RUN TIMER

The cumulative compressor run timer circuit gets its name from the fact that the timer motor is permitted to operate only when the compressor is in operation and the thermostat is closed. The schematic for this circuit is shown in Figure 28–6. Notice that the

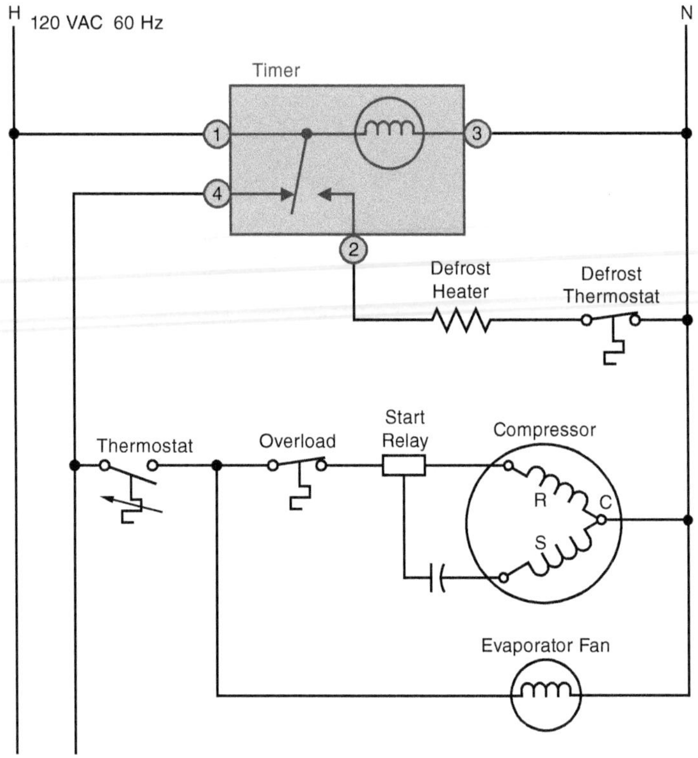

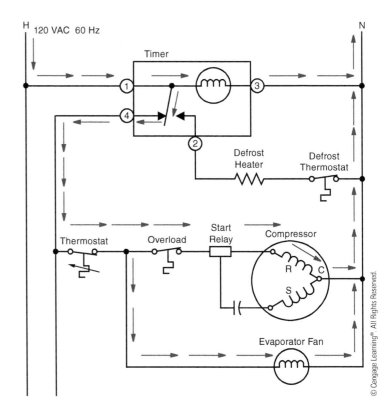

▶ **Figure 28–3**
Schematic of a defrost timer used in a
continuous run circuit.

▶ **Figure 28–4**
Current path during cooling operation.

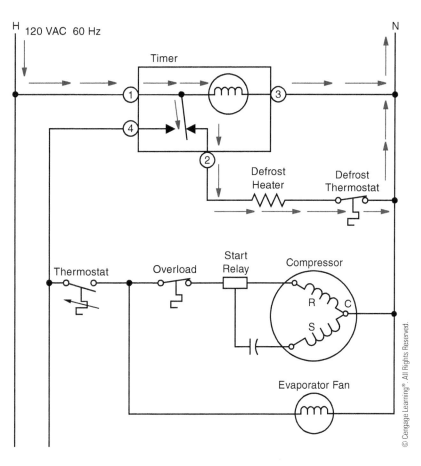

▶ **Figure 28-5**
Current path during defrost
operation.

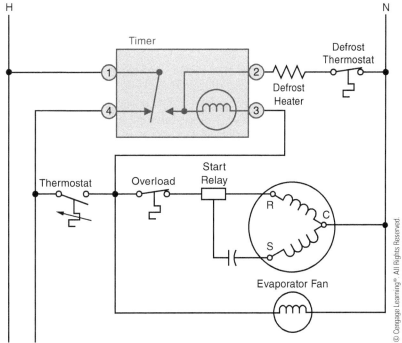

▶ **Figure 28-6**
Defrost timer connected in
a cumulative compressor run circuit.

H ← → Timer → Defrost Heater N

Figure 28–7
Current path during the cooling cycle.

pigtail lead of the clock motor has been connected to terminal 2 instead of terminal 1. Figure 28–7 shows the current path during compressor operation. The timer contact is making connection between terminals 1 and 4. This permits power to be applied to the thermostat. When the thermostat contact closes, current is permitted to flow through the compressor motor, the evaporator fan motor, and the defrost timer motor. In this circuit, the timer motor is connected in series with the defrost heater. The operation of the timer motor is not affected, however, because the impedance of the timer motor is much greater than the resistance of the heater. For this reason almost all the voltage of this circuit is dropped across the timer motor. The impedance of the timer motor also limits the current flow through the defrost heater to such an extent that it does not become warm.

Figure 28–8 shows the current path through the circuit when the defrost cycle has been activated. Notice in this circuit that the defrost heater is connected directly to the power line. This permits the

heater to operate at full power and melt any frost accumulation on the evaporator. There is also a current path through the timer motor and run winding of the compressor motor. In this circuit, the timer motor is connected in series with the run winding of the compressor. As before, the impedance of the timer motor is much greater than the impedance of the run winding of the compressor. This permits almost all the voltage in this circuit to be applied across the timer motor. At the end of the defrost cycle, the timer contact returns to its normal position and the compressor is permitted to operate.

TESTING THE TIMER

An ohmmeter can be used to check the continuity of the contacts and the motor winding. However, to really test the timer for operation takes time. The cam can be manually turned to the position so that the defrost cycle is turned on. This can be checked with a voltmeter to determine when full circuit voltage is applied to terminal 2. It is then necessary to

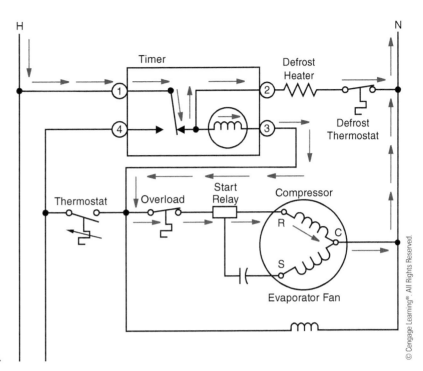

H   N

Timer

Defrost
Heater

Defrost
Thermostat

Start
Relay

Thermostat   Overload   Compressor

R
C
S

Evaporator Fan

▶ **Figure 28–8**
Current path during the defrost cycle.

wait long enough for the timer to open the contact to the defrost heater and reconnect the compressor circuit. If the thermostat is closed, the compressor will start when the timer contact changes position. This test shows that the timer motor is operating and that the contact does change position.

COMMERCIAL DEFROST TIMERS

Many large commercial refrigeration units often use a separate **timer clock** to control the defrost cycle. This has several advantages over the previously discussed defrost timer. When this method is used, the timer clock is connected directly across the power line, as shown in Figure 28–9. This separates the operation of the timer from the operation of the compressor. In this way, the defrost cycle can be started during periods when the unit is in minimum use.

Timers of this type, Figure 28–10, generally have two timed settings. One determines the time of day or night the timer turns on. The second setting determines how long the timer is permitted to remain on. The timer shown in this example can be started on even-numbered hours of the day or night. The center knob sets how long the contacts are energized before they return to their normal position. Once turned on, the contacts can be set to remain in their energized position for a minimum of 2 minutes to a maximum of 120 minutes.

This timer has a separate timer release solenoid incorporated into its design. When the timer release solenoid is energized, it causes the contacts to return to their normal reenergized position immediately. This permits the action of some type of external limit switch, such as a temperature or pressure switch, to terminate the defrost cycle.

P/N S814100 SUPCO
COMMERCIAL DEFROST CONTROL
ELECTRIC HEAT DEFROSTING

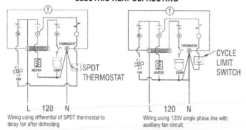

Wiring using differential of SPDT thermostat to delay fan after defrosting

Wiring using 120V single phase line with auxiliary fan circuit.

CAUTION - RISK OF ELECTRICAL SHOCK HAZARD!
DISCONNECT ALL POWER AT MAIN PANEL BEFORE REMOVING THE INSULATOR. MORE THAN ONE DISCONNECT SWITCH MAY BE REQUIRED TO DE-ENERGIZE THE DEVICE FOR SERVICING. MAKE SURE TO REPLACE THE INSULATOR AFTER WIRING.

USE COPPER CONDUCTORS ONLY
SUPPLY CONNECTIONS USE THE GAUGE SPECIFIED, SUITABLE FOR 75°C (167°F).

MINIMUM GAUGE WIRE	MAXIMUM LOAD
8	40 AMPS
10	30 AMPS
12	20 AMPS
14	10 AMPS

INSTALLING AND OPERATING DIRECTIONS
PLACE START PINS IN OUTER (24 HOUR) DIAL AT THE TIME OF DAY THAT THE SWITCH CONTACTS ARE TO BE REVERSE FROM SHOWN BELOW WHEN THE DIAL PINS ARE OPPOSITE TIME POINTER.
CAUTION: LEAVE AT LEAST 1 HOLE BETWEEN EACH ADJACENT PIN.
TO SET BACK UP DEFROST TERMINATION: PUSH DOWN AND ROTATE POINTER ON INSIDE (2 HOUR) DIAL UNTIL IT IS OPPOSITE THE DESIRED TIME.

TO SET THE TIME OF DAY: GRASP THE KNOB IN THE CENTER OF THE INNER (2 HOUR) DIAL AND ROTATE IT IN A COUNTER CLOCKWISE DIRECTION. THIS WILL REVOLVE THE OUTER DIAL. LINE UP THE CORRECT TIME OF DAY ON THE OUTER DIAL WITH THE TIME POINTER. **DO NOT TRY TO SET THE TIME CONTROL BY GRASPING THE OUTER DIAL. ROTATE THE INNER DIAL ONLY!**
TO REMOVE THE TIMER MOVEMENT FROM CASE: PRESS THE SPRING CLIP AT RIGHT CENTER OF CASE AND LIFT THE TIMER MOVEMENT OUT.
TO REPLACE THE TIMER MOVEMENT IN CASE: INSERT TABS ON THE LEFT SIDE OF THE TIME SWITCH PLATE INTO HOLES INSIDE OF THE CASE. PUSH DOWN AND SNAP INTO PLACE.
FOR REPLACEMENT OF THIS TIMER CONTACT
SUPCO @ 800 - 333 - 9125.

MAXIMUM CONTACT RATING

40 AMPS	NON - INDUCTIVE	120 VAC
2 HP		120 VAC
690 VA	PILOT DUTY	120 VAC

TIMING MOTOR: 120V 60 Hz

SEALED UNIT PARTS CO., INC.
PO BOX 21, 2230 LANDMARK PLACE, ALLENWOOD, NJ 08720 USA
Phone: 732-223-6644 • Fax: 732-223-1617
www.supco.com • info@supco.com

▶ **Figure 28-9**
Schematic diagram of a commercial defrost timer.

▶ **Figure 28-10**
Commercial defrost timer.

SUMMARY

- Frost-free appliances use a defrost timer to control the operation of the defrost cycle.
- The function of the defrost timer is to disconnect the compressor from the circuit and connect a resistive heater located near the evaporator at regular intervals.
- The defrost timer is a cam-operated timer powered by a small single-phase synchronous motor similar to an electric clock motor.
- Two basic connections for a defrost timer are:

    A. The continuous run timer.

    B. The cumulative compressor run timer.
- The motor of the continuous run timer is connected directly to the power line and runs at all times.
- The motor of the cumulative compressor run timer is connected in such a manner that it runs only during the time the compressor motor is in operation.
- Commercial defrost timers often use a separate time clock to control the defrost cycle.

KEY TERMS

cam
continuous run timer
cumulative compressor
  run timer

defrost timer
evaporator
resistive heating element
timer clock

REVIEW QUESTIONS

1. What type of motor is used to operate the timer?
2. Why is one of the motor leads brought outside the timer?
3. Name two ways of connecting the defrost timer.
4. What function does the defrost heater perform?
5. To which terminal is the pigtail lead of the timer motor connected if the timer is to operate continuously?

UNIT 29 ▶ The Thermostat

OBJECTIVES

After studying this unit, the student should be able to:

▷ Define a thermostat

▷ Describe the operation of bimetal thermostats

▷ Discuss the operation of contact-type and mercury-contact thermostats

▷ Discuss the operation of heating and cooling thermostats

▷ Describe the operation of the fan switch

▷ Discuss the operation of the heat anticipator and the cooling anticipator

▷ Discuss the operation of line voltage, programmable, staging, and differential thermostats

▷ Connect a thermostat in a circuit

Thermostats are temperature-sensitive switches. They use a variety of methods to sense temperature, and can be found with different **contact** arrangements. Some thermostats are designed to be used with low-voltage systems, generally 24 volts; and others are designed to be connected directly to line voltage and operate motors and heating units. The advantage of low-voltage thermostats is that they are more economical and safer to use inside the home.

BIMETAL THERMOSTATS

One of the most common methods of sensing temperature is with a bimetal strip. When used as the temperature-sensing element of a thermostat, the bimetal strip is generally bent in a spiral that resembles a clock spring. If a contact is attached to the end of the strip and another contact is held stationary, a thermostat is formed, Figure 29–1. A small permanent magnet is used to provide a snap action for the contacts.

This type of thermostat is inexpensive and has the advantage of not having to be mounted in a level position. The greatest enemy of an open-contact thermostat is dirt. This is especially true for thermostats designed for low-voltage operation. If poor thermostat contact is suspected, the contacts should be cleaned. This can be done with a strip of hard paper, such as typing paper, and alcohol. Soak a strip of hard paper in alcohol and place the strip between the contacts. Close the contacts and draw the strip through the closed contacts. This will generally remove any accumulation of dirt and oil. After cleaning, the contacts should be buffed to remove any alcohol residue. This can be done by drawing a piece of dry hard paper through the contacts several times. This type of thermostat is shown in Figure 29–2.

To avoid the problem of dirty contacts, the contacts of some thermostats are enclosed inside a glass tube, Figure 29–3. Because the contacts are enclosed in glass, they are stationary. The bimetal strip is attached to a permanent magnet instead of a contact. In the case of a double-acting thermostat, there are two magnets attached to the bimetal strip. When the magnet is close enough to the glass tube, it is attracted to the metal contacts and causes the contacts to close with a snap action.

▶ **Figure 29–2**
Open-contact thermostat.

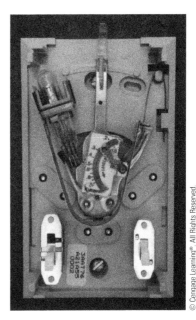

▶ **Figure 29–3**
Contacts are enclosed inside a glass tube.

MERCURY CONTACT THERMOSTAT

Another type of contact used with the bimetal type of thermostat is the mercury contact. In this type of thermostat, a small pool of mercury is sealed inside

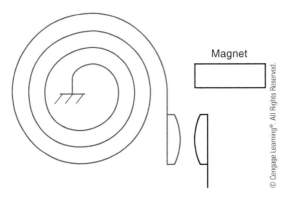

▶ **Figure 29–1**
Contacts operated by a bimetal strip.

Magnet

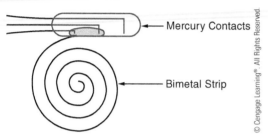

Figure 29–4
Mercury contacts.

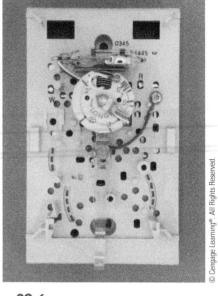

Figure 29–6
Mercury thermostat.

Figure 29–5
Mercury-type thermostat.

a glass container. A set of contacts is also sealed in the glass. Most mercury-type contacts are made to be single-pole double-throw, which means there are a common terminal, a normally open terminal, and a normally closed terminal. Figure 29–4 illustrates this type of contact. Notice in this example that the pool of mercury makes connection with the common terminal, located in the center, and the normally closed terminal. If the glass bulb is tilted in the other direction, the mercury will flow to the opposite end and make connection between the common terminal and the normally open contact.

The mercury contact has the advantage of being sealed in glass and not subjected to dirt and oil. When this type of contact is used with a bimetal strip, it is generally mounted as shown in Figure 29–5. This type of thermostat uses the weight of the mercury to provide a snap action for the contact instead of a magnet. When the bimetal strip has turned far enough to permit the mercury to flow from one end of the glass bulb to the other, the weight of the mercury prevents any spring action of the bimetal strip from snapping the contact open. A mercury thermostat, however, must be mounted in a level position if it is to operate properly. This

can sometimes be a problem in homes that do not remain level. A mercury-type thermostat is shown in Figure 29–6.

Mercury-type thermostats are particularly sensitive to the way they are mounted. If they are to operate correctly, they must be mounted level. The manufacturer's instructions generally show how the thermostat should be checked with a leveling device.

HEATING AND COOLING THERMOSTATS

Many thermostats are designed to be used for both heating and cooling applications. This can be done with thermostats that contain both a normally open and a normally closed contact. A simple schematic diagram of this type of thermostat is shown in Figure 29–7. Notice the thermostat contact is a single-pole double-throw type. The selector switch is a double-pole type. The dashed line indicates mechanical intertie. With the selector switch in the position shown, the thermostat is being used for heating. If the selector switch is changed, the

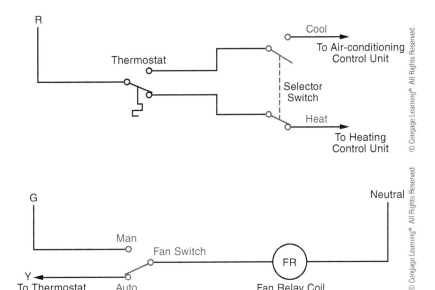

 Figure 29–7
Dual operation of a thermostat.

▶ **Figure 29–8**
Fan switch.

bottom movable contact will break connection with its stationary contact, and the top movable contact will make connection with its stationary contact. Notice that changing this switch will also change the sense of the thermostat. In the heating position, the thermostat activates the heating unit when the contact closes because of a decrease in temperature. In the cooling position, the thermostat activates the air-conditioning unit when the contact makes connection because of an increase in temperature.

THE FAN SWITCH

Many thermostats are designed to permit manual control of the blower fan. This is done to permit the blower fan to be operated separately. Some people find it desirable to operate the blower fan continuously to provide circulation of air throughout the building. This is especially true for buildings equipped with electronic air cleaners (precipitators) or for buildings that must remove undesirable elements such as smoke in an office building or night club. A schematic diagram of this type of circuit is shown in Figure 29–8. The **fan switch** is a single-pole double-throw switch. When the switch is in one position, it permits the fan relay to be controlled

by the thermostat. If the switch is thrown in the opposite direction, the fan relay is connected directly to the control voltage.

THE HEAT ANTICIPATOR

The **heat anticipator** is a small resistance heater located near the bimetal strip. The function of this heater is to slightly preheat the bimetal strip and prevent overrun of the heating system. For example, many heating systems, such as fuel oil or gas, operate by heating a metal container called a heat exchanger. When the temperature of the heat exchanger reaches a high enough level, a thermostatically controlled switch causes the blower to turn on and blow air across the heat exchanger. The moving air causes heat to be removed from the heat exchanger to the living area. When the thermostat is satisfied, the heating unit is turned off. The blower will continue to operate, however, until the excess heat has been removed from the heat exchanger.

Now assume that the thermostat has been set for a temperature of 75°F. If the heating unit is permitted to operate until the temperature reaches 75°F, the final temperature of the living area may be from

3° to 5° higher than the thermostat setting by the time enough heat has been removed from the heat exchanger to cause the blower to turn off.

If the heat anticipator has been properly set, however, it will cause the thermostat to turn off several degrees before the room temperature has reached the thermostat setting. This permits the excess heat of the heat exchanger to raise the temperature to the desired level without overrunning the thermostat setting.

The setting of the heat anticipator is controlled by a sliding contact. There are markings such as .2, .25, .3, .35, and .4. The sliding contact is generally set at the number that corresponds to the current rating of the control system. The current rating can generally be located in the service information or on the control unit itself. The heat anticipator does not have to be set at that position, however. The service technician should set it to operate the unit for longer or shorter periods depending on the desires of the customer.

THE COOLING ANTICIPATOR

A device that operates in a similar manner to the heat anticipator is the **cooling anticipator**. The cooling anticipator is a resistive heating element that operates in an opposite sense to the heat anticipator. The cooling anticipator operates while the thermostat contacts are open and the air-conditioning unit is not running. The cooling anticipator heats the thermostat slightly and causes it to close its contacts before the ambient temperature increases enough to close them.

The circuit shown in Figure 29–9 displays the current path of the heat anticipator for a heating and cooling thermostat during the heating cycle. In this mode of operation, current flows through the heat anticipator while the thermostat contact is closed and the heating unit is in operation.

In Figure 29–10, the thermostat has been switched to the cooling mode. Notice that a current path exists through the cooling anticipator when the

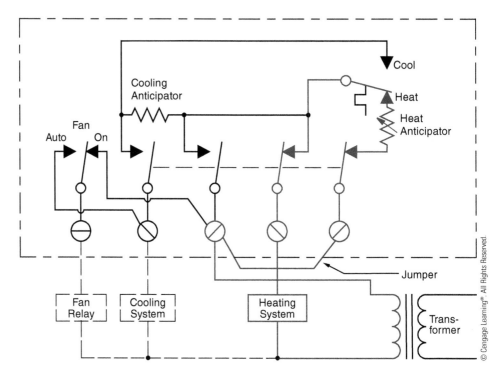

▶ **Figure 29–9**
Current path for heat anticipator.

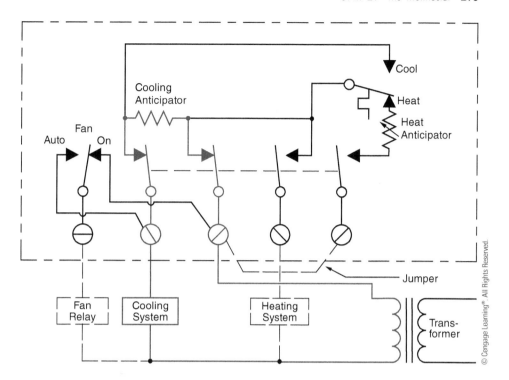

▶ **Figure 29–10**
Current path for
cooling anticipator.

thermostat contact is open and the air-conditioning unit is not in operation. When the thermostat contact closes, a low-resistance path exists around the cooling anticipator. This stops the flow of current through it while the air-conditioning unit is in operation.

LINE VOLTAGE THERMOSTATS

Line voltage thermostats are generally used to control loads such as blower fans and heating elements. This means that the thermostat must contain contacts that are capable of handling the current needed to operate these loads without an intervening relay. This type of thermostat is shown in Figure 29–11. Many of these thermostats use the pressure of refrigerant in a sealed system to sense temperature, Figure 29–12. When the temperature increases, the pressure in the system increases also. The increase in pressure causes the bellows to expand. When the bellows expands far enough, it activates a set of spring-loaded contacts.

Although line voltage thermostats can be used for many applications, they do not contain a heat anticipator and are not as accurate as low-voltage thermostats.

▶ **Figure 29–11**
Line voltage thermostat.

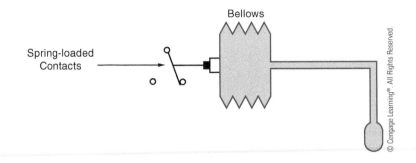

▶ **Figure 29-12**
Thermostat contacts are operated by pressure.

PROGRAMMABLE THERMOSTATS

The term *programmable* is a catch word that has taken on many meanings. In the case of thermostats, the term *programmable* generally means that a thermostat can be set to operate at different temperature settings at different times. **Programmable thermostats** range in complexity from units that use a simple time clock to units that are operated by integrated circuits (ICS) and permit the temperature to be set to any desired level at any desired time. A programmable thermostat is shown in Figure 29–13. This thermostat uses a quartz-operated time clock and two separate thermostat units. The time clock is used to operate a switch. The setting of the clock determines the position of the switch at any particular time. A schematic for this type of circuit is shown in Figure 29–14. Notice that

▶ **Figure 29-13**
Programmable thermostat.

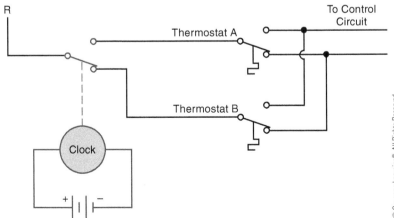

▶ **Figure 29-14**
Schematic for a programmable thermostat.

the position of the clock-operated switch determines which thermostat is used to control the system. To understand the operation of this system, assume that thermostat A has been set for a temperature of 95°F, and thermostat B is set for 75°F. The time clock has been set to permit thermostat A to control the air-conditioning system when there is no one in the residence. One hour before people are to return home, the time clock changes the contact and thermostat B is used to control the system. Because thermostat B has been set for 75°F, the residence will have been cooled to that temperature when the people arrive.

The programmable thermostat can reduce energy consumption by maintaining a desired temperature only during the hours the dwelling is occupied. The temperature can be maintained at an uncomfortable level the rest of the time, which permits the air-conditioning unit to operate much less.

Electronic Programmable Thermostats

Many programmable thermostats are solid state and do not contain mechanical contacts. These units generally employ electronic devices such as thermistors to sense temperature. They can be programmed for several different conditions or events. Many permit the thermostat to be programmed differently for each day. The temperature can be set differently for different times of the day. Most electronic thermostats provide a variety of functions and options that are set by the homeowner. An electronic thermostat is shown in Figure 29–15.

STAGING THERMOSTATS

Staging thermostats are similar to programmable thermostats in that they contain two separate sets of contacts. Unlike the programmable thermostat, however, the staging thermostat contains only one bimetal strip, which is used to control the action of both sets of contacts. One set of contacts is designed to operate slightly behind the other set. A good example of how a staging thermostat is used can be found in a heat-pump system. Assume that the first contact is used to operate the compressor relay, and the second contact is used to operate the contactor, which controls the electric resistance heating strips. When the temperature decreases, the first thermostat contact closes and connects the compressor to the line. If the compressor is able to provide enough heat to the dwelling, the second contact will never make connection. If, however,

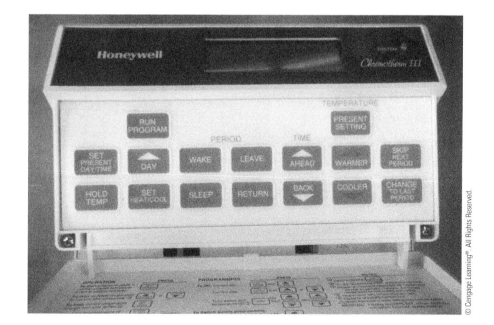

▶ **Figure 29–15**
Electronic thermostat.

the outside temperature is low enough that the heat pump cannot provide enough heat, the second thermostat contact will close and permit the electric heat strip to operate.

THE DIFFERENTIAL THERMOSTAT

The **differential thermostat** is used primarily with solar-powered heating systems. A differential thermostat is shown in Figure 29–16. This thermostat uses two separate temperature sensors and is activated by the difference of temperature between them. A solar hot-water system is shown in Figure 29–17. A solar collector is used to heat the water. A storage tank stores the heated water and acts as a heat exchanger for the domestic hot water for the home. The system is controlled by the differential thermostat. When the temperature of the collector becomes greater than the temperature of the water in the storage tank by so many degrees, the thermostat activates the pump motor. The pump motor circulates water from the storage tank to the collector, and from the collector back to the tank. When the temperature of the collector is within a certain amount of the water temperature, the thermostat turns the pump off. In this way, water is circulated through the collector only when the collector is at a higher temperature

than the stored water. A common setting for the differential thermostat is 20 and 5. This means that the thermostat will turn the pump on when the collector is 20° hotter than the stored water, and turn the pump off when the collector is only 5° hotter than the stored water.

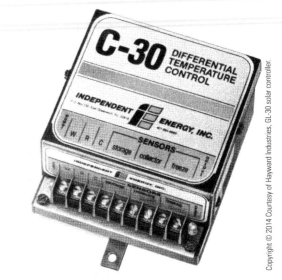

▶ **Figure 29–16**
Differential thermostat.

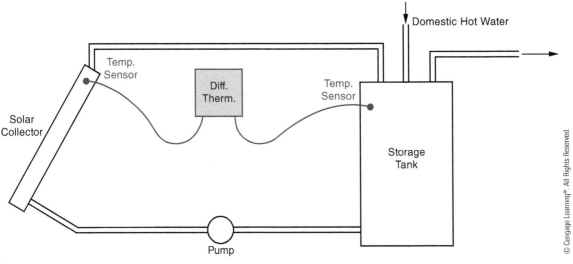

▶ **Figure 29–17**
Solar hot-water system controlled by a differential thermostat.

Some differential thermostats provide extra features, such as **antifreeze protection**. Antifreeze protection turns the pump on and circulates warm water through the collector when its temperature is near freezing. This does cool the warmed water, but cooling the water is generally preferred to damaging the collector. Some solar systems used a separate water supply for the collector. These systems use a mixture of antifreeze and water in the collector loop to avoid freezing problems.

THERMOSTAT TERMINAL IDENTIFICATION

Thermostats generally contain letters that are used to identify the terminal connections. The most common letters are R, G, Y, W, B, and O. The letters stand for common colors of thermostat wire.

R = Red
G = Green
Y = Yellow
W = White
B = Blue or Black
O = Orange

The B and O terminals are seldom used. The B terminal connects to a heating damper and the O terminal connects to a cooling damper. The other terminals connect as follows:

R – One side of the control transformer (generally 24 volts)
G – Fan relay
Y – Compressor relay
W – Heating relay

Some heating/cooling thermostats use R_C and R_H instead of R. When these thermostats are used for both heating and cooling, terminals R_C and R_H are connected with a jumper, Figure 29–18.

Heat pump thermostats generally contain terminals not found on common heating/cooling thermostats. Figure 29–19 illustrates the terminal connections for a common heat pump thermostat.

A listing of thermostat terminals and their meanings follows:

A   General purpose (could be anything)
B   Damper (heating) or Reversing solenoid on heat pumps (heating)
C   Unswitched side, Class 2 power
DF   Defrost

E   Emergency heat relay on heat pumps
G   Fan
K1   Switched side, Second source—Class 2 power
K2   Unswitched side, Second source—Class 2 power
L   Indicator circuits or system monitors
O   Damper (cooling) or reversing valve on heat pump (cooling)
R   Switched side, Class 2 power (single source)
RC   Switched side, Class 2 power, Cooling side
RH   Switched side, Class 2 power, Heating side
T   Outdoor thermostat
TT   One side, Class 2 circuit-switch—Heat
TT   Other side, Class 2 circuit-switch—Heat
W   Heating
W1   1st Stage heating
W2   2nd Stage heating
W3   3rd Stage heating
X   Lock-out reset
Y   Cooling
Y1   1st Stage cooling
Y2   2nd Stage cooling
Y3   3rd Stage cooling

NOTE: *Class 2 power sources are generally low voltage and low power. They are not capable of supplying enough power to create a fire hazard or a shock hazard.*

Some manufacturers do not use the letters R, G, Y, and W, but use letters V, F, C, and H. The V terminal stands for voltage and is the same as the R terminal. The F terminal stands for "fan" and is the same as the G terminal. Terminal C stands for "compressor" and is the same as the Y terminal, and the H stands for "heat" and corresponds to the W terminal. A chart illustrating the thermostat terminal identification for different manufacturers is shown in Figure 29–20.

TESTING THE THERMOSTAT

It is possible to perform a simple test to determine if a problem exists in a thermostat. A jumper wire is used to bypass parts of the thermostat. A basic heating and cooling thermostat is shown in Figure 29–21. To perform the test, use a jumper

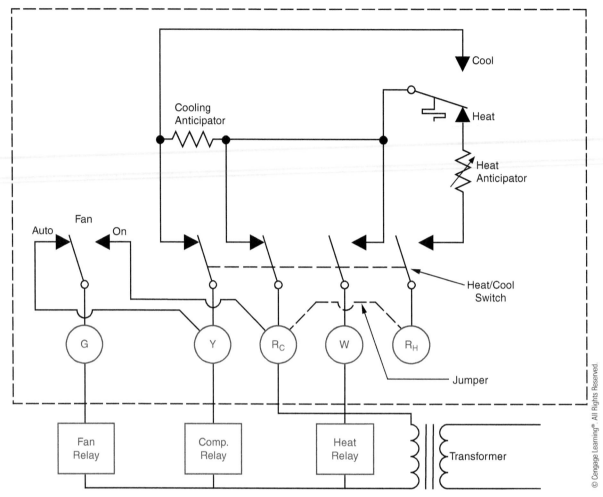

▶ Figure 29–18
Thermostat terminal identification for a common heating/cooling thermostat.

wire to make connection from the power input terminal, R_C or R_H to the G, Y, and W terminals one at a time, Figure 29–22. When connection is made from one of the R terminals and the W terminal, the heating relay should energize. When connection is made from one of the R terminals and the Y terminal, the compressor relay should energize, and when connection is made from one of the R terminals and the G terminal, the fan relay should energize.

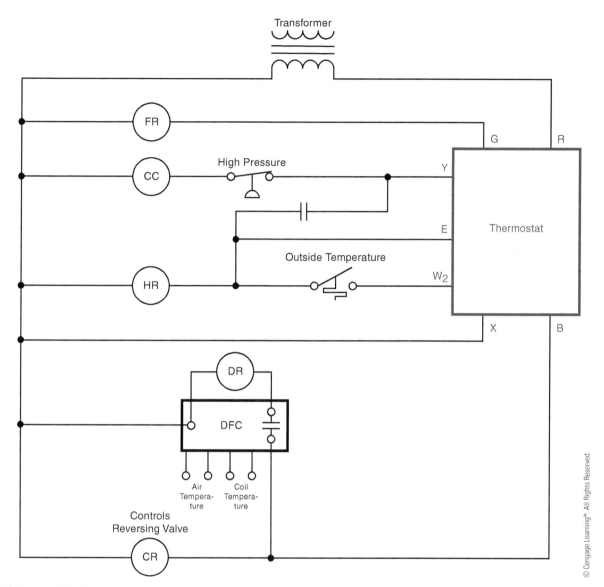

Figure 29–19
Typical thermostat connection for a heat pump.

MANUFACTURER	TYPE	COMMON	COOLING	HEATING	FAN
Cam-Stat	T17	R	Y	W	G
Cont. Corp.	360	R	Y	W	G
General	T91	V	C	H	F&G
General	T199	R	Y	W	G
Honeywell	T834	R	Y	W	G
Honeywell	T87	R	Y	W	G
White-Rogers	IF56	R	Y	W	G

▶ **Figure 29–20**
Typical thermostat markings.

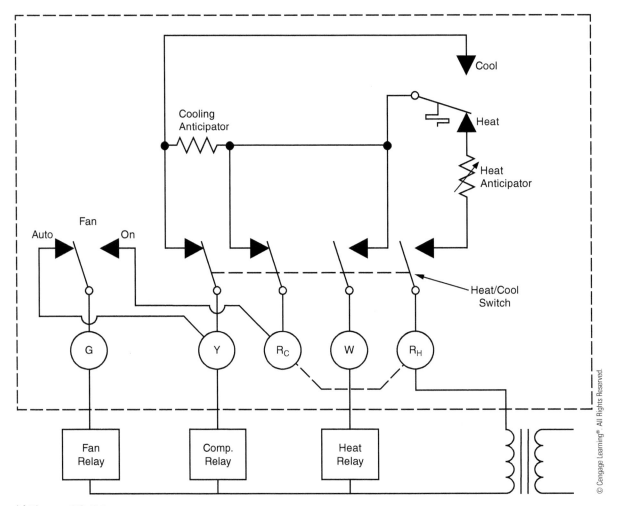

▶ **Figure 29–21**
Basic heating and cooling thermostat.

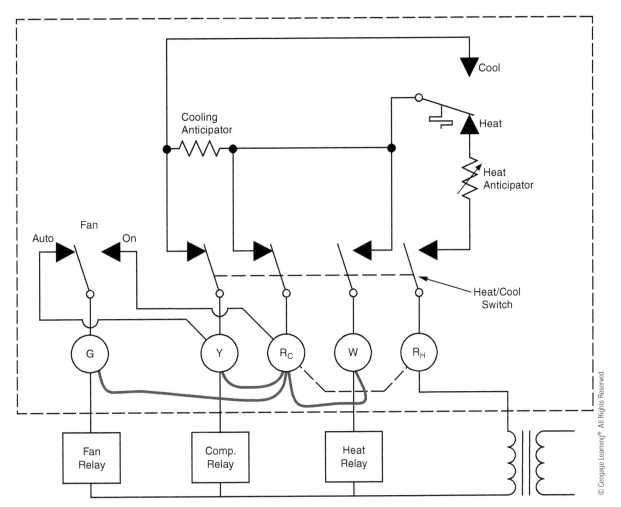

Figure 29–22
Testing a thermostat.

SUMMARY

- ⊙ Thermostats are temperature-sensitive switches.
- ⊙ Most thermostats are intended to be used on low-voltage control systems.
- ⊙ A bimetal strip is often used to sense the temperature.
- ⊙ Bimetal strips are generally bent in a spiral and resemble a clock spring. This is done to permit a longer length to fit into a small area.
- ⊙ Contact thermostats have one stationary contact and one movable contact. The movable contact is attached to the bimetal strip.
- ⊙ Contact-type thermostats use a small permanent magnet to provide a snap action when the contact opens or closes.

▶ Mercury-type thermostats operate by enclosing a pool of mercury inside a glass envelope. The glass envelope is attached to a bimetal strip.

▶ The weight of the mercury provides the snap action for the contacts when they open or close.

▶ Some thermostats are designed to be used as both a heating and cooling thermostat.

▶ The heat anticipator is a small resistive heater located near the bimetal strip. Its purpose is to open the contacts before the temperature actually reaches the thermostat setting.

▶ The cooling anticipator is a small resistive heater located near the bimetal strip of an air-conditioning thermostat. It operates when the thermostat contacts are open and causes them to close before the ambient temperature becomes high enough to close the contacts.

▶ Some thermostats contain a fan switch, which permits the blower fan to be operated independently of the heating or cooling system.

▶ Line voltage thermostats are made with large contacts and can be used to connect a load to the line.

▶ Programmable thermostats contain more than one set of contacts and can be set to operate at different temperature settings at different times.

▶ Staging thermostats contain more than one set of contacts and are generally used with heat-pump systems. The first set of contacts is used to turn on the compressor, and the second set of contacts is used to turn on the strip heaters.

▶ KEY TERMS

antifreeze protection	fan switch	staging thermostat
contact	heat anticipator	thermostat
cooling anticipator	line-voltage thermostat	
differential	programmable	
thermostat	thermostat	

▶ REVIEW QUESTIONS

1. What is a thermostat?
2. What is the advantage of an open-contact thermostat?
3. What is the disadvantage of an open-contact thermostat?
4. What is the advantage of a mercury thermostat?
5. What is used to provide a snap action for the contacts in an open-contact type of thermostat?
6. What is used to provide a snap action for the mercury thermostat?

7. What method of sensing temperature is often used with line voltage thermostats?

8. What is a programmable thermostat?

9. What is the advantage of the programmable thermostat?

10. What is a differential thermostat?

11. What are differential thermostats generally used to control?

12. What is antifreeze protection in reference to a differential thermostat?

13. What is the advantage of a low-voltage thermostat over a line voltage thermostat?

14. What is the purpose of the heat anticipator?

15. How is the setting of the heat anticipator generally determined?

16. To what does the G terminal on a typical heating/cooling thermostat connect?

17. Some thermostats contain terminals marked R_C and R_H. What must be done if this thermostat is used for both heating and cooling?

18. A thermostat has a terminal marked C. To what does this terminal connect?

Pressure Switches

OBJECTIVES

After studying this unit, the student should be able to:

- ▷ Describe the operation of high-pressure switches
- ▷ Describe the operation of low-pressure switches
- ▷ Make connection of a high-pressure switch
- ▷ Make connection of a low-pressure switch

High- and low-pressure switches are used to sense the amount of pressure in an air-conditioning and refrigeration system. They are used to disconnect the compressor from the power line if the pressure should become too high or too low. Most of the pressure switches used for air conditioning are operated by a **bellows**. A tube is attached to one end of the bellows, and the other end is connected to the discharge or suction side of the compressor, depending on which type of pressure switch is used.

THE HIGH-PRESSURE SWITCH

Figure 30–1 illustrates the operation of a **high-pressure switch**. The bellows is connected to the discharge side of the compressor via the tube. As the pressure of the system increases, the bellows

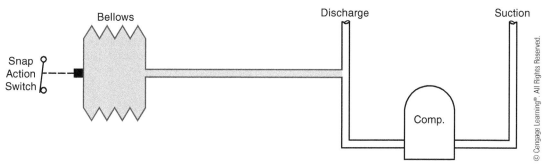

▶ **Figure 30–1**
Pressure switch connected to sense high pressure.

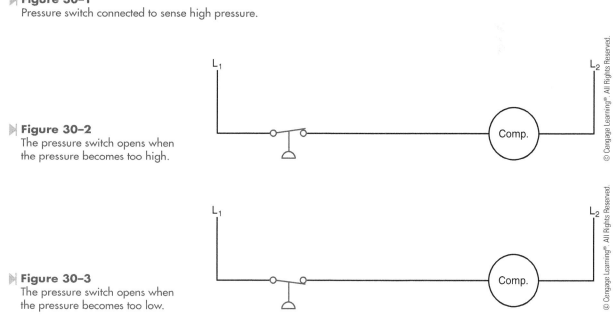

▶ **Figure 30–2**
The pressure switch opens when the pressure becomes too high.

▶ **Figure 30–3**
The pressure switch opens when the pressure becomes too low.

expands. The bellows is used to activate a spring-loaded normally closed switch. If the pressure should become too great, the bellows will expand far enough to open the switch. The normally closed pressure switch is connected in series with the compressor circuit shown in Figure 30–2. The pressure switch may be connected in series with the compressor or in series with the compressor control relay, depending on the type of control circuit.

THE LOW-PRESSURE SWITCH

The **low-pressure switch** is very similar in construction to the high-pressure switch. The low-pressure switch, however, is connected to the low-pressure, or suction, side of the compressor. The low-pressure switch is used to disconnect the compressor from the circuit if the pressure on the suction side should become too low. Figure 30–3 illustrates this type of circuit. The low-pressure switch is a normally open, held-closed switch. The switch is held in the closed position by the pressure of the system. If the pressure should drop low enough, the switch will open and disconnect the compressor from the circuit. As with the high-pressure switch, the low-pressure switch can be used to disconnect the compressor from the line or disconnect the compressor relay, depending on the type of control circuit. A low-pressure switch is shown in Figure 30–4.

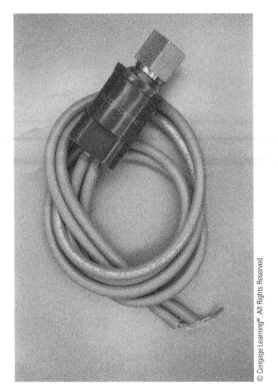

▶ **Figure 30–4**
Low-pressure switch.

▶ **Figure 30–5**
Adjustable pressure switch.

CONSTRUCTION

Most of the pressure switches used for commercial or industrial systems are **adjustable**, Figure 30–5. This feature allows the service technician to use the switch on different systems. For example, the pressure settings are different for different types of refrigerant. Figure 30–6 shows a table of common pressure settings for high- and low-pressure switches used with different types of refrigerant.

A **dual-pressure switch** is shown in Figure 30–7. This switch incorporates both high- and low-pressure switches in the same housing. Some manufacturers use a **nonadjustable** type of pressure switch. This switch is commonly found on central units designed for residential use. This

PRESSURE SWITCH SETTINGS				
Type of Refrigerant	High Pressure Cut out	Cut in	Low Pressure Cut out	Cut in
12	275	145	15	35
22	380	300	38	68
500	280	200	22	46

▶ **Figure 30–6**
Pressure switch settings.

type of switch is shown in Figure 30–8. The advantage of this type of switch is that it is inexpensive. When replacing this type of pressure switch, however, it must be matched to the refrigerant system. High- and low-pressure switches are shown in the schematic in Figure 30–9.

Figure 30–7
Dual-pressure switch.

Figure 30–8
Nonadjustable pressure switch.

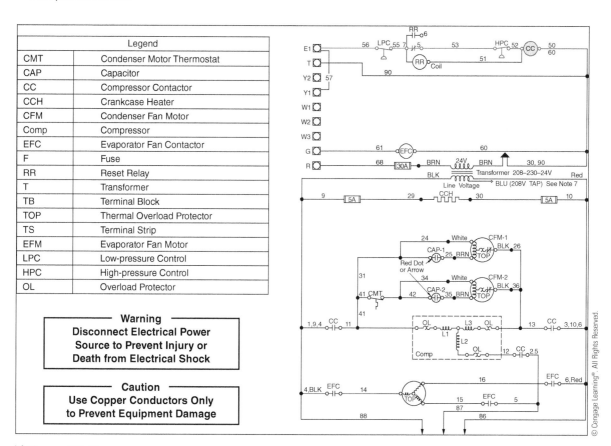

Legend	
CMT	Condenser Motor Thermostat
CAP	Capacitor
CC	Compressor Contactor
CCH	Crankcase Heater
CFM	Condenser Fan Motor
Comp	Compressor
EFC	Evaporator Fan Contactor
F	Fuse
RR	Reset Relay
T	Transformer
TB	Terminal Block
TOP	Thermal Overload Protector
TS	Terminal Strip
EFM	Evaporator Fan Motor
LPC	Low-pressure Control
HPC	High-pressure Control
OL	Overload Protector

Warning
Disconnect Electrical Power
Source to Prevent Injury or
Death from Electrical Shock

Caution
Use Copper Conductors Only
to Prevent Equipment Damage

Figure 30–9
High- and low-pressure switches are used to disconnect the compressor contactor.

SUMMARY

⊙ High- and low-pressure switches are used to sense the pressure in an air-conditioning and refrigeration system.

⊙ High- and low-pressure switches generally use a bellows, which expands and contracts with a change of pressure. The bellows operates a set of contacts.

⊙ High-pressure switches are normally closed switches that open when the pressure becomes too high. The switch is connected in series with the coil of the compressor relay.

⊙ Low-pressure switches are normally open, held-closed switches that open when the pressure drops below a certain level. Low-pressure switches are connected in series with the coil of the compressor relay also.

⊙ Some pressure switches can be adjusted and some cannot.

KEY TERMS

adjustable
bellows

dual-pressure switch
high-pressure switch

low-pressure switch
nonadjustable

REVIEW QUESTIONS

1. What device is used to construct most of the pressure switches used in the air-conditioning field?
2. What type of contact is used with a high-pressure switch?
3. What type of contact is used with a low-pressure switch?
4. Where in the refrigerant system is the high-pressure switch connected?
5. Where in the refrigerant system is the low-pressure switch connected?

▷ TROUBLESHOOTING QUESTIONS

Refer to the schematic shown in Figure 30–9 to answer the following questions.

1. Is the compressor in this unit operated by a single-phase or a three-phase motor?

   A. Single-phase

   B. Three-phase

2. In order for the compressor contactor to energize, the thermostat (not shown) must make contact between which terminals?

   A. R and G

   B. R and T

   C. R and E1

   D. R and Y2

3. Are the condenser fan motors single-phase or three-phase?

   A. Single-phase

   B. Three-phase

4. In order to energize the EFC contactor coil, connection should be made between which terminals?

   A. E1 and G

   B. T and G

   C. Y1 and G

   D. R and G

5. Is the evaporator fan motor single-phase or three-phase?

   A. Single-phase

   B. Three-phase

UNIT 31 ▷ The Flow Switch

OBJECTIVES

After studying this unit, the student should be able to:

▷ Describe the operation of a flow switch

▷ Draw the standard NEMA symbol for a flow switch

▷ Connect a flow switch in a circuit

The type of **flow switch** used in air-conditioning systems senses the flow of air instead of the flow of liquid. The flow switch is often referred to as a **sail switch** because it operates on the principle of a sail. The flow switch is constructed from a snap-action microswitch. A metal arm is attached to the microswitch. A piece of thin metal or plastic is connected to the metal arm. The thin piece of metal or plastic has a large surface area and offers resistance to the flow of air. When a large amount of airflow passes across the sail, enough force is produced to cause the metal arm to operate the contacts of the switch. A flow switch is shown in Figure 31–1.

The flow switch is used to give a positive indication that the evaporator or condenser fan is operating before the compressor is permitted to operate. The airflow switch is the only positive method of indicating that the fan is actually in operation. For example, in the circuit shown in Figure 31–2, the thermostat controls the operation of a **control relay (CR)**. When the thermostat closes its contacts, CR coil energizes. This causes all CR contacts to close. When the first CR contact closes, **CFM (condenser fan motor) relay**

energizes and starts the condenser fan motor. When the second CR contact closes, **EFM (evaporator fan motor) relay** coil energizes and starts the evaporator fan motor. The third CR contact cannot energize the **compressor relay coil**, however, because it is interlocked with CFM and EFM relays. The compressor relay coil can be energized only after the condenser fan and evaporator fan relay coils have energized.

The idea behind this type of control is to ensure that the compressor cannot be started until both the condenser and evaporator fans are operating. This control circuit, however, does not fulfill that requirement. This circuit does not sense whether the fans are actually operating. It does sense whether the relay coils, which control those fan motors, are energized. This circuit cannot detect whether a fan motor is not operating or a belt is broken between the motor and the fan.

The circuit shown in Figure 31–3 has been modified from the circuit in Figure 31–2. Notice in this circuit that the normally open CFM and EFM contacts connected in series with the compressor relay have been replaced with airflow switches **CFS (condenser flow switch)** and **EFS (evaporator flow switch)**. These switches are operated by the force of air created by the condenser fan or the evaporator fan. In this circuit, the compressor can be started only after the condenser and evaporator fans are actually operating. If the circuit is in operation and one of the fans should stop, the compressor

▶ **Figure 31–1**
Airflow switch.

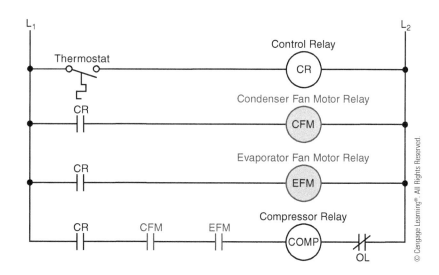

▶ **Figure 31–2**
Compressor is interlocked with condenser and evaporator fan relays.

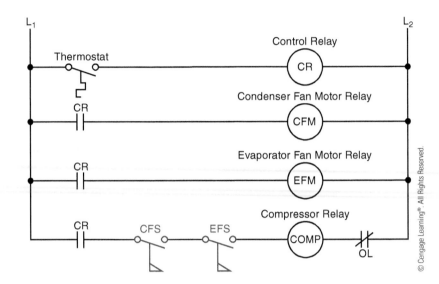

▶ **Figure 31–3**
Compressor is interlocked with airflow switches.

▶ **Figure 31–4**
Flow switch senses the increase of air pressure inside the duct when the blower is in operation.

relay will be disconnected from the circuit, thereby disconnecting the compressor motor from the circuit.

Another type of flow switch is shown in Figure 31–4. This switch is actually a pressure switch that detects the increase in air pressure inside the duct when the blower is in operation. The switch can be adjusted for more or less sensitivity. The switch contacts connect in the same manner as shown in Figure 31–3.

SUMMARY

- ▷ Some flow switches sense the flow of liquid, and some are made to sense the flow of air.
- ▷ Airflow switches are sometimes referred to as sail switches.
- ▷ Airflow switches are generally used to ensure that air is moving across the evaporator coil before the compressor is permitted to start.
- ▷ Flow switch contacts are generally normally open and are connected in series with the coil of the compressor relay.

KEY TERMS

CFM (condenser fan motor) relay

CFS (condenser flow switch)

compressor relay coil

CR (control relay)

EFM (evaporator fan motor) relay

EFS (evaporator flow switch)

flow switch

sail switch

REVIEW QUESTIONS

1. What is a common name for the airflow switch?

2. What function does the airflow switch perform in a circuit?

3. What is interlocking in a control circuit?

UNIT 32 ▷

The Humidistat

OBJECTIVES

After studying this unit, the student should be able to:

▷ Discuss the operation of a humidistat
▷ Discuss how the humidistat is connected into the control circuit
▷ Connect a humidistat in a circuit

The control of humidity can be very important in some heating and air-conditioning systems. Some industries such as mills that knit polyester and nylon fibers must maintain a constant humidity because these materials contract and expand with a change of humidity. The control of humidity is also important in heating systems. The amount of humidity in the air has a great effect on the comfort of the living area. If the humidity is to be maintained at a constant level, some device must be used to detect the amount of humidity and then operate some type of control.

The **humidistat** is a device that can sense the amount of humidity in the air and activate a set of contacts if the humidity should become too high or too low. The two most common materials used to sense humidity are **hair** and **nylon**. The materials contract and expand with a change in the amount of humidity in the air. A humidistat using hair as the sense element is shown in Figure 32–1.

If a humidifier is used in a central-heating or air-conditioning system, it is generally operated only when the blower is in operation. For this reason, some means is used to interlock the humidifier with the blower. The circuit shown in Figure 32–2 uses a humidistat to control the operation of a solenoid coil. The solenoid coil operates a valve that supplies water to the humidifier. Notice that the solenoid coil is interlocked with an airflow switch. The coil can be energized only when the sail switch indicates there is airflow in the system. Some controls use a combination of a humidistat and a sail switch. Nylon strips are used to sense the amount of humidity in the air, and a plastic sail is used to sense the flow of air in the duct.

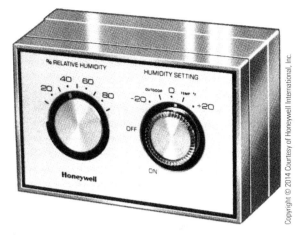

▶ **Figure 32–1**
Humidistat.

▶ **Figure 32–2**
Humidifier is interlocked with an airflow switch.

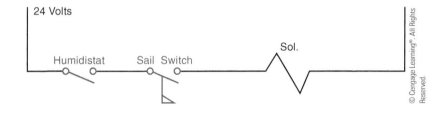

▶ SUMMARY

- ⊙ A humidistat is used to control the amount of humidity in the air.
- ⊙ The amount of humidity in the air greatly affects the comfort of the living area.
- ⊙ The humidistat is a device that senses the amount of humidity in the surrounding air and controls a set of contacts.
- ⊙ The two most common materials used to sense humidity are hair and nylon.
- ⊙ A humidifier used in a central air-conditioning system is generally permitted to operate only when the blower is in operation.

▶ KEY TERMS

hair
humidistat
nylon

▷ REVIEW QUESTIONS

1. What is a humidistat?

2. What are the two most common materials used to sense humidity?

3. What type of control is often used to interlock the humidifier with the blower?

UNIT 33

Fan-Limit Switches

OBJECTIVES

After studying this unit, the student should be able to:

▷ Discuss the operation of a fan-limit switch

▷ Describe different types of fan and limit switches

▷ Draw the schematic symbol for a limit switch

▷ Test a fan or limit switch with an ohmmeter

▷ Connect a fan-limit switch in a circuit

The blower fan of a heating system is generally not permitted to operate until the heat exchanger reaches a high enough temperature to ensure that cold air will not be delivered into the living area. **Fan switches** are generally operated by a bimetal strip that closes a set of contacts when the temperature of the heat exchanger reaches a high enough level. A fan switch of this type is shown in Figure 33–1. The control shown on the face of the switch is used to determine the temperature at which the fan will turn off. The temperature at which the switch contacts will close is determined by the manufacturer. Longer operation time of the fan generally increases the overall efficiency of the heating unit because more of the heat is delivered to the living area and less escapes to unheated areas. Some people, however, do not like the cooler air being delivered by the blower at the end of a heating cycle. For this reason, the switch can be set to turn off sooner and prevent this problem.

Some fan switches are designed with large enough contacts to permit them to control the operation of the blower motor without a fan relay. Other fan switches have small auxiliary contacts and are used to control the coil of a fan relay. The circuit in Figure 33–2 shows a fan switch being used to control a blower motor. Notice the schematic symbol is the same as a thermostat. This symbol is used because it is a thermally activated switch.

Another type of fan switch is shown in Figure 33–3. This type of switch does not sense the heat of a heat exchanger to close a set of contacts.

This switch is basically a timer. It uses a small resistance heater that is controlled by the thermostat. The heater causes a bimetal strip to bend. When the strip has bent far enough, it closes a set of spring-loaded contacts and connects the motor to the line. A schematic of this relay is shown in Figure 33–4. Notice that this switch has two sections that are isolated from each other. The 24-volt section is connected to the heating element. The 120-volt section is connected to the switch contacts. When connecting this type of fan switch, care must be taken not to connect the terminals to the wrong voltage. The advantage of this type of switch is that it can be used to replace almost any type of thermally operated fan switch because it can be mounted almost anywhere.

Another type of fan switch is shown in Figure 33–5. This switch is used to control the speed of a fan motor as opposed to turning it on or off. This switch is more common to an air-conditioning system than a heating system and is used to control the speed of the condenser fan. Some systems decrease the speed of the condenser fan if the temperature of the condenser drops too low. A low condenser temperature causes a low head pressure. This switch is basically a single-pole double-throw switch,

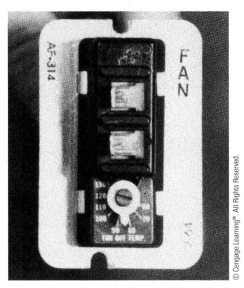

Figure 33–1
Fan switch.

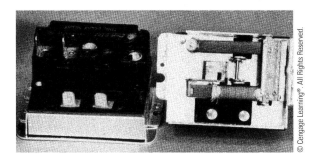

Figure 33–3
Time-delay fan switch.

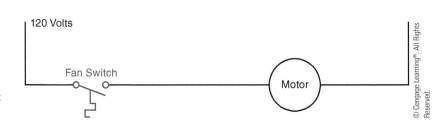

Figure 33–2
Fan switch controls the operation of the motor.

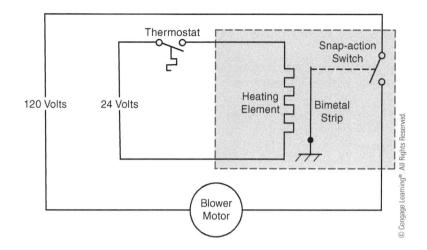

▶ **Figure 33–4**
Timer used as a fan switch.

▶ **Figure 33–5**
Bimetal fan-speed switch.

Figure 33–6. When the condenser temperature is low, the motor is connected to low-speed operation. When the condenser temperature increases, the motor is connected for high-speed operation.

LIMIT SWITCHES

One type of **limit switch** is used as a safety cutoff switch for a heating system. Limit switches are generally a bimetal-operated switch. Two types of limit switches are shown in Figure 33–7. Limit switches contain a normally closed contact, which is connected in series with the system control. They are not, however, connected in series with the blower fan, Figure 33–8. Notice that the blower motor can operate independently of the burner control. This permits the blower to cool down the heating unit if the high-limit switch should open and turn the blower off.

▶ **Figure 33–6**
Condenser speed-control switch.

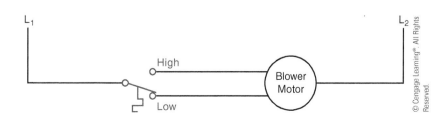

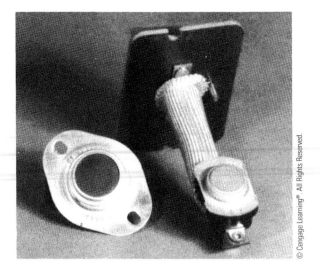

▶ **Figure 33–7**
High-limit switches.

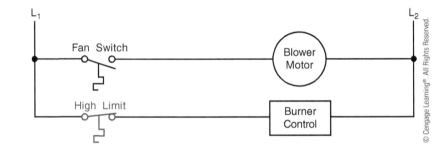

▶ **Figure 33–8**
Limit switch is connected in series
with the burner control.

FAN-LIMIT SWITCH

The **fan-limit switch** contains both a fan
switch and a high-limit switch in one housing,
Figure 33–9. This type of switch uses a bimetal
strip formed in the shape of a spiral. When the
bimetal is heated, it causes a cam to rotate. When
the cam has turned far enough, the normally open
fan switch closes and connects the blower fan to
the line, Figure 33–10. If the system is operating
properly, the blower fan prevents the temperature
of the heat exchanger from rising high enough to
open the high-limit switch contacts. If the blower
fan does not operate, however, the temperature of
the heat exchanger will increase enough to open
the normally closed limit switch. This will cause the
burner to turn off and prevent damage to the heating system.

▶ **Figure 33–9**
Fan-limit switch.

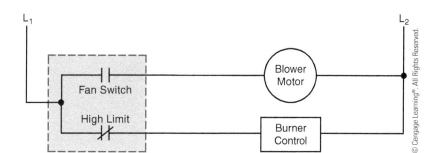

▶ **Figure 33–10**
Schematic for fan-limit switch.

▷ SUMMARY

- ▷ The blower fan of a heating system is generally not permitted to operate until the heat exchanger has reached a high enough temperature.
- ▷ Fan switches are generally operated by a bimetal strip that closes a set of contacts.
- ▷ Some fan switches can be adjusted to operate longer or turn off sooner.
- ▷ Some fan switches are designed with large enough contacts to control the operation of the blower motor without the use of a fan relay.
- ▷ Some fan and limit switches are constructed as one unit, and others are constructed as separate units.
- ▷ Limit switches are generally used to stop the operation of the heating system if the temperature should become too high.

▷ KEY TERMS

fan-limit switch
fan switches
limit switch

▷ REVIEW QUESTIONS

1. Why do some fan switches permit setting the temperature at which the switch will turn on and off?

2. What type of sensing device do most fan switches use to determine when the temperature is high enough to start the blower fan?

3. What type of contact arrangement is used for switches that control the speed of a condenser fan motor?

4. What is the most common use for a high-limit switch?

5. Why is the blower fan not connected in series with the limit switch?

The Oil-Pressure Failure Switch

OBJECTIVES

After studying this unit, the student should be able to:

≫ Discuss why the oil-pressure failure switch is necessary on large air-conditioning systems

≫ Discuss differential pressure

≫ Describe the operation of the time-delay circuit

≫ Connect an oil-pressure failure switch for operation on a 120- or 240-volt system

Many of the larger air-conditioning units use a forced-oil system for the compressor instead of a splash system. When a forced-oil system is used, an **oil-pressure failure switch** is often employed to protect the compressor from insufficient oil pressure. The oil-pressure failure switch actually contains several control functions in the same unit. These functions include a **differential pressure switch**, a timer, and a set of **control contacts**.

THE DIFFERENTIAL PRESSURE SWITCH

The actual oil pressure in a compressor is the difference in pressure between the suction pressure and the discharge pressure of the oil pump. For example, if the suction pressure is 35 psi, and the oil pump discharge pressure is 65 psi, the actual amount of oil pressure in the system is 30 psi (65 − 35 = 30). If the oil-pressure failure switch is to be used to measure the actual oil pressure in the system, it must be able to measure the difference between these two pressures. This is accomplished with a differential pressure switch. Figure 34–1 illustrates the operation of an oil-pressure failure switch. Notice in this illustration that two bellows are employed. One bellows is connected to the suction side of the compressor. The other bellows is connected to the oil pump discharge. If the oil pressure is low, the bellows connected to the suction side of the compressor forces the differential pressure switch to remain closed. If the oil pressure increases high enough above the pressure of the suction line, the oil pressure bellows will provide enough force to overcome the force of the suction line bellows and open the differential pressure switch. Notice that the differential pressure

switch remains closed until there is enough oil pressure to open it.

THE TIME-DELAY CIRCUIT

The differential pressure switch is used to control the **time-delay circuit**. The time-delay circuit consists of a **current-limiting resistor**, a resistance heating element, and a bimetal strip. Most oil pressure failure switches are designed to be used on 120- or 240-volt connections. This selection is made possible by the value of the current-limiting resistor. Notice that this resistor is center tapped. If 240 volts is to be applied to the circuit, the full value of the current-limiting resistor is connected in series with the heater. If the circuit is 120 volts, the line is connected to the center tap position of the current-limiting resistor. Because the value of the resistor

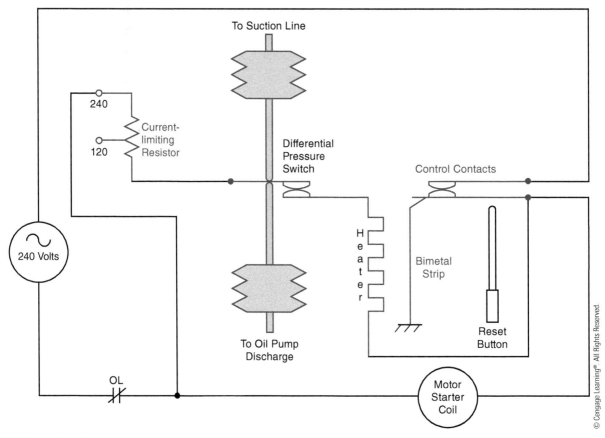

▷ **Figure 34–1**
Schematic of oil-pressure failure switch.

is cut in half for the 120-volt connection, the same amount of current will flow through the heater for either line voltage.

The resistance heater is used to heat the bimetal strip. If the heater is permitted to operate long enough, the bimetal strip will warp away, and the control contact will open. The time-delay circuit is necessary to permit the compressor to operate long enough for oil pressure to build up in the system. When the oil pressure reaches a high enough level, the differential pressure switch opens and disconnects the heater from the circuit. This stops the warping action of the bimetal strip and the control contacts do not open.

THE CONTROL CONTACTS

Notice that the control contacts are connected in series with the motor starter coil to the compressor. If the control contacts should open, the circuit to the motor starter will be broken and the compressor will be disconnected from the line. Notice that the control contacts provide power to the heater of the timer. If the control contacts should open, power cannot be applied to the heater circuit until the contacts are closed. Once the contacts have opened, they must be manually reset by the reset button.

SPECIFICATIONS

The normal usable oil pressure for most reciprocating compressors is generally between 35 and 45 psi. The differential pressure control permits the cut-in and cut-out points to be set. A common setting for this type of switch is cut in at 18 psi and cut out at 12 psi. This means that the differential pressure switch contacts open when the oil pressure becomes 18 psi greater than the pressure of the suction line and close when the oil pressure drops to a point that it is only 12 psi above the suction line pressure. The amount of time delay is set by the manufacturer and is generally about 2 minutes. An oil-pressure failure switch is shown in Figure 34–2.

Figure 34–2
Oil-pressure failure switch.

▶ SUMMARY

- ⊙ Many large air-conditioning compressors use a forced-oil system instead of a splash system.
- ⊙ The oil-pressure failure switch is used to protect the compressor from insufficient oil pressure.
- ⊙ The oil-pressure failure switch senses the difference in pressure between the oil pump and suction pressure of the compressor.
- ⊙ A time-delay circuit is used to stop the compressor if the oil pressure has not reached a sufficient level within a certain time.
- ⊙ A current-limiting resistor permits the oil-pressure failure switch to be connected to 120 or 240 volts.
- ⊙ The control contacts of the oil-pressure failure switch are connected in series with the coil of the compressor starter relay.
- ⊙ Most reciprocating compressors operate with an oil pressure of 35 to 45 psi.
- ⊙ Oil-pressure failure switches are generally set to cut in at about 18 psi and to cut out at about 12 psi.

KEY TERMS

control contacts
current-limiting
resistor

differential pressure
switch

oil-pressure failure
switch
time-delay circuit

REVIEW QUESTIONS

1. How can the actual amount of useful oil pressure in a compressor be found?

2. What is the function of the current-limiting resistor?

3. Why is the current-limiting resistor center tapped?

4. Does a high enough oil pressure open the differential pressure switch contacts or close them?

5. What is the function of the heater?

6. Explain the sequence of events that take place if the oil pressure does not become great enough to disconnect the heater circuit.

7. What is the cut-in point?

8. What is the cut-out point?

9. Is the timer circuit connected in series with the motor starter coil?

10. Are the overload contacts connected in series with the motor starter coil?

UNIT 35 ▷

Solenoid Valves

OBJECTIVES

After studying this unit, the student should be able to:

▷ Define a solenoid

▷ Properly connect the inlet and outlet of a solenoid valve

▷ Describe the operation of a four-way, or reversing, valve

▷ Make the proper electrical connections for a solenoid valve

A **solenoid valve** is an electrically operated valve. These valves are used to control the flow of gases or liquids. They range in complexity from a simple on–off valve to four-way reversing valves used on heat-pump systems. A simple plunger-type of solenoid valve is shown in Figure 35–1. This type of valve is often used to control the flow of gas or liquid in an air-conditioning system. The plunger of the solenoid is used to lift the valve off its seat. The valve is held closed by a spring when it is in its normal, or de-energized, position. When the coil is energized, the plunger lifts the valve off the seat, and liquid or gas is permitted to flow from the **inlet** to the **outlet**. When the coil is de-energized, the spring returns the valve to the seat and stops the flow of liquid or gas.

Notice that the valve is marked with an inlet and outlet side. The inlet is connected to the side of the system with the highest pressure. In this way, the

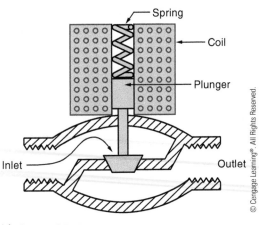

> **Figure 35–1**
> Solenoid valve.

pressure of the system is used to help keep the valve closed. If the valve should be reversed and pressure applied to the outlet side, the pressure of the system could be enough to overcome the tension of the spring and lift the plunger off the seat. This would cause the valve to leak.

THE REVERSING VALVE

A very common solenoid valve used in the air-conditioning field is the **four-way valve**, or **reversing valve**. Reversing valves are used to change the direction of flow of refrigerant in a heat-pump system. Figure 35–2 shows the direction of refrigerant flow when the heat-pump unit is in the cooling cycle. Notice that the high-pressure gas leaving the compressor enters the reversing valve. It is then directed to the outside coil being used as the condenser during the cooling cycle. Liquid refrigerant flows from the outside coil to the **metering device**, where it is changed to a low-pressure liquid. The low-pressure liquid then enters the inside coil, where it attracts heat from the inside air and changes to a gas. It then flows to the reversing valve. The reversing valve directs the flow to the accumulator and back to the compressor.

If the unit is now to be used for heating, the flow of refrigerant must be reversed through the system, Figure 35–3. Notice that the flow of hot, high-pressure gas is still from the discharge side of the compressor to the reversing valve. In this example, however, the flow of high-pressure gas is directed to the inside coil, which is now being used as the condenser. Liquid refrigerant leaves the inside coil and flows to the metering valve. The refrigerant is changed into a low-pressure liquid after going through the metering valve and flowing to the outside coil. Heat is then added to the liquid from the surrounding outside air. The gas then flows to the reversing valve, the accumulator, and back to the suction line of the compressor.

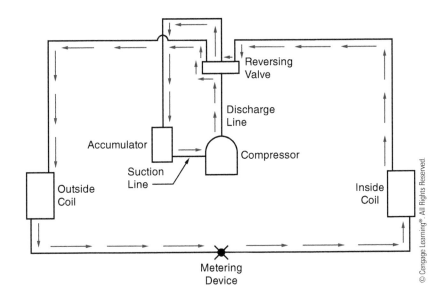

> **Figure 35–2**
> Refrigerant flow during the cooling cycle.

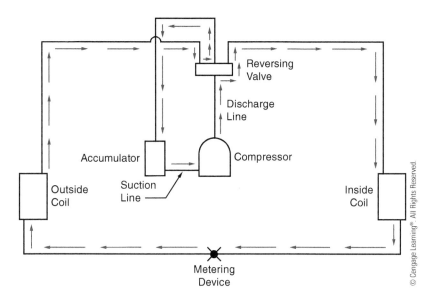

Figure 35–3
Refrigerant flow during the heating cycle.

Notice in both examples that the direction of refrigerant flow from and to the compressor is the same. The reversing valve was used to change the direction of flow.

REVERSING VALVE OPERATION

The four-way reversing valve is actually two valves that operate together. There is a main valve that actually controls the flow of refrigerant in the system, and a **pilot valve** that controls the operation of the main valve, Figure 35–4. The force needed to operate the main valve is provided by the compressor. The valve shown in Figure 35–4 has a sliding valve body that is used to control the flow of refrigerant through the system. In the illustration shown, the system is being used in the cooling cycle. Notice that on each side of the valve there is a small passage called an **orifice**. The orifice provides a path for a very small amount of refrigerant to flow. Notice also that there is a small capillary tube connected from each end of the valve body to the pilot valve and a third capillary tube connected from the pilot valve to the suction line of the compressor. In the position shown, the plunger of the pilot valve is blocking the capillary from the left side of the main valve body. The capillary tube

connected to the right side of the main valve is connected to the suction side of the compressor through the pilot valve. With the plunger of the pilot valve in this position, a high pressure is formed on the left side of the main valve and a low pressure is formed on the right side. The high pressure created on the left side of the main valve forces the main valve to slide to the right. With the main valve in this position, the discharge line of the compressor is connected to the outside coil and the inside coil is connected to the suction side of the compressor.

If the **solenoid coil** is energized, the plunger of the pilot valve will change to the position shown in Figure 35–5. The plunger now blocks the capillary tube connected to the right side of the main valve body. The capillary tube connected on the left side of the main valve is now connected to the suction line of the compressor through the pilot valve. This causes a high pressure to be created on the right side of the main valve and a low pressure on the left side. The high pressure forces the main valve to slide to the left. When the reversing valve is in this position, the discharge side of the compressor is connected to the inside coil, and the suction line is connected to the outside coil. The unit is now in the heating cycle.

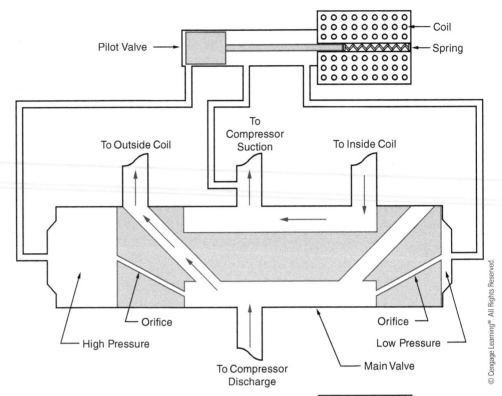

> **Figure 35–4**
> Reversing valve
> set for the
> cooling cycle.

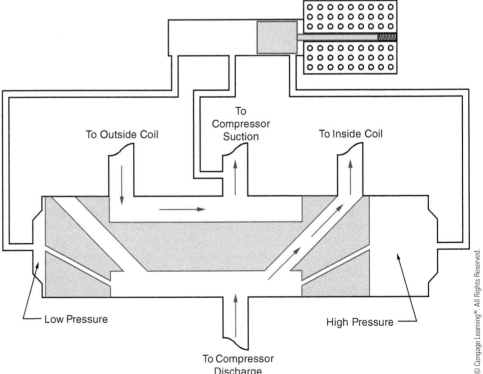

> **Figure 35–5**
> Reversing valve
> set for the
> heating cycle.

The valve illustrated in this example shows the valve is in the cooling cycle when the solenoid is de-energized and in the heating cycle when energized. This has been standard for many years for heat-pump systems. Now, however, some manufacturers are reversing this procedure. Some reversing valves are made in such a manner that when the valve is de-energized, the unit is in the heating cycle. This was done so that valve failure would result in the unit being in the heating cycle, the feeling being that heat is necessary to life, whereas air-conditioning is not. A four-way reversing valve is shown in Figure 35–6.

Figure 35–6
Reversing valve.

SUMMARY

- A solenoid valve is an electrically operated valve.
- The inlet and outlet side of a valve should never be reversed.
- Four-way valves are used to reverse the flow of refrigerant in a heat-pump system.
- Reversing valves are actually two valves that operate together.
- The solenoid coil of a four-way valve actually operates a pilot valve that is used to operate the main valve.
- The force needed to operate the main valve is provided by the compressor.

KEY TERMS

four-way valve

inlet

metering device

orifice

outlet

pilot valve

reversing valve

solenoid coil

solenoid valve

REVIEW QUESTIONS

1. What is a solenoid valve?
2. Why is it important not to reverse the connection of the inlet and outlet side of a solenoid valve?
3. What is used to cause the plunger to close when the solenoid coil is de-energized?
4. What is the function of a four-way reversing valve?
5. What is the function of the pilot valve?
6. What is the function of the main valve?
7. What is actually used to change the position of the main valve from one setting to another?

The Short-Cycle Timer

OBJECTIVES

After studying this unit, the student should be able to:

- ≫ Define the condition known as short cycling
- ≫ List reasons why short cycling occurs
- ≫ Describe the construction of a short-cycle timer
- ≫ Describe the circuit operation of a short-cycle timer
- ≫ Connect a short-cycle timer in the circuit

Short cycling is a condition that occurs when the compressor is restarted immediately after it has been turned off. This causes the compressor to restart against a high head pressure. Trying to restart a compressor in this manner can result in damage to the compressor, motor winding, or at the very least, open a circuit breaker or overload relay. After a compressor has been turned off, enough time should be permitted to pass to allow the pressure in the system to equalize before the compressor is restarted.

CAUSES OF SHORT CYCLING

Short cycling can be caused by several situations. For example, a loose thermostat wire can result in a bad connection that will make the compressor alternately start and stop. A momentary interruption of the power line can cause the compressor to stop and then restart when power is restored. People can also cause short cycling. Assume, for example, that the air-conditioning system is in operation. Now assume that someone changes the thermostat setting and causes the thermostat contact to open and stop the compressor. Now assume that the person changes his mind and again changes the thermostat so that the compressor tries to restart. Regardless of the reason or causes of short cycling, it should be avoided whenever possible.

THE SHORT-CYCLE TIMER

The **short-cycle timer** is a cam-operated, motor driven, on-delay timer. A photograph of a short-cycle timer is shown in Figure 36–1. This timer is used in conjunction with a relay generally referred to as a holding relay. The timer contains a set of double-pole double-throw contacts (DPDT). A basic schematic of a short-cycle timer circuit is shown in Figure 36–2. Notice the two sets of double-throw contacts labeled A and B. The dashed line

▶ **Figure 36–1**
Short-cycle timer.

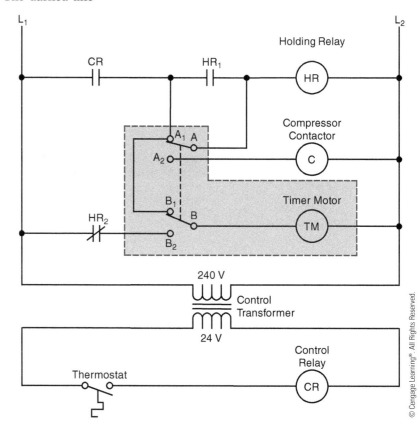

▶ **Figure 36–2**
Basic schematic of a short-cycle timer circuit.

between the contacts indicates mechanical connection so that they operate together. Notice also the holding relay labeled HR. The circuit is controlled by the operation of the thermostat.

CIRCUIT OPERATION

To understand the operation of this circuit, refer to the schematic shown in Figure 36–3. The arrows indicate the paths for current flow. Notice there is a path from line L_1 through the primary of the control transformer and back to L_2. This provides 24 volts for the operation of the control circuit. When the thermostat contacts close, a circuit is provided through the coil of the control relay (CR).

This causes contact CR to close and provides a current flow path to the short-cycle timer. Notice there are two paths of current flow at the timer. The current enters the A_1 terminal and flows to the B_1 contact terminal. The current can then flow through the contact to terminal B and then to the **timer motor**. There is also a current path from terminal A_1 to A. The current can then flow from A to the **holding relay (HR)**.

The holding relay energizes and changes both HR contacts, as shown in Figure 36–4. The now closed HR_1 contact is used as a holding contact. Notice that current is also flowing through the timer motor. The timer motor is geared to permit a delay of about 3 minutes before the contacts change position.

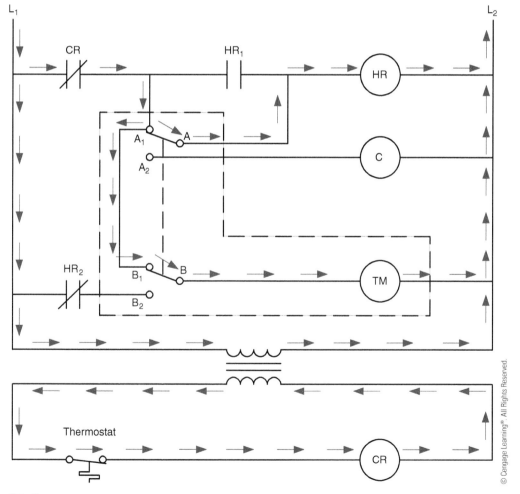

▷ **Figure 36–3**
The thermostat energizes the control relay.

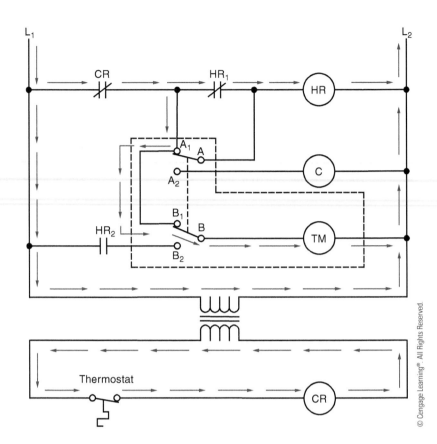

▶ **Figure 36–4**
The holding relay and the timer
motor are energized.

Figure 36–5 illustrates the operation of the circuit when the timer contacts change position. Notice that current now flows through the closed CR contact and the closed HR contact to contact A. Current can now flow to contact A_2 and then to the compressor contactor. When the compressor contactor energizes, it connects the compressor to the line. Notice that contact B_2 is connected to the now open HR_2 contact. Because there is no current flow to the timer motor, the timing operation is stopped for as long as the thermostat maintains a circuit to CR relay.

After the thermostat has been satisfied, it will reopen and de-energize CR relay coil. This causes CR contact to open and de-energize HR relay coil. When HR relay de-energizes, HR_2 contacts will again close and current flow is provided through the B_2 and B, contact to the timer motor, Figure 36–6. The timer motor now operates and resets the contacts to their original position, as shown in Figure 36–2. The circuit is now ready for another operation sequence. If the thermostat momentarily opens and then recloses, or if there is a momentary loss of power, the holding relay will de-energize and the timer will have to time out before the compressor can be reconnected to the line. About 1 minute is required for the timer to reset.

ELECTRONIC SHORT-CYCLE TIMERS

Many short-cycle timers use electronic components to affect a time delay. These units are relatively simple to install and set. All logic functions are contained within the timer, eliminating the need for a holding relay. Electronic short-cycle timers can be obtained that operate as **delay-on-make** or as **delay-on-break**.

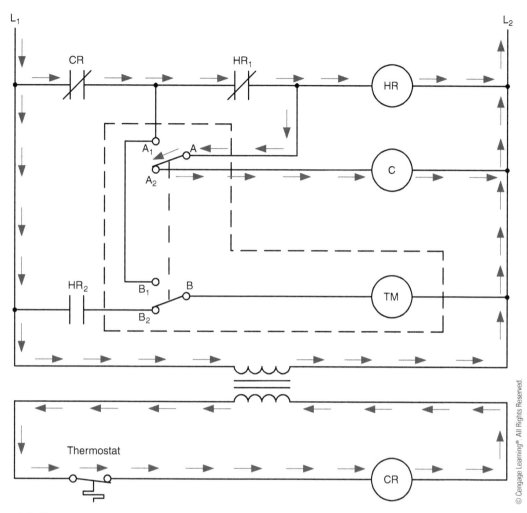

▶ **Figure 36–5**
The compressor contactor energizes when the timer contacts change position.

The Delay-on-Make Short-Cycle Timer

Delay-on-make timers are basically on-delay timers. They connect in series with the coil of the compressor contactor, Figure 36–7. When the thermostat contacts close and apply power to the timer, the timer starts timing for the period of time set on the timer. After the time-delay period, power is applied to the coil of the compressor contactor and the compressor starts. The delay-on-make timer is not as popular as the delay-on-break timer because the

delay-on-make timer does not permit the compressor to start when the thermostat contacts are first closed. A delay-on-make short-cycle timer is shown in Figure 36–8.

The Delay-on-Break Short-Cycle Timer

The delay-on-break short-cycle timer begins its timing sequence when the thermostat contacts open instead of close. This permits the delay-on-break timer to start the compressor instantly when the

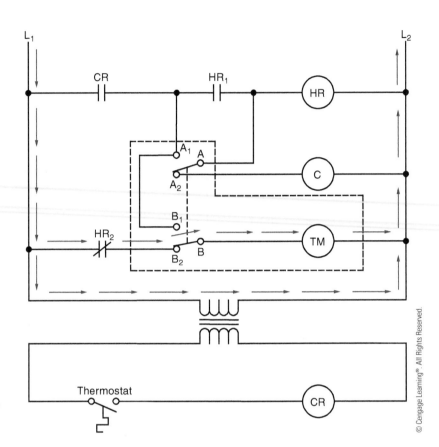

▶ **Figure 36–6**
When the thermostat de-energizes,
current is provided to the timer
motor to reset the contacts.

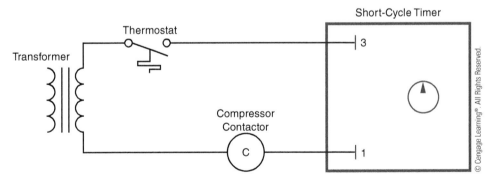

▶ **Figure 36–7**
Delay-on-make short-cycle timer connection.

▶ Figure 36–8
Delay-on-make short-cycle timer.

▶ Figure 36–10
Delay-on-break short cycle timer.

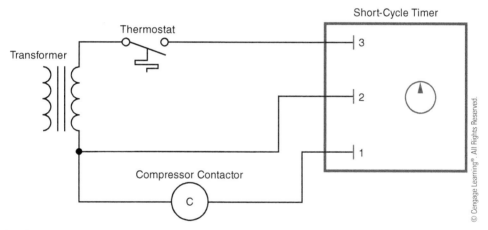

▶ Figure 36–9
Delay-on-break short-cycle timer connection.

thermostat contacts close. If the power is interrupted, the timer must time out before power can be reapplied to the compressor contactor. The delay-on-brake short cycle time has three connection terminals instead of two, Figure 36–9. A delay-on-break short-cycle timer is shown in Figure 36–10.

Some delay-on-break timers do not require a third terminal. The timer shown in Figure 36–11 is connected in the same manner as the delay-on-make timer. The connection for this time is the same as that shown in Figure 36–7. The short-cycle timer shown in Figure 36–11 can be operated on voltages that range from 12 to 240 VAC. If the timer is to be connected to 120 or 240 volts, the wire loop must be cut.

It should be noted that some electronic thermostats have a fixed internal time delay that prevents the compressor from starting for some period of time after power to the compressor contactor is interrupted, Figure 36–12. The power interruption could be caused by the thermostat disconnecting power to the compressor contactor or by a power failure. Thermostats of this type have a built-in delay-on-break short-cycle timer function.

▶ **Figure 36–11**
Some delay-on-break timers do not require connection to a third terminal.

▶ **Figure 36–12**
Some electronic thermostats have a fixed delay time.

Line Monitors and Short-Cycle Timers

Some electronic controls combine more than one function in a single unit. The control shown in Figure 36–13 is a combination line monitor and short-cycle timer for a single-phase unit. The control monitors the line voltage for a high or low condition. A 5-second time delay is built into the unit in the event a voltage fault is encountered. This prevents the control from opening the circuit to the coil of the compressor contactor in the event of a momentary line fluctuation that can occur as a result of a heavy load being connected to the power line. The voltage drop due to the starting of a large motor is a good example of a momentary voltage fluctuation.

This control also acts as a short-cycle timer. There are two control knobs on the control unit. One is set to the line voltage value the control is connected to and the other sets the delay time of the short-cycle timer. Voltage values of 95 to 135 volts AC or 190 to 260 volts AC can be selected. Control voltage can range from 18 to 240 volts AC. A schematic connection diagram for this control is shown in Figure 36–14.

A three-phase line monitor and short-cycle timer control is shown in Figure 36–15. This unit performs the same basic function as the single-phase unit except that it monitors three line voltages. Like

▶ **Figure 36–13**
Single-phase line monitor and short-cycle timer.

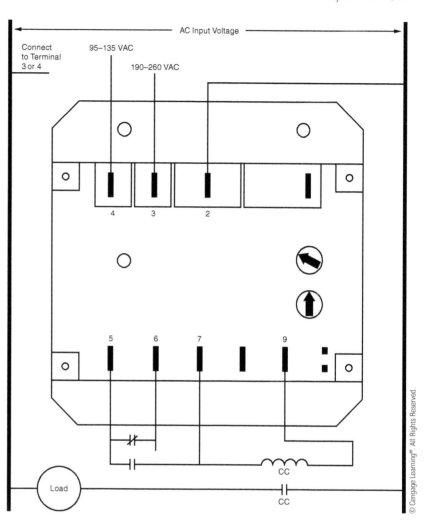

Figure 36–14
Connection for a single-phase line monitor and short-cycle timer.

Figure 36–15
Three-phase line monitor and short-cycle timer.

Figure 36–16A
The monitor is designed to plug into an eight-pin tube socket.

Figure 36–16B
Voltage adjustment for the three-phase line monitor.

the single-phase control, the three-phase control has settings for different voltage ranges and delay times for the short-cycle timer.

Another three-phase line monitor is shown in Figure 36–16A and Figure 36–16B. This monitor does not provide short-cycle protection. Its function is to monitor the line voltages and disconnect

the compressor from the line if a problem should exist. The monitor is designed to plug into an eight-pin tube socket. A set of contacts are connected in series with the coil of the compressor contactor. In the event of a line problem, the contacts open and the contactor disconnects the compressor from the line.

SUMMARY

- ⊙ Short cycling is a condition that occurs when a compressor is restarted immediately after it has been stopped.
- ⊙ Short cycling can cause the compressor to try to restart against a high head pressure and damage the compressor motor.
- ⊙ A short-cycle timer can be installed to prevent short cycling.
- ⊙ The short-cycle timer is a cam-operated, motor driven, on-delay timer.
- ⊙ The short-cycle timer is controlled by the thermostat.
- ⊙ Some electronic thermostats contain a delay-on-break short-cycle timer function.

KEY TERMS

delay-on-break   holding relay (HR)   short-cycle timer
delay-on-make   short cycling   timer motor

REVIEW QUESTIONS

1. What is short cycling?

2. What is used to provide the timing operation for the short-cycle timer?

3. What type of contacts are used in the short-cycle timer?

4. How many and what type of contacts must the holding relay have?

5. What does the dashed line drawn between the two sets of timer contacts represent?

Methods of Sensing Temperature

OBJECTIVES

After studying this unit, the student should be able to:

- ▷ List two factors that determine the amount of expansion that occurs when metal is heated
- ▷ Describe the construction and use of a bimetal strip
- ▷ Describe how a bimetal strip can be used to construct a thermometer and to operate a set of contacts
- ▷ Describe the operation of a thermocouple
- ▷ List two factors that determine the amount of voltage produced by a thermocouple
- ▷ Discuss the operation of a resistance temperature detector (RTD)
- ▷ Discuss the operation of thermistors
- ▷ Describe methods to measure temperature

In the air-conditioning and refrigeration field, the ability to sense and measure temperature is of great importance. There are numerous methods used to sense the temperature. In fact, there has probably been more emphasis on the ability to measure temperature than any other quantity. This unit deals with some of these methods.

EXPANSION OF METAL

A very common and reliable method for sensing temperature is by the **expansion** of metal. It has been known for many years that metal expands when heated. The amount of expansion is proportional to two things:

1. The type of metal used
2. The amount of heat

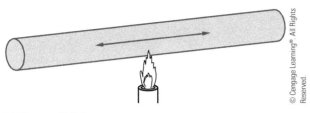

▶ **Figure 37–1**
Metal expands when heated.

Consider the metal bar shown in Figure 37–1. When the bar is heated, its length expands. When the metal is permitted to cool, it contracts. Although the amount of the movement due to contraction and expansion is small, a simple mechanical principle can be used to increase the amount of movement, Figure 37–2.

The metal bar is mechanically held at one end. This permits the amount of expansion to be in only one direction. When the metal is heated and the bar expands, it pushes against the mechanical arm. A small movement of the bar causes a great amount of movement in the mechanical arm. This increased movement of the arm can be used to indicate the temperature of the bar, or it can be used to operate a switch as shown. It should be understood that illustrations are used to convey a principle. In actual practice, the switch shown in Figure 37–2 would be spring loaded to provide a "snap" action for the contacts. Electrical contacts must never be permitted to open or close slowly. This produces poor contact pressure and causes the contacts to burn or cause erratic operation of the equipment they are intended to control. A device that uses this principle is one type of starting relay known as the hot-wire relay, which was covered in an earlier chapter.

Another very common device that operates on the principle of expansion and contraction of metal is the mercury thermometer. Mercury is a metal that remains in a liquid state at room temperature. If the mercury is confined in a narrow glass tube, as shown in Figure 37–3, it will rise up the tube as it expands due to heat. If the tube is calibrated correctly, it provides an accurate measurement for temperature.

▶ **Figure 37–3**
A mercury thermometer operates by the expansion of metal.

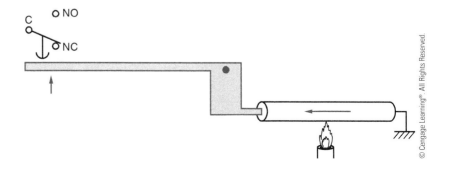

▶ **Figure 37–2**
Expanding metal operates a set of contacts.

THE BIMETAL STRIP

The bimetal strip is another device that operates by the expansion of metal. It is probably the most common heat-sensing device used in the production of thermometers and thermostats. The bimetal strip is made by bonding two dissimilar types of metal together, Figure 37–4. Because these metals are not alike, they have different expansion rates. This difference causes the strip to bend or wrap when heated, Figure 37–5.

The bimetal strip is often formed into a spiral shape, as shown in Figure 37–6. The spiral permits a longer bimetal strip to be used in a small space. The longer the strip is, the more movement will be

▶ **Figure 37–4**
A bimetal strip.

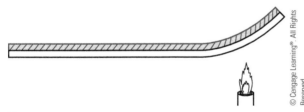

▶ **Figure 37–5**
A bimetal strip warps with a change of temperature.

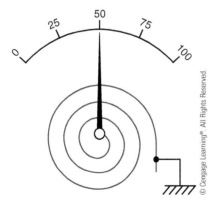

▶ **Figure 37–6**
A bimetal strip used as a thermometer.

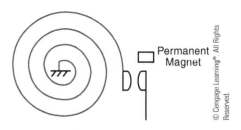

▶ **Figure 37–7**
A bimetal strip used to operate a set of contacts.

produced by a change of temperature. If one end of the strip is mechanically held and a pointer is attached to the center of the spiral, a change in temperature will cause the pointer to rotate. If a calibrated scale is placed behind the pointer, it becomes a thermometer. If the center of the spiral is held and a contact is attached to the end of the bimetal strip, it becomes a thermostat. As stated previously, electrical contacts cannot be permitted to open or close slowly. This type of thermostat uses a small permanent magnet to provide a snap action for the contact, Figure 37–7. When the moving contact reaches a point that is close to the stationary contact, the magnet attracts the metal strip and causes a sudden closing of the contacts. When the bimetal strip cools, it attempts to pull itself away from the magnet. When the force of the bimetal strip becomes strong enough, it overcomes the force of the magnet and the contacts snap open. This type of thermostat is inexpensive and has been used in homes for many years.

THE THERMOCOUPLE

The **thermocouple** is made by joining two dissimilar metals together at one end. When the joined end of the thermocouple is heated, a voltage is produced at the opposite end, Figure 37–8. The amount of voltage produced is proportional to:

1. The types of metals used to produce the thermocouple.
2. The difference in temperature of the two junctions.

The chart in Figure 37–9 shows common types of thermocouples. The metals the thermocouples are constructed from are shown, as well as their normal temperature range.

The amount of voltage produced by a thermocouple is small, generally on the order of millivolts (1 millivolt = 0.001 volt). The polarity of the voltage of a thermocouple is determined by the temperature. For example, a type "J" thermocouple produces zero volts at about 32°F (Fahrenheit). At temperatures above 32°, the iron wire is positive and the constantan wire is negative. At a temperature of 300°F, this thermocouple produces

a voltage of about 7.9 millivolts. At a temperature of −300°F, it produces a voltage of about −7.5 millivolts. This indicates that at temperatures below 32°F, the iron wire becomes the negative lead and the constantan wire becomes the positive-voltage lead.

Because thermocouples produce such low voltages, they are often connected in series, as shown in Figure 37–10. This series connection permits the voltages to add and produce a higher output voltage. This connection is known as a **thermopile**.

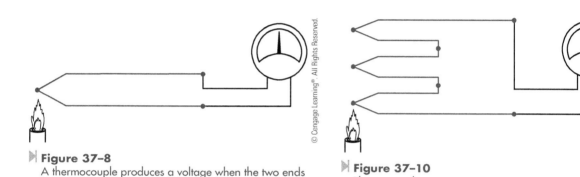

© Cengage Learning®. All Rights Reserved.

◢ **Figure 37–8**
A thermocouple produces a voltage when the two ends are at different temperatures.

◢ **Figure 37–10**
Thermocouple.

© Cengage Learning®. All Rights Reserved.

TYPE	MATERIAL		DEGREES F	DEGREES C
J	Iron	Constantan	−328 to +32 +32 to +1432	−200 to 0 0 to 778
K	Chromel	Alumel	−328 to +32 +32 to +2472	−200 to 0 0 to 1356
T	Copper	Constantan	−328 to +32 +32 to +752	−200 to 0 0 to 400
E	Chromal	Constantan	−328 to +32 +32 to +1832	−200 to 0 0 to 1000
R	Platinum 13% Rhodium	Platinum	+32 to +3232	0 to 1778
S	Platinum 10% Rhodium	Platinum	+32 to +3232	0 to 1778
B	Platinum 30% Rhodium	Platinum 6% Rhodium	+992 to +3352	533 to 1800

© Cengage Learning®. All Rights Reserved.

◢ **Figure 37–9**
Thermocouple chart.

RESISTANCE TEMPERATURE DETECTORS

The **resistance temperature detector (RTD)** is made of platinum wire. The resistance of platinum changes greatly with temperature. When platinum is heated, its resistance increases at a very predictable rate. This makes the RTD an ideal device for measuring temperature very accurately. RTDs are used to measure temperatures that range from −328 to +1166°F (−200 to +630°C). RTDs are made in different styles to perform different functions. Figure 37–11 illustrates a typical RTD used as a probe. A very small coil of platinum wire is encased inside a copper tip. Copper is used to provide good thermal contact. This permits the probe to be very fast acting. The chart in Figure 37–12 shows resistance versus temperature for a typical RTD probe.

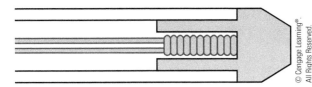

© Cengage Learning®. All Rights Reserved.

▶ **Figure 37–11**
Resistance temperature detector.

DEGREES C	RESISTANCE
0	100
50	119.39
100	138.5
150	157.32
200	175.84
250	194.08
300	212.03
350	229.69
400	247.06
450	264.16
500	280.93
550	297.44
600	313.65

© Cengage Learning®. All Rights Reserved.

▶ **Figure 37–12**
Temperature and resistance for a typical RTD.

The temperature is given in degrees Celsius (°C) and resistance is given in ohms.

THERMISTORS

The term *thermistor* is derived from the words *thermal* and *resistor*. Thermistors are actually thermally sensitive semiconductor devices. There are two basic types of thermistors. One type has a **negative temperature coefficient (NTC)** and the other has a **positive temperature coefficient (PTC)**. A thermistor that has a negative temperature coefficient decreases its resistance as the temperature increases. A thermistor that has a positive temperature coefficient increases its resistance as temperature increases. The NTC thermistor is the most widely used.

Thermistors are highly nonlinear devices. For this reason, they are difficult to use for measuring temperature. Devices that measure temperature with a thermistor must be calibrated for the particular type of thermistor being used. If the thermistor is ever replaced, it has to be an exact replacement or the circuit will no longer operate correctly. Because of their nonlinear characteristic, thermistors are often used as set point detectors as opposed to actual temperature measurement. A set point detector is a device that activates some process or circuit when the temperature reaches a certain level. For example, assume a thermistor has been placed inside the stator of a motor used to operate a compressor. If the motor becomes overheated, the windings of the motor could be severely damaged or destroyed. The thermistor can be used to detect the temperature of the windings. When the resistance of the thermistor falls to a certain level NTC type, a set of contacts connected in series with the motor starter coil of the compressor opens. When the compressor motor starter de-energizes, the compressor is disconnected from the power line. Thermistors can be operated in temperatures that range from about −100 to +300°F.

THE PN JUNCTION

Another device that has the ability to measure temperature is the **PN junction**, or diode. The diode is becoming a very popular device for measuring temperature because it is accurate and linear.

When a silicon diode is used as a temperature sensor, a constant current is passed through the diode. Figure 37–13 shows this type of circuit. In this circuit, resistor R1 limits the current flow through

the transistor and the sensor diode. The value of R1 also determines the amount of current that will flow through the diode. Diode D1 is a 5.1-volt zener diode used to produce a constant voltage between the base and emitter of the PNP transistor. Resistor R2 limits the amount of current flow through the zener diode and the base of the transistor. Diode D2 is a common silicon diode. It is being used as the temperature sensor for the circuit. If a digital voltmeter is connected across the diode, a voltage drop between 0.8 and 0 volt can be seen. The amount of the voltage drop is determined by the temperature of the diode.

If the diode is subjected to a lower temperature, say by touching it with a piece of ice, it will be seen that the voltage drop of the diode will increase. If the temperature of the diode is increased by holding it between two fingers or bringing a hot soldering iron near it, its voltage drop will decrease. Notice that the diode has a negative temperature coefficient. As its temperature increases, its voltage drop becomes less. The circuit shown in Figure 37–14

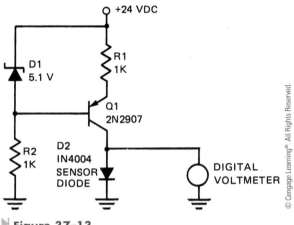

▶ Figure 37–13
Constant current generator.

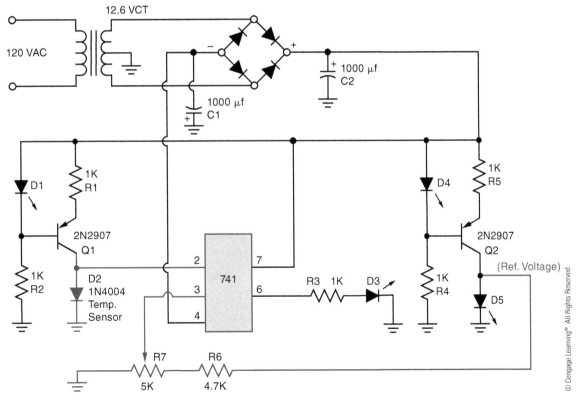

▶ Figure 37–14
Set level detector for temperature.

can be used as a set point detector. The operation of the circuit is as follows.

A bridge rectifier and a center-tapped transformer are used to produce an above- and below-ground power supply. If ground is considered as zero volts, the positive output of the bridge will be positive with respect to ground, and the negative output of the bridge will be negative with respect to ground. Capacitors C1 and C2 are used to filter the DC output voltage of the rectifier. Notice that capacitor C1 has its positive lead connected to ground and C2 has its negative lead connected to ground. The positive output of the rectifier will produce a voltage that is about +9 volts compared to ground, and the negative output will produce a voltage that is about −9 volts compared to ground.

Diode D1 is a light-emitting diode connected in the forward direction. In this circuit, the LED is used as a low-voltage zener diode. Because the LED has a constant voltage drop of about 1.7 volts, it can be used to provide a constant voltage. Resistor R1 limits the current flow through the diode and the sensor resistor. Resistor R2 limits current flow through the LED and the base of the transistor. Notice this is the same constant current generator circuit shown in Figure 37–13, with the exception of the LED's being used as the zener diode.

Transistor Q2, resistors R5 and R4, and LED D4 form another constant current generator circuit. Notice this generator is connected to an LED, D5. In this circuit, D5 is used to provide a low-voltage reference source for the operational amplifier. When a light-emitting diode is connected to a constant current source, its voltage drop is very stable. This makes it an ideal choice when a steady reference voltage is needed. Resistors R6 and R7 are used to form a voltage divider. Resistor R5 is a 5000-ohm variable resistor that has a voltage drop across its entire resistance of about 1 volt. The wiper tap of this resistor is connected to the noninverting input of the 741. Because resistor R5 has a voltage drop of only 1 volt across its resistance, the full range of the wiper will adjust the voltage applied to the noninverting input between 1 volt

and 0. This is done to make adjustment of the detector circuit easy. Because the voltage drop of diode D2 will never be greater than 0.8 volt, resistor R7 can adjust the entire range over which the detector can operate.

Diode D3 is used as an output indicator. When the output is low, D3 will be turned off. When the output of the op amp goes high, D3 will be turned on. Diode D3 is used only as an indicator in this circuit. The output of the op amp could be used to operate the input of a transistor or a solid-state relay. The transistor or relay could be used to operate almost anything desired. Resistor R3 limits the current flow through D3.

To understand the operation of this circuit, assume that resistor R7 has been adjusted to a point that the output of the op amp is off or low. This means that the voltage applied to the inverting input, pin 2, is more positive than the voltage set at pin 3. If the temperature of diode D2 is increased, its voltage drop will decrease. When the temperature of the sensor diode becomes high enough, its voltage drop will be less than the voltage set at the noninverting input. When the voltage applied to pin 3 becomes more positive than the voltage applied to pin 2, the output of the op amp will go high or turn on. Adjustment of resistor R7 permits the detector to be used over a wide range of temperatures.

EXPANSION DUE TO PRESSURE

Another common method of measuring temperature is by the increase of pressure of some chemicals. Refrigerant, for example, increases pressure as temperature increases. If a simple bellows is connected to a line containing refrigerant, Figure 37–15, the bellows will expand as the pressure inside the sealed system increases. When the surrounding temperature decreases, the pressure inside the system decreases, and the bellows contracts. When the bellows is made to operate a set of contacts, it is generally referred to as a **bellows-type thermostat**.

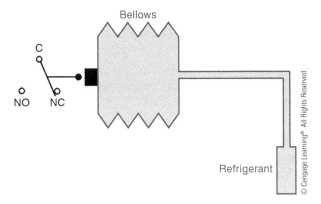

Bellows

C

NO   NC

Refrigerant

Figure 37–15
Bellows contracts and expands with a change of refrigerant pressure.

SUMMARY

- A common method of sensing temperature is by the expansion of metal.
- Two factors that determine the amount of expansion that will occur when metal is heated are:
  A. The type of metal used.
  B. The temperature of the metal.
- A common device that operates on the principle of expansion of metal is the mercury thermometer.
- A bimetal strip is constructed by bonding two types of metal together that expand at a different rate.
- The sensitivity of a bimetal strip is proportional to its length.
- Bimetal strips are often wound into a spiral to permit a longer strip to be used in a small space.
- Thermocouples produce a voltage when heated at one end.
- Thermocouples are made by joining two dissimilar metals together at one end.
- The amount of voltage produced by a thermocouple is determined by:
  A. The types of metals used.
  B. The difference in temperature of the two junctions.
- Resistance temperature detectors (RTDs) are made of platinum wire.
- The resistance of platinum changes at a very predictable rate as the temperature changes.
- Thermistors are devices that rapidly change their resistance with a change of temperature.
- The resistance of a thermistor with a positive temperature coefficient will increase with an increase of temperature.

⊙ The resistance of a thermistor with a negative temperature coefficient decreases with an increase of temperature.

⊙ Because of thermistors' ability to rapidly change resistance with a change of temperature, they are generally used as temperature-sensitive switches.

⊙ A PN junction can be used to sense temperature by passing a constant current through it and detecting the voltage drop across the junction.

⊙ When a constant current is passed through a PN junction, its voltage drop is proportional to the temperature.

⊙ A metal bellows connected to a sealed refrigerant line can be used to sense temperature because the pressure in the sealed system will be proportional to the temperature.

KEY TERMS

bellows-type
   thermostat
expansion
negative temperature
   coefficient (NTC)

PN junction
positive temperature
   coefficient (PTC)
resistance temperature
   detector (RTD)

thermocouple
thermopile

REVIEW QUESTIONS

1. Should a metal bar be heated or cooled to make it expand?

2. What type of metal remains in a liquid state at room temperature?

3. How is a bimetal strip made?

4. Why are bimetal strips often formed into a spiral shape?

5. Why should electrical contacts never be permitted to open or close slowly?

6. What two factors determine the amount of voltage produced by a thermocouple?

7. What is a thermopile?

8. What do the letters RTD stand for?

9. What type of wire are RTDs made of?

10. What material is a thermistor made of?

11. Why is it difficult to measure temperature with a thermistor?

12. If the temperature of an NTC thermistor increases, will its resistance increase or decrease?

13. How can a silicon diode be made to measure temperature?

14. Assume that a silicon diode is being used as a temperature detector. If its temperature increases, will its voltage drop increase or decrease?

15. What is an above- and below-ground power supply?

Gas Burner Controls

OBJECTIVES

After studying this unit, the student should be able to:

- Discuss the functions of a gas burner control
- Discuss the pilot light method of igniting a gas burner
- Discuss high-voltage spark ignition
- Describe the operation of a thermocouple
- Describe the operation of a "fire eye" and "flame rod" flame sensor
- Discuss the operation of the main control valve

The primary function of a gas burner control is to ensure that gas is not permitted to enter the system if it cannot be ignited in a safe manner. An accumulation of gas is extremely explosive and must be avoided. Several methods of igniting the main burner can be employed. The two most common in use today are the pilot light and high-voltage spark ignition.

PILOT LIGHT

Probably the oldest method of automatically igniting the main burner is with a **pilot light**. A pilot light is a small gas flame that burns continuously near the main burner. When gas is permitted to flow to the main burner, the pilot light ignites the fuel. If the pilot light should not be in operation when the gas is permitted to flow to the main burner, an accumulation of gas could result in an explosion. Therefore, the control system must have some means of sensing the presence of the pilot flame. If the pilot flame is not present, the main gas valve goes into safety shutdown and does not permit gas to be supplied to the main burner.

HIGH-VOLTAGE SPARK IGNITION

Many of the newer gas-operated appliances and heating systems use an **electric arc** to ignite the gas flame. This system uses less energy than a pilot light because it does not depend on a gas flame being present at all times. The electric arc is used only during the actual ignition sequence. When electric arc ignition is used, the gas-control system must be different also. Instead of sensing the presence of a pilot flame, the control system turns on the electric ignitor and permits gas to flow. If a flame is not detected in a short period of time, the control system turns off the flow of gas.

FLAME SENSORS

There are several methods used to sense the presence of a gas flame. One of the most common is with the use of a thermocouple. The thermocouple is a device that produces a voltage when heated. If the thermocouple is inserted in the gas flame, as shown in Figure 38–1, a voltage will be produced. The voltage produced by the thermocouple is used to create a current flow through the coil of a solenoid. The current produces a magnetic field that holds the valve open. As long as the solenoid receives enough current, a valve is held open, and gas is permitted to

flow to the main burner. If the pilot light goes out, no voltage will be produced, and the pilot valve will stop the flow of gas. It should be noted that the thermocouple has the ability to produce enough current to hold the valve open, but it cannot produce enough current to reopen the valve if it is closed. The pilot valve must be opened manually by pushing the pilot button located on the main valve. It should also be noted that some controls of this type are actually thermopiles and not thermocouples. Recall that a thermopile is a series connection of several thermocouples used to produce a higher voltage. When replacing a thermocouple, care must be taken to use the proper type. A thermocouple is shown in Figure 38–2.

Another type of flame sensor uses pressure. This control is similar to the pressure type of thermostat. A refrigerant-filled bulb is located in the pilot flame. When the refrigerant is heated, a pressure is produced that holds the pilot safety valve open.

Another type of gas flame sensor is the **"fire eye"** or **"flame eye."** The fire eye is a gas-filled tube that has a very high resistance in its normal state and will not conduct electricity. A gas flame contains ultraviolet (UV) radiation. The ultraviolet radiation of the gas flame causes the gas in the fire eye to ionize and conduct electricity. Notice that this type of sensor detects the light of a flame and not the heat. This type of control is generally used to sense

▶ **Figure 38–2**
Thermocouple and pilot burner.

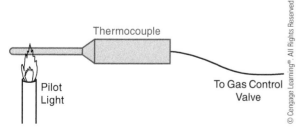

▶ **Figure 38–1**
Thermocouple senses pilot flame.

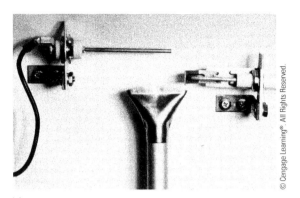

▶ Figure 38–3
Flame rod, burner head, and high-voltage ignition electrode.

the presence of the main burner flame instead of the pilot flame. Fire eye detectors are generally used with timers that turn the gas supply off if a flame is not detected within a certain time after a call for heat.

The **"flame rod"** is another sensor that is generally used to detect the presence of flame at the main burner. The flame rod operates by using the gas flame as a conductor of electricity. A gas flame contains many ionized particles that will conduct electricity in a similar manner to some types of vacuum tubes. When the flame rod is inserted in a flame, a current path exists between the rod and the metal of the burner head itself. As long as there is a flame, there can be a flow of electricity between the rod and the burner head. If the flame should be extinguished, the flow of electricity will stop. Figure 38–3 shows a photograph of a flame rod, a small burner head, and an electric spark ignitor.

CONTROL VALVES

The gas **control valve** is the real heart of the gas heating system. Control valves control the flow of gas to the main burner and the pilot light, if used. Many of them contain an internal **pressure regulator**, which maintains a constant pressure to the main burner. A simple gas control valve is shown in Figure 38–4. This illustration is used to show the basic principle of operation. This type of valve uses a thermocouple to detect the presence of a pilot flame.

Notice that a spring is used to close the valve if the thermocouple should stop producing current for the solenoid coil. Also notice that a solenoid coil is used to open the main valve when the thermostat calls for heat. Different valves use different methods of opening the main valve. Some valves use a small electric heater to heat a bimetal strip that opens the main valve. Others use a small heater to cause a metal rod to expand and open the main valve. Regardless of the method used, all control valves perform the same basic function.

The schematic in Figure 38–5 shows a basic control circuit for a gas heating system. Notice that the fan and high-limit switch are connected in the 120-volt line ahead of the control transformer. When the thermostat closes, 24 volts AC is applied to the control valve. This permits the valve to open and supply gas to the main burner. Notice that this control valve uses a thermocouple to sense the presence of a pilot light. If the pilot light should go out, the pilot valve will close and gas flow to the burner will stop.

The schematic in Figure 38–6 shows a control circuit that uses a high-voltage spark ignitor. Notice that a fan-limit switch is connected ahead of the 24 volts control transformer. This is the same as the other type of control. In this circuit, however, when the thermostat calls for heat, 24 volts AC is applied to a direct spark ignition control module. When the control module receives a call for heat, it turns on the main control valve and provides about 15,000 volts to the ignition electrode. The module also starts an electronic timer at the same time. When the gas is ignited at the burner head, a current flows from the flame rod to the base of the burner. This completes a circuit through the ground wire back to the control module. This flow of current is used to turn off the timer and electric ignitor. As long as a flame is present and the thermostat calls for heat, the main valve is permitted to remain open. If the flame should go out, however, current flow between the flame rod and burner ground will be broken and the timer and electric ignitor will be started. If a flame is not established in a predetermined time, the main valve will be turned off and the flow of gas stopped. Some systems are equipped with an alarm relay that is turned on by a solid-state relay when the control module senses an unsafe condition.

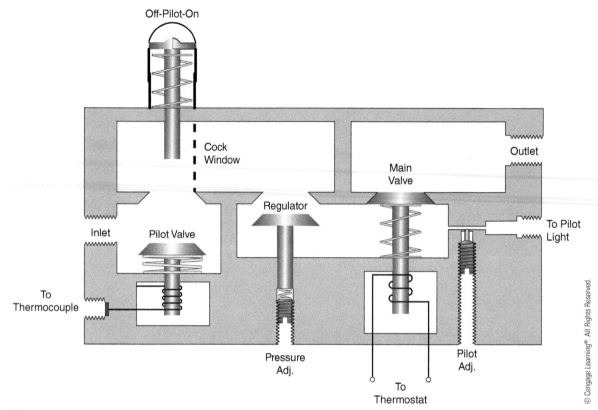

▶ **Figure 38–4**
Basic gas control valve.

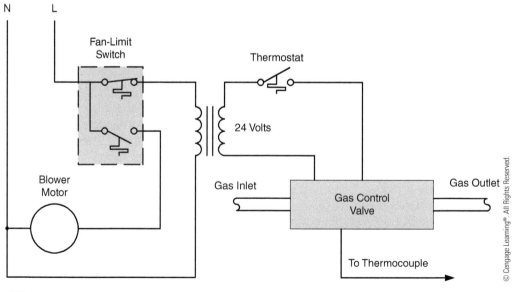

▶ **Figure 38–5**
Basic gas control system.

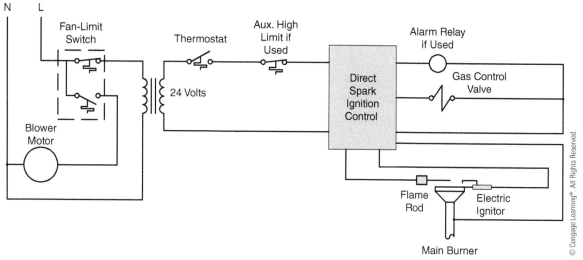

▶ **Figure 38–6**
Electric spark ignition control.

▶ **Figure 38–7**
Direct spark ignition control module.

▶ **Figure 38–8**
Main gas valve used with electric ignition.

If a flame is not established when the thermostat calls for heat, the timer will shut down the system in the same manner. A direct ignition control module is shown in Figure 38–7, and a gas control valve used with an electric spark ignition system is shown in Figure 38–8.

SUMMARY

⊙ The primary function of a gas burner control is to ensure that gas is not permitted to enter the system if it cannot be ignited in a safe manner.

⊙ The oldest method of automatically igniting a gas burner is with a pilot light.

⊙ A thermocouple is used to sense the presence of a pilot light before gas is permitted to flow to the main burner.

⊙ The pilot light heats the thermocouple, which supplies an electric current to hold a solenoid valve open in the main control valve.

⊙ High-voltage spark ignition is used on some gas appliances to ignite the main burner.

⊙ The "fire eye" type of flame sensor changes resistance in the presence of the light of the flame of the main burner.

⊙ A "flame rod" senses the presence of a gas flame by using ionized particles in the flame as a conductor.

⊙ The gas control valve controls the flow of gas to the main burner.

KEY TERMS

control valve
electric arc

"fire eye" or "flame eye"
"flame rod"

pilot light
pressure regulator

REVIEW QUESTIONS

1. What is the purpose of a pilot light?

2. Why is it necessary to be certain that the gas is ignited at the main burner on a call for heat by the thermostat?

3. What is a thermocouple?

4. What is a thermopile?

5. Why must the pilot control valve be reset manually if it should open?

6. Explain how a "fire eye" works.

7. Explain the operation of a "flame rod."

8. What is the common amount of voltage applied to an electric spark ignitor?

9. Why must a ground wire be connected between the direct spark ignition control module and the burner head?

10. What is the advantage of electric spark ignition over pilot light ignition?

UNIT 39 ▷ Oil Burner Controls

OBJECTIVES

After studying this unit, the student should be able to:

▷ Discuss the operation of an oil-fired heating system

▷ Describe electric ignition of a gun-type oil heating system

▷ Discuss the operation of the primary control

▷ Describe the operation of the CAD cell

Some of the controls on an oil-fired heating system are basically the same as the controls on a gas-fired system. The fan and limit controls are very similar and in some cases the same. The major part of an oil-fired control system is the **primary control**. The primary control's function is to ensure that when the thermostat calls for heat, the flame will be established within a predetermined amount of time. This is to prevent an accumulation of oil vapor in the combustion chamber. If a large amount of oil is formed in the combustion chamber and ignited, an explosion could occur.

IGNITION

A **gun-type** oil furnace is ignited by an electric arc. Two electrodes are located near the nozzle. When the thermostat calls for heat, the primary control connects the ignition transformer to the 120-volt AC power line. The transformer steps the 120 volts up to 10,000 volts. The 10,000 volts is connected to two electrodes. This causes an arc to be produced between the two electrodes. The air produced by the combustion fan motor causes the arc to be blown in a horseshoe shape, as shown in Figure 39–1. This arc is used to ignite the oil. The electrodes are adjusted in such a manner that they do not enter into the oil spray produced by the nozzle. Only the horseshoe-shaped arc is permitted to contact the oil spray. If the electrodes are adjusted too far in front of the nozzle, they may touch the spray, which will cause them to burn and soot. If they are adjusted too far behind the nozzle, the arc will not contact the oil spray. This will cause the furnace to start hard and have delayed ignition.

PRIMARY CONTROL

The schematic of a typical primary control is shown in Figure 39–2. Notice that this control employs several solid-state components in its operation. These components are:

1. The **silicon bilateral switch (SBS)**.
2. The triac.
3. The **cadmium sulfide cell (CAD cell)**.

In this circuit, the gate lead of the SBS has been left disconnected. This permits the SBS to operate in a manner very similar to that of a diac. When the voltage applied to the SBS reaches a high enough level, assume 5 volts for this example, it will turn on and conduct current to the gate of the triac. This will permit the triac to turn on.

CAD CELL

The CAD cell is a device that changes its resistance in accordance with the amount of light it is exposed to. When the CAD cell is in darkness, it will have a very high resistance of several hundred thousand ohms. When it is in light, its resistance will decrease to about 50 ohms.

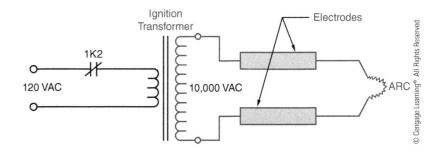

Figure 39–1
Electric arc ignites oil-fired furnace.

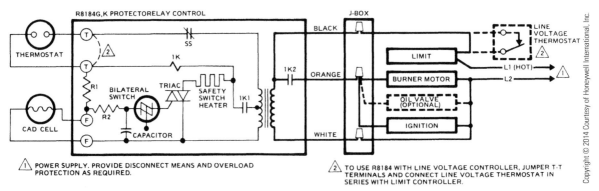

Figure 39–2
Internal schematic and typical hookup for R8184G.

CIRCUIT OPERATION

To help in understanding how this circuit works, it will be shown in different stages of operation. In the circuit shown in Figure 39–3, the thermostat has just called for heat. The arrows are used to show the path of current flow through the circuit. The current leaves one side of the step-down transformer and flows through the thermostat contacts. The current then flows through resistor R1. Because the CAD cell is in darkness, it has a very high resistance. This causes most of the voltage to be dropped at the junction point of R1 and R2. Because the voltage at this point is greater than 5 volts, the SBS will turn on and conduct current to the gate of the triac. When the triac turns on, current is permitted to flow through relay coil 1K, the safety switch heater, the triac, and back to the transformer. Notice that coil 1K is connected in series with the safety switch heater at this time.

Figure 39–4 illustrates the operation of the circuit when relay coil 1K energizes. Notice that both contacts 1K1 and 1K2 are shown closed. When contact 1K2 closes, 120 volts is connected to the burner motor and the ignition transformer. When contact 1K1 closes, a different current path for the relay coil and safety heater is provided to the center tap of the transformer. Relay coil 1K and the safety switch heater are no longer connected in series. Notice that one current path is through the thermostat, and 1K relay coil. The current path through the SBS and triac gate is still provided because the oil flame has not been ignited as yet and the CAD cell is still in darkness.

A second current path is provided through the triac and safety switch heater. If, for some reason, ignition

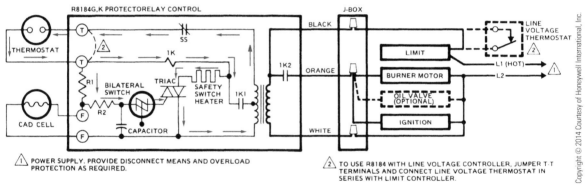

Figure 39–3
Internal schematic and typical hookup for R8184G after thermostat has called for heat.

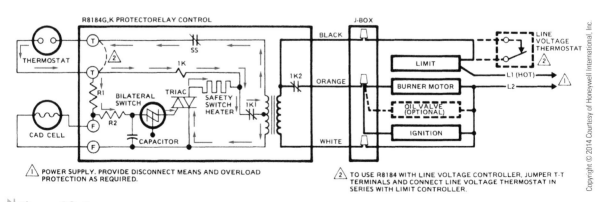

Figure 39–4
Internal schematic and typical hookup for R8184G when relay coil 1K energizes.

should not occur, current will continue to flow through the triac and safety switch heater. This will eventually cause the bimetal contact SS to open and disconnect the thermostat circuit. If this should happen, it is necessary to manually reset the primary control with the reset button located on the control unit.

In Figure 39–5, the circuit is shown in its normal operating condition after ignition. Notice that current is still permitted to flow through the 1K relay coil to keep it energized. The triac, however, has been turned off. When ignition occurs, the CAD cell "sees" the light of the flame. This causes its resistance to drop to a low value. When this happens,

the voltage drop at the junction of resistors R1 and R2 becomes very low. Because there is now less than 5 volts, the SBS is turned off and no current is conducted to the gate of the triac. This stops the current flow through the safety switch heater and the circuit continues to operate until the thermostat is satisfied.

A photograph of a CAD cell used as the flame detector in an oil furnace is shown in Figure 39–6. A primary control unit for an oil furnace is shown in Figure 39–7. A burner assembly with pump, burner motor, primary control, and ignition transformer is shown in Figure 39–8.

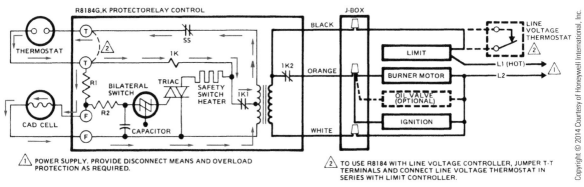

▷ **Figure 39–5**
Internal schematic and typical hookup for R8184G in normal operating condition after ignition.

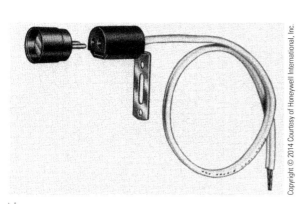

▷ **Figure 39–6**
CAD cell flame detector.

▷ **Figure 39–7**
Primary control.

Figure 39–8
Oil burner assembly
with pump, burner mo-
tor, primary control, and
ignition transformer.

SUMMARY

- ⊙ The primary control controls the major operation of an oil-fired heating system.
- ⊙ The primary control must ensure that ignition is established in the combustion chamber to prevent an accumulation of oil vapor.
- ⊙ Oil-fired systems are generally ignited by a high-voltage transformer producing an arc across two electrodes located near the nozzle.
- ⊙ Most primary controls contain solid-state components.
- ⊙ A CAD cell located in a position close to the flame is used to sense burner igniting by detecting the light of the flame.
- ⊙ A CAD cell is a solid-state device that lowers its resistance in the presence of light.

KEY TERMS

cadmium sulfide cell (CAD cell)
gun-type
primary control
silicon bilateral switch (SBS)

▶▶ REVIEW QUESTIONS

1. What is the function of the ignition transformer?
2. How much voltage is supplied to the electrodes?
3. Are the electrodes permitted to enter into the oil spray?
4. What does enter into the oil spray to cause ignition?
5. What device is controlled by the operation of the triac?
6. What solid-state device controls the flow of gate current to the triac?
7. Does the CAD cell have a high resistance or low resistance when in the presence of light?
8. How would the circuit operate if the CAD cell were in the presence of light when the thermostat called for heat?

SECTION 6

Troubleshooting Using Control Schematics

UNIT 40

Introduction to Troubleshooting

OBJECTIVES

After studying this unit, the student should be able to:

- Use a voltmeter to measure voltage across circuit components
- Use an ohmmeter to measure continuity and component resistance
- Use an ammeter to measure circuit current
- Explain the hopscotch method of troubleshooting
- Discuss the use of current transformers
- Discuss safety considerations when using CTs

Troubleshooting is probably the most important part of a service technician's job. Good troubleshooting ability will save hours of valuable time and money. Unfortunately, troubleshooting can be one of the most confusing aspects of the job if the technician does not know the basics. Different technicians use different methods. Most adopt a procedure they are comfortable with and understand. Some basic questions that should be considered when troubleshooting any type of system are listed here:

- What is the system supposed to do?
- How does it do it?
- What is it doing that it should not be doing, or what is it not doing that it should be doing?

Knowing the answer to these basic questions will generally point you in the right direction for determining the problem.

Troubleshooting electric circuits generally involves the use of electric **measuring instruments**. The most common are the voltmeter, ammeter, and ohmmeter. Understanding how each of these instruments functions and how to employ them is one of the keys to developing good troubleshooting skills. The general use of each will be discussed in this unit.

THE VOLTMETER

Recall that one definition of voltage is electrical pressure. The voltmeter indicates the amount of potential between two points, in much the same way a pressure gauge indicates the pressure difference between two points. In the circuit shown in Figure 40–1, assume that a voltage of 120 volts exists between L1 and N. If the leads of a voltmeter were to be connected between L1 and N, the meter would indicate 120 volts. Now assume that the leads of the voltmeter are connected across the lamp, Figure 40–2.

QUESTION 1: Assuming the lamp filament is good, would the voltmeter shown in Figure 40–2 indicate 0 volts, 120 volts, or some voltage value between 0 and 120?

ANSWER: The voltmeter would indicate 0 volts. In the circuit shown in Figure 40–2, the switch and lamp are connected in series. One of the basic rules for a series circuit is that the voltage drop across all components equals the applied voltage. The voltage drop across each component is proportional to the amount of resistance of the component and the amount of current flow. In the circuit shown in Figure 40-2, there is no current flow because the switch is open. Because no current can flow through the lamp, there can be no voltage drop.

QUESTION 2: If the voltmeter probes were to be moved across the switch, as shown in Figure 40–3, would the meter indicate 0 volts, 120 volts, or some value between 0 and 120 volts?

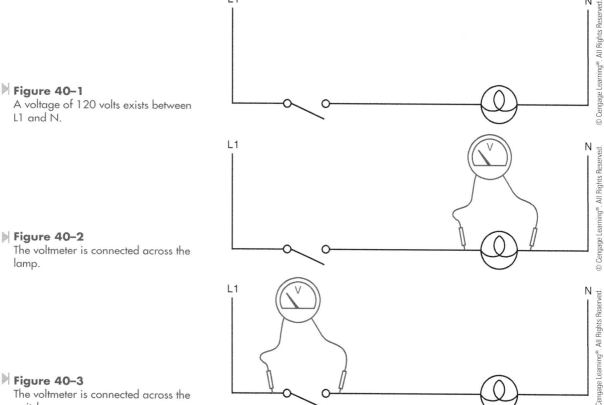

▶ **Figure 40–1**
A voltage of 120 volts exists between L1 and N.

▶ **Figure 40–2**
The voltmeter is connected across the lamp.

▶ **Figure 40–3**
The voltmeter is connected across the switch.

ANSWER: The voltmeter would indicate 120 volts. Because the switch is an open circuit, the resistance is infinite at this point, which is millions of times greater than the resistance of the lamp filament. Remember that voltage is electrical pressure. The only current flow necessary to measure voltage is the current flow through the meter itself. In this circuit, the only current path is through the resistance of the voltmeter and the lamp filament, Figure 40–4. If the probes of the voltmeter were to be connected to a wall outlet, the meter would indicate 120 volts, but there would be no current flow except through the meter itself.

QUESTION 3: If the total or applied voltage in a series circuit equals the voltage drop across each component, why is all the voltage drop across the voltmeter resistor and none across the lamp filament?

ANSWER: There is some voltage drop across the lamp filament because the current of the voltmeter is flowing through it. The voltage drop across the lamp filament, however, is so small as compared with the voltage drop across the voltmeter resistance that it is generally considered to be zero. Assume the lamp filament to have a resistance of 50 ohms. Now assume the voltmeter is a digital voltmeter with a resistance of 10 million ohms. The total circuit resistance is 10,000,050 Ω. The total circuit current is 0.000011999 amp (120/10,000,050), or about 12 microamperes. The voltage drop across the lamp filament is 0.0006 volt or 0.6 millivolt (50 $\times$ 12 μA).

QUESTION 4: Assume that the filament of the lamp is open or burned out. Would the voltmeter in Figure 40–3 indicate 0 volts, 120 volts, or some value between 0 and 120 volts?

ANSWER: The voltmeter would indicate 0 volts. If the lamp filament is open or burned out, a current path for the voltmeter does not exist and the voltmeter would indicate zero, Figure 40–5. In order for the voltmeter to indicate voltage, it would have to be connected across L1 and N so that a complete circuit through the meter would exist, Figure 40–6.

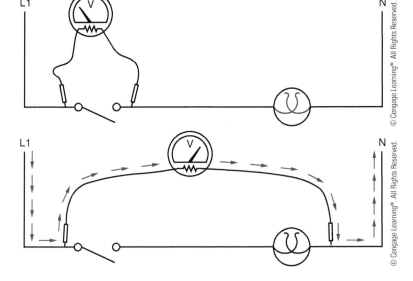

▶ **Figure 40-4**
Current flows through the meter and lamp filament.

▶ **Figure 40-5**
If the lamp filament is open, no current can flow.

▶ **Figure 40-6**
A complete circuit must exist through the meter to measure voltage.

TROUBLESHOOTING WITH THE VOLTMETER

Before it is possible to troubleshoot a circuit, the technician must have an understanding of how the circuit operates normally. The circuit shown in Figure 40–7 is a basic control circuit for a compressor. The circuit operates as follows:

1. When the thermostat contacts close, power is provided to the FR (fan relay) coil.

2. This causes both FR load contacts to close and connect the fan motor to the 240-volt line.

3. The moving air of the fan causes flow switch FL to close.

4. When switch FL closes, power is provided to the compressor contactor (CC).

5. Both CC load contacts close and supply power to the compressor motor.

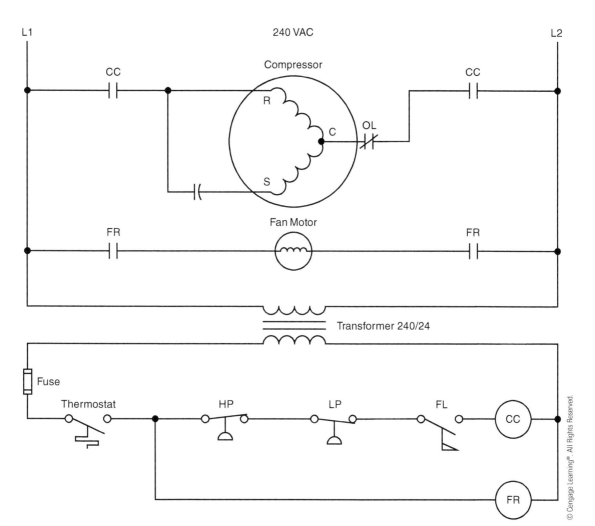

▶ **Figure 40–7**
Basic compressor circuit.

6. High-pressure and low-pressure switches connected in series with the compressor contactor provide protection for the compressor.

7. When the thermostat opens, the circuit to both the compressor contactor and fan relay is broken disconnecting the compressor and fan from the line.

Assume that a problem has developed with the unit. The service technician is told that the air conditioner will not work. The first test to be made is to determine whether control voltage is available from the secondary of the transformer. This can be done by checking for 24 volts from the thermostat to CC, Figure 40–8. For the purpose of this example, it will be assumed that the voltmeter indicated 24 volts.

The next step is to attempt to operate the unit. Many service technicians use a jumper to short the thermostat contacts. When a jumper is used to short components in a control circuit, a **fused jumper**

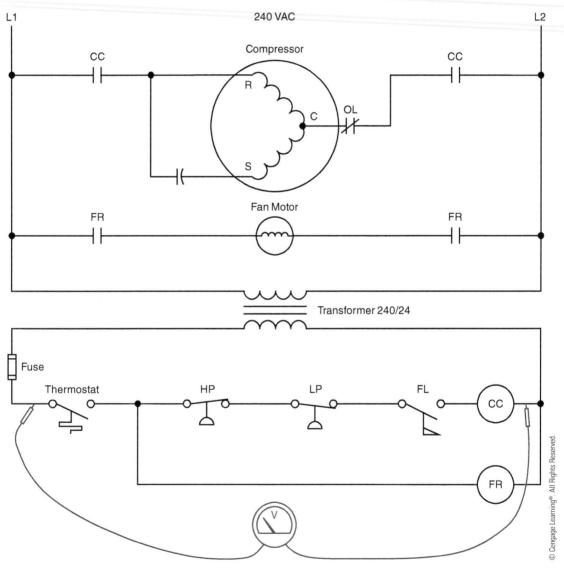

▶ **Figure 40–8**
Testing the transformer voltage.

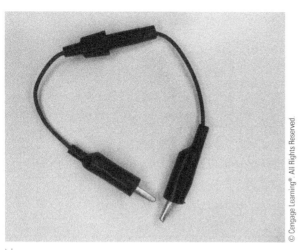

▶ **Figure 40–9**
Fused jumper for shorting control components.

is recommended, Figure 40–9. The jumper contains a small amp value fuse such as 3 or 4 amperes. If the jumper should be accidentally connected across power, the fuse will blow instantly. Assume then that the thermostat was jumped; the fan motor started, but the compressor did not.

The next step is to determine what could be the problem. Looking at the schematic, make mental notes of what could cause the compressor not to start.

1. CC contactor is defective.
2. Flow switch FL did not close.
3. The low-pressure switch, LP, is open.
4. The high-pressure switch, HP is open.
5. CC load contacts are burned out and not connecting power to the compressor.
6. The compressor overload is open.
7. The compressor start capacitor is bad.
8. The compressor motor is bad.

The next logical step is probably to determine whether voltage is being applied to the coil of the compressor contactor. This can be done by jumping the thermostat and checking across CC coil with a voltmeter, Figure 40–10.

For this example, it is assumed that the voltmeter indicated that there was no voltage applied to contactor coil CC.

THE HOPSCOTCH METHOD

A very common troubleshooting method is called the **hopscotch method**. As the name implies, you jump from one component to another until the open part of the circuit is found. In the example in Figure 40–10, the voltmeter is connected across the coil of contactor CC. To use the hopscotch method of troubleshooting, leave one voltmeter probe connected to one side of the transformer and connect the other probe on the other side of the next component in line, Figure 40–11. If the voltmeter indicates 24 volts, it means that the flow switch is open and preventing the compressor contactor from energizing. If the voltmeter indicates 0 volts, it means that there is still an open condition somewhere else in the circuit that is preventing the voltmeter from receiving a flow or current. The 0 volt reading does not mean that contact FL is closed. Contact FL could be open, but there is something else open in the circuit ahead of it. In this example, it will be assumed that the voltmeter indicates 0 volts.

The next step is to hopscotch to the next component, which is the low-pressure switch, Figure 40–12. If the voltmeter indicates a voltage of 24 volts, it is an indication that the low-pressure switch is open. If the voltmeter indicates 0 volts, the voltmeter probe should be moved across the next component in line. For this example it will be assumed that the voltmeter indicates a voltage of 24 volts. The next step is to determine whether the switch is defective or the system is low on refrigerant.

THE OHMMETER

The ohmmeter is generally used in two primary ways:

1. To measure the amount of resistance in a circuit
2. To test a circuit for continuity

Assume that a service technician is sent on a service call. The only information given is that the air conditioner will not run. Using the same circuit as in the previous example, assume that the technician places a jumper across the thermostat and discovers that the condenser fan operates, but the

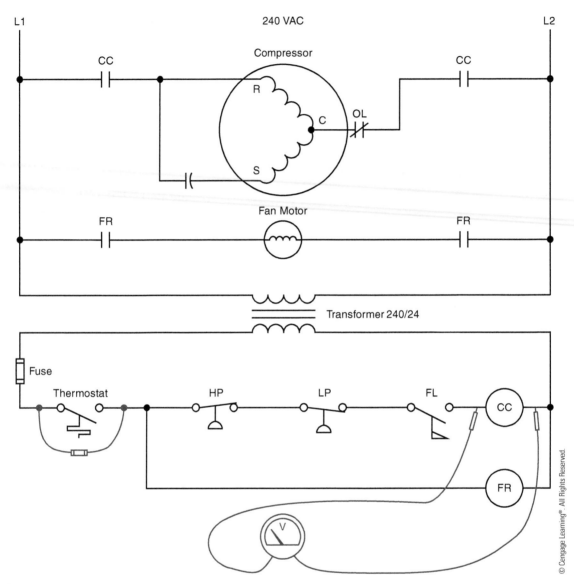

▶ **Figure 40–10**
Measuring voltage across the compressor contactor.

compressor does not. Now assume that he discovers that the compressor contactor is energized. The first step is to test the voltage across the run/start and common terminals of the compressor, Figure 40–13. It is assumed that the voltmeter indicates a value of 240 volts.

The next step is to make mental notes of what could cause this problem.

1. The compressor windings are open.
2. The overload relay is tripped.

Assume that the overload is checked and found not to be tripped. The next step is to check the compressor run and start windings for continuity and resistance. This is done by connecting one probe of an ohmmeter to the common terminal of the compressor and

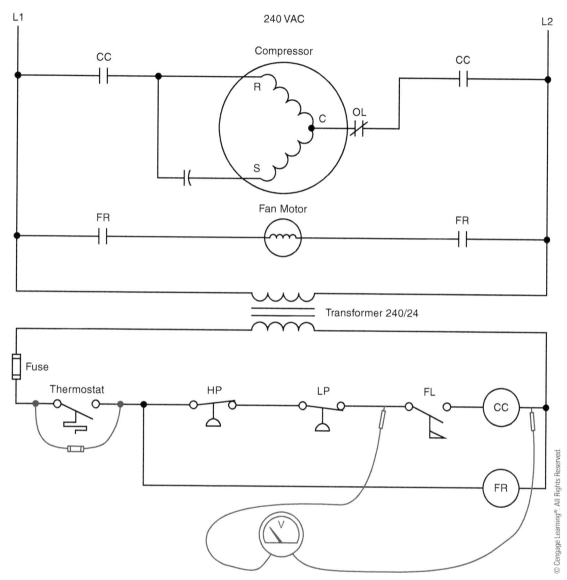

L1   240 VAC   L2

> **Figure 40–11**
> The voltmeter is connected to the next component in line.

the other terminal to the run and start terminals, one at a time, Figure 40–14. The ohmmeter should indicate continuity between the common terminal and the run terminal, and continuity between the common terminal and the start terminal. If it does not, it is an indication that the winding is open.

The resistance reading generally gives some indication as to the state of the winding, although trying to determine whether a winding is shorted with an ohmmeter is a guess at best. The actual resistance of the winding is determined by the size and type of compressor motor and will probably not be known

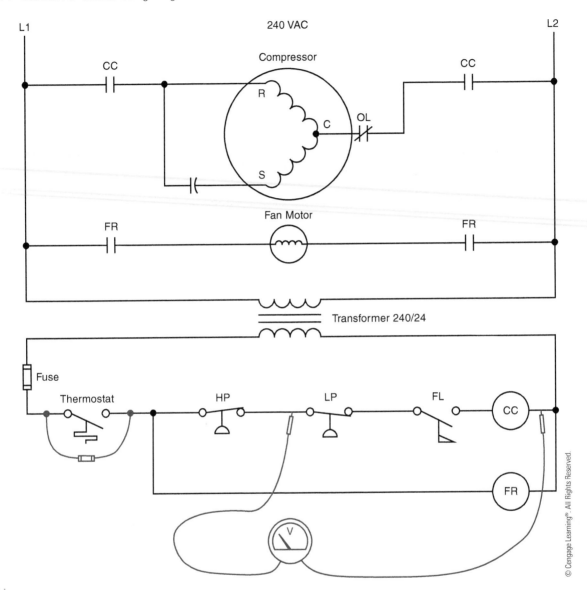

▶ **Figure 40–12**
The next component in line is tested.

by the service technician, but some general guidelines can be followed.

1. If the resistance of either winding is extremely low as compared with the other, there is a good possibility that the winding is shorted.
2. The start winding should be a little more resistive than the run winding.

The run and start windings should also be tested to ensure that they are not grounded. This can be done by connecting one side of the ohmmeter probe to the case of the compressor and testing for continuity to each winding, Figure 40–15. The ohmmeter should indicate infinite resistance if the winding is not grounded.

Another type of ohmmeter, called a megohmmeter or "megger," is often used to test the insulation of

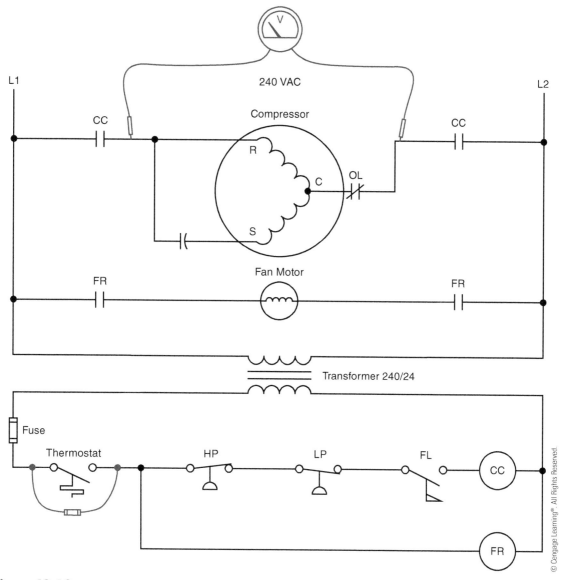

▶ **Figure 40–13**
Testing voltage to the compressor.

wire, Figure 40–16. The megohmmeter is designed to measure resistance in the range of millions of ohms. It is generally used to test the insulation of wire and also produces a much higher voltage than a standard ohmmeter. Often, the insulation around wire appears to be good when tested with an ohmmeter, but breaks down when it is subjected to a high voltage. Ohmmeters generally use from 1.5 to 9 volts to supply current to the circuit being tested. Meggers generally

use from 500 to 1000 volts to supply current to the circuit being tested. This higher voltage will often reveal problems that a low voltage will not.

USING AN AMMETER

The ammeter is used to measure the actual amount of electricity flowing in a circuit. This can be extremely valuable when trying to determine whether

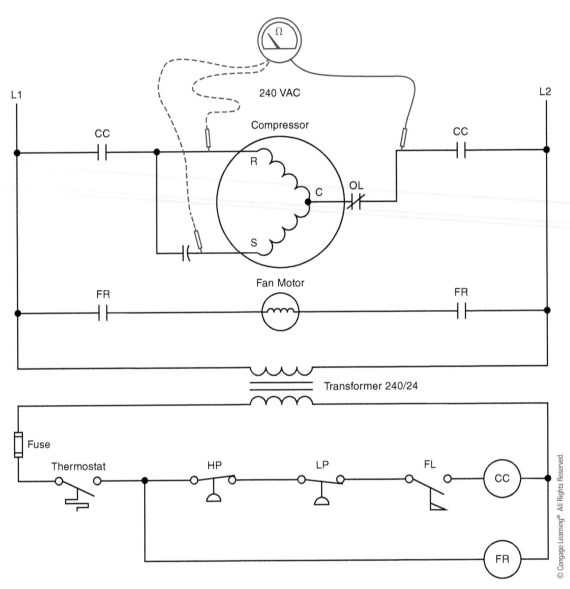

▶ **Figure 40–14**
An ohmmeter is used to test the windings for continuity.

something is actually operating or not. Assume that you are troubleshooting an electric furnace. Also assume that the furnace has three stages rated at 5 kW each. The stages are generally sequenced so that they come on one at a time. A very fast method of checking the furnace is to connect a clamp-on ammeter to the incoming line, turn on the heat, and watch the readings on the ammeter, Figure 40–17.

Because each element has a power rating of 5000 watts, the ampere draw of each element can be determined using Ohm's law.

$$I = \frac{P}{E}$$

$$I = \frac{5000}{240}$$

$$I = 20.8 \text{ amperes}$$

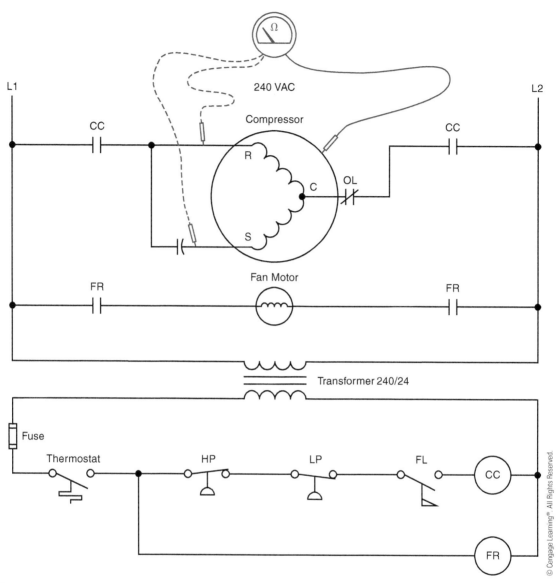

Figure 40-15
Each winding is tested for a grounded condition.

When the furnace is energized, the ammeter should indicate a current flow of approximately 21 amperes. After a few minutes the current should increase to approximately 42 amperes (20.8 + 20.8 = 41.6), and after another delay the current should increase to approximately 62.5 amperes (20.8 × 3 = 62.4). A voltmeter could be used to determine whether voltage is being applied to each element, but unless the power was turned off and the heating elements disconnected and tested with an ohmmeter, you could not be certain an element was not burned open. The ammeter permits a quick check of the unit so that it can be confirmed that each element is operating.

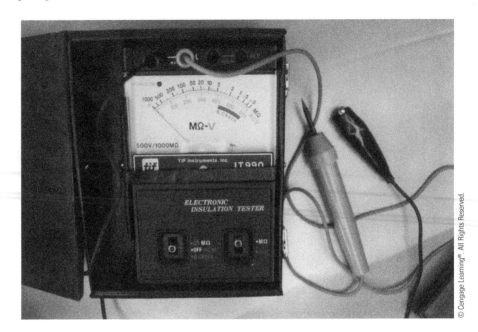

Figure 40–16
A megohmmeter for testing insulation resistance.

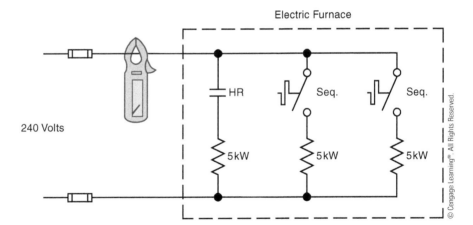

Figure 40–17
Testing an electric furnace with an ammeter.

THE SHOTGUN METHOD

As stated previously, most technicians adopt their own troubleshooting methods that are developed with time and experience. The *shotgun method* involves testing the circuit at various locations to determine trouble areas rather than following a step-by-step procedure as outlined in the hopscotch method. In this example, the circuit to be tested is a central air-conditioning and electric heating system. Probably the best place to start is at the thermostat because it is readily accessible.

1. Check the power supply. To do this, set the fan switch to ON or MAN to see whether the blower fan turns on. If it does, you have determined that the 24-volt supply is working. If the blower does not turn on, the problem could be the thermostat, fan relay, blower motor, run capacitor, 24-volt transformer, or main power supply to the transformer. At this point, it can be determined whether the problem is with the inside unit or the outside unit.

2. Test the thermostat. Remove the thermostat from its base, and check the wires connected to the thermostat base with a voltmeter to determine whether 24 volts are available. If 24 volts are available, use a fused jumper to test the circuit components controlled by the thermostat. Connect one lead to the power terminal (R), and make connection to each of the other terminals to determine whether there is a response. If there is a response to each of the terminals, the thermostat is defective.

3. If there was not a response to a particular circuit component, replace the thermostat on the base and check that component, starting with the power supply. In this example, assume that the air-conditioning unit did not respond.

4. Check the 240-volt power supply to the unit. This can be checked at the breaker, disconnect switch, or main contactor, depending on which is most accessible. In this example it is assumed that power is present at the main contactor.

   a. Check the output of the main contactor to determine whether power is being supplied to the compressor. If not, check the 24-volt supply to the coil of the main contactor. If 24 volts is supplied to the coil, the contactor is defective.

5. If 24 volts is not present at the coil of the contactor, check the thermostat wires where they enter the outside unit. If power is not present, check the wiring between the thermostat and the outside unit.

6. If 24 volts is present at the unit, check any components between the 24-volt supply and the coil of the contactor. Components such as high-pressure switches, low-pressure switches, and so on, are connected in series with the low-voltage circuit.

   Now assume that instead of no response at the outside unit, the condenser fan started, but the compressor did not.

7. If 240 volts is available at the output of the main contactor, check all components, such as run and start capacitors, between the contactor and the compressor.

8. If all components between the contactor and compressor are good, check the power supplied to the compressor terminals. If power is present at the compressor terminals, disconnect power to the outside unit by opening the disconnect switch or circuit breaker.

9. Disconnect the power terminals connected to the compressor. Use an ohmmeter and check between each terminal to determine whether there is an open circuit. Also check between each terminal and the compressor case to determine whether there is a grounded circuit. Note: It is possible for the motor windings to be shorted and not be open or grounded. Shorted windings will cause the motor to draw an excessive amount of current or may not permit the compressor to start when power is supplied. An ohmmeter generally does not reveal this condition.

10. If the ohmmeter indicates an open circuit in the compressor, note whether the compressor is hot to the touch. If so, the internal overload may be open. It cannot be determined whether the compressor winding is open or if the internal overload is open until the compressor cools. This overload cannot be bypassed. If the compressor is hot, it may take hours for the overload to reset, depending on the temperature of the compressor, the ambient temperature, and whether the compressor is located in direct sunlight. The only way to know whether the compressor is defective or whether another problem caused the overload to open is to wait until the overload resets. It is recommended to leave the power disconnected to the outside unit until the compressor cools and allows the overload to reset. This will allow the technician to observe whether the compressor restarts or not.

11. Some of the circumstances that can cause the internal overload to open are:

    • Defective windings in the compressor, causing it to draw excessive current.
    • A stuck compressor.
    • A brief power interruption, such as a loss of power or someone opening the thermostat contacts and reclosing them.

- Lack of airflow across the condenser and compressor. This can be caused by a dirty condenser or anything blocking air to the condenser. The condenser fan can also be defective and thus prevented from obtaining full speed.
- Low voltage supplying the compressor.
- Overcharge of refrigerant, causing high head pressure. This would cause the compressor to draw excessive current.
- Low charge of refrigerant. The compressor could overheat because it depends on cool vapor returning from the evaporator to help cool the motor.
- Very high ambient temperature and being exposed to direct sunlight.

12. If the compressor eventually restarts, check the current draw of the unit and compare this reading to the nameplate current rating. If the current draw is greater than the full-load-amp (FLA) draw listed on the nameplate, determine if the problem is a defective compressor or one of the other causes listed.

SUMMARY

- Before it is possible to troubleshoot a circuit, the technician must have an understanding of how the circuit operates.
- The three main instruments used for troubleshooting are the voltmeter, ohmmeter, and ammeter.
- Power must always be disconnected from the circuit before using an ohmmeter.
- A complete circuit must exist before a voltmeter will indicate voltage.
- An ammeter is used to determine whether current is actually flowing through the circuit.
- The hopscotch method of troubleshooting involves starting at one of the circuits and moving from component to component until the problem is discovered.

KEY TERMS

fused jumper
hopscotch method
measuring instruments

REVIEW QUESTIONS

1. A voltmeter is connected across the terminals of an electric heating element. The voltmeter indicates a voltage of 240 volts. Is this a true test to determine whether the heating element is operating?

2. What electrical measuring instrument should be used to determine whether the heating element in question 1 is operating?

3. It is suspected that the high-pressure switch in a control circuit is open. Explain the steps in testing this component with an ohmmeter.

4. Refer to Figure 40–10. Assume that the voltmeter indicates a value of 24 volts, but the compressor contactor is not energized. What is the most likely problem?

    A. A jumper wire was not placed around the thermostat contacts.

    B. The flow switch is open.

    C. The coil of the CC contactor is open.

    D. The coil of the CC contactor is shorted.

5. Refer to the circuit in Figure 40–7. Assume that when a jumper is placed around the thermostat contacts, the fan motor starts, the compressor contactor energizes, but the compressor motor does not start. Which of the following could *not* cause this problem?

    A. The flow switch is not closing.

    B. The CC load contacts are defective.

    C. The compressor overload relay is open.

    D. The compressor start capacitor is defective.

UNIT 41 ▶

Room Air Conditioners

OBJECTIVES

After studying this unit, the student should be able to:

▶ Identify circuit components on a schematic diagram

▶ Analyze a schematic diagram to determine how the circuit functions

In the previous units, basic symbols and rules for reading a schematic diagram have been covered. In actual practice, however, schematics do not always look like the classic textbook examples. Many schematic diagrams use a **legend** to aid in understanding. A legend is a list that shows a symbol or notation and gives the definition of that symbol or notation. The legend that will be used with the schematics presented in this unit is shown in Figure 41–1.

SCHEMATIC 1

The first circuit to be discussed is shown in Figure 41–2. First find the **major components** shown on the schematic: the switch, fan motor, compressor, capacitor, overload, and thermostat. Notice that these components may not be shown exactly as you would

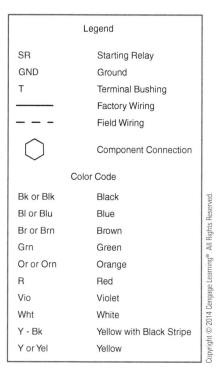

▶ **Figure 41–1**
Schematic legend.

expect. Notice the overload symbol, for example. The symbol used is the same as the symbol for an overload heater discussed earlier in the text. There are, however, no overload contacts shown. The schematic is indicating the use of a small, single-phase bimetal overload unit that acts as both heater and overload contact.

Next, find and examine the fan motor. Notice that this motor has several windings that indicate that it is used for multispeed operation. Notice also that there is no capacitor connected to this motor. The small winding shown separate is the start winding. Because there is no start or run capacitor shown, this motor is a resistance-start induction-run motor. Notice that the white wire is connected from the motor to B terminal of the capacitor and then to one lead of the service chord. This indicates that the white wire is common to the other windings. Now trace the connection of the red, blue, and black wires. The red wire is connected to the LO speed position on the switch; the blue wire is connected to the MED speed position; and the black wire is connected to the HI speed position.

Next, examine the compressor. Notice that two windings are shown. Each winding is connected

to a terminal. One terminal is labeled with an R to represent the run winding. The middle terminal is labeled with an S, which represents the start winding terminal, and the third terminal is labeled with a C, which indicates the common terminal. Trace the common terminal through the overload and thermostat to terminal C on the switch. Notice that the thermostat and overload are connected in series with the compressor. Now trace the run lead of the compressor. Notice that it is connected to the B terminal of the capacitor. This shows that the run winding is connected to the common side of the service chord. Now trace the start lead to the A side of the capacitor. Notice that the capacitor is connected in series between the common side of the service chord and the start winding. The thermistor connected across the capacitor terminals is used to decrease the capacitance connected in series with the compressor after the compressor is in operation. Recall that a thermistor is a temperature-sensitive resistor. This thermistor has a negative temperature coefficient, which means that it will have a very high resistance when it is cool. When its temperature increases, its resistance decreases. When the compressor is first started, the thermistor is cool because no current has been flowing through it. This causes its resistance to be much greater than the capacitive reactance of the capacitor. The full amount of the capacitor is now connected in series with the start winding.

As current flows through the thermistor, its temperature begins to increase. This causes a decrease in its resistance, which permits more current to flow. As the resistance of the thermistor decreases, the effect of the capacitor on the motor decreases also. The effect is very similar to having a compressor that has both a start and run capacitor in the circuit for starting, and then disconnecting the start capacitor and permitting the motor to operate with the run capacitor only.

The last component to be discussed is the switch. Notice that it is not shown with internal electrical connections. There is a legend at the bottom of the schematic, however, that shows which terminals are connected when the switch is set in different positions. In the LO position, for example, terminal L is connected to both the LO fan speed position and the C position, which permits the thermostat to control the compressor.

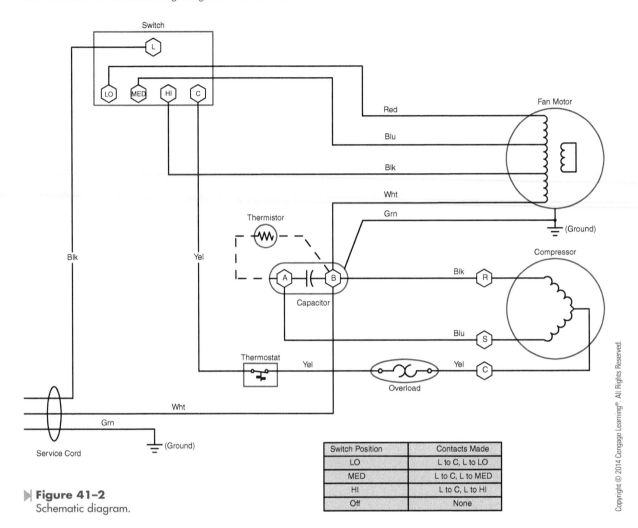

Switch Position	Contacts Made
LO	L to C, L to LO
MED	L to C, L to MED
HI	L to C, L to HI
Off	None

▶ **Figure 41–2**
Schematic diagram.

SCHEMATIC 2

The second schematic to be discussed is shown in Figure 41–3. This schematic is for another room-type air conditioner, but it has some added components. This unit is used to provide heat as well as cooling. An **electric resistance heating element** is used to provide heat in cool weather. Notice also the addition of the start capacitor and start relay. The thermostat in this circuit is **double-acting** instead of a single-pole single-throw. This permits the same thermostat to be used for both heating and cooling. Notice also that the run capacitor is different. This capacitor is actually

two capacitors contained in the same case. The junction point between the two capacitors is connected to one side of the service chord. The fan motor in this unit is different also. Notice that this fan used a run capacitor connected in series with the start winding of the motor at all timers. This motor is a permanent split-capacitor motor. Notice that this motor has two speeds instead of three.

The start capacitor is connected in parallel with the run capacitor to increase the starting torque of the compressor. The resistor shown connected across the terminals of the start capacitor is a relatively high value of fixed resistance used to discharge the capacitor when it is disconnected from

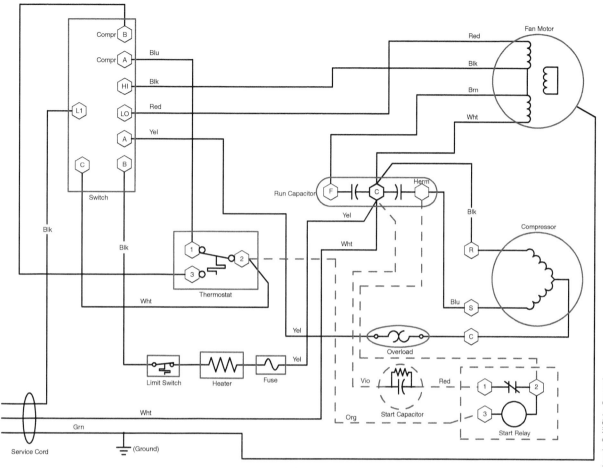

Switch Position	Contacts Made
Fan	H1 to L1
HI Cool	Compr A and H1 to L1, A to C
LO Cool	Compr A and LO to L1, A to C
HI Heat	Compr B and HI to L1, B to C
LO Heat	Compr B and LO to L1, B to C
Off	None

▶ **Figure 41–3**
Room air conditioner circuit.

the circuit. Notice the start relay. The start capacitor is connected in series with the normally closed contact. This is a potential starting relay, which senses the voltage induced in the start winding and opens the contact when the motor reaches about 75% of its full speed.

In this circuit, the switch is the main controller. For example, trace the circuit when the switch is placed in the high cool position. The legend at the bottom of the schematic indicates that when the switch is in the HI cool position, terminals COMPR A and HI are connected to terminal L1, and terminal A is connected to terminal C. When HI is

connected to L1, the fan motor will operate in its high speed. When terminal COMPR A is connected to L1, a current path is provided to terminal 1 of the thermostat. Terminal 2 is connected to terminal C of the switch. Because switch terminal A is connected to switch terminal C, power is connected to the compressor motor through the thermostat contact. When the thermostat is connected in this manner, an increase in temperature causes the thermostat contacts to close and a decrease in temperature causes them to open.

Now assume that the switch has been set to the low heat position. The switch legend indicates that terminals COMPR B, and LO are connected to L1, and terminal B is connected to terminal C. When LO is connected to L1, the fan motor operates in the low-speed position. When terminal COMPR B is connected to L1, power is provided to terminal 3 of the thermostat. Terminal 2 is connected to terminal C of the switch. Because switch terminal B is connected to the resistance heater through the high-limit switch and fuse, the thermostat controls the operation of the heater. When the thermostat is connected in this manner, a decrease in temperature causes the thermostat contacts to close, and an increase in temperature causes them to open.

SUMMARY

- ⊙ Legends are sometimes used with schematic diagrams to aid in understanding.
- ⊙ A legend is a list of symbols and/or notations and gives the definition of these symbols and/or notations.
- ⊙ When using a schematic to interpret the operation of a circuit, it is generally helpful to identify the major components in the circuit first.

KEY TERMS

double-acting
electric resistance heating element
legend
major components

REVIEW QUESTIONS

1. What is a legend?
2. Refer to Figure 41–2. What would be the action of this circuit if the overload relay should burn open?
3. What purpose does the thermistor connected in parallel with the capacitor serve?
4. In Figure 41–2, what switch connections are made when the switch is in the HI position?
5. In Figure 41–3, why is the thermostat switch as a single-pole double-throw?
6. In Figure 41–3, what do the dashed lines showing connection between the start capacitor and start relay to other parts of the circuit mean?
7. In Figure 41–3, what color wire is connected between terminal 2 of the thermostat and terminal C of the switch?
8. What color wire is connected between terminal 2 of the thermostat and the start relay?
9. In Figure 41–3, if no continuity is shown when one lead of an ohmmeter is connected to switch terminal A and the other is connected to terminal C of the compressor, what does it mean?
10. In Figure 41–3, to what two points should the terminals of an ohmmeter be connected to check the continuity of the resistance heater circuit?

TROUBLESHOOTING QUESTIONS

Refer to the schematic shown in Figure 41–3 to answer the following questions.

1. Assume that the switch has been set in the HI HEAT position. Now assume that the thermostat controls the operation of the electric heating element, but does not control the operation of the fan motor. Which of the following could cause this condition?

A. The thermostat is defective.

B. The high-limit switch is stuck in the closed position.

C. The switch is not making connection between contacts B and C.

D. There is nothing wrong with the unit. This is normal operation for this unit.

2. When the switch is set in the LO COOL position, the unit operates normally. When the switch is set in the HI COOL position, the compressor operates, but the fan motor does not. Which of the following could cause this condition?

A. The fan motor winding between the red and black wire is open.

B. The fan motor winding between the black and brown wire is open.

C. The switch is not making connection between terminals COMPR A and HI.

D. The capacitor section between terminals C and F is defective.

3. When the switch is set in HI COOL or LO COOL position, the unit operates normally. When the switch is set in HI HEAT or LO HEAT position, the fan motor operates normally, but the unit does not provide any heating. Which of the following could cause this condition?

A. The thermostat is not making connection between 2 and 3.

B. The limit switch is open.

C. The fuse link is open.

D. All of the above.

4. When the switch is set in the LO COOL position, the unit operates normally. When the switch is set in the HI COOL position the fan motor operates, but the compressor does not operate. Which of the following could cause this condition?

A. The overload unit is open.

B. The switch is not making contact between terminals A and C.

C. The potential starting relay is defective.

D. The thermostat is defective.

5. Assume that the switch has been set in the LO COOL position, and the fan motor operates normally. When the thermostat contact closes between terminals 1 and 2, the compressor hums, but does not start. An ohmmeter test of the compressor is as follows:

R to C 2 ohms

S to C 6 ohms

R to S 8 ohms

R to case infinity ohms

S to case infinity ohms

Which of the following would *not* cause this problem?

A. Overload is open.

B. The capacitor between terminals C and HERM is defective.

C. The start capacitor is defective.

D. The potential starting relay is defective.

A Commercial Air-Conditioning Unit

OBJECTIVES

After studying this unit, the student should be able to:

⯈ Recognize electrical components from the symbols on the schematic

⯈ Discuss the operation of a commercial air-conditioning unit

⯈ Interpret a three-phase schematic diagram

In this unit, a commercial air-conditioning system will be discussed. The legend for this schematic is shown in Figure 42–1. The schematic to be discussed is shown in Figure 42–2. Notice that this control system contains several devices not normally found in a residential system. The compressor, for example, is operated by a **three-phase squirrel-cage induction motor**. It can be seen that the motor is three phase by the wye connection of the stator winding. It can be determined that the motor is a squirrel cage because it has no external resistors that would be used for the rotor circuit of a wound rotor induction motor. There is also no DC circuit that would be required to excite the rotor of a three-phase synchronous motor.

The condenser fan motor is a single-phase permanent split-capacitor motor. Notice that the condenser fan motor is connected in parallel with two

LEGEND			
C	– Contactor	SC	– Start Capacitor
CC	– Cooling Compensator	SR	– Start Relay
CH	– Crankcase Heater	ST	– Start Thermistor
Comp or Compr	Compressor	TC	– Thermostat Cooling
		TD	– Time Delay
		Therm	– Thermostat
CPCS	– Compressor Protection Control System	TM	– Timer Motor
CR	– Control Relay	Tran or Trans	Transformer
CT	– Current Transformer		
FC	– Fan Capacitor		
FM	– Fan Motor	⬠	– Component Connection (Marked)
FS	– Fan Switch	○	– Component Connection (Unmarked)
FT	– Fan Thermostat	⌐Ω⌐	– Field Splice
HC	– Heating Control	—•—	– Splice
HPS	– High-pressure Switch	—→	– Plug
HR	– Holding Relay	>—	– Receptacle
IFM	– Indoor Fan Motor	——	– Factory Wiring
IFR	– Indoor Fan Relay	▬▬▬	– Field Power Wiring
IP	– Internal Protector	– – –	– Field Ground Wire
LPS	– Low-pressure Switch	– – –	– Field Control Wiring (NEC Class II)
OL	– Overload	– – – –	– Alternate Wiring
QT	– Quad Terminal	▬▬▬	– Indicates Common Potential (Does not represent wire)
R	– Resistor		–
RC	– Run Capacitor		–
Recep	– Receptacle		–
Res	– Bleed Resistor		–

▶ **Figure 42–1**
Schematic legend.

lines of the compressor. When contactor C energizes, both C contacts close and connect both the compressor and condenser fan motors to the line.

The **crankcase heater** is shown directly below the condenser fan motor and is connected to terminals 21 and 23. Notice the crankcase heater is energized at all times. As long as power is connected to the circuit, the crankcase heater will be energized.

The control transformer contains two primary windings and two secondary windings. This transformer can be connected to permit a 460- or 230-volt connection to the primary, and the secondary can provide 230 or 115 volts. In the circuit shown, the primary winding is connected in series, which permits 460 volts to be connected to it. The secondary winding is also connected in series, which provides an output voltage of 230 volts.

The 230-volt circuit is used to operate a short-cycle timer circuit. This is the same circuit that was discussed in Unit 35.

The 24-volt circuit is shown at the bottom of the schematic. Notice that only the secondary of the transformer is shown. This is indicating that its power can be derived from almost anywhere. The primary of this transformer could be connected to a 120-volt circuit inside the building. This circuit contains the high- and low-pressure switches. If one of them opens, it will have the same effect as opening the thermostat.

Notice that the **indoor fan relay (IFR)** is shown, but the fan motor is not. In a commercial location, there may actually be several fans operated by the IFR relay. In practice, the IFR relay may be used to control the coils of other relays, which connect the fan motors to the line.

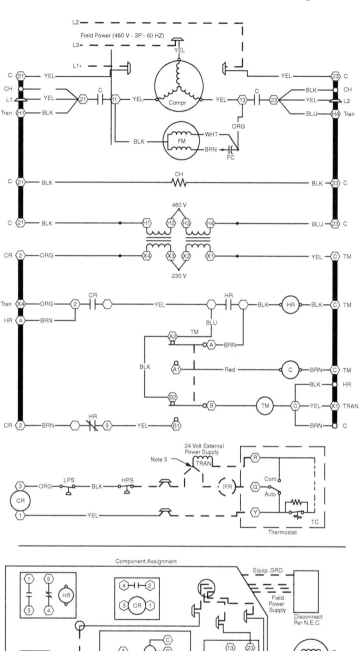

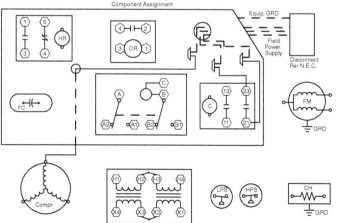

Figure 42–2
Schematic diagram.

The thermostat is a single-pole single-throw contact. The resistor shown connected around the thermostat contact represents the heat anticipator. A switch is also provided that permits the indoor fan to be operated automatically or manually.

The last item shows the component arrangement. This is used to aid the service technician in locating the different control components in the system.

SUMMARY

- ⊙ The compressor in this example circuit is powered by a three-phase squirrel-cage motor.
- ⊙ The windings in the compressor motor are connected in a wye configuration.
- ⊙ The condenser fan motor is a single-phase permanent split-capacitor motor.
- ⊙ The control transformer contains two primary and two secondary windings.
- ⊙ The thermostat, low-pressure switch, and high-pressure switch are connected to a 24-volt system.
- ⊙ The control system for this unit contains a short-cycle timer.

KEY TERMS

crankcase heater
indoor fan relay (IFR)
three-phase squirrel-cage induction motor

REVIEW QUESTIONS

1. What do the letters CC mean on a control schematic?
2. What does CPCS mean on a schematic?

   Refer to Figure 42–2 for the following questions.

3. If it is desired to change the voltage controlling the short-cycle timer from 230 volts to 115 volts, what transformer leads should be connected together?
4. Assume the system has stopped operation. A voltmeter is connected across the LPS switch terminals, and it indicates 24 volts. The voltmeter is then connected across the HPS switch, and it indicates 0 volts. Which switch is stopping the operation of the circuit?
5. When the system is operating normally, how much voltage should be seen across the CR relay coil?

TROUBLESHOOTING QUESTIONS

Refer to the schematic shown in Figure 42–2 to answer the following questions.

1. Referring to the schematic in its present state, will the compressor start with the thermostat in the closed position? Explain your answer.

A. Yes

B. No

2. What voltage is used to operate the short-cycle timer and compressor relay?

A. 460 VAC

B. 230 VAC

C. 120 VAC

D. 24 VAC

3. What relay controls the operation of the condenser fan motor?

A. IFR

B. HR

C. C

D. CR

4. Which of the following components is not shown on the schematic?

A. Compressor motor

B. Condenser fan motor

C. Thermostat

D. Evaporator fan motor

5. Assume that the unit is in operation and suddenly stops. A voltmeter test reveals the following information:

Voltage across L_1, L_2, and L_3 = 460 VAC

Voltage across X_1 to X_4 of the control transformer = 230 VAC

Voltage across CR coil = 24 VAC

Voltage across terminals 2 and 4 of the CR contact = 230 VAC

Voltage across coil C = 0

What is the most probable cause of trouble? Explain your answer.

A. CR coil is open.

B. CR contacts are stuck closed.

C. Coil HR of the short-cycle timer is open.

D. Coil C is open.

Heat-Pump Controls

OBJECTIVES

After studying this unit, the student should be able to:

▷ Describe the operation of a heat pump
▷ Discuss the function of a double-acting thermostat in a heat-pump system
▷ Discuss the operation of a sequence timer
▷ Describe the operation of the defrost thermostat and timer

A heat pump is a device that provides both heating and air conditioning within the same unit. In the cooling cycle, the outside heat exchange unit is used as the condenser and the inside heat exchanger is used as the evaporator. When the heat pump is used for heating, the reversing valve reverses the flow of refrigerant in the system and the outside heat exchanger becomes the evaporator. The inside heat exchanger becomes the condenser. Heat pumps also contain some type of backup heating system that is used when the outside temperature is too low to make heat transfer efficient. The most common type of backup heat is electric-resistance heat.

Heat pumps contain other control devices that are generally used only with heat-pump equipment, such as two-stage thermostats, sequence relays, and defrost timers.

TWO-STAGE THERMOSTATS

The two-stage thermostat is a thermostat that contains two separate mercury contacts. It is similar to the programmable thermostat except that the two mercury contacts cannot be set independently of each other. The mercury contacts of the two-stage thermostat are so arranged that one contact makes connection slightly ahead of the other. For example, assume the heat pump is being used in the heating mode. Now assume that the temperature drops. One of the contacts will make connection first. This contact turns on the compressor, and heat is provided to the living area. If the compressor can provide enough heat to raise the temperature to the desired level, the second mercury contact does not make connection. If the compressor cannot provide the heat needed, the second mercury contact will close and turn on the electric-resistance heating elements to provide extra heat to the living area.

THE SEQUENCE TIMER

The **sequence timer** is an on-delay timer used to connect the heating elements to the line in stages instead of all at once. Most sequence timers contain two or three contacts and are operated by a small heating element that heats a bimetal strip. When the bimetal strip becomes hot enough, it snaps from one position to another and closes the two contacts. A photograph of this type of timer is shown in Figure 43–1.

Copyright © 2014 Courtesy Sensata Technologies, Inc.

▷ **Figure 43–1**
Sequence relay.

DEFROST TIMER

When the heat pump is used in the heating mode of operation, it removes heat from the air and delivers it inside the living area. This means that the outside heat exchanger is being used as the evaporator, and cold refrigerant is circulated through it. Any moisture in the air can cause frost to form on the coil and reduce the airflow through it. This will reduce the efficiency of the unit. For this reason, it is generally necessary to defrost the outside heat exchanger. Defrosting is done by disconnecting the condenser or outside fan motor and reversing the flow of refrigerant through the coil. This causes the unit to temporarily become an air conditioner and warm refrigerant is circulated through the coil.

Before the defrost cycle can be activated, two separate control conditions must exist. The **defrost thermostat**, located on the outside heat exchanger, must be closed; and the defrost timer must permit the defrost cycle to begin. A schematic diagram of a basic defrost control circuit is shown in Figure 43–2. Notice that the defrost timer is connected in parallel with the compressor. This means that the timer can operate only when the compressor is in operation. Notice also that the defrost cycle energizes the **reversing valve solenoid**. This means that this unit is in the heating mode when the solenoid is de-energized.

Notice the defrost timer contains two contacts, DT_1 and DT_2. DT_1 is normally open and DT_2 is normally closed. The defrost relay (DFR) contains three contacts. DFR_1 is normally closed and is connected in series with the outside fan motor. DFR_2 is normally open and is connected in parallel with contact DT_1. Contact DFR_3 is normally open and is connected to the reversing valve solenoid.

The schematic shown in Figure 43–3 illustrates the condition of the circuit when the defrost cycle first begins. Notice that the defrost timer (DT) has caused contact DT_1 to close, but contact DT_2 has not opened. The contacts of the defrost timer are operated by two separate cams. The cams are so arranged that contact DT_1 will close before DT_2 opens.

The schematic in Figure 43–4 illustrates the condition of the circuit immediately after the defrost relay has energized. Notice that all DFR contacts

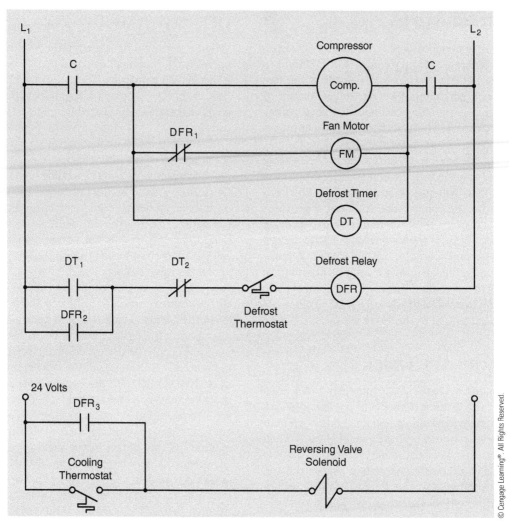

▶ **Figure 43–2**
Defrost cycle circuit.

have changed position. DFR_1 contact is now open and the outside fan motor has been disconnected from the circuit. DFR_2 contact is closed and is used as a holding contact around contact DT_1. DFR_3 contact is closed and provides current to the reversing valve solenoid to reverse the flow of refrigerant in the system.

The schematic shown in Figure 43–5 illustrates the condition of the circuit after contact DT_1 reopens. Notice that contact DFR_2 maintains a current flow path around the now open DT_1 contact and the defrost cycle is permitted to continue. The unit will remain in the defrost cycle until the defrost thermostat is satisfied and opens the circuit, or the defrost timer causes the DT_2 contact to open. When this occurs, the system will change back to its original condition, shown in Figure 43–2.

Electronic Defrost Timers

Many units now employ an electronic defrost timer similar to the one shown in Figure 43–6. This timer has a fixed 10-minute defrost time. The time interval between defrost cycles can be set by

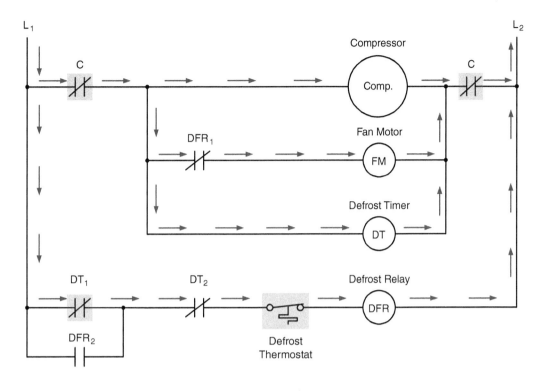

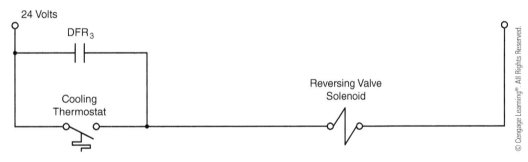

▶ Figure 43–3
Initial circuit operation.

moving a jumper lead to the proper pin. The time interval can be set for 30, 60, or 90 minutes. Two pins marked TEST can be shorted to reduce the defrost time by a factor of 2, 5, and 6 seconds. This permits the service technician to test the unit without waiting a long period of time. The HOLD terminal permits the timer to accumulate time only during the time that the compressor is in operation. The connection diagram for this module is shown in Figure 43–7.

THE FULL SYSTEM SCHEMATIC

A schematic for a residential heat-pump unit is shown in Figure 43–8. The legend for the schematic is shown in Figure 43–9. Notice in Figure 43–8 that the schematic is divided into three main sections. One section shows the **outdoor compressor controls**. The second section shows the **indoor resistance heat** and blower-fan controls, and the third section shows the **low-voltage controls**.

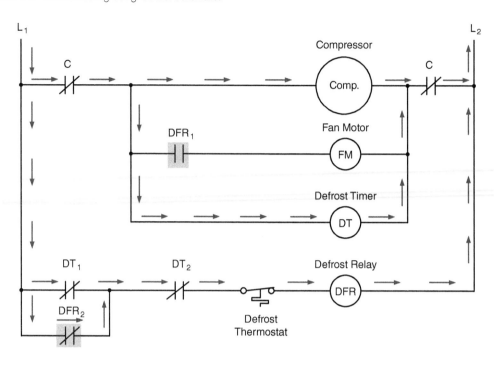

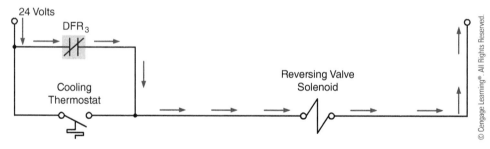

▶ **Figure 43–4**
Defrost relay energizes.

To begin the study of this control system, locate the low-voltage section of the schematic. It is divided into three sections. One section is located directly below the blower fan motor. Notice the 24-volt transformer used to provide needed power. Now locate the **terminal board** directly to the left of the control transformer. The terminal board shows terminal connections marked inside hexagon-shaped figures. Starting at the top and going down they are R, G, O, Y, and so on. Now locate the second control section directly under the outdoor unit schematic. Notice

the terminal board contains some of the same letter connection points as the other terminal board. Now locate the thermostat. Notice the thermostat contains **terminal markings** that are the same as the other two boards. These terminal markings are used to aid in tracing the circuit. For example, locate the terminal marked Y on the thermostat. Now, locate the terminal marked Y on the board closest to the control transformer. Finally, locate the terminal marked Y on the terminal board located under the outdoor unit. If the wires are traced, it will be seen

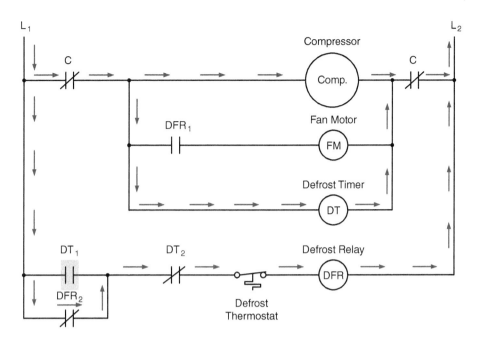

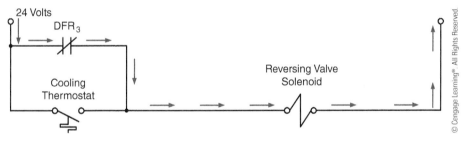

Figure 43–5
Defrost timer opens DT_1 contact.

Figure 43–6
Electronic defrost timer module.

that all of the terminals marked Y are connected together. This is true for all the other terminals that are marked with the same letter. Terminal markings are often used to help simplify a schematic by removing some to the connecting wires. The circuit shown in Figure 43–10 is very similar to the schematic in Figure 43–8 except that the terminal markings are used instead of connecting wires.

Notice the use of the two-stage thermostat in the schematic shown in Figure 43–8. The thermostat is so constructed that contact H1 will close before H2. Also, contact CO will close before C1. Also notice that the thermostat has an emergency heat position that permits the switch to override the thermostat and the electric resistance heaters to be connected in the circuit on a call for heat. Notice also that

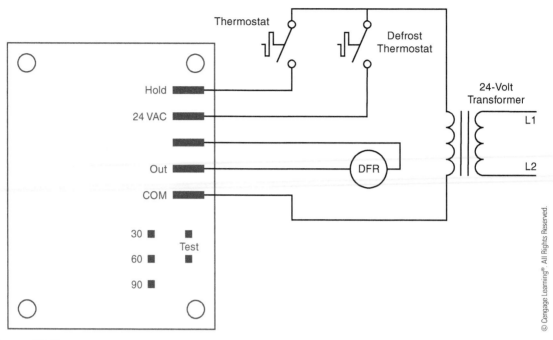

▶ **Figure 43–7**
Connection for a typical electronic defrost timer.

this control uses an outdoor thermostat (ODT). The ODT senses the outside temperature and permits sequence relays 1 and 2 to operate only if the outside temperature is below a certain level.

Locate the three sequence relay coils, SEQ1, SEQ2, and SEQ3. Trace the operation of the circuit when sequence relay 1 is energized. The two SEQ1 contacts located in the resistance heat section close and connect the heating elements to the line. A third SEQ1 contact is connected in series with SEQ2 timer. When this contact closes, SEQ2 timer can begin operation provided the outdoor thermostat is closed. When this timer completes its time sequence, all SEQ2 contacts close. The two SEQ2 contacts located in the resistance heat section connect the second bank of resistance heaters to the line. The

third SEQ2 contact permits current to flow to SEQ3 timer. At the end of its time cycle, the two SEQ3 contacts located in the resistance heater section close and connect the third bank of heaters to the line.

Locate the blower fan motor. Notice that this is a multispeed fan motor. Only two of the speeds are used, however. High speed is used when the fan control relay is energized by the thermostat. Notice that when the normally open FR contacts close, high speed is connected to the line. Also notice that the normally closed FR contacts are connected to the first SEQ1 contact. When SEQ1 contact closes, the second fan speed is connected to the line.

Now locate the defrost timer and defrost relay. Trace the action of the circuit as described earlier in this unit. Notice that the DFR contact located between

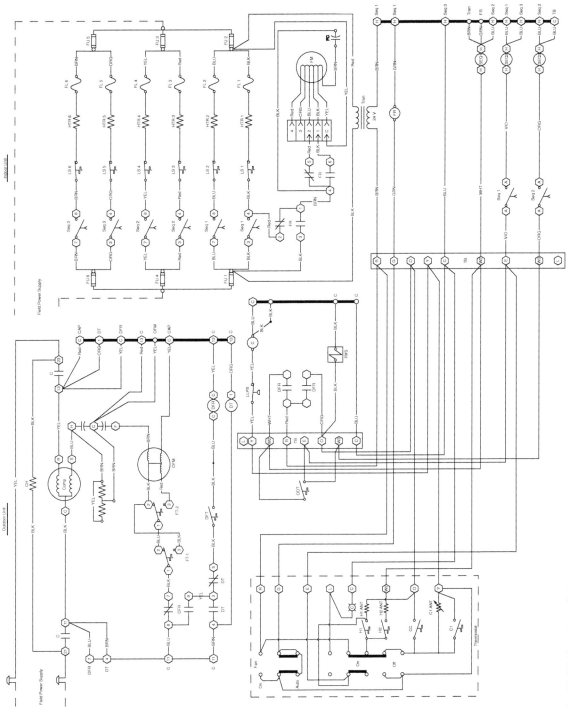

▶ **Figure 43–8**
Heat-pump circuit.

LEGEND

Ant.	–	Anticipator
C	–	Contactor
Co	–	First-stage Cooling Thermostat
C1	–	Second-stage Cooling Thermostat
Cap.	–	Capacitor
CH	–	Crankcase Heater
Comp	–	Compressor
DFR	–	Defrost Relay
DFT	–	Defrost Thermostat
DT	–	Defrost Timer
Em Ht	–	Emergency Heat
EHR	–	Emergency Heat Relay
FC	–	Fan Capacitor
FL	–	Fuse Link
FT	–	Fan Thermostat
Fu	–	Fuse
H1	–	First-stage Heating Thermostat
H2	–	Second-stage Heating Thermostat
Htr	–	Heater
IFM or **FM**	–	Indoor Fan Motor
IFR or **FR**	–	Indoor Fan Relay
LLPS	–	Liquid Line Pressure Switch

LS	–	Limit Switch
ODT	–	Outdoor Thermostat
OFM	–	Outdoor Fan Motor
PI	–	Plug
QT	–	Quad Terminal
RC	–	Run Capacitor
RVS	–	Reversing Valve Solenoid
SC	–	Start Capacitor
Seq	–	Sequencer
SR	–	Start Relay
ST	–	Start Thermistor
TB	–	Terminal Board
Tran	–	Transformer

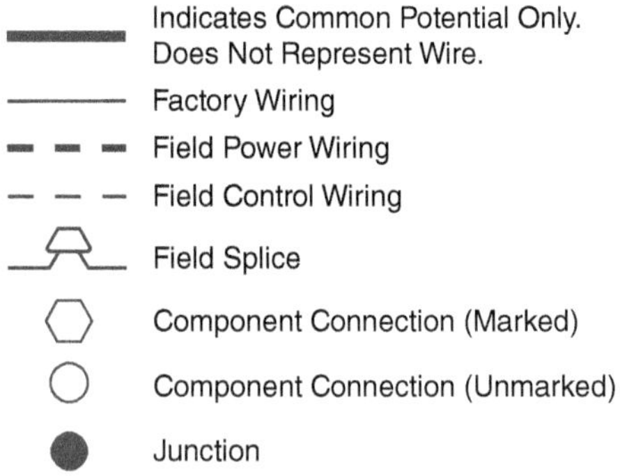

━━━	Indicates Common Potential Only. Does Not Represent Wire.
─────	Factory Wiring
▬ ▬ ▬	Field Power Wiring
─ ─ ─	Field Control Wiring
⌂	Field Splice
⬡	Component Connection (Marked)
◯	Component Connection (Unmarked)
●	Junction

▶ **Figure 43–9**
Schematic legend.

terminals 4 and 6 is used to override the outdoor thermostat. This permits the resistance heaters to be used during the defrost cycle, preventing cold air from being blown in the living areas during the time the unit is operating as an air conditioner.

Now locate the two outside fan-speed control thermostats labeled FT1 and FT2. Notice that when FT1 is in the position shown, it permits FT2 to operate the fan in either high speed or low speed. When FT1 changes position and makes connection between terminals 1 and 3, it connects the fan in the high-speed position, and FT2 has no control over the speed.

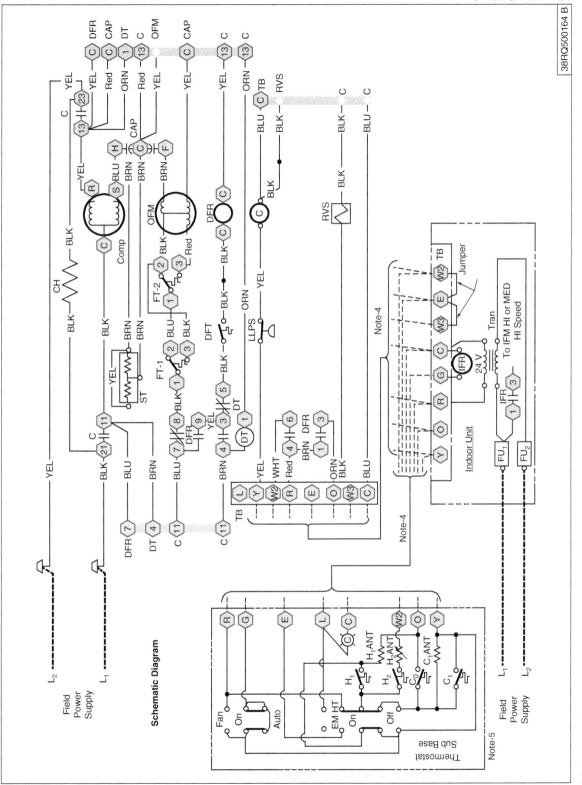

Figure 43–10
Terminal identification is used to simplify schematic.

▷ SUMMARY

- ⊙ A heat pump is a device that provides both heating and cooling within the same unit.
- ⊙ Heat-pump systems contain a backup heating system that is used when the outside ambient temperature is too low to provide enough heat transfer.
- ⊙ The double-acting thermostat contains two separate sets of contacts.
- ⊙ The first contact of a double-acting thermostat turns the heat-pump compressor on. The second contact turns the strip heaters on.
- ⊙ Sequence timers are used to turn the strip heaters on in stages instead of all at one time.
- ⊙ The defrost timer is used to reverse the flow of refrigerant at periodic intervals to melt any accumulation of frost on the outside evaporator coil.

▷ KEY TERMS

defrost thermostat
indoor resistance heat
low-voltage controls

outdoor compressor controls
reversing valve solenoid
sequence timer

terminal board
terminal markings

▷ REVIEW QUESTIONS

1. What is the purpose of terminal markings?
2. What two control components must be in a closed position before a heat pump is permitted to go into the defrost cycle?
3. The thermostat shows a small pilot light connected between terminals L and C. What condition of the thermostat turns this light on?
4. What is the purpose of the outdoor thermostat?
5. What is the operating voltage of the reversing valve solenoid?

▷ TROUBLESHOOTING QUESTIONS

Refer to the schematic shown in Figure 43–8 and Figure 43–10 to answer the following questions.

1. Assume that the unit is set in the heating mode and that thermostat contact H_1 closes. Also assume that when H_1 closed, the indoor fan motor and outside fan motors started, but the compressor motor did not. Which of the following could cause this condition?

   A. The liquid line pressure switch is open.

   B. Coil C is open.

   C. The compressor motor is defective.

   D. All of the above are true.

2. If the unit is set in the heating mode and thermostat contact H_1 closes, what is the normal action of the unit?

A. The heat strips, indoor fan motor, outdoor fan motor, and compressor turn on.

B. The indoor fan motor, outdoor fan motor, and compressor turn on.

C. The heat strips and indoor fan motor turn on.

D. The heat strips, outdoor fan motor, indoor fan motor, compressor, and reversing valve turn on.

3. Assume that cooling thermostat C_0 is closed and C_1 is open. What is the normal operating sequence of the unit?

A. The indoor fan motor, outdoor fan motor, and compressor turn on.

B. Only the indoor fan motor turns on.

C. The indoor fan motor, outdoor fan motor, reversing valve, and compressor turn on.

D. The reversing valve turns on.

4. Assume that fuses FL_1 and FL_2 connected to strip heaters HTR-1 and HTR-2 are blown. Now assume that thermostat HT_1 closes and starts the heat-pump compressor, but the outside temperature is too low to permit the heat pump to sufficiently heat the dwelling and thermostat H_2 closes also. What will be the action of the unit?

A. The compressor will continue to run, but the heat strips will not activate because they are staged and must operate in proper sequence.

B. The heat-pump compressor stops operating because thermostat H_2 prevents both the compressor and heat strips from being turned on at the same time.

C. The compressor continues to operate and after some period of time sequencer 1 will time out and turn on sequencer 2 energizing heating elements HTR-3 and HTR-4. If enough time elapses, sequencer 2 will time out and turn on sequencer 3, which energizes heating elements HTR-5 and HTR-6.

D. The compressor will eventually be disconnected from the line due to high pressure because the inside blower cannot operate in the heating mode if the first bank of strip heaters fails to operate.

5. The schematic shows a small lamp connected between terminals C and L on the thermostat. What is this lamp used to indicate?

A. The compressor is in operation.

B. The compressor motor has failed.

C. The unit is in the defrost mode.

D. The thermostat has been set for emergency heat.

Packaged Units: Electric Air Conditioning and Gas Heating

OBJECTIVES

After studying this unit, the student should be able to:

- Discuss the operation of an air-conditioning system that operates in conjunction with a gas heating system
- Interpret the schematic diagram for the air-conditioning system
- Interpret the schematic diagram for the heating system

The schematic shown in Figure 44–1 is for a unit that contains both electric air conditioning and gas heating in the same package. This drawing shows both a connection diagram and a schematic diagram of this unit.

This diagram shows mainly the heating and blower controls. At the bottom of the schematic diagram is a component labeled **condensing unit**. This is the only reference to the air-conditioning compressor and condenser fan on this schematic. This is not uncommon for a packaged unit.

THE COOLING CYCLE

The thermostat shows four terminal connections. The terminal labeled R is connected to one side of the 24-volt control transformer. When the thermostat is in the cooling position, an increase in temperature

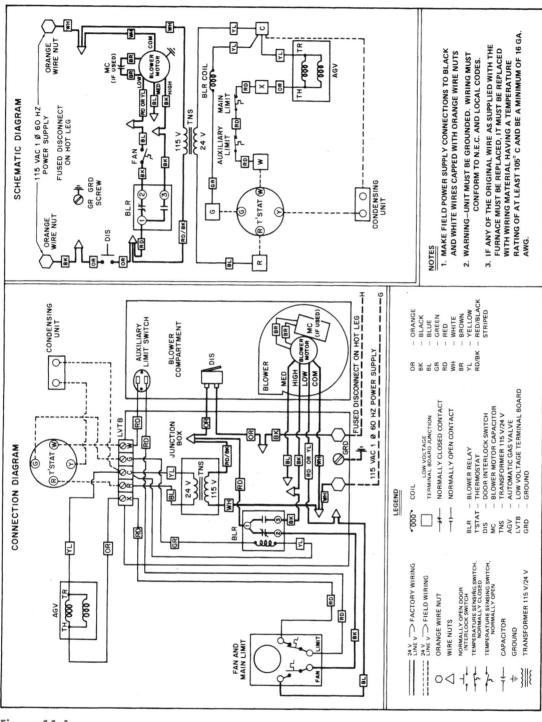

▷ **Figure 44–1**
Schematic for an air-conditioning and heating package unit.

will cause terminal R to make connection with terminals G and Y. When power is applied to terminal Y, a circuit is completed to the condensing unit. The other side of the condensing unit is connected to terminal C, which completes the circuit back to the control transformer. This starts the air-conditioning compressor and condenser fan.

Terminal G of the thermostat is connected to the blower relay coil (BLR). When BLR coil energizes, both BLR contacts change position. The normally closed contact opens and prevents the possibility that power can be applied to the low-speed terminal of the blower fan motor. The normally open contact closes and connects power to the high-speed terminal of the motor. Notice that the indoor blower fan operates in the high-speed position when the air-conditioning unit is started.

THE HEATING CYCLE

When the thermostat is in the heating position, a decrease of temperature will cause the thermostat to make connection between terminals R and W. This permits a circuit to be completed through the automatic gas valve (AGV). When the AGV is energized, gas is permitted to flow to the main burner, where it is ignited by the pilot light. Two high-limit contacts are connected in series with the automatic gas valve. One is labeled **auxiliary limit**, and the other is labeled **main limit**. The wiring diagram shows the main limit to be located in the fan-limit switch. The auxiliary limit switch is in a separate location. Both of these switches are normally closed and are shown to be temperature activated. The schematic also shows that an increase in temperature will cause them to open. Because both are connected in series with the AGV, the circuit will be broken to the valve if either one opens.

In the heating cycle, the indoor blower fan is controlled by the fan switch. The fan switch is temperature activated. After the gas burner has been turned on, the temperature of the furnace increases. When the temperature has risen to a high enough level, the fan switch will close and connect the low-speed terminal of the blower motor to the power line. Notice that the fan switch is connected in series with the normally closed BLR contact. When BLR relay is de-energized, the fan switch is permitted to control the operation of the blower fan. The blower fan relay

permits the fan to operate in low speed when the unit is in the heating cycle, and in high speed when the unit is in the cooling cycle.

THE DOOR INTERLOCK SWITCH

The **door interlock** is shown on the schematic as a normally open pushbutton labeled (DIS). The function of this switch is to permit the unit to operate only when the furnace door is closed. When the door is opened, the 120-volt power supply is broken to the unit. Most door interlock switches are so designed that they are actually a two-position switch. When the door is open, the switch can be pulled out. This causes the switch to make connection so the unit can be serviced.

ELECTRONIC CONTROL OF BLOWER MOTOR AND COMPRESSOR LOCK-OUT

Many units employ electronic control for some of their functions. Troubleshooting for electronic circuit boards is generally accomplished by determining whether the board has the proper input information to obtain an output. If the inputs and outputs are correct, the board is good and the problem lies somewhere else in the circuit. The circuit board shown in Figure 44–2 is designed to control the blower for a cooling unit with gas or electric heat. The circuit board also provides lock-out protection for the compressor.

▶ **Figure 44–2**
Blower motor and compressor lock-out control board.

LOCK-OUT PROTECTION

Lock-out protection involves the use of a high-impedance relay that becomes connected in series with the compressor contactor coil in the event of a problem with the compressor circuit. In the circuit shown in Figure 44–3, the coil of the **lock-out relay** is connected in parallel with the normally closed safety switches used to help protect the compressor. Under normal conditions, the high-pressure switch, low-pressure switch, and the low evaporator temperature switch are closed and provide a complete circuit to the compressor

contactor coil. Because the lock-out relay coil is connected in parallel with these switches, almost no voltage is dropped across the LOR coil, and it remains turned off. If one of the switches should open, however, the lock-out relay coil becomes connected in series with the compressor contactor coil, Figure 44–4. Because the impedance of the lock-out relay coil is much higher than the compressor contactor coil, almost all of the 24 volts is dropped across the LOR coil, and very little voltage is across the CC coil. The compressor contactor cannot energize because of the low voltage applied to the coil.

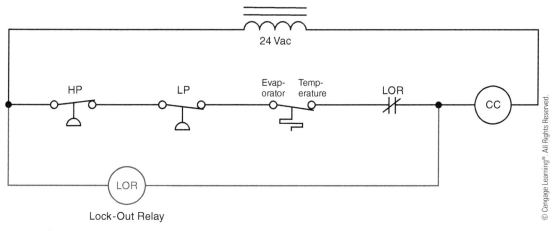

▶ **Figure 44–3**
Basic lock-out circuit.

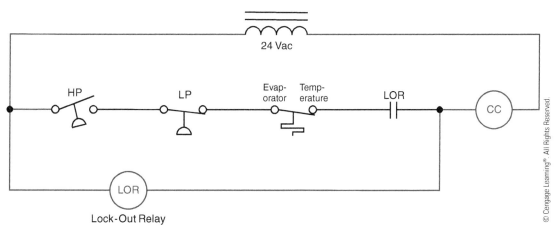

▶ **Figure 44–4**
A circuit exists through the LOR coil and CC coil.

The normally closed LOR contact connected in series with the compressor contactor coil opens. If the high-pressure switch should close, the now open LOR contact prevents the compressor contactor from energizing. The circuit will remain in this condition until the control power is turned off and the lock-out relay coil de-energizes. The line voltage connections for the blower control and lock-out relay circuit board are shown in Figure 44–5. The circuit board and basic controls are shown in Figure 44–6.

CIRCUIT OPERATION

When the heating thermostat contact closes, a circuit exists to the time-delay relay (TDR) and draft motor relay (DMR), Figure 44–7. Power is also supplied to the ignition control module. The draft motor controls the centrifugal switch connected to the thermostat input of the ignition control module. Its function is used to ensure that gas is not to supplied to the burner unless the draft motor is in operation.

In Figure 44–8, it is assumed that the centrifugal switch has closed and permitted the gas burner to ignite. It is also assumed that the time-delay relay has permitted TDR contacts to close. When the TDR contacts close, a current path is provided to the coil of the K3 relay, causing all K3 contacts to change position. When the normally open K3 contact closes, a current path is provided to the coil of the 2M contactor. When 2M energizes, the blower motor turns on, Figure 44–5. The circuit will continue to operate in this manner until the thermostat contacts reopen.

OPERATION OF THE COMPRESSOR LOCK-OUT RELAY

The compressor lock-out relay can operate only when the thermostat is set in the cooling mode. In Figure 44–9 it is assumed that the thermostat has been set for the cooling mode and the thermostat contact is closed. The main current path is through the normally closed K1 contact, low-pressure switch, high-pressure switch, and evaporator temperature switch to the coil of 1M contactor. The current takes this path because of the high impedance of coil K1. When coil 1M energizes, the compressor and condenser fan start, Figure 44–5. There is also a current path through the thermostat fan switch, the normally closed K3 contact, and 2M coil. When the 2M coil energizes, the evaporator fan motor starts. The circuit continues to operate in this manner until the thermostat contact opens or some other problem occurs.

Now assume that the low-pressure switch opens, Figure 44–10. The open circuit caused by the open low-pressure switch now connects coil K1 in series with coil 1M. Because coil K1 has a much higher impedance than 1M, most of the voltage is across K1 and not 1M, causing the K1 relay to energize and the 1M contactor to de-energize. The normally closed K1 contacts open and the normally open K1 contacts close. When the normally open K1 contacts close, a current path is provided to the compressor lock-out indicator. The now open K1 contact prevents the compressor from restarting if the low-pressure switch should reclose. The circuit will remain in this condition until the control power is interrupted.

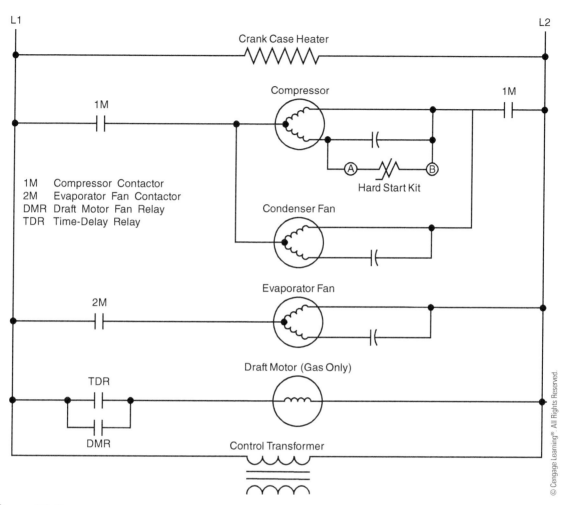

▷| **Figure 44–5**
Line voltage circuit.

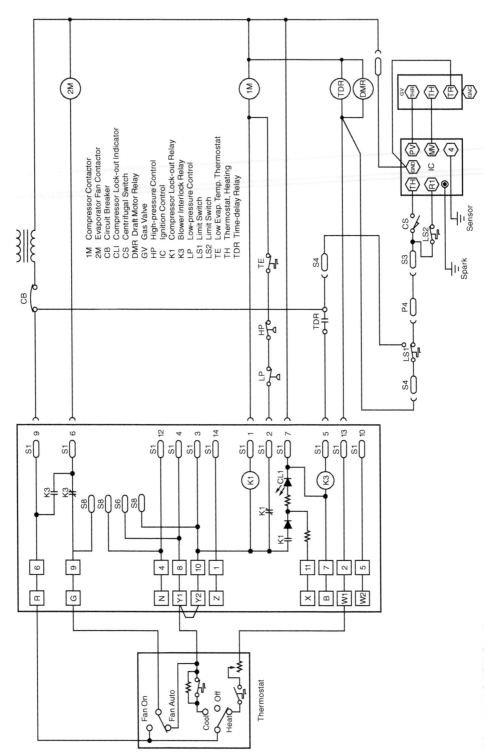

1M Compressor Contactor
2M Evaporator Fan Contactor
CB Circuit Breaker
CLI Compressor Lock-out Indicator
CS Centrifugal Switch
DMR Draft Motor Relay
GV Gas Valve
HP High-pressure Control
IC Ignition Control
K1 Compressor Lock-out Relay
K3 Blower Interlock Relay
LP Low-pressure Control
LS1 Limit Switch
LS2 Limit Switch
TE Low Evap. Temp. Thermostat
TH Thermostat, Heating
TDR Time-delay Relay

▶ **Figure 44–6**
Control circuit for a cooling unit and gas heat.

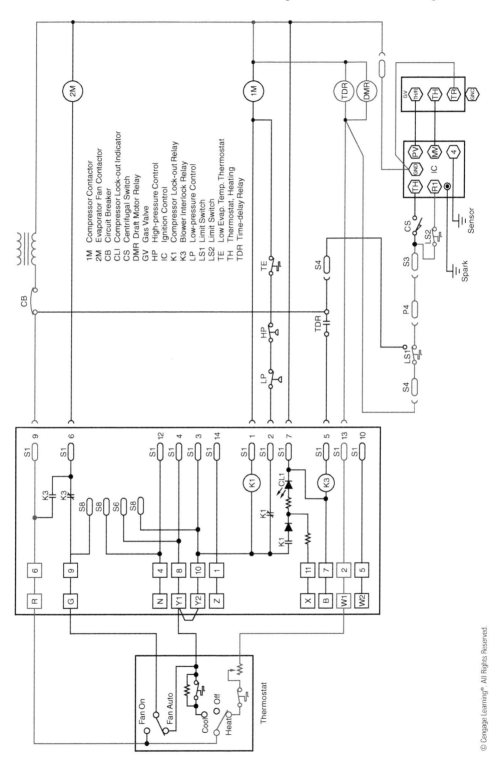

1M Compressor Contactor
2M Evaporator Fan Contactor
CB Circuit Breaker
CLI Compressor Lock-out Indicator
CS Centrifugal Switch
DMR Draft Motor Relay
GV Gas Valve
HP High-pressure Control
IC Ignition Control
K1 Compressor Lock-out Relay
K3 Blower Interlock Relay
LP Low-pressure Control
LS1 Limit Switch
LS2 Limit Switch
TE Low Evap. Temp. Thermostat
TH Thermostat, Heating
TDR Time-delay Relay

Figure 44–7
The heating thermostat contact closes.

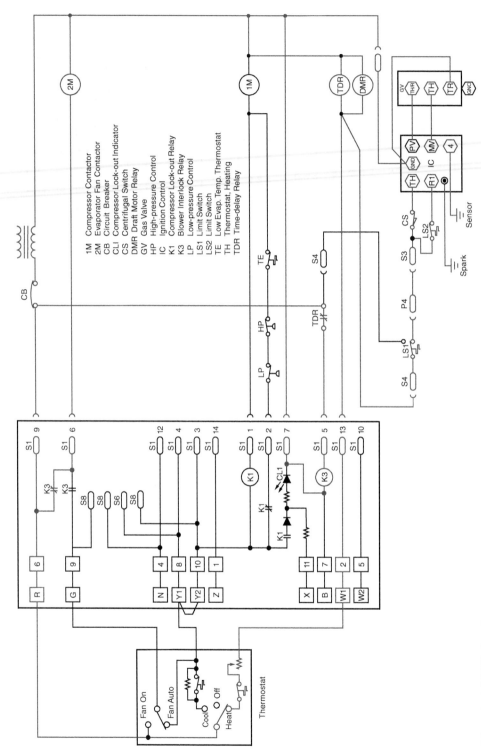

1M Compressor Contactor
2M Evaporator Fan Contactor
CB Circuit Breaker
CLI Compressor Lock-out Indicator
CS Centrifugal Switch
DMR Draft Motor Relay
GV Gas Valve
HP High-pressure Control
IC Ignition Control
K1 Compressor Lock-out Relay
K3 Blower Interlock Relay
LP Low-pressure Control
LS1 Limit Switch
LS2 Limit Switch
TE Low Evap. Temp. Thermostat
TH Thermostat, Heating
TDR Time-delay Relay

▶ **Figure 44–8**
The blower motor turns on after a time delay.

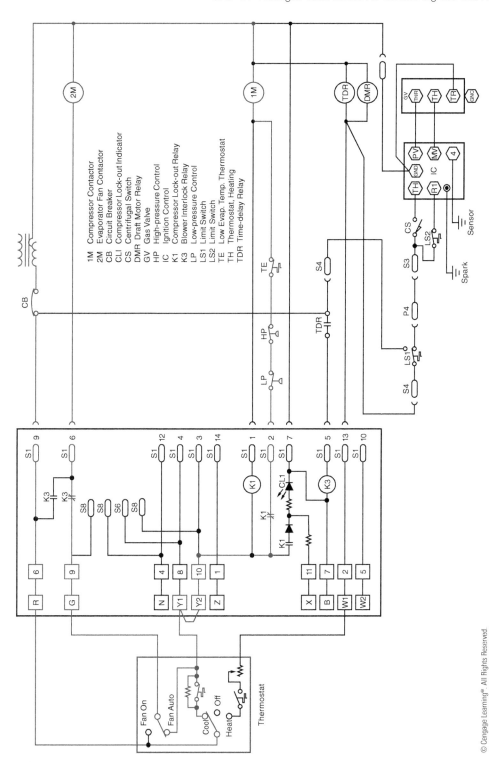

◗ **Figure 44–9**
Circuit during normal cooling cycle.

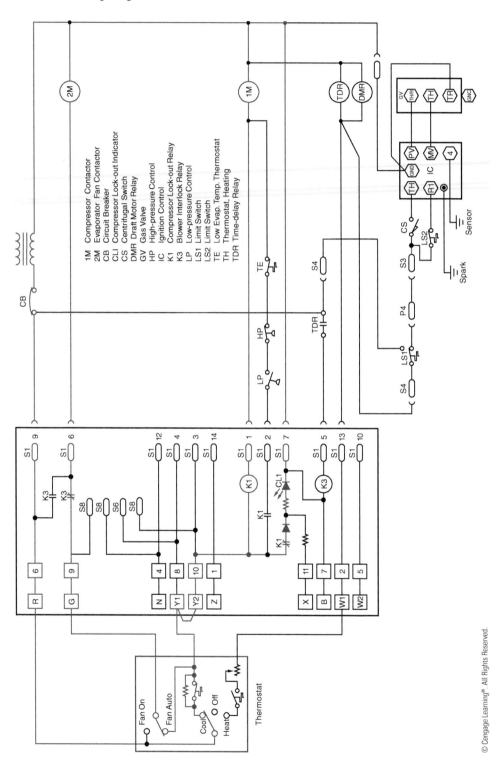

1M	Compressor Contactor
2M	Evaporator Fan Contactor
CB	Circuit Breaker
CLI	Compressor Lock-out Indicator
CS	Centrifugal Switch
DMR	Draft Motor Relay
GV	Gas Valve
HP	High-pressure Control
IC	Ignition Control
K1	Compressor Lock-out Relay
K3	Blower Interlock Relay
LP	Low-pressure Control
LS1	Limit Switch
LS2	Limit Switch
TE	Low Evap. Temp. Thermostat
TH	Thermostat, Heating
TDR	Time-delay Relay

▶ **Figure 44–10**
The lock-out relay has energized.

SUMMARY

- ⊙ The schematic diagram in this example shows mostly the operation of the heating system.
- ⊙ The thermostat in this diagram has four terminals labeled R, G, Y, and C.
- ⊙ The thermostat is connected to the 24-volt side of the transformer.
- ⊙ The heating system contains two high-limit switches.
- ⊙ Both of the high-limit switches are connected normally closed and are connected in series with the automatic gas valve.

KEY TERMS

auxiliary limit	door interlock	main limit
condensing unit	lock-out relay	

REVIEW QUESTIONS

Refer to the schematic shown in Figure 44–1 for the following questions.

1. The unit operates normally in the cooling cycle. When the unit is switched to the heating cycle, the gas burner will not ignite. List four possible problems.

2. The blower fan operates normally in the cooling cycle. In the heating cycle, however, the fan will not operate. List three possible problems.

3. The unit will not operate in the heating or cooling cycle. A voltage check shows that there is no low voltage for operation of the control circuit. List three possible problems.

4. A lock-out protection circuit involves connecting the coil of the lock-out relay in series with the coil of the compressor contactor. When this circuit is energized, why is most of the voltage across the lock-out relay coil and very little across the compressor contactor coil?

5. Once the lock-out relay has been energized, what must be done to reset the circuit?

TROUBLESHOOTING QUESTIONS

Refer to the schematic shown in Figure 44–1 to answer the following questions. Note that this drawing is a combination of both a schematic and a wiring diagram. Also note that most of the control shown is of the heating system. The air-conditioning part of this unit is referred to as "condensing unit" on the diagram. The thermostat contact arrangement is not shown in this diagram, but recall that in the cooling mode, terminals G and Y make connection with terminal R when the thermostat contact closes. In the heating mode, terminal R makes connection with terminal W when the thermostat contact closes.

1. When the unit is in the heating mode, what controls the operation of the blower motor?

   A. Thermostat

   B. Fan relay

   C. Fan switch

   D. Main limit switch

2. When the unit is used in the heating mode, which blower fan speed is used?

   A. HIGH

   B. MED

   C. LOW

3. Assume that when in the heating mode, the gas burner starts when the thermostat calls for heat, but the blower motor will not run and eventually the burner is turned off by the main limit switch. Now assume that when the fan switch on the thermostat is moved to the manual position, the blower motor begins operating. List three conditions that could cause this problem.

   A. _____

   B. _____

   C. _____

4. When the unit is in the cooling mode, which thermostat terminal controls the operation of the blower motor?

   A. R

   B. W

   C. Y

   D. G

5. Describe the operation of the door interlock switch.

   A. The door interlock switch prevents the unit from operating in either the heating or the cooling mode. If the unit is in operation and the door is opened, the unit will stop operation. If the unit is not operating, it cannot start in either the heating or the cooling mode.

   B. The door interlock switch prevents operation in the heating mode only.

   C. The door interlock switch prevents operation in the cooling mode only.

   D. The door interlock switch prevents the unit from starting in either the heating or the cooling mode, but if the unit is in operation when the door is opened, it will continue to operate until the thermostat is satisfied.

To answer the following questions, refer to the schematic diagram in Figure 44–6.

6. Assume that the thermostat is set in the heating mode and that the heating thermostat contact closes. Now assume that the DMR relay coil is defective. Will the gas burner operate? Explain your answer.

7. Which relay controls the operation of the 2M contactor when the unit is in the heating mode?

8. What does the CLI indicator signify?

   A. Incandescent lamp

   B. Neon lamp

   C. Light-emitting diode

   D. The print does not indicate the type of light-producing device.

SECTION 7

Ice Maker and Refrigeration Controls

Household Ice Makers

OBJECTIVES

After studying this unit, the student should be able to:

- ▷ Discuss the operation of a compact-type ice maker
- ▷ Describe the operation of the water valve and flow washer
- ▷ Discuss the sequence of events that takes place during each stage of the ice maker's operation
- ▷ Discuss the operation of a flex tray ice maker

Ice makers can be divided into two major categories, household and commercial. Unlike commercial units, household ice makers do not recirculate water. They fill a tray or mold and the water is allowed to freeze. Various methods are used to sense when the water has been frozen and to eject the ice from the tray.

Commercial ice makers generally recirculate the water during the freeze cycle. The one reason for this is that pure water freezes faster than water containing impurities and minerals. The ice formed is more pure and clearer in color. This does not apply to flaker-type machines, however. Flaker or crushed-ice machines use an auger to scrape ice off an evaporator after the water has been frozen.

Some cube-type machines freeze water in the shape of the cube, and others freeze water as a slab. The slab-type machines use a grid of cutter wires to cut the frozen slab into cubes.

HOUSEHOLD ICE MAKERS

One of the most widely used household ice makers is the **compact**, Figure 45–1. Although a newer model has been introduced, many of these original units are still in operation. The basic operation of this unit is as follows:

1. An electric solenoid valve, Figure 45–2, turns on and fills the tray or mold with water. The valve contains a **flow washer** that meters the amount of water. The washer is designed to work with pressures that range between 15 and 100 psi. The length of time the water is permitted to flow is controlled by a cam operated by a small electric motor. The time can be adjusted by moving the water solenoid switch closer to or farther away from the cam. The amount of water needed to fill the mold is approximately 135 cc, or 4 oz. It should be noted that insufficient water causes the thermostat to cool too quickly, causing the ice maker to eject hollow cubes.

2. A thermostat senses when the water is frozen. It is mounted directly on the mold by a spring clip. The thermostat controls the start of the ejection and refill cycle.

3. When the thermostat contact closes, it turns on the **mold heater** and motor. The motor operates the timing cam and ejector blades. The ice maker is so designed that the **ejector blades** can stall against the ice cubes without causing harm to the motor or mechanical parts. When the heater has warmed the mold sufficiently, the ice cubes are pushed out by the ejector blades.

4. During the ejection cycle, the shutoff arm raises and lowers. The shutoff arm senses the height of ice in the holding bin. If the bin is not full, the arm returns to its original position, and the ice maker is permitted to eject ice cubes again after they have been frozen. If the holding bin is full, however, the arm cannot return to its normal position, and the next ejection cycle cannot begin. The ice maker can be manually turned off by raising the shutoff arm above its normal range of travel.

The ice maker normally permits the ejector blades to make two revolutions before the thermostat reopens its contact and allows the process to stop at the end of the cycle. If the ejector blades make only one revolution, the ice cubes will be left on top of the blades instead of being dumped into the holding bin. This is not a problem, however, because the cubes will be dumped at the beginning of the next ejection cycle. Near the end of the cycle the mold is refilled with water.

▶ **Figure 45–1**
Early model of a compact ice maker.

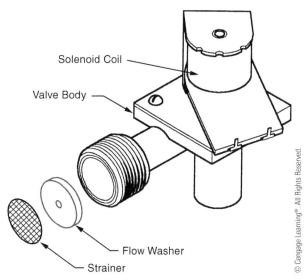

Solenoid Coil

Valve Body

Flow Washer

Strainer

▶ **Figure 45–2**
Water valve.

OPERATION OF THE CIRCUIT

The basic circuit for the compact ice maker is shown in Figure 45–3. The circuit is shown during the freeze cycle. It is assumed that the mold has been filled with water and the thermostat contact is open. The shutoff arm is in its normal position, indicating that the holding bin is not full. Note the position of the ejector blade and the shutoff arm.

Figure 45–4 shows the circuit at the beginning of the ejection cycle. At this time, the thermostat has cooled sufficiently for its contact to close. A current path now exists through the mold heater and motor, and the ejector blades begin to turn.

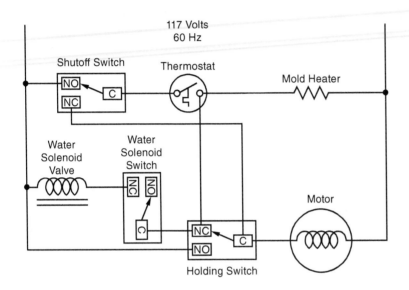
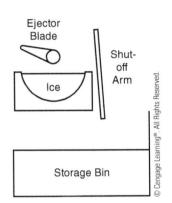

▶ **Figure 45–3**
Basic circuit for the compact ice maker.

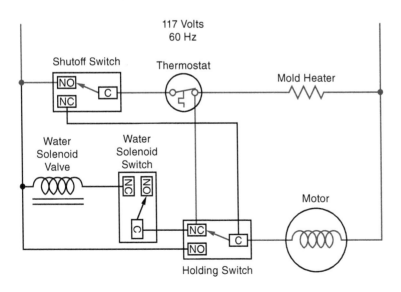
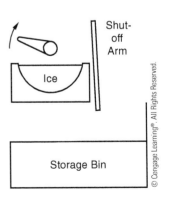

▶ **Figure 45–4**
Beginning of the ejection cycle.

As the motor turns the ejector blades and timing cam, the **holding switch** changes position, and the shutoff arm begins to rise, Figure 45–5. The function of the holding switch is to maintain the circuit until the cam returns to the freeze, or off, position.

In Figure 45–6, the timing cam causes the shut-off arm to rise and fall, making the shutoff switch change position. When the ejector blades reach the ice in the mold, the motor will stall until the ice cubes are thawed loose by the mold heater. Notice that the circuit to both the heater and the motor has

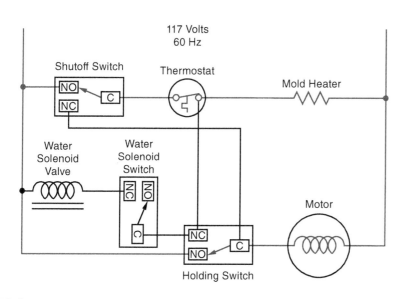
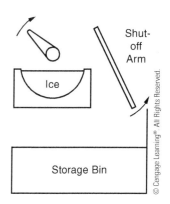

▶ **Figure 45–5**
The shutoff arm begins to rise.

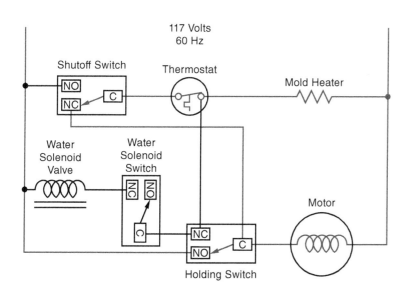
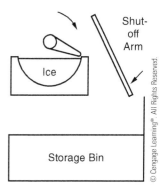

▶ **Figure 45–6**
The shutoff switch changes position.

been maintained by the holding switch. Note that it is possible for the thermostat to open its contact at any point in this process. If this should occur, power is turned off to the mold heater but maintained to the motor by the holding switch.

Near the completion of the first revolution of the ejector blades, the timing cam closes the water solenoid switch, Figure 45–7. Although the water solenoid switch is now closed, current cannot flow through the coil. As long as the thermostat contact is closed, the same voltage potential is applied to both sides of the water solenoid coil. Because there is no potential difference across the coil, no current can flow, and the water valve does not open to permit water flow into the mold.

At the end of the first revolution, Figure 45–8, the shutoff arm and ejector blades have returned to their normal position, and the timing cam has reset

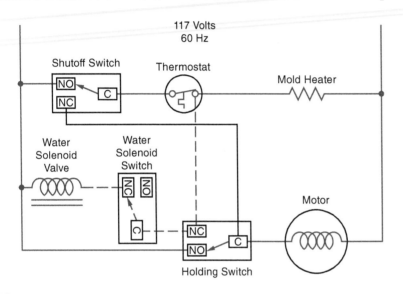
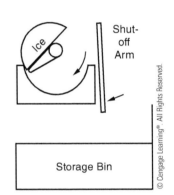

▶ Figure 45–7
The timing cam closes the water solenoid switch.

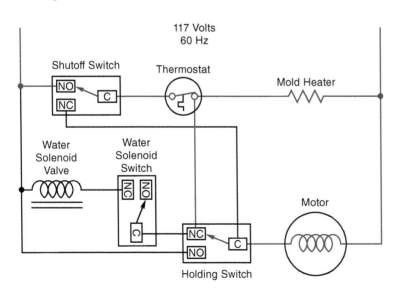
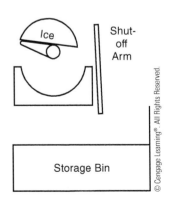

▶ Figure 45–8
End of the first revolution.

all cam-operated switches back to their normal position. Notice, however, that the thermostat contact has remained in the closed position, permitting the second revolution to begin.

After the timing cam has rotated a few degrees, the holding switch again closes to maintain a current path to the motor and mold heater, Figure 45–9. The shutoff arm rises and changes the position of the shutoff switch. The continued

rotation of the ejector blades dumps the ice into the holding bin.

During the second revolution, the increased temperature from the mold heater causes the thermostat contact to reopen, which de-energizes the heater, Figure 45–10. The holding contact, however, provides a continued current path to the motor. If the storage bin is full, the shutoff arm will not return to its normal position, and the shutoff switch will not be reset.

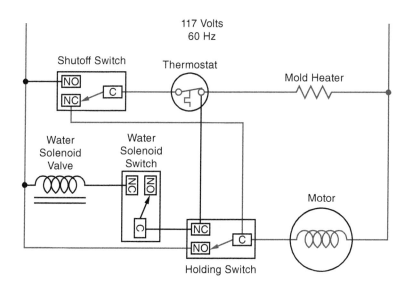
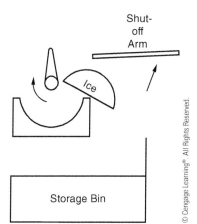

▶ **Figure 45–9**
Beginning of the second revolution.

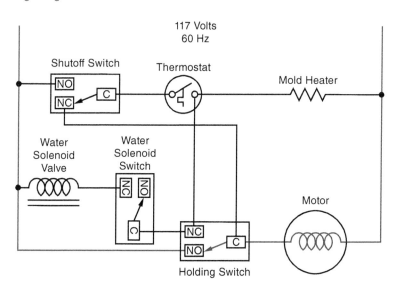
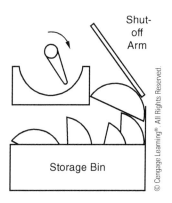

▶ **Figure 45–10**
Middle of the second revolution.

Near the completion of the second revolution, Figure 45–11, the timing cam again closes the water solenoid switch. A current path now exists through the solenoid coil and the mold heater. Although the solenoid coil and mold heater are now connected in series, the impedance of the solenoid coil is much higher than that of the heater. This permits most of the voltage, about 105 to 110 volts, to be applied across the coil, causing it to energize and open the water valve.

The cycle ends when the timing cam reopens the water solenoid and holding switch, Figure 45–12. If the storage bin is full, as shown in this illustration, a new ejection and refill cycle cannot begin until sufficient ice has been removed from the storage bin to permit the shutoff arm to return to its normal position.

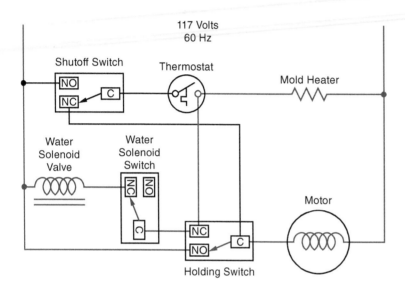

▶ **Figure 45–11**
End of the second revolution.

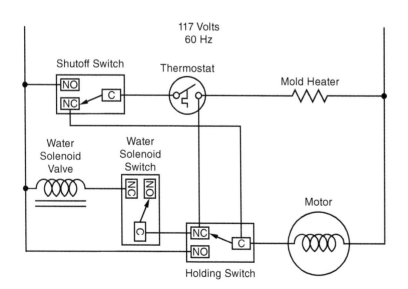

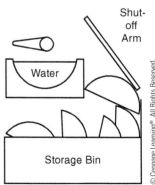

▶ **Figure 45–12**
End of the cycle.

THE NEW MODEL COMPACT ICE MAKER

Although the new model compact ice maker, Figure 45–13, is very similar in design and operating principle to the original version just discussed, there are some significant differences. Some of these differences are listed here:

1. The ejector blades on the newer model stop at a different position, as shown in Figure 45–14. Also shown in Figure 45–14 is the position of the ejector blades when different actions occur during the ejection cycle.

2. The ejector blades make only one revolution instead of two during the ejection cycle.

3. Most of the new models have an external water level adjustment knob, Figure 45–13. Turning the knob moves a set of contacts in relation to its contact ring, permitting the fill time to be longer or shorter.

4. On the original model compact ice maker, the gear located on the front of the unit could be turned manually to advance the ice maker through different parts of the cycle. This gear should *never* be turned on the newer model. To do so will cause damage to the ice maker. The front gears of both the original and the newer compact ice makers are shown in Figure 45–15.

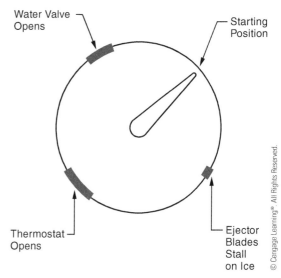

▶ **Figure 45–14**
Ejector blade positions on new model compact ice maker.

▶ **Figure 45–15**
Front gears of the original (left) and newer model (right) compact ice makers.

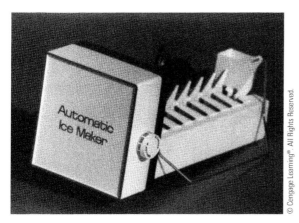

▶ **Figure 45–13**
New compact ice maker.

5. The new model compact provides **test points** on the plate located behind the front cover, Figure 45–16. It is possible to test different parts of the electrical circuit using a voltmeter and ohmmeter. The letters indicate the following test points:
   - N = Neutral side of the line
   - L = L1 (HOT) side of the line
   - M = Motor connection

- H = Heater connection
- T = Thermostat connection
- V = Water valve connection

6. Probably the greatest difference lies in the electrical circuit itself. In this model, copper strips are laminated on an insulated plate

located on the back side of the drive gear. When the motor turns, it turns these copper strips also. Contacts ride against these copper strips and make or break connection to operate the circuit. A diagram of the copper strips and contacts is shown in Figure 45–17.

The basic electrical circuit for this unit is shown in Figure 45–18. Please note that the contact points A, B, C, and D correspond to the contacts shown in Figure 45–17. At this point, connection is made between contacts B and C.

When the thermostat reaches approximately 17°F, its contact closes and produces a current path to both the motor and the heater, as shown in Figure 45–19. The motor begins to turn both the ejector blades and the copper strips located on the back of the main gear. At some point, contact between points B and C is broken and contact between points C and D is made, as shown in Figure 45–20. The ejector blades then stall against the ice. A current path is maintained to the motor between points C and D, and a current path is maintained to the heater by the closed thermostat contact.

After the surface of the ice has been thawed by the heater, the ejector blades will begin to turn again. After the ejector blades have rotated approximately 180°, the thermostat contact opens,

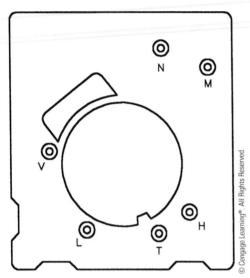

▶ **Figure 45–16**
Test points.

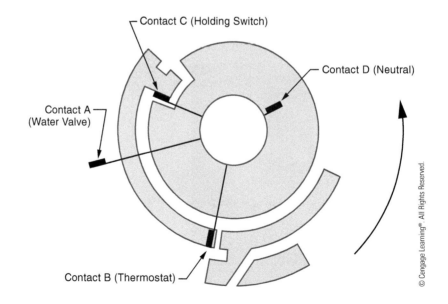

▶ **Figure 45–17**
Rotary switch located
on back of drive gear.

Contact C (Holding Switch)

Contact D (Neutral)

Contact A
(Water Valve)

Contact B (Thermostat)

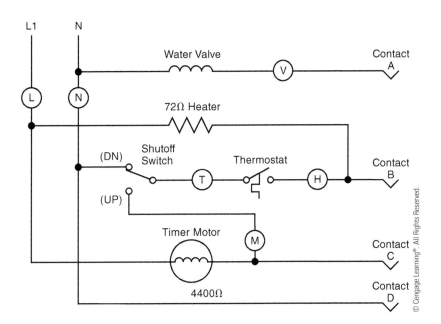

Figure 45–18
Basic schematic diagram for
new model of Whirlpool compact
ice maker.

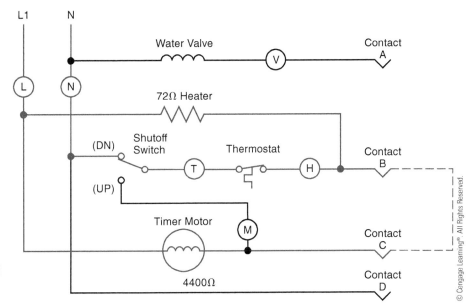

Figure 45–19
A current path is provided
through the heater and
timer motor.

Figure 45–14. As the blades continue to turn, the shutoff arm rises and lowers, and the copper strips advance until connection is made between contacts A and B, Figure 45–21. This provides a current path through the mold heater to the water solenoid valve. Because the coil of the solenoid has a much higher impedance than the mold heater, most of the line voltage will be dropped across the valve, causing it to open and refill the mold. The ejector blades will continue to turn until they reach the end of the cycle and the circuit returns to its original condition, as shown in Figure 45–18.

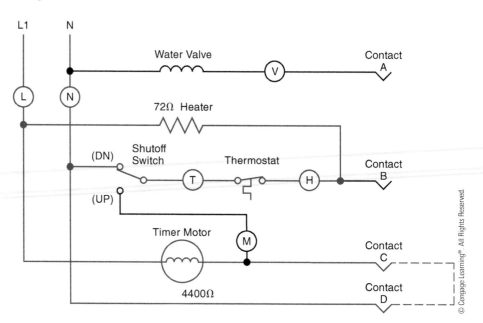

Figure 45–20
The holding contact maintains the timer motor circuit.

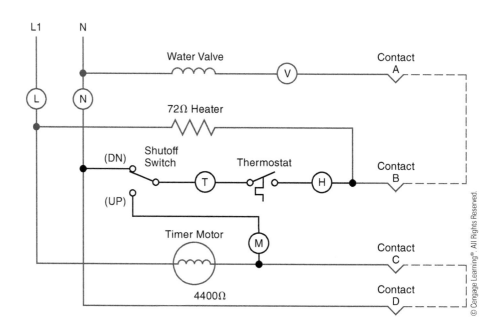

Figure 45–21
Water valve energizes.

SUMMARY

- ⊙ Ice makers can be divided into two major categories: household and commercial.
- ⊙ An electric solenoid valve permits the tray to be filled with water at the proper time.
- ⊙ A flow washer is used to meter the amount of water flow to the tray.
- ⊙ The flow control washer is designed to operate with pressures that range between 15 and 100 psi.
- ⊙ The amount of water needed to fill the mold or tray is approximately 135 cc, or 4 oz.
- ⊙ A thermostat mounted to the mold senses when the water is frozen.
- ⊙ A mold heater is used to slightly melt the cubes so the ejector blades can push them out of the mold.
- ⊙ The shutoff arm senses when the storage bin is full of cubes.
- ⊙ The shutoff arm can be used to manually prevent starting another cycle.

KEY TERMS

compact   holding switch
ejector blades   mold heater
flow washer   test points

REVIEW QUESTIONS

1. Ice makers are divided into what two major categories?

2. What is the advantage of continually recirculating the water during the ice-making process?

3. What component controls the amount of water flow into the original compact ice maker?

4. In the original compact ice maker, what method is used to sense when the water has been frozen?

5. In the original compact ice maker, what controls the start of the ejection and refill cycle?

6. How can the original compact ice maker be manually turned off?

7. How many revolutions does the ejector blades of the original compact ice maker normally make during the ejection cycle?

8. What is the function of the holding switch in the original compact ice maker circuit?

9. Concerning the newer type compact ice maker, what method is used to change the contacts labeled A, B, C, and D in the schematic diagram shown in Figure 45–18?

10. Can the gear of the new type compact ice maker be rotated to manually advance the operation of the ice maker?

11. How many revolutions do the ejector blades of the newer type compact ice maker make during the ejection cycle?

Commercial Ice Makers

OBJECTIVES

After studying this unit, the student should be able to:

▷ Discuss the basic design of different types of commercial ice makers

▷ Discuss the operation of a cube-type and flaker-type ice maker

▷ Discuss the control system of a cube-type and flaker-type ice maker

Commercial ice makers are designed to produce large quantities of ice and are generally found in restaurants, cafeterias, motels, and hotels. Some ice makers produce cubes and others produce flaked ice. The first commercial-type ice maker to be discussed is manufactured by Scotsman Company. The basic components of this unit are shown in Figure 46–1. Notice that this unit can be equipped with either an air-cooled or a water-cooled condenser. The water-cooled unit operates much more quietly.

This unit produces ice by cascading water over a metal plate used as the evaporator, Figure 46–2. A water pump provides continuous circulation of water when the compressor is operating. A water distributor, located at the top of the plate, provides an even flow of water over the entire surface of the plate. Excess water is caught by a trough at the bottom of the plate and is returned to a sump where

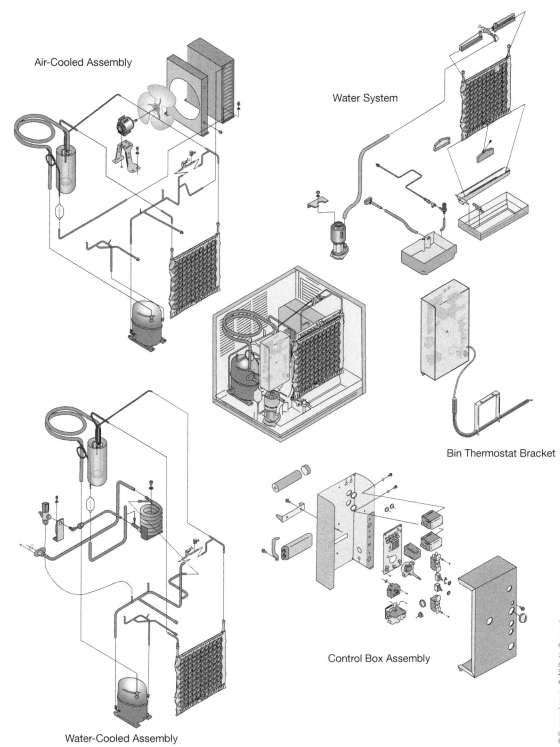

Air-Cooled Assembly

Water System

Bin Thermostat Bracket

Water-Cooled Assembly

Control Box Assembly

▶ **Figure 46–1**
Scotsman cube-type ice maker.

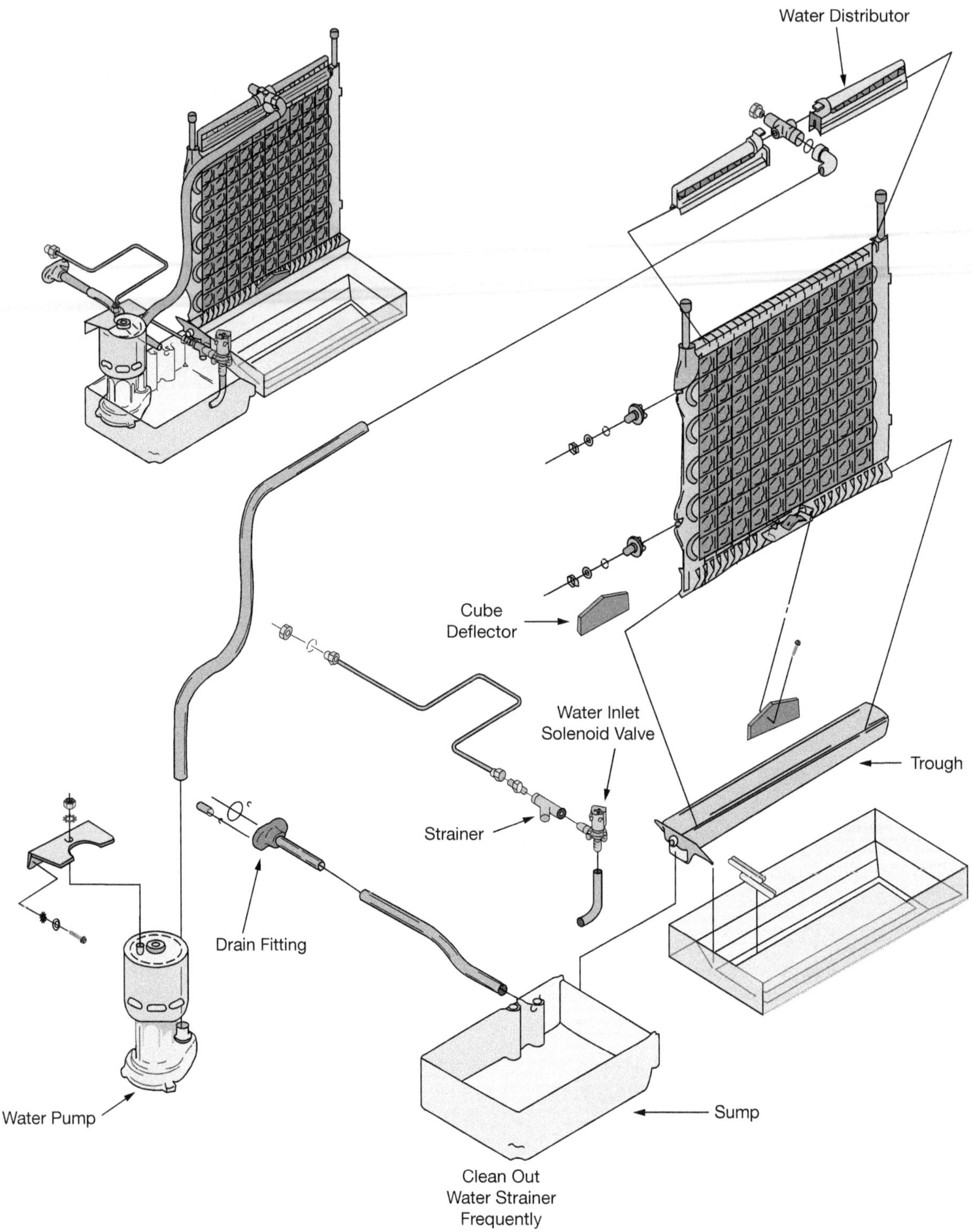

Water Distributor

Cube
Deflector

Water Distributor

Water Inlet
Solenoid Valve

Trough

Strainer

Drain Fitting

Water Pump

Sump

Clean Out
Water Strainer
Frequently

▶| **Figure 46–2**
Water cascades over a metal plate to produce ice.

it is recirculated by the pump. Continuous circulation of water produces a clearer ice because pure water freezes faster than impure water. This is not to say that water can be purified by circulating it. The purification is a result of the freezing process. Basically, the water freezes before the impurities and minerals have a chance to freeze. The water, minus the impurities and minerals, is frozen to the evaporator plate, and the impurities and minerals are returned to the sump.

After the ice has formed, the harvest cycle begins. At the beginning of this cycle, a **hot-gas solenoid valve** opens and permits high-pressure hot gas to

be diverted to the evaporator plate. This high-pressure gas warms the plate and thaws the surface of the ice that is in contact with it. The combination of the warm plate and the cascading water loosens the ice cubes so that they drop away from the plate and fall into the storage bin below. During the harvest cycle, a **water solenoid valve** opens and permits fresh water to flow into the sump. This not only refills the sump but also flushes impurities out the overflow drain.

The basic electrical schematic diagram for this machine is shown in Figure 46–3. There are two manual switches in this circuit. One is the master

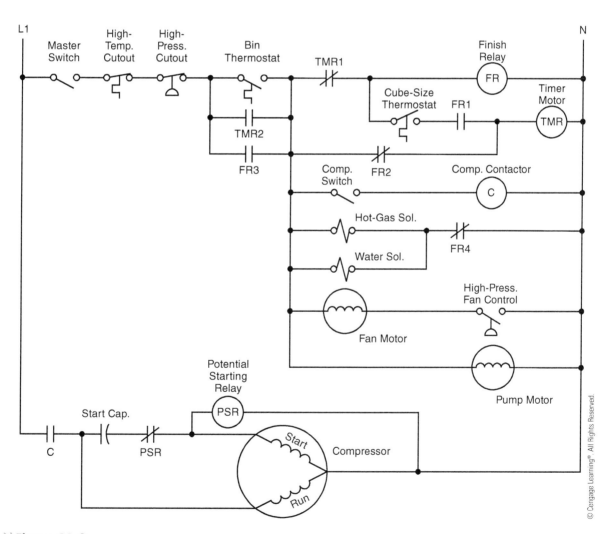

▶ **Figure 46–3**
Basic schematic diagram of a commercial ice cube maker.

switch, and can be used to disconnect power to the entire control circuit. The second is connected in series with the compressor contactor relay coil. This switch can be used to turn off the compressor separately. Two safety switches, the high-temperature cutout and the high-pressure cutout, are connected in series with the master switch. If either of these switches opens, they will disconnect power to the control circuit.

This circuit also contains two thermostats, the **bin thermostat** and the **cube-size thermostat**. The sensor for the bin thermostat is located in the ice storage bin and senses the level of ice. When the ice reaches a high enough level, it touches the sensor and causes the contact to open. This stops the operation of the ice maker at the end of the harvest cycle. The sensor for the cube-size thermostat is mounted on the evaporator plate. When the evaporator plate reaches a low enough temperature, the thermostat contact closes and completes a circuit to the timer motor. The timer contains a set of cam-operated contacts and is used to complete the freeze and harvest cycle.

This circuit also contains a fan motor controlled by a pressure switch that senses the pressure on the high side of the compressor. This fan motor is used only on units with an air-cooled condenser. Because this fan motor is controlled by a pressure switch, it may cycle on and off during the unit's operation.

For a better understanding of this circuit, it is shown at the beginning of the freeze cycle, Figure 46–4. It is assumed that the master switch and the compressor switch have been closed and that the bin thermostat contact is closed. A circuit is now completed to the finish relay, FR, causing all FR contacts to change position. The FR3 contact serves as a holding contact to permit completion of the cycle in the event the bin thermostat contact should open. The FR1 contact has closed to permit a current path to the timer motor when the cube-size thermostat closes. The compressor contactor coil is energized, which closes its contact and connects the compressor to the line. The pump motor is energized causing water to flow over the evaporator plate. It is also assumed that the high-pressure fan control switch is closed, permitting the condenser fan motor to operate. Notice, however, that the FR4 contact has opened to prevent the hot-gas solenoid and the water solenoid from operating.

After the circuit has operated in this condition for some period of time, the evaporator plate becomes cold enough to permit the cube-size thermostat to close as shown in Figure 46–5. This completes a circuit to the timer motor. The timer is used to complete the cycle in the event the bin thermostat should open.

After the timer has operated for some length of time, the timer contacts will change position as shown in Figure 46–6, starting the harvest cycle. The TMR2 contact closes to maintain a current path around the bin thermostat contact, and the TMR1 contact opens and de-energizes coil FR. When coil FR de-energizes, all of its contacts return to their normal position. The FR2 contact recloses and maintains a current path to the timer motor, permitting it to complete the cycle. When the FR4 contact recloses, the hot-gas solenoid and water solenoid valves open. As hot gas is circulated through the evaporator plate, it warms and permits the cube-size thermostat contact to reopen. The circuit will continue to operate in this condition until the timer completes the cycle and resets both TMR contacts. At this point, the freeze cycle will begin again if the bin thermostat is still closed.

FLAKER-TYPE ICE MAKERS

Flaker-type ice makers produce ice continuously as opposed to harvesting ice cubes at certain intervals. Flaked ice has a soft, flaky texture and is often preferred by restaurants. A basic diagram of a flaker-type ice maker is shown in Figure 46–7. The water supply from the building enters the water reservoir. A float valve maintains a constant water level in the reservoir.

Water from the reservoir enters the bottom of the **freezer assembly**. The freezer assembly is the evaporator of the refrigeration unit. The freezer assembly is basically a hollow tube surrounded by a cylindrical container. Refrigerant is used to cool the hollow tube. A stainless steel **auger** is placed inside the hollow tube. The motor drive assembly turns the auger. As water enters the bottom of the freezer assembly, it is frozen into ice and carried upward by the auger. When the ice reaches the top, the flared end of the auger presses excess water out of the ice before it is extruded, or flaked out, through

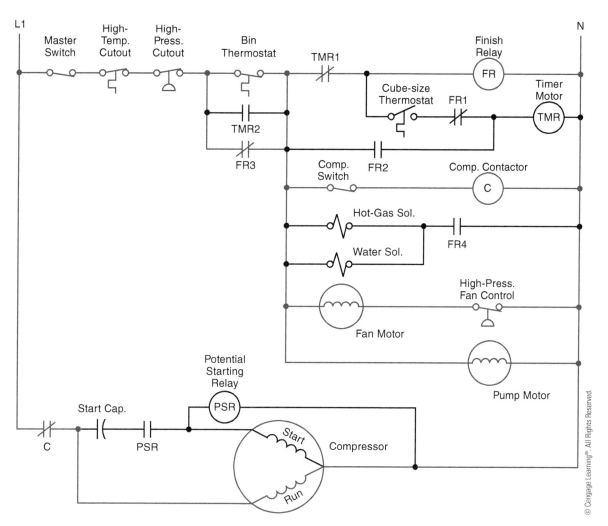

Figure 46–4
Beginning of freezing cycle.

the ice spout. A **nylobraid tube** carries the ice to the ice storage bin. When enough ice accumulates, it touches the sensor bulb of the bin thermostat. The bin thermostat contact then opens and the compressor is disconnected from the line, but the auger drive motor continues to operate for approximately 1 to 2 minutes. This permits the auger to clear the ice out of the freezer unit before it stops operating.

The basic schematic diagram for the flaker-type ice maker is shown in Figure 46–8. Notice that the auger drive motor contains two separate centrifugal switches, one normally closed and the other normally open. The normally closed switch connects the motor start winding to the line when the motor is started. The normally open centrifugal switch controls the coil of the compressor contactor. The compressor contactor can be energized only when the auger drive motor operates within a certain speed range. If ice becomes compacted in the freezer unit, it will cause the auger drive motor to slow down. If the speed of the auger drive motor is reduced below a certain point, the centrifugal switch connected in

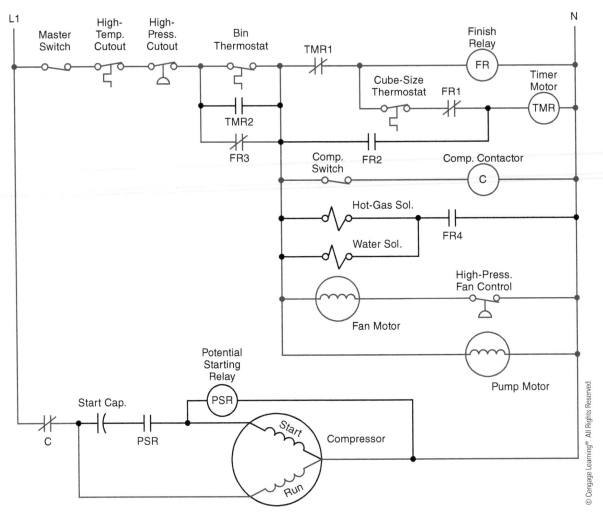

▶| **Figure 46–5**
The cube-size thermostat closes to complete a circuit to the timer motor.

series with the compressor contactor coil will open. If this should happen, the compressor turns off, but the auger delay pressure control switch permits the auger to continue operating for approximately 1½ minutes. If the ice is cleared sufficiently in that length of time, the auger drive motor speed will increase and permit the centrifugal switch to reclose and start the compressor.

The auger delay pressure control switch is a single-pole double-throw pressure switch connected in the low side of the refrigeration system. When the

system is turned off, and the pressures have equalized in the system, the low-side pressure is high enough to hold the switch in the position shown in Figure 46–8. When the compressor starts, the low-side pressure begins to decrease. When it has decreased to 20 **psig (pounds per square inch gauge)**, the contacts change position. They will remain in the changed position until the low-side pressure increases to 32 psig.

The bin thermostat senses the level of ice in the storage bin and normally controls the operation of

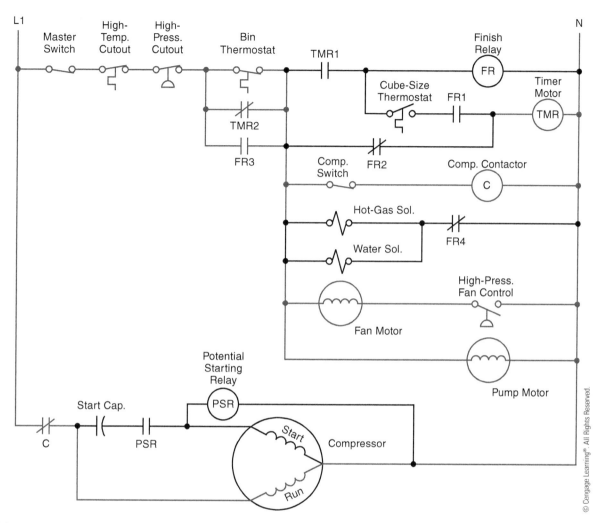

> **Figure 46–6**
Harvest cycle.

the ice maker. A low water-pressure switch is connected to the water supply line. If the water pressure drops below 5 psig, the switch contacts will open. They reclose when the water pressure reaches 20 psig. The low head-pressure switch can interrupt operation of the compressor if the head pressure should become too low.

A master switch disconnects power to the entire control circuit. The spout switch can also disconnect power to the entire circuit in case the ice becomes compacted in the nylobraid tube and spout. If the spout switch becomes tripped, it must be manually reset.

This unit utilizes two condenser fan motors. One motor is mounted at the bottom of the condenser and the other is mounted at the top. The bottom fan motor is connected in parallel with the compressor and will operate any time the compressor is in operation. The top fan motor is controlled by a pressure switch that senses the high side of the refrigeration system. If the pressure becomes high enough, the

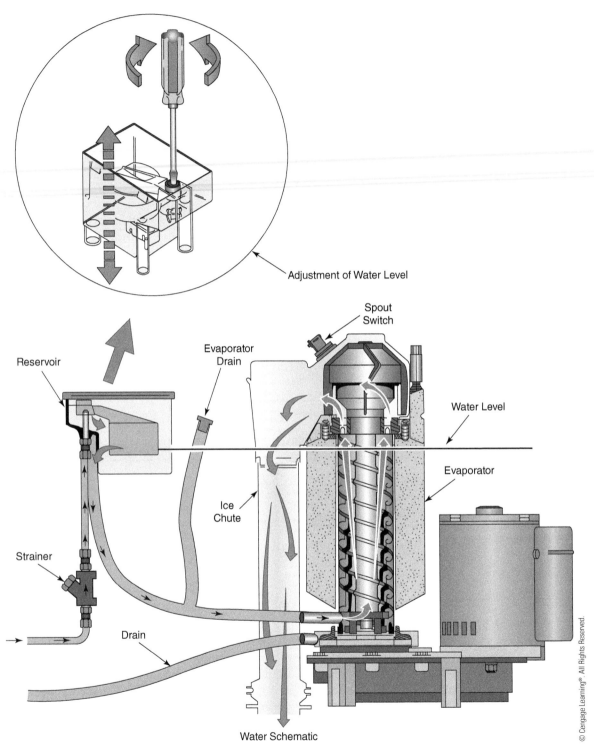

Adjustment of Water Level

Spout Switch

Reservoir

Evaporator Drain

Water Level

Ice Chute

Evaporator

Strainer

Drain

Water Schematic

▶| **Figure 46–7**
Water schematic.

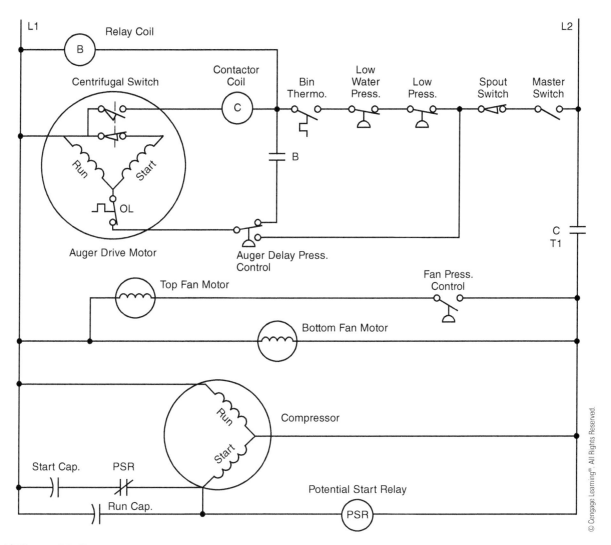

▶ **Figure 46–8**
Basic schematic of flaker-type ice maker.

switch contact will close and start the top condenser fan motor.

OPERATION OF THE CIRCUIT

The circuit in Figure 46–9 shows the initial start-up of the ice maker. It is assumed that the master switch and the bin thermostat switch are closed. A circuit is first completed through coil B. This closes contact B and completes a circuit through the auger delay pressure control switch to the auger drive motor. The normally closed centrifugal switch contact completes a circuit to the start winding and permits the auger drive motor to start.

The normal running mode of the circuit is shown in Figure 46–10. In this phase of operation, it is assumed that the auger drive motor is operating at the proper speed, and the centrifugal switch

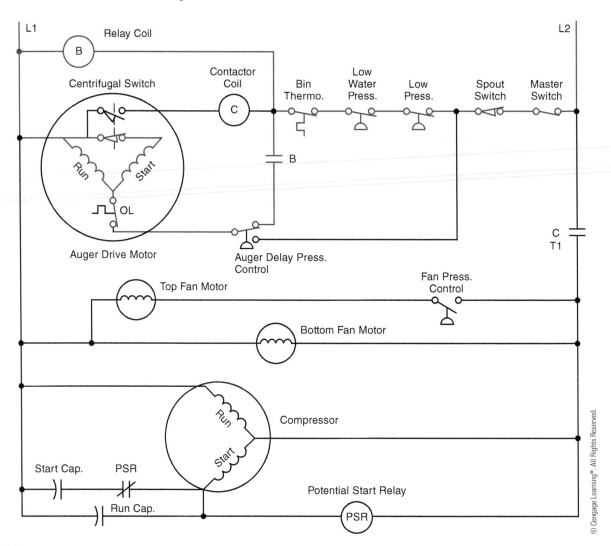

▶ **Figure 46-9**
Initial start sequence.

connected in series with the compressor contactor has closed and permitted the compressor to start. The suction pressure has dropped low enough to permit the auger delay pressure control switch contacts to change position. The bottom condenser fan motor is in operation, and the top fan motor may or may not be operating depending on the pressure on the high side of the refrigeration system.

In the schematic shown in Figure 46–11, it is assumed that the bin thermostat has been satisfied

and has opened its contact. This opens the circuit to coils B and C and stops the operation of the compressor. The auger drive motor will continue to operate until the pressure on the low side of the refrigeration system increases enough to reset the auger delay pressure switch. The circuit will then be back in its original, de-energized position.

The circuit in Figure 46–12 shows the operation of the circuit when the auger drive motor slows down enough to cause the centrifugal switch in

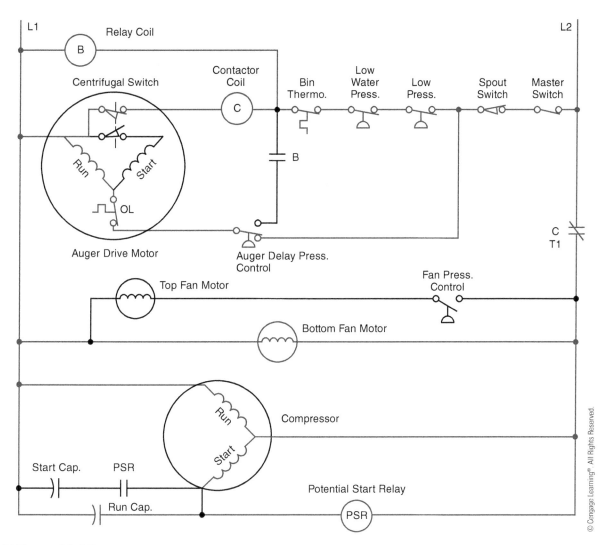

▶ Figure 46–10
Normal ice-making mode.

series with the compressor contactor to open. If this occurs, the compressor will be disconnected from the line and stop operating. A current path is maintained through the auger drive motor and auger delay pressure switch. Notice also that a current path is maintained through relay coil B. If the auger drive motor speed does not increase sufficiently before the auger delay switch changes position, the closed B contact will provide a current path through the reset auger delay switch and permit the auger drive motor to continue operation.

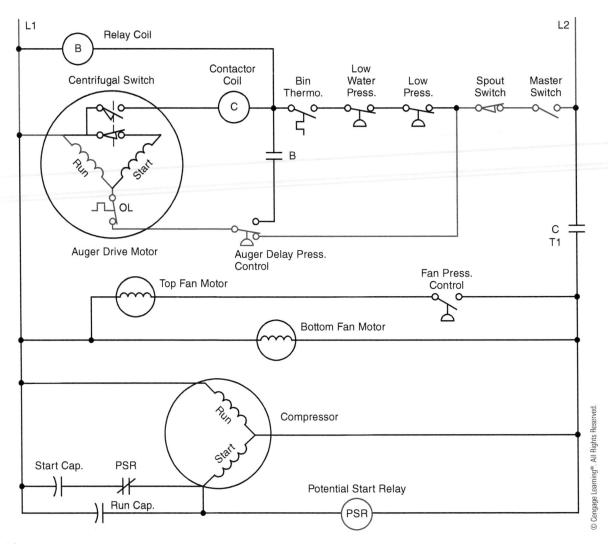

Figure 46–11
The bin thermostat stops the ice-making process.

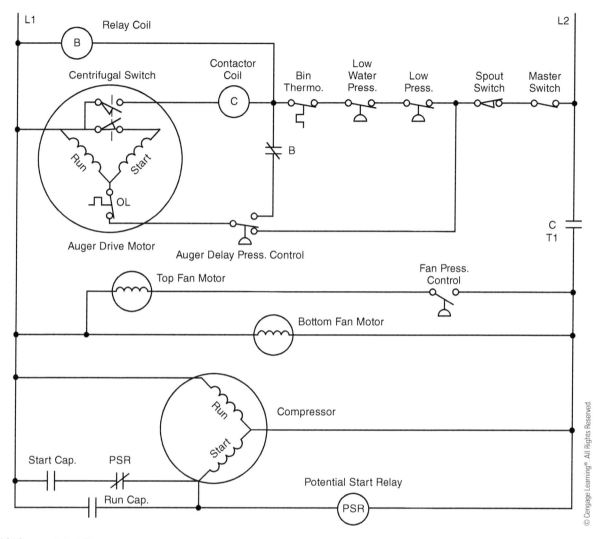

▶ **Figure 46–12**
Auger becomes overloaded and disconnects the compressor.

▶ SUMMARY

- ⊙ Commercial ice makers are designed to produce large quantities of ice.
- ⊙ Continuous circulation of water produces a clearer ice because pure water freezes faster than impure water.
- ⊙ At the beginning of the harvest cycle of the cube-type ice maker, a hot-gas solenoid valve permits high-pressure gas to be diverted to the evaporator plate. This warms the plate and thaws the surface of the ice.

⊙ During the harvest cycle, a water solenoid valve opens and permits fresh water to flow into the sump.

⊙ The cube-type ice maker contains two thermostats: the bin thermostat and the cube-size thermostat.

⊙ Flaker-type ice makers produce ice continuously.

⊙ In the flaker-type ice maker, a float valve maintains a constant water level in the water reservoir.

⊙ The flaker-type ice maker uses a hollow tube as the evaporator.

⊙ An auger is used to carry ice up the freezer assembly.

⊙ A thermostat located inside the bin of a flaker-type ice maker senses when the bin is full.

⊙ The auger motor of the flaker-type ice maker contains two centrifugal switches.

▷ KEY TERMS

auger	flaker-type	nylobraid tube
bin thermostat	freezer assembly	psig (pounds per square inch gauge)
cube-size thermostat	hot-gas solenoid valve	water solenoid valve

▷ REVIEW QUESTIONS

1. Concerning the Scotsman cube-type ice maker, what device is used to cause the water to flow evenly over the surface of the evaporator plate?

2. Concerning the Scotsman cube-type ice maker, what method is used to thaw the surface of the ice in contact with the evaporator plate during the harvest cycle?

3. Concerning the Scotsman cube-type ice maker, what are the two safety switches used to disconnect power from the control circuit?

4. What device is used to sense the level of ice cubes in the storage bin of the Scotsman cube-type ice maker?

5. What electrical component starts the operation of the timer motor in the Scotsman cube-type ice maker?

6. Concerning the Scotsman flaker-type ice maker, what device is used to carry the ice to the top of the evaporator tube?

7. How is excess water pressed out of the ice before it is ejected into the storage bin of the flaker-type ice maker?

8. Explain the operation of the auger delay switch used in the flaker-type ice maker.

9. Concerning the flaker-type ice maker, why is it desirable to have the auger drive motor continue to operate for some period of time after the compressor has stopped operation?

10. What controls the operation of the bottom condenser fan motor in the flaker-type ice maker?

11. What electrical component is used to stop the operation of the compressor if the auger should become overloaded?

12. Concerning the Scotsman flaker-type ice maker, which safety switch must be manually reset if it trips?

UNIT 47 ▷

Refrigeration Controls

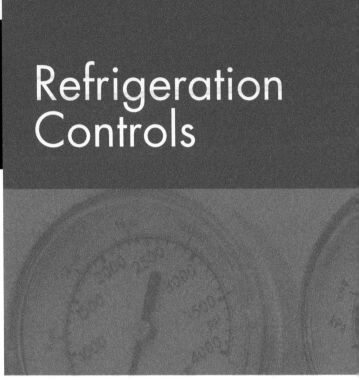

OBJECTIVES

After studying this unit, the student should be able to:

- ▷ Discuss differences between refrigeration and air-conditioning control systems
- ▷ Discuss problems of low head pressures in a refrigeration system
- ▷ Discuss ways of improving the operation of refrigeration systems
- ▷ Explain condenser flooding
- ▷ Discuss the use of shutters and dampers
- ▷ Discuss fan cycling

Refrigeration and air-conditioning systems are essentially the same in that they both involve removing heat from the surrounding air. The differences that occur are in the amount of heat removed and the operating environment. Air-conditioning systems operate at a higher temperature and are used for comfort cooling and humidity control. They generally operate only in the warmer months, and the condenser units are located outside the structure being cooled.

Refrigeration systems are intended to produce colder temperatures and generally operate throughout the year. Some are intended to produce temperatures that range from 35°F to 45°F (1.6°C to 7.2°C) for food storage, and others produce temperatures of 0°F (−17.7°C) or lower for hard freezing. Probably the greatest difference between air-conditioning and refrigeration, as far as controls are concerned, lies in the fact that refrigeration systems operate at lower temperatures and must employ some method for defrosting the evaporator.

Many refrigeration systems, such as open freezers in supermarkets, are intended to operate inside an air conditioned building. It is this operating environment that can create some special problems. The cold ambient air temperature in winter or the cool air inside an air conditioned building can cause the compressor head pressure to drop below a point such that the pressure differential between the high and low side of the system is insufficient for the unit to operate efficiently. When this is the case, some method must be employed to raise the temperature of the condenser and permit the compressor head pressure to increase. Some common ways of accomplishing this are fan cycle control, shutters, and condenser flooding. Refrigeration and air-conditioning units that employ water-cooled condensers control the flow of cooling water to maintain head pressure.

CONDENSER FLOODING

Condenser flooding is accomplished by placing a pressure-operated valve in the refrigerant line between the condenser and metering valve. More than one method can be employed to accomplish this. Flooding the condenser with liquid refrigerant

has the effect of covering the condenser with a plastic blanket. This causes an increase in condenser temperature and a corresponding increase in head pressure. To accomplish condenser flooding, the unit must contain enough liquid refrigerant to flood the condenser. This calls for a large charge of refrigerant and some means of storing it. Units intended to use condenser flooding contain a receiver to hold the excess refrigerant.

Nonadjustable Head Pressure Valve

Figure 47–1 illustrates the connection of a nonadjustable head pressure control valve. A line drawing of the valve is shown in Figure 47–2. The valve's main port is between the condenser and receiver. As long as receiver pressure remains above a certain level, the bypass between discharge and receiver portions of the valve is closed. If the receiver pressure should drop, such as would be the case with low ambient temperature, the spring-loaded valve overcomes the receiver pressure, and hot gas begins to flow through the discharge portion of the valve. Low receiver pressure also causes the valve to decrease the flow from the condenser, causing refrigerant to back up in the condenser. This has the effect of decreasing the surface area of the condenser, causing an increase in temperature and a corresponding

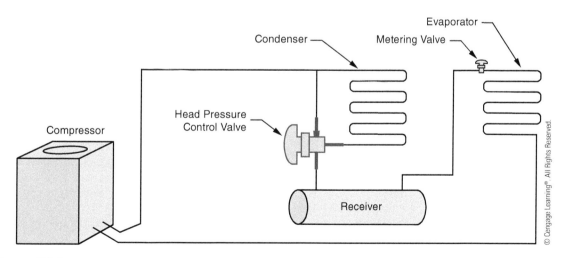

▶ **Figure 47–1**
The head pressure control valve meters the flow of refrigerant through the condenser.

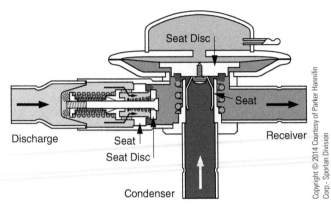

▶ **Figure 47–2**
Operation of a nonadjustable head pressure control valve.

increase in pressure. This valve maintains an almost constant pressure and function well in temperatures of up to about –40°F (–40°C). A nonadjustable head pressure control valve is shown in Figure 47–3.

Adjustable Head Pressure Valve

Adjustable head pressure systems generally require the use of two valves. One valve **opens on rise of inlet** pressure (ORI) and the other **opens on rise of differential** pressure (ORD). A basic piping connection for the adjustable head pressure system is shown in Figure 47–4. An illustration of the ORI valve is shown in Figure 47–5. The ORI valve is an inlet pressure-regulating valve that responds to changes in the condenser pressure. Note that both the inlet and outlet pressures are against the seat disc. One tends to cancel the effects of the other. In warm weather the condenser pressure is greater than the receiver pressure, which causes the valve to open and permit refrigerant to flow through the condenser.

The ORD valve provides a bypass around the condenser. An illustration of the ORD valve is shown in Figure 47–6. The compressor head pressure acts to open the ORD valve and the receiver pressure and

▶ **Figure 47–3**
A nonadjustable head pressure valve.

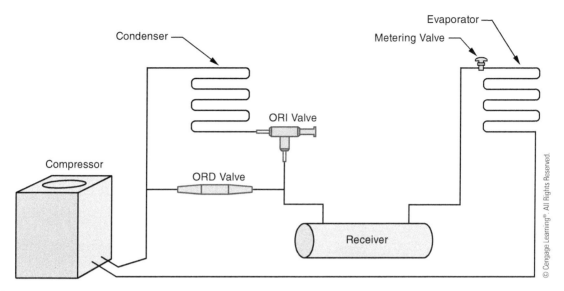

▶ **Figure 47–4**
An adjustable head pressure system requires two valves.

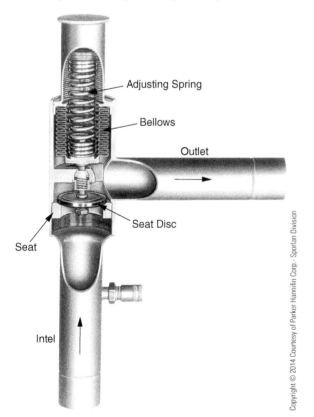

▶ **Figure 47–5**
The open on rise of inlet pressure (ORI) valve can be adjusted.

spring tends to keep the valve closed. If the receiver pressure should drop due to low ambient temperature, two actions take place:

1. The ORI valve begins to close and reduces the flow of refrigerant through the condenser.
2. The ORD valve begins to open and permits the hot gas to bypass the condenser and flow to the receiver.

The reduced flow of refrigerant through the condenser causes an increase in temperature and pressure. An ORI valve is shown in Figure 47–7, and an ORD valve is shown in Figure 47–8.

FAN CYCLE CONTROL

Another common method of increasing head pressure is by controlling the amount of air across the condenser. Air-cooled condensers of large refrigeration units and air conditioners employ a fan to increase airflow across them. The increased airflow permits a greater amount of heat to be removed. During periods that the ambient temperature is so low that the head pressure drops below a certain amount, a fan cycle switch can be used to disconnect power to the condenser fan, Figure 47–9. The switch shown in Figure 47–9 is pressure

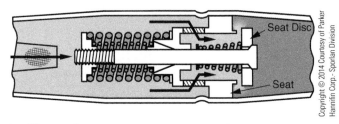

▶ **Figure 47–6**
The open on rise of differential (ORD) valve operates on the difference in pressure between the compressor head pressure and receiver pressure.

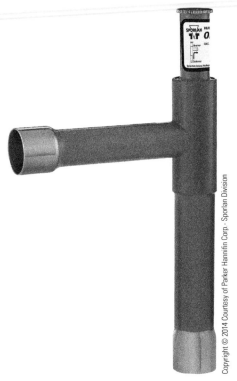

▶ **Figure 47–7**
An open on rise of inlet pressure (ORI) valve.

operated. It is connected in the high side of the unit, Figure 47–10. The switch can be adjusted for the amount of pressure required to turn the fan on and off. A typical setting for refrigerant R12 is 125 psi turn off and 175 psi turn on. At pressures greater than 175 psi, a set of electrical contacts inside the switch closes and connects the condenser fan to the power line. If the high side pressure drops below 125 psi, the contacts will open and turn off the condenser fan. The pressure differential prevents rapid cycling of the condenser fan motor. The fan cycle switch does not hinder operation during warm months and provides good operation during cold months.

The fan cycle switch is relatively inexpensive and can be added to an existing system with very little trouble. Generally no alterations to the piping system are required. There is one potential problem with this type of head pressure control. The pressure differential between fan turn on and turn off can cause erratic operation of the expansion valve.

▶ **Figure 47–8**
An open on rise of differential pressure (ORD) valve.

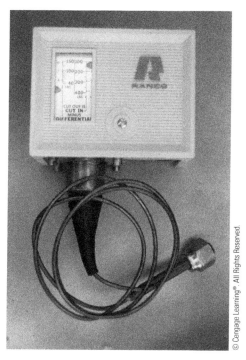

▶ **Figure 47–9**
Pressure-operated fan cycle switch.

Units that employ more than one condenser fan often have one fan controlled by pressure and the others controlled by temperature-sensitive switches. The fans is set to turn on or off in stages. One temperature switch, for example, may turn a fan off at 75°F (23.9°C) and another switch may turn a fan off at 65°F (18.3°C). This helps maintain a more constant head pressure. When temperature switches are used, the temperature-sensing element is generally connected to the liquid line.

Variable-Speed Control

Another type of fan cycle control employs a solid-state device called a triac to control motor speed, Figure 47–11. The triac has the ability to control the output voltage applied to the motor. Refer to Unit 56 for more information concerning the operation of a triac. Some controls vary the output voltage applied to the condenser fan in accord with the temperature of the liquid refrigerant line, and others sense ambient temperature. A thermistor is used to sense the

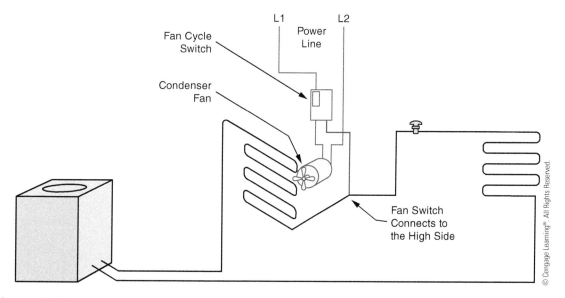

▶ **Figure 47–10**
A pressure-operated fan cycle switch controls the condenser fan to control head pressure.

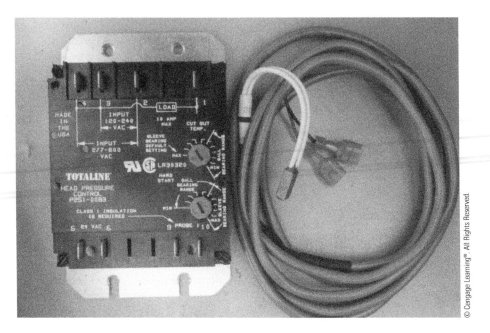

▶ **Figure 47–11**
Variable-speed fan cycle control.

temperature. A decrease of temperature causes the unit to reduce the voltage applied to the condenser fan motor causing it to slow down. As the temperature increases the output voltage increases permitting the motor to increase speed. A chart illustrating typical voltage and temperature relationships for a control that senses liquid line temperature is shown in Figure 47–12. The voltage/temperature relationship can be changed to some degree by changing the type thermistor used to sense the temperature. Notice on the chart that at temperatures below about 75°F the motor is turned off and at temperatures above about 115°F full output voltage is applied to the motor. The variable-speed control helps eliminate some of the pressure differential problems encountered with fan cycle controls that simply turn on or off. A typical connection diagram for this type control is shown in Figure 47–13.

SHUTTERS

Another method for controlling head pressure is with the use of shutters, which can be opened or closed to control the airflow across the condenser.

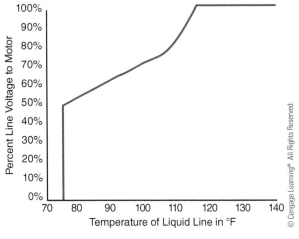

▶ **Figure 47–12**
Typical output voltage and temperature curve for a variable speed fan cycle control.

They can be installed on the inlet or outlet side of the condenser fan. A pressure-operated piston is used to open the shutters and permit more air to flow. The piston is connected to the high side of

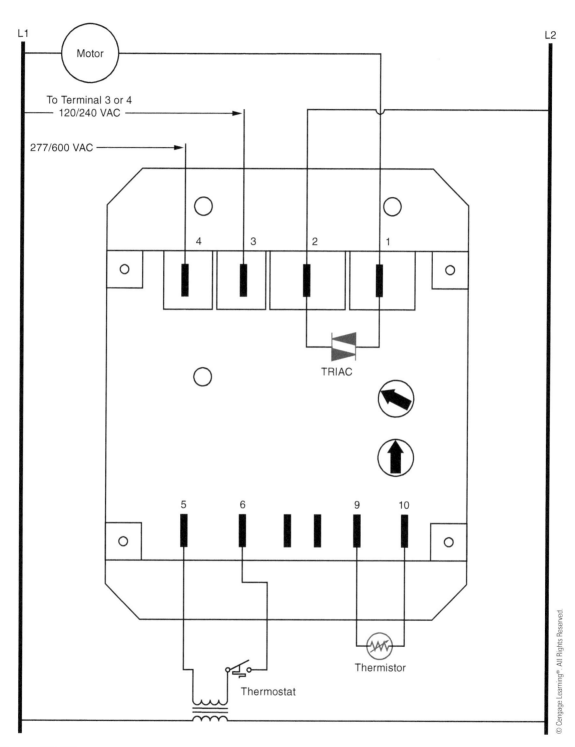

▶ Figure 47–13
Basic connection for a variable speed fan control.

the system in much the same way as the pressure-operated fan cycle switch. If pressure decreases, the shutters close to restrict airflow. If the pressure rises, the piston pushes against a mechanical rod, causing the shutters to open and permit more airflow.

An advantage of shutters is that they open and close slowly, maintaining an even head pressure. If a condenser contains multiple fans, one fan is generally equipped with a shutter, and the other fans are cycled on or off with pressure-activated or temperature-activated switches.

SUMMARY

- ⊙ Refrigeration systems differ from air-conditioning systems in the temperature ranges in which they operate, and refrigeration systems require some method of defrosting the evaporator.
- ⊙ Refrigeration systems can experience difficulties due to low head pressure in cold weather.
- ⊙ Head pressure can be increased by limiting the airflow across the condenser.
- ⊙ Condenser flooding has the effect of covering the condenser with a plastic blanket.
- ⊙ To employ condenser flooding, the unit must have a large charge of refrigerant and a place to store it.
- ⊙ Condenser flooding is generally accomplished by connecting a pressure-operated valve between the condenser and the expansion valve.
- ⊙ Fan cycle control is accomplished by turning the condenser fan on or off in relation to pressure or temperature.
- ⊙ Variable speed fan cycle controls reduce the voltage applied to the motor if the liquid line temperature or ambient air temperature decreases below a certain point.
- ⊙ Shutters can be used to decrease airflow across the condenser.
- ⊙ Shutters generally employ a pressure-operated piston to control the opening and closing of the shutters.

KEY TERMS

opens on rise of differential (ORD)
opens on rise of inlet (ORI)

REVIEW QUESTIONS

1. Explain why low ambient temperatures can cause problems with refrigeration systems.
2. What can be done to make the unit act as if the condenser has been covered with a plastic blanket?

3. Referring to the chart in Figure 47–12, what is the approximate percent of output voltage at a temperature of 100°F (37.7°C)?

4. Referring to the chart in Figure 47–12, at about what temperature does the control turn the condenser fan off?

5. Name two requirements that must be met to use condenser flooding to control head pressure.

6. What device is used to operate the shutters on most shutter systems?

7. Does an increase in pressure cause the shutters to open or close?

8. What problem can be caused by cycling the condenser fan on and off to control head pressure?

9. What device is used to sense liquid line temperature with the variable speed fan cycle control described in this unit?

10. What solid state device does the variable speed control unit employ to control the voltage to the condenser fan motor?

SECTION 8

Solid-State Devices

UNIT 48 ▷ Resistors and Color Codes

OBJECTIVES

After studying this unit, the student should be able to:

▷ Discuss different types of resistors

▷ Determine the resistance of a resistor, using the color code

▷ Test a resistor to determine whether it is within its rated tolerance

Resistors are among the most common components found in electrical circuits. It is sometimes necessary for a technician to be able to determine the value of a resistor in a circuit. Some resistors are intended to carry large amounts of current and produce heat, such as the resistors in electric heating systems, small space heaters, and the burners of an electric range, Figure 48–1. These resistors are generally made from a special type of wire called nichrome. **Nichrome** wire is about 65 times more resistive than copper and can be operated at very high temperatures.

Wire wound resistors are made from nichrome wire also. These resistors are often made by winding nichrome wire on a hollow porcelain tube, Figure 48–2. As a general rule, the ohmic value and wattage rating is written on the resistor. Wire wound resistors of this type should be mounted vertically and not horizontally. The

▶| **Figure 48–1**
Electric heating element.

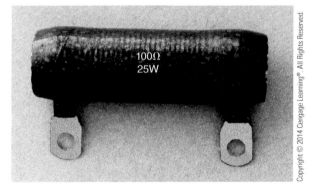

100Ω
25W

▶| **Figure 48–2**
Wire wound resistor.

hollow portion of the resistor acts as a chimney to permit air to circulate through the resistor, Figure 47–3. People often employ lightbulbs as small heaters to protect well pumps during periods of cold weather. The problem with lightbulbs is that they have a bad habit of burning out during the coldest nights of the year. A better solution is to use a wire wound resistor instead of a lightbulb. A resistor with a value of 120 ohms produces 120 watts of heat when connected to 120 volts. If the resistor were rated at 150 watts or more, it would probably never burn out.

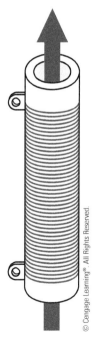

▶| **Figure 48–3**
Mounting the resistor vertically permits air to flow through the hollow opening.

Wire wound resistors generally have the ohmic value and power rating written on the resistor. If the ohmic value is not written on the resistor, an ohmmeter can be used to determine its value.

Color Code

Small **fixed resistors** with ratings of ⅛ to 2 watts are generally marked with **bands of color** to indicate their ohmic value and tolerance. The size of the resistor indicates its **wattage rating**, Figure 48–4. Resistors can have from three to five bands of color. Most have four bands. The colors are used to indicate a numeric value. The chart in Figure 48–5 lists the colors and their corresponding number value. Resistors with a tolerance of ±20% contain three bands of color. Resistors with a tolerance of ±10%, ±5%, and ±2% contain four bands of color, and resistors with a tolerance of ±1% and some special purpose resistors contain five bands of color. When determining the ohmic value and tolerance of a resistor with three or four bands of color, the first two bands represent numbers, the third band is the multiplier, and the fourth band

COLOR	NUMBER VALUE
Black	0
Brown	1
Red	2
Orange	3
Yellow	4
Green	5
Blue	6
Violet	7
Gray	8
White	9
Tolerance	
No fourth band	20%
Silver fourth band	10%
Gold fourth band	5%
Red fourth band	2%
Brown fifth band	1%
Special Multipliers	
Gold third band	0.10
Silver third band	0.01

▶ **Figure 48–5**
Colors represent numeric values.

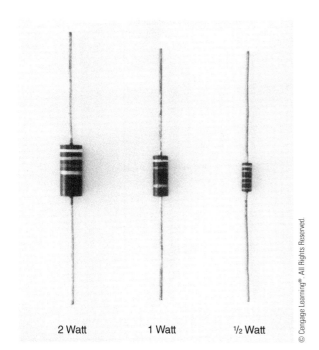

▶ **Figure 48–4**
The size of a fixed resistor indicates its wattage value.

2 Watt   1 Watt   ½ Watt

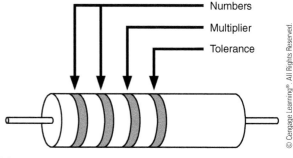

Numbers
Multiplier
Tolerance

▶ **Figure 48–6**
Four-band resistor.

indicates the tolerance, Figure 48–6. If the resistor has a tolerance of ±10%, the fourth band is silver. The fourth band is gold for a resistor with a tolerance of ±5%, and red for a resistor with a tolerance of ±2%.

Assume a resistor has color bands of yellow, violet, orange, and gold, Figure 48–7. The first two bands represent numbers. Yellow is 4 and violet is 7. The third band is the multiplier. Add the number of

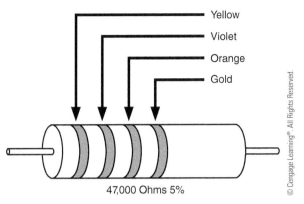

Figure 48–7
The resistor has a value of 47,000 ohms with a tolerance of 5%.

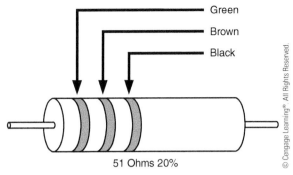

51 Ohms 20%

Figure 48–8
The resistor has a value of 51 ohms and a tolerance of 20%.

zeros indicated by the color. Orange is three, so add 3 zeros. The number becomes 47,000. The resistor has a value of 47,000 ohms with a tolerance of ±5%.

Now assume that a resistor has color bands of green, brown, black, Figure 48–8. Green is 5, brown is 1, and black is 0. The first two colors are numbers 51 and the third band color is black, which is zero. This means there is no multiplier. The resistor has an ohmic value of 51 ohms. Because there is no third band the resistor has a tolerance of ±20%.

Resistors that have a value less than 10 ohms use gold and silver in the third band as multipliers. When a resistor has a third band of gold, it means to multiply the first two numbers by 0.1 or divide the first two numbers by 10. If the third band is silver, multiply the first two numbers by 0.01 or divide the first two numbers by 100. Assume a resistor has colors of blue, gray, gold, and red, Figure 48–9. Blue is 6, gray is 8, and gold means to divide by 10. The resistor has an ohmic value of 6.8 ohms. The red fourth band indicates a tolerance of ±2%.

Tolerance

The **tolerance** indicates the limits of ohmic value. Assume that a resistor is marked 1000 ohms with a tolerance of ±10%. To determine whether this resistor is within its tolerance rating, find 10% of the rated value (1000 × 0.10 = 100 Ω). The resistor will be in tolerance if its value is between 1100 and 900 ohms (1000 + 100 = 1100 and 1000 – 100 = 900).

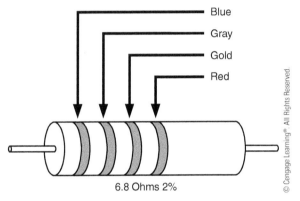

6.8 Ohms 2%

Figure 48–9
The resistor has a value of 6.8 ohms with a tolerance of 2%.

Standard Resistance Values

Fixed resistors are generally manufactured in standard values. The higher the tolerance value, the fewer resistance values available. Standard resistor values for different tolerances are listed in the chart shown in Figure 48–10. In the column under 10% only 12 values of resistance are listed. These standard values, however, can be multiplied by factors of 10. Notice that one of the standard values listed is 33 Ω. There are also standard values in 10% resistors of 0.33; 3.3; 330; 3300; 33,000; and 330,000, 3,300,00 Ω. Notice there is no listing for a value 32 or 34 ohms. They do not exist as a standard value. The 2% and 5% column lists 24

STANDARD RESISTANCE VALUES

0.1%, 0.25%, 0.5%	1%	0.1%, 0.25%, 0.5%	1%	0.1%, 0.25%, 0.5%	1%	0.1%, 0.25%, 0.5%	1%
10.0	10.0	17.8	17.8	31.6	31.6	56.2	56.2
10.1	–	18.0	–	32.0	–	56.9	–
10.2	10.2	18.2	18.2	32.4	32.4	57.6	57.6
10.4	–	18.4	–	32.8	–	58.3	–
10.5	10.5	18.7	18.7	33.2	33.2	59.0	59.0
10.6	–	18.9	–	33.6	–	59.7	–
10.7	10.7	19.1	19.1	34.0	34.0	60.4	60.4
10.9	–	19.3	–	34.4	–	61.2	–
11.0	11.0	19.6	19.6	34.8	34.8	61.9	61.9
11.1	–	19.8	–	35.2	–	62.6	–
11.3	11.3	20.0	20.0	35.7	35.7	63.4	63.4
11.4	–	20.3	–	36.1	–	64.2	–
11.5	11.5	20.5	20.5	36.5	36.5	64.9	64.9
11.7	–	20.8	–	37.0	–	65.7	–
11.8	11.8	21.0	21.0	37.4	37.4	66.5	66.5
12.0	–	21.3	–	37.9	–	67.3	–
12.1	12.1	21.5	21.5	38.3	38.3	68.1	68.1
12.3	–	21.8	–	38.8	–	69.0	–
12.4	12.4	22.1	22.1	39.2	39.2	69.8	69.8
12.6	–	22.3	–	39.7	–	70.6	–
12.7	12.7	22.6	22.6	40.2	40.2	71.5	71.5
12.9	–	22.9	–	40.7	–	72.3	–
13.0	13.0	23.2	23.2	41.2	41.2	73.2	73.2
13.2	–	23.4	–	41.7	–	74.1	–
13.3	13.3	23.7	23.7	42.2	42.2	75.0	75.0
13.5	–	24.0	–	42.7	–	75.9	–
13.7	13.7	24.3	24.3	43.2	43.2	76.8	76.8
13.8	–	24.6	–	43.7	–	77.7	–
14.0	14.0	24.9	24.9	44.2	44.2	78.7	78.7
14.2	–	25.2	–	44.8	–	79.6	–
14.3	14.3	25.5	25.5	45.3	45.3	80.6	80.6
14.5	–	25.8	–	45.9	–	81.6	–
14.7	14.7	26.1	26.1	46.4	46.4	82.5	82.5
14.9	–	26.4	–	47.0	–	83.5	–
15.0	15.0	26.7	26.7	47.5	47.5	84.5	84.5
15.2	–	27.1	–	48.1	–	85.6	–
15.4	15.4	27.4	27.4	48.7	48.7	86.6	86.6
15.6	–	27.7	–	49.3	–	87.6	–
15.8	15.8	28.0	28.0	49.9	49.9	88.7	88.7
16.0	–	28.4	–	50.5	–	89.8	–
16.2	16.2	28.7	28.7	51.1	51.1	90.9	90.9
16.4	–	29.1	–	51.7	–	92.0	–
16.5	16.5	29.4	29.4	52.3	52.3	93.1	93.1
16.7	–	29.8	–	53.0	–	94.2	–
16.9	16.9	30.1	30.1	53.6	53.6	95.3	95.3
17.2	–	30.5	–	54.2	–	96.5	–
17.4	17.4	30.9	30.9	54.9	54.9	97.6	97.6
17.6	–	31.2	–	55.6	–	98.8	–

STANDARD RESISTANCE VALUES *continued*							
2%, 5%	10%	2%, 5%	10%	2%, 5%	10%	2%, 5%	10%
10	10	18	18	33	33	56	56
11	–	20	–	36	–	62	–
12	12	22	22	39	39	68	68
13	–	24	–	43	–	75	–
15	15	27	27	47	47	82	82
16	–	30	–	51	–	91	–

▶ **Figure 48–10**
Standard resistance values.

standard values and the 1% column lists 96 values. All of the values listed can be multiplied by factors of 10 to obtain other resistance values. Resistors with tolerance ratings of 0.1%, 0.25%, and 0.5% generally have the resistance value printed on the resistor.

1% Value Resistors

Notice in the chart shown in Figure 48–10 that the resistor values listed in the 1% column have three numeric numbers instead of two as is the case with 2%, 5%, and 10% values. Because resistors with a tolerance of 1% use three numbers instead of two, a five-band resistor must be used to indicate their value, Figure 48–11. Assume a five-band resistor has color bands of brown, blue, yellow, orange, and brown, Figure 48–12. The first three bands represent numbers: brown = 1, blue = 6, and yellow = 4. The fourth band is orange, which means that you would multiply by 1000 or move the decimal three places; 16.4 becomes 16,400 ohms. The fifth band is brown, which indicates a tolerance value of ±1%.

Other Fifth-Band Colors

Some resistors have five bands of color that are not 1% resistors. These are generally military markings. A resistor with a fifth band of yellow or orange is a reliability rating. The military often needs resistors that have been tested for reliability. It has long been known that if a resistor can operate within its tolerance for some number of hours without failure that the chances of failure become much smaller. The military often employs companies to test resistors for some period of time by operating them in a circuit and then checking the value to see whether the resistor has

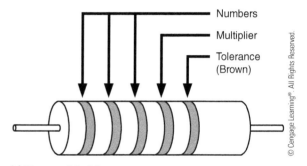

Numbers
Multiplier
Tolerance (Brown)

▶ **Figure 48–11**
Five-band resistor.

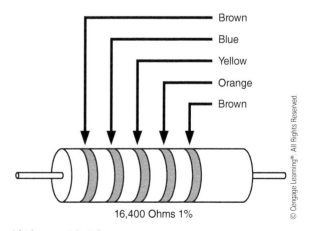

Brown
Blue
Yellow
Orange
Brown

16,400 Ohms 1%

▶ **Figure 48–12**
A 1% resistor has five color bands.

remained within its tolerance rating. Resistors with a fifth band of yellow are rated reliable enough for space flight equipment. An orange fifth band indicates the resistor is reliable enough for use in missile systems. A white fifth band indicates the leads are solderable.

VARIABLE RESISTORS

A variable resistor is a resistor whose values can be changed or varied over a range. Variable resistors can be obtained in different case styles and power ratings. Figure 48–13 illustrates how a variable resistor is constructed. In this example, a resistive wire is wound in a circular pattern, and a sliding tap makes contact with the wire. The value of resistance can be adjusted between one end of the resistive wire and the sliding tap. If the resistive wire has a total value of 100 ohms, the resistor can be set between the values of 0 and 100 ohms.

A variable resistor with three terminals is shown in Figure 48–14. This type of resistor has a wiper arm inside the case that makes contact with the resistive element. The full resistance value is between the two outside terminals and the wiper arm is connected to the center terminal. The resistance between the center terminal and either of the two outside terminals can be adjusted by turning the shaft, which changes the position of the wiper arm. The resistive material for this type of variable resistor is a carbon compound similar to that

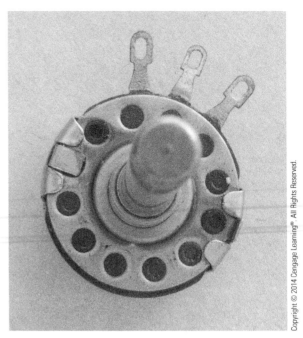

▶ **Figure 48–14**
A variable resistor with three terminals.

used to make fixed carbon resistors. Wire wound variable resistors of this type can be obtained also, Figure 48–15. The advantage of the wire wound type is a higher power rating.

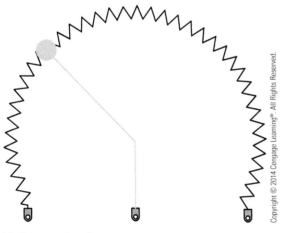

▶ **Figure 48–13**
Basic construction of a variable resistor.

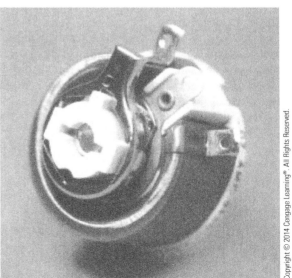

▶ **Figure 48–15**
Wire wound variable resistor.

The resistor shown in Figure 48–14 can be adjusted from its minimum to maximum value by turning the control approximately 3/4 of a turn. In some types of electrical equipment this range of adjustment may be too course to allow for sensitive adjustments. When this becomes a problem, a multi-turn resistor can be used, Figure 48–16. Multi-turn variable resistors operate by moving the wiper arm with a screw of some number of turns. They generally range form 3 turns to 10 turns. If a 10 turn variable resistor is used, it will require 10 turns of the control knob to move the wiper from one end of the resistor to the other instead of 3/4 of a turn.

Variable Resistor Terminology

Variable resistors are known by several common names. The most popular name is *pot*, which is shortened from the word potentiometer. Another common name is rheostat. A rheostat is actually a variable resistor that has only two terminals instead of three, but three terminal variable resistors are often referred to as rheostats also. A potentiometer

describes how a variable resistor is used rather than some specific type of resistor. The word potentiometer comes from the word "potential" or voltage. A potentiometer is a variable resistor used to provide a variable voltage as shown in Figure 48–17. In this example, one end of a variable resistor is connected to +12 volts, and the other end is connected to ground. The middle terminal or wiper is connected to the + terminal of a voltmeter and the − lead is connected to ground. If the wiper is moved to the upper end of the resistor, the voltmeter will indicate a potential of 12 volts. If the wiper is moved to the bottom, the voltmeter will indicate a value of 0 volts. The wiper can be adjusted to provide any value of voltage between 12 and 0 volts.

Schematic Symbols

Electrical schematics use symbols to represent the use of a resistor. Unfortunately, the symbol used to represent a resistor is not standard. Figure 48–18 illustrates several schematic symbols used to represent both fixed and variable resistors.

▶ **Figure 48–16**
Multi-turn variable resistor.

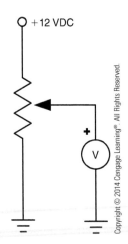

Figure 48–17
A variable resistor used to provide a variable voltage is called a potentiometer.

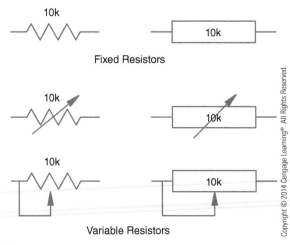

Figure 48–18
Resistor symbols.

SUMMARY

- ⊙ Wire wound resistors are generally used for applications requiring a high power rating.
- ⊙ Nichrome wire is generally employed in the construction of wire wound resistors.
- ⊙ Electric heating elements are generally made of nichrome wire.
- ⊙ Resistors in sizes ranging from ⅛ to 2 watts generally have their values marked with bands of color.
- ⊙ The size of a resistor is generally an indication of its power rating.
- ⊙ A resistor color code is generally used to indicate a resistor's ohmic value and tolerance.
- ⊙ Resistors that have only three bands of color are rated at ±20%.
- ⊙ Resistors with a tolerance rating of ±2%, ±5%, and ±10% have four color bands.
- ⊙ Resistors with a tolerance rating ±1% have a fifth brown band.
- ⊙ The fifth band of some military resistors indicates reliability.
- ⊙ The value of variable resistors can be adjusted.

KEY TERMS

bands of color	tolerance	wattage rating
fixed resistors	variable resistors	wire wound
nichrome		

REVIEW QUESTIONS

1. What type of resistor is generally used when a high power rating is needed?

2. What type of wire is generally used in the construction of resistors intended to be operated at high temperatures?

3. A resistor is marked orange, orange, orange, and gold. An ohmmeter indicates that the resistor value is 34,700 ohms. Is this resistor within its tolerance rating?

4. What would be the color bands for a 1000-ohm resistor with a tolerance of $\pm 2\%$?

5. What would be the color bands for a resistor valued at 365,000 ohms?

6. A resistor has color bands of yellow, orange, gold, gold. What is the value and tolerance of this resistor?

7. What color bands would be found on a resistor with an ohmic value of 510 Ω and a tolerance of $\pm 10\%$?

8. Should a wire wound resistor with a hollow core be mounted vertically or horizontally?

9. A wire wound resistor has a value of 100 Ω and a power rating of 150 watts. If this resistor is connected to 120 volts, will its power rating be exceeded?

10. A circuit requires a resistor with a value of 5000 ohms. What is the closest standard value of a 5% resistor that can be used in this circuit?

UNIT 49 ▷ Semiconductor Materials

OBJECTIVES

After studying this unit, the student should be able to:

▷ Discuss the atomic structure of conductors, insulators, and semiconductors

▷ Discuss how a P-type material is produced

▷ Discuss how an N-type material is produced

Many of the air-conditioning controls are operated by solid-state devices as well as magnetic and mechanical devices. If a service technician is to install and troubleshoot control systems, he or she must have an understanding of electronic devices as well as relays.

Solid-state devices, such as diodes and transistors, are often referred to as **semiconductors**. The word *semiconductor* refers to the type of material solid-state devices are made of. To understand how solid-state devices operate, one must study the atomic structure of conductors, insulators, and semiconductors.

CONDUCTORS

Conductors are materials that provide an easy path for electron flow. Conductors are generally made from materials that have large, heavy atoms. This is why

Valence Electron —

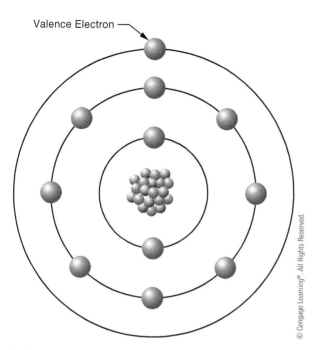

© Cengage Learning®. All Rights Reserved.

▶ **Figure 49–1**
Atom of a conductor.

most conductors are metals. The best electrical conductors are silver, copper, and aluminum. Conductors are materials that have only one or two valence electrons in their atom, Figure 49–1. An atom that has only one valence electron makes the best electrical conductor because the electron is loosely held in orbit and is easily given up for current flow.

INSULATORS

Insulators are generally made from lightweight materials that have small atoms. In the atoms of an insulating material, their outer orbits are filled or almost filled with valence electrons. This means an insulator has seven or eight valence electrons, Figure 49–2.

Because an insulator has its outer orbit filled or almost filled with valence electrons, the electrons are tightly held in orbit and not easily given up for current flow.

SEMICONDUCTORS

Semiconductors, as the name implies, are materials that are neither good conductors nor good insulators. Semiconductors are made from materials that have four valence electrons in their outer orbit, Figure 49–3.

The most common semiconductor materials used in the electronics field are **germanium** and

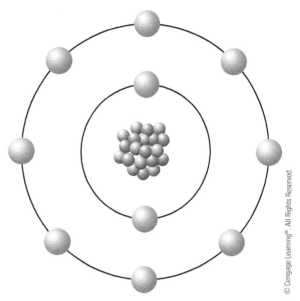

© Cengage Learning®. All Rights Reserved.

▶ **Figure 49–2**
Atom of an insulator.

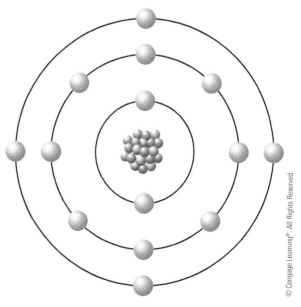

© Cengage Learning®. All Rights Reserved.

▶ **Figure 49–3**
Atom of a semiconductor.

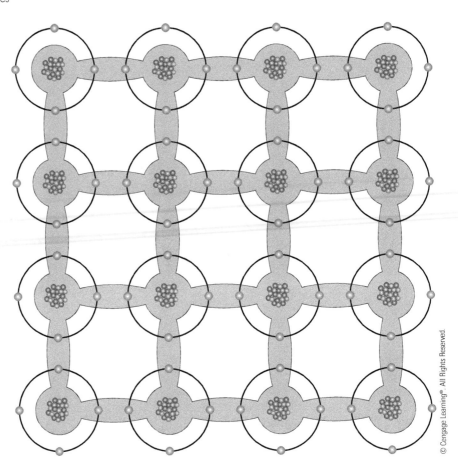

▶ **Figure 49–4**
Lattice structure of a pure
semiconductor material.

silicon. Of these two materials, silicon is used more often because of its ability to withstand heat. When semiconductor materials are refined into a pure form, the molecules arrange themselves into a crystal structure that has a definite pattern, Figure 49–4. A pattern such as this is known as a **lattice structure**.

A pure semiconductor material such as silicon has no special properties and does little more than make a poor conductive material. If a semiconductor material is to become useful for the production of solid-state components, it must be mixed with an impurity. When the semiconductor material is mixed with an impurity that has only three valence electrons, such as idium or gallium, the lattice structure also becomes different, Figure 49–5. When a material that has only three valence electrons is mixed with a pure semiconductor, a hole is left in the material when the lattice structure is formed. This hole is caused by the lack of an electron where one should be. Because the material now has a lack of electrons, it is no longer electrically neutral. Electrons are negative particles. Because a hole is in a place where an electron should be, the hole has a positive charge. This semiconductor material now has a net positive charge and is therefore known as a **P-type material**.

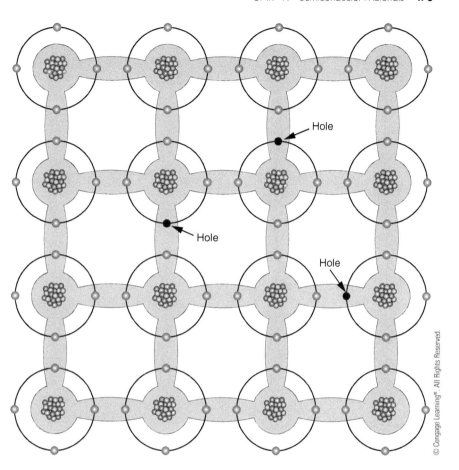

Hole

Hole

Hole

▶ **Figure 49–5**
Lattice structure of a P-type material.

When a semiconductor material is mixed with an impurity that has five valence electrons, such as arsenic or antimony, the lattice structure will have an excess of electrons, Figure 49–6. Because electrons are negative particles, and there are more electrons in the material than there should be, the material has a net negative charge. This material is referred to as an **N-type material** because of its negative charge.

All solid-state devices are made from a combination of P- and N-type materials. The type of device formed is determined by how the P- and N-type materials are connected or joined together. The number of layers of material and the thickness of various layers play an important part in determining what type of device will be formed. For instance, the diode is often called a PN junction because it is made by joining together a piece of P-type and a piece of N-type material, Figure 49–7.

The transistor, on the other hand, is made by joining three layers of semiconductor material, Figure 49–8. Regardless of the type of solid-state device being used, it is made by the joining together of P- and N-type materials.

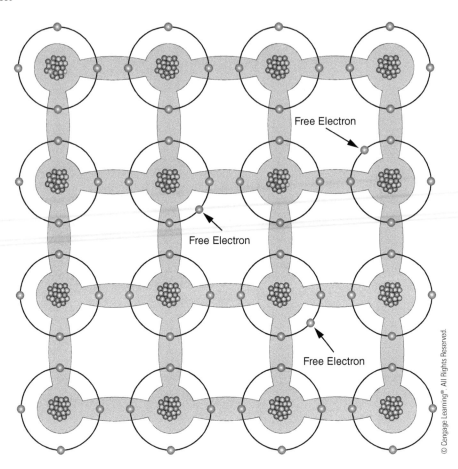

Figure 49–6
Lattice structure of an N-type
material.

Figure 49–7
PN junction.
© Cengage Learning®. All Rights Reserved.

Figure 49–8
Transistor.
© Cengage Learning®. All Rights Reserved.

SUMMARY

- ⊙ Conductors are materials that provide an easy path for electron flow.
- ⊙ The best conductors are silver, copper, and aluminum.
- ⊙ Valence electrons are electrons located in the outermost orbit or shell of an atom.
- ⊙ Conductors are made from materials that generally contain one or two valence electrons.

- Insulators are materials that do not conduct electricity easily.
- Insulators are made from materials that generally contain seven or eight valence electrons.
- Semiconductors are materials that contain four valence electrons.
- The two most common semiconductor materials are germanium and silicon.
- Silicon is used more often than germanium because it can withstand more heat.
- P-type material is made by combining a material that has three valence electrons with a pure semiconductor material.
- A P-type semiconductor material has an excess of holes in its structure.
- N-type material is made by combining a material that has five valence electrons with a pure semiconductor material.
- N-type semiconductor material has an excess of electrons in its structure.

KEY TERMS

germanium   N-type material   semiconductors
lattice structure   P-type material   silicon

REVIEW QUESTIONS

1. How many valence electrons are contained in a material used as a conductor?
2. How many valence electrons are contained in a material used as an insulator?
3. What are the two most common materials used to produce semiconductor devices?
4. What is a lattice structure?
5. How is a P-type material made?
6. How is an N-type material made?
7. What type of semiconductor material can withstand the greatest amount of heat?
8. All solid-state components are formed from combinations of P- and N-type materials. What factors determine what kind of components will be formed?

UNIT 50 ▷

Diodes

As stated previously, solid-state devices are made by combining P- and N-type materials together. The device produced is determined by the number of layers of material used, the thickness of the layers of material, and the manner in which the layers are joined together. Hundreds of different electronic devices have been produced since the invention of solid-state components.

It is not within the scope of this text to cover even a small portion of these devices. The devices to be covered by this text have been chosen because of their frequent use in the air-conditioning industry as opposed to communications or computers. These devices are presented from a straightforward, practical viewpoint, and mathematical explanation is used only when necessary.

The PN junction is often referred to as the **diode**. The diode is the simplest of all electronic devices.

It is made by joining together a piece of P-type material and a piece of N-type material. Refer to Figure 50–1. The schematic symbol for a diode is shown in Figure 50–2. The diode operates like an electric check valve in that it will permit current to flow through it in only one direction. If the diode is to conduct current, it must be **forward biased**. The diode is forward biased only when a positive voltage is connected to the **anode** and a negative voltage is connected to the **cathode**. If the diode is **reverse biased**, the negative voltage connected to the anode and the positive voltage connected to the cathode, it will act like an open switch, and no current will flow through the device.

One thing the service technician should be aware of when working with solid-state circuits is that the explanation of the circuit is often given assuming conventional current flow as opposed to electron flow. *The **conventional current flow theory** assumes that current flows from positive to negative as opposed to the **electron flow theory***, *which states that current flows from negative to positive*. Although it has been known for many years that current flows from negative to positive, many of the electronic circuit explanations assume a positive to negative current flow. There are several reasons for this. For one, ground is generally negative and considered to be 0 volts in an electronic circuit. Any voltage above or greater than ground is positive. Most people find it is easier to think of something flowing downhill or from some point above to some point below. Another reason is that all the arrows in an electronic schematic are pointed in the direction of conventional current flow. The diode shown in Figure 50–2 is forward biased only when a positive voltage is applied to the anode and a negative voltage is applied to the cathode. If the conventional current flow theory is used, current will flow in the direction the arrow is pointing. If the electron theory of current flow is used, current must flow against the arrow.

A common example of the use of the conventional current flow theory is the electrical systems of automobiles. Most automobiles use a negative ground system, which means the negative terminal of the battery is grounded. The positive terminal of the battery is considered to be the "HOT" terminal,

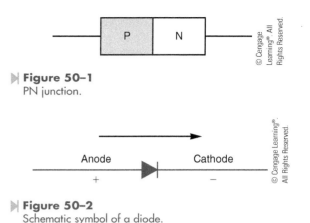

Figure 50–1
PN junction.

Anode Cathode
+ −

Figure 50–2
Schematic symbol of a diode.

and it is generally assumed that current flows from the "HOT" to ground. This explanation is offered in an effort to avoid confusion when troubleshooting electronic circuits.

Diode Ratings

Several ratings concerning diodes can be obtained from a manufacturer's data sheet. However, the two ratings that are generally of concern to the technician in the field are the peak inverse voltage (PIV) rating and the current rating. The peak inverse voltage rating describes the amount of voltage the diode can withstand when reverse biased without breaking down and permitting current to flow through it in the reverse direction. This rating may also be listed as (PRV) peak reverse voltage. Note that the rating lists the *peak* voltage rating. If the diode is connected into an alternating current circuit, it may be necessary to determine the RMS voltage rating of the diode. This can be accomplished by multiplying the RMS rating by 1.414.

EXAMPLE 1: Will a diode with a PIV rating of 150 volts be damaged if it is connected in a 120-volt AC circuit?

$$120 \times 1.414 = 169.68 \text{ V}$$

SOLUTION: Yes, the diode will be damaged. An RMS voltage of 120 volts has a peak value of 169.68 volts.

The forward voltage rating may or may not be listed. The forward voltage rating is generally determined by the semiconductor material used to construct the diode. Most diodes are made from silicon and have a forward voltage rating of 0.6 to 0.7 volt. Diodes made from germanium exhibit a forward voltage rating of 0.3 to 0.4 volt. Although germanium diodes have less voltage drop in the forward direction, silicon is generally preferred because it can withstand a much greater amount of heat without being damaged.

The second diode rating of major concern is the forward current rating. This rating indicates the amount of current the diode can safely conduct in the forward direction. Diodes can be obtained in different case styles and sizes determined by the amount of current the diode must conduct. The greater the current, the more heat the diode must dissipate. If the diode has a forward voltage drop of 0.6 volt and a current of 1 ampere, the diode must dissipate 0.6 watt of heat.

$$P = E \times I \quad P = 0.6 \times 1 \quad P = 0.6 \text{ watt}$$

If the diode has a current flow of 25 amperes and a forward voltage drop 0.6 volt, it must dissipate 15 watts of heat.

$$P = E \times I \quad P = 0.6 \times 25 \quad P = 15 \text{ watts}$$

Diodes in different sizes and case styles are shown in Figure 50–3.

TESTING THE DIODE

The diode can be tested with an ohmmeter. When the leads of an ohmmeter are connected to a diode, the diode should show continuity in only one direction. For example, assume that when the leads of an ohmmeter are connected to a diode, it shows continuity. If the leads are reversed, the ohmmeter should indicate an open circuit. If the diode shows continuity in both directions, it is shorted. If the ohmmeter indicates no continuity in either direction, the diode is open. To test the diode, follow this two-step procedure:

1. Connect the ohmmeter leads to the diode. Notice whether or not the meter indicates continuity through the diode, Figure 50–4.

2. Reverse the diode connection to the ohmmeter, Figure 50–5. Notice whether or not the meter indicates continuity through the diode. The ohmmeter should indicate continuity through the diode in only one direction.

NOTE: *If continuity is not indicated in either direction, the diode is open. If continuity is indicated in both directions, the diode is shorted.*

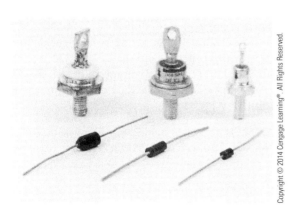

▶ **Figure 50–3**
Diodes in different case styles and sizes.

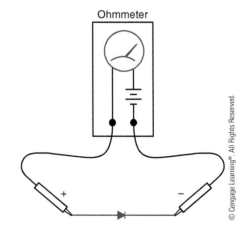

▶ **Figure 50–4**
Testing a diode.

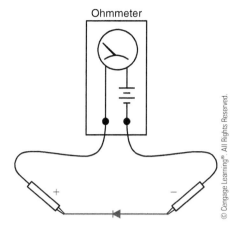

▶| **Figure 50–5**
A diode connected in the reverse direction.

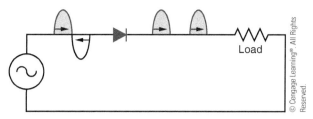

▶| **Figure 50–6**
Half-wave rectifier.

RECTIFIERS

Diodes can be used to perform many jobs, but their most common use in industry is to construct a **rectifier**. A rectifier is a device that changes or converts AC voltage into DC voltage. The simplest type of rectifier is known as the **half-wave rectifier**. Refer to the circuit shown in Figure 50–6. The half-wave rectifier can be constructed with only one diode, and gets its name from the fact that it will rectify only half of the AC waveform applied to it. When the voltage applied to the anode is positive, the diode is forward biased, and current can flow through the diode and load resistor, and back to the power supply. When the voltage applied to the anode becomes negative, the diode is reverse biased, and no current will flow.

Because the diode permits current to flow through the load in only one direction, the current is DC.

Diodes can be connected to produce full-wave rectification, which means both halves of the AC waveform are made to flow in the same direction. One type of **full-wave rectifier** is known as the **bridge rectifier** and is shown in Figure 50–7. Notice the bridge rectifier requires four diodes for construction.

To understand the operation of the bridge rectifier, assume that point X of the AC source is positive and point Y is negative. Current will flow to point A of the rectifier. At point A, diode D4 is reverse biased and D1 is forward biased. The current will flow through diode D1 to point B of the rectifier. At point B, diode D2 is reverse biased, so the current must flow through the load resistor to ground. The current returns through ground to point D of the rectifier. At point D, both diodes D4 and D3 are forward biased, but current will not flow from positive to positive. Therefore, the current will flow through diode D3 to point C of the bridge, and then to point Y of the AC source, which is negative at this time. Because current flowed through the load resistor during this half cycle, a voltage is developed across the resistor.

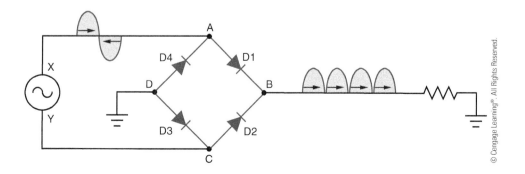

▶| **Figure 50–7**
Bridge rectifier.

Now assume that point Y of the AC source is positive and X is negative. Current will flow from point Y to point C of the rectifier. At point C, diode D3 is reverse biased and diode D2 is forward biased. The current will flow through diode D2 to point B of the rectifier. At point B, diode D1 is reverse biased, so the current must flow through the load resistor to ground. The current flows from ground to point D of the bridge. At point D, both diodes D3 and D4 are forward biased. As before, current will not flow from positive to positive, so the current will flow through diode D4 to point A of the bridge and then to point X, which is now negative. Because current flowed through the load resistor during this half cycle, a voltage is developed across the load resistor. Notice that the current flow was in the same direction through the resistor during both half cycles.

Most of industry operates on three-phase power instead of single-phase. Six diodes can be connected to form a three-phase bridge rectifier, which will change three-phase AC voltage into DC voltage.

TWO-DIODE-TYPE FULL-WAVE RECTIFIER

Another type of full wave can be constructed using only two diodes. The advantage of this rectifier is that it has a voltage drop across only two diodes instead of four. In order to construct a two-diode full-wave rectifier, a center-tapped transformer must be employed, Figure 50–8. Each diode is connected across one half of the secondary winding, and therefore, is connected to half the voltage of the secondary. If the transformer has a secondary voltage of 24 volts, the rectifier actually works on 12 volts.

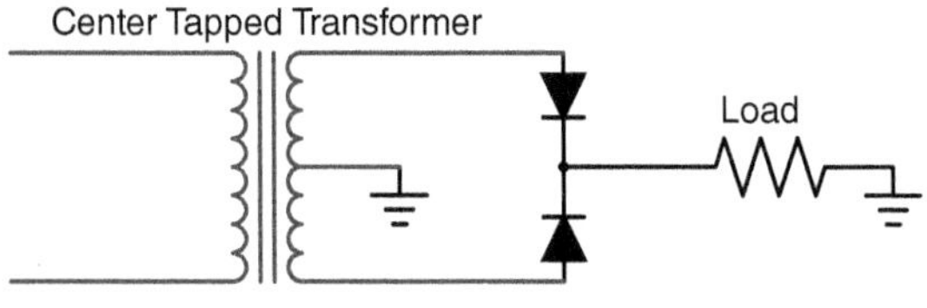

The DC output voltage would be approximately the same as a bridge rectifier connected to 12 volts.

THREE-PHASE RECTIFIER

Most of industry operates on three-phase power instead of single-phase. There are two types of rectifiers that can be used to rectify three-phase alternating current into direct current. One is the half wave and the other is the full wave. The half-wave three-phase rectifier requires that the three-phase connection be a four-wire system that is derived from a wye connection with a center tap, Figure 50–9. One advantage of the three-phase half-wave rectifier is that it produces much less ripple than a single-phase full-wave rectifier, Figure 50–10. Note that the voltage never drops back to zero. This produces a higher average voltage output than the single-phase full-wave rectifier.

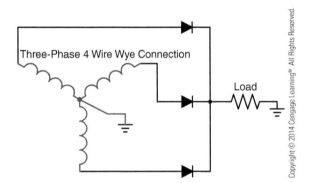

▶ **Figure 50–9**
Three-phase half-wave rectifier.

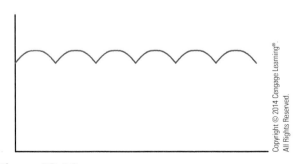

▶ **Figure 50–10**
Three-phase half-wave rectifiers have less ripple than single-phase full-wave rectifiers.

▶ **Figure 50–8**
Two-diode-type full-wave rectifier.

Center Tapped Transformer

Load

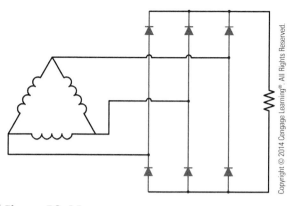

▶ **Figure 50–11**
Three-phase bridge rectifier.

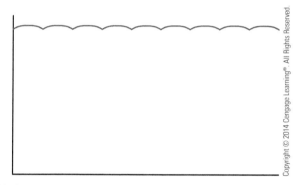

▶ **Figure 50–12**
The three-phase bridge rectifier produces one-half the ripple of a three-phase half-wave rectifier.

The second type of three-phase rectifier is the bridge type, which produces full-wave rectification. Three-phase bridge rectifiers require six diodes. An advantage of the bridge type rectifier is that it does not require a three-phase four-wire system. The bridge type rectifier can be connected to a delta or wye three-phase connection, Figure 50–11. An advantage of the three-phase full-wave rectifier as compared to the three-phase half-wave rectifier is that it produces one-half the amount of ripple, Figure 50–12.

AVERAGE VOLTAGE VALUES

The difference between peak and RMS voltage was discussed in a previous unit. The peak value is the maximum voltage an AC waveform will reach

during one half cycle, and the RMS value is the value of voltage that can produce the same amount of power as an equivalent value of direct current. When making this comparison, it is assumed that the DC voltage is derived from a constant source such as a battery. Although rectifiers do convert AC voltage into DC voltage, the output is not a constant value due to the amount of ripple produced by different types of rectifiers. The output voltage of a rectifier is called the *average* value because a DC voltmeter indicates the average voltage determined by the peak value and the amount of ripple. For a single-phase half-wave rectifier the average value is 0.3185 the peak value or 0.45 the RMS value.

EXAMPLE 2: A single-phase half-wave rectifier is connected to a 24-volt transformer. What is the DC output voltage?

SOLUTION: One method is to change the 24-volt RMS value into peak and then multiply by 0.3185.

$$24 \times 1.414 = 33.936$$

$$33.936 \times 0.3185 = 10.8 \text{ volts}$$

The second method is to multiply the RMS value by 0.45.

$$24 \times 0.45 = 10.8 \text{ volts}$$

The average value for a single-phase full-wave rectifier can be determined by multiplying the peak value by 0.637 or the RMS value by 0.9.

Because three-phase rectifiers produce much less ripple, the average output voltage will be greater than that for a single-phase rectifier. To determine the average output voltage for a three-phase half-wave rectifier, multiply the peak value by 0.827 or the RMS value by 1.169. To determine the average voltage value for a three-phase full-wave rectifier, multiply the peak value by 0.955 or the RMS value by 1.35.

IDENTIFYING DIODE LEADS

When the diode is to be connected in a circuit, there must be some means of identifying the anode and the cathode. Diodes are made in different case styles so there are different methods of identifying the leads.

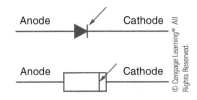

Figure 50–13
Lead identification of a plastic case diode.

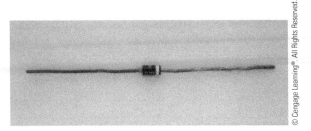

Figure 50–14
Junction diode.

Large stud-mounted diodes often have the diode symbol printed on the case to show proper lead identification. Small plastic case diodes often have a line or band around one end of the case, Figure 50–13. This line or band represents the line in front of the arrow on the schematic symbol of the diode. An ohmmeter can always be used to determine the proper lead identification if the polarity of the ohmmeter leads is known. The positive lead of the ohmmeter must be connected to the anode to make the diode forward biased. A junction diode is shown in Figure 50–14.

ZENER DIODES

Zener diodes are special diodes that have a low reverse voltage breakdown rating. Zener diodes generally have a reverse breakdown voltage of about 3 to 200 volts. They are intended to be operated with current flowing through them in the reverse direction. The symbol for a zener diode is shown in Figure 50–15.

When the reverse breakdown voltage of a diode is exceeded, the current suddenly begins to flow with almost no restriction. This is generally referred to as an **avalanche** condition. Zener diodes are sometimes referred to as avalanche diodes. This avalanche current is referred to as the **zener region**, Figure 15–16. Note in Figure 50–16 that the reverse voltage drop is relatively constant over a wide range of current. Diodes are not harmed when they operate in the zener region as long as the electrical values of their current or watts are not exceeded.

Because the voltage drop across the zener diode is constant, it is generally used as a voltage regulator. Any device or devices connected in parallel with the zener will have the same voltage applied. An example circuit is shown in Figure 50–17. The zener diode has a reverse voltage rating of 12 volts and a power rating of 5 watts. Resistor R_1 is used to limit the current in the circuit. Resistors R_2 and R_3 represent the load for the circuit. Note the supply voltage must be greater than the zener voltage for the circuit to operate.

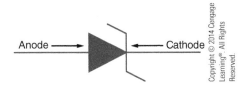

Figure 50–15
Schematic symbol of a zener diode.

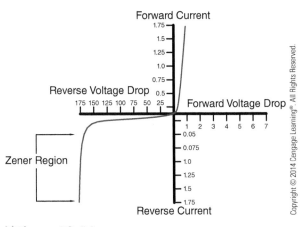

Figure 50–16
The current avalanches in the reverse direction.

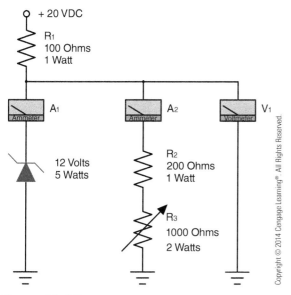

+ 20 VDC

R₁
100 Ohms
1 Watt

A₁ Ammeter

A₂ Ammeter

V₁ Voltmeter

12 Volts
5 Watts

R₂
200 Ohms
1 Watt

R₃
1000 Ohms
2 Watts

▶ **Figure 50–17**
Zener diode used as a voltage regulator.

Resistor R_1 and the zener diode form a series circuit to ground. Because the zener has a constant voltage drop of 12 volts, resistor R_1 has a voltage drop of 8 volts, (20 volts − 12 volts = 8 volts). Resistor R_1 therefore permits a total current flow of 0.08 ampere (80 milliamperes) in this circuit, (8 volts/100 Ω = 0.08 A).

The load circuit, R_2 and R_3, is connected in parallel with the zener diode. Because the load

is in parallel with the zener, the voltage dropped across the zener will be applied to the load. If the zener maintains a constant voltage drop of 12 volts, the load will have a constant applied voltage of 12 volts. The maximum current that can flow through the load is 0.06 ampere (60 milliamperes), (12 volts/200 Ω = 0.06 A). Note that the value of R_1 was chosen to ensure that there would be sufficient current to operate the load. In order for a load current of 60 mA to flow, the value of R_3 is adjusted to 0 ohms. At this point, ammeter A_2 indicates a flow of 60 mA, and ammeter A_1 indicates 20 mA. The values of A_1 and A_2 always add to equal the maximum current that can flow in the circuit, (60 mA + 20 mA = 80 mA). Voltmeter V_1 indicates a voltage of 12 volts.

Now assume that resistor R_3 is adjusted to a value of 200 ohms. The load now has a resistance of 400 ohms (200 Ω + 200 Ω = 400 Ω). Ammeter A_2 now indicates a current flow of 30 mA (12 volts/400 Ω = 0.030). Ammeter A_1, however, now indicates a current flow 50 mA (50 mA + 30 mA = 80 mA). Voltmeter V_1 indicates a value of 12 volts.

Zener diodes can be tested in the same manner as rectifier diodes provided the ohmmeter voltage is less than the reverse voltage value of the zener. For example, an ohmmeter that provided an output voltage of 6 volts could not be used to test a zener diode with a reverse voltage rating of 5 volts. The zener diode would appear to be shorted even if it were not.

▶ SUMMARY

⊙ The PN junction is formed by joining a piece of P-type and a piece of N-typesemiconductor material together.

⊙ The diode operates like an electronic check valve in that it will permit current to flow through it in only one direction.

⊙ The diode can be used to change alternating current into direct current.

⊙ A half-wave rectifier rectifies only one-half of the AC waveform into DC.

⊙ A full-wave rectifier rectifies both halves of the AC waveform into DC.

⊙ The diode can be tested with an ohmmeter by connecting it first one way and then the other. Current should flow through it in only one direction.

⊙ The conventional current flow theory states that current flows from positive to negative.

▶ The electron flow theory states that current flows from negative to positive.

▶ The PIV rating indicates the amount of voltage the diode can withstand in the reverse biased condition.

▶ The forward current rating indicates the amount of current the diode can conduct in the forward direction.

▶ The output voltage of a rectifier is called the average value.

▶ Zener diodes are commonly used as voltage regulators.

▶ Zener diodes are operated in the reverse direction as compared to a junction diode.

▶ KEY TERMS

anode	diode	rectifier
bridge rectifier	electron flow theory	reverse biased
cathode	forward biased	zener diode
conventional current flow theory	full-wave rectifier	zener region
	half-wave rectifier	

▶ REVIEW QUESTIONS

1. What is the PN junction more commonly known as?

2. On a plastic case diode, how are the leads identified?

3. Explain how a diode operates.

4. Explain the difference between the conventional current flow theory and the electron flow theory.

5. Explain the difference between a half-wave rectifier and a full-wave rectifier.

6. Explain how to test a diode with an ohmmeter.

7. A single-phase full-wave bridge rectifier is connected to 240 volts AC. Assuming that 240 is the RMS value, what is the average DC voltage?

8. What type of rectifier require six diodes to construct?

9. What is the most common application for a zener diode?

UNIT 51

Light-Emitting Diodes (LEDs) and Photodiodes

OBJECTIVES

After studying this unit, the student should be able to:

▷ Discuss the operation of a light-emitting diode

▷ Compute the resistance needed for connecting an LED into a circuit

▷ Connect an LED in a circuit

▷ Discuss the differences between light-emitting diodes and photodiodes

▷ Draw the schematic symbols for LED and photodiodes

Light-emitting diodes (LEDs) are among the most common devices found in the electrical and electronics fields. They are used as indicator lights in many types of equipment. They have an extremely long life when operated within their ratings because there is no filament to burn out. LEDs are constructed by joining special semiconductor materials together that emit photons when power is applied. The color produced is determined by the types of materials used. LED colors are generally IR (infrared), red, green, yellow, orange, and blue. The basic light-emitting diode is formed by joining gallium arsenide (GaAs) or gallium phosphide (GaP) with some other material. These two solutions can be combined to form a solid solution known as gallium arsenide phosphide (GaAsP). Different colors are produced by adding other compounds, called dopants, such as zinc selenide (ZnSe) or silicon carbide (SiC). The

Color	Wavelength (nm)	Voltage	Semiconductor Material
Red	610–760	1.63–2.03	Aluminum gallium arsenide (AlGaAs) Gallium arsenide phosphide (GaAsP) Aluminum gallium indium phosphide (AlGaInP) Gallium(III) phosphide (GaP)
Orange	590–610	2.03–2.10	Gallium arsendide phosphide (GaAsP) Aluminium gallium indium phosphide (AlGaInP) Gallium(III) phosphide (GaP)
Yellow	570–590	2.10–2.18	Gallium arsendide phosphide (GaAsP) Aluminium gallium indium phosphide (AlGaInP) Gallium(III) phosphide (GaP)
Green	500–570	1.9–4	Indium gallium nitride (InGaN) / Gallium(III) nitride (GaN) Gallium(III) phosphide (GaP) Aluminium gallium indium phosphide (AlGaInP) Aluminium gallium phosphnide (AlGaP)
Blue	450–500	2.48–3.7	Zinc selenide (ZnSe) Indium gallium nitride (InGaN) Silicon carbide (SiC) as substrate
Violet	400–450	2.76–4	Indium gallium nitride (InGaN)
Purple	Depends on type	2.48–3.7	Dual blue/red LEDs Blue LED with red phosphor or White LED with purple plastic
Infared	>760	<1.9	Gallium arsendie (GaAs) Aluminium gallium arsenide (AlGaAs)
Ultraviolet	<400	3.1–4.4	Diamond (235 nm) Boron nitride (215 nm) Aluminium nitride (AlN) (210 nm) Aluminium gallium nitride (AlGaN) Aluminium gallium indium nitride (AlGaInN)
White	Broad spectrum	Up to 3.5	Blue/UV diode with yellow phosphor

Figure 51–1
The color of an LED is determined by the material it is made from.

chart in Figure 51–1 shows different-colored LEDs, the wavelength of light in nanometers, and the materials used to construct the diode.

LED CHARACTERISTICS

The electrical characteristics of light-emitting diodes vary considerably from those of the common junction or rectifier diode. Junction diodes have a forward voltage drop of about 0.7 volt for silicon and 0.4 volt for germanium. LEDs have a forward voltage drop of about 1.7 volts or greater, depending on the material the diode is made of. Most light-emitting diodes are operated at about 20 mA or less current. A chart showing the typical

forward voltage drop of different diodes is shown in Figure 51–2. Junction diodes typically have a PIV rating of 100 volts or greater, but LEDs have a typical PIV rating of about 5 volts. For this reason, when light-emitting diodes are used in applications where they are intended to block any amount of reverse voltage, they are connected in series with a junction diode.

Testing LEDs

Light-emitting diodes can be tested in a manner similar to that of testing a junction diode. The LED is a rectifier and should permit current to flow through it in one direction only. When testing an

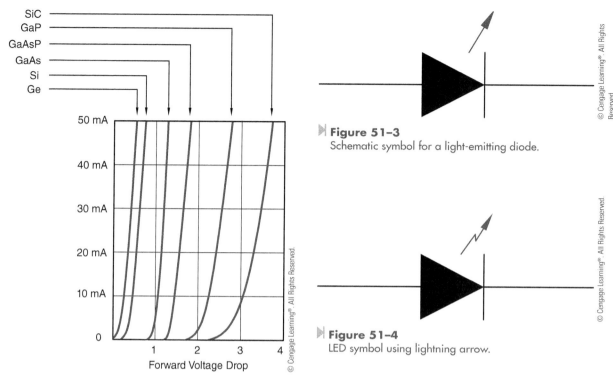

▶ **Figure 51–2**
Forward voltage and current characteristics of diodes.

▶ **Figure 51–3**
Schematic symbol for a light-emitting diode.

▶ **Figure 51–4**
LED symbol using lightning arrow.

LED with an ohmmeter, it must be capable of supplying enough voltage to overcome the forward conduction voltage of about 1.7 volts or higher. The meter, however, must not supply a voltage that is higher than the reverse breakdown voltage. The schematic symbol for a light-emitting diode is shown in Figure 51–3. Some symbols use a straight arrow, as shown in Figure 51–3, and others use a lightning arrow, as shown in Figure 51–4. The lightning arrow symbol is employed to help prevent the arrow from being confused with a lead attached to the device. The important part of the symbol is that the arrow is pointing away from the diode. This indicates that light is being emitted or given off by the diode.

LED Lead Identification

Light-emitting diodes are housed in many different case styles. Regardless of the case style, however, there is generally some method of identifying which lead is the cathode and which is the anode. The case of most LEDs will have a flat side that is located closer to the cathode lead, Figure 51–5. Also, the cathode lead is generally shorter.

Seven-Segment Displays

A very common device that employs the use of light-emitting diodes is the seven-segment display, Figure 51–6. The display actually contains eight LEDs, each segment plus the decimal point. Common cathode displays have all the cathodes connected together to form a common point. The display is energized by connecting a more positive voltage to the anode lead of each segment. Common anode displays are energized by connecting the appropriate cathode lead to a more negative voltage (generally ground). The seven-segment display can be used to display any number from 0 to 9.

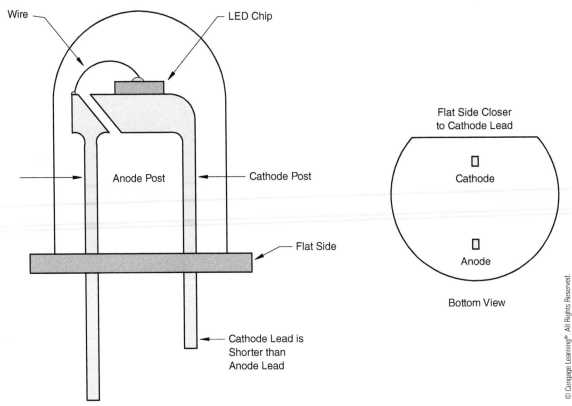

Wire

LED Chip

Anode Post

Cathode Post

Flat Side

Cathode Lead is Shorter than Anode Lead

Flat Side Closer to Cathode Lead

Cathode

Anode

Bottom View

▶ **Figure 51–5**
Identifying the leads of an LED.

Connecting the LED in a Circuit

When used in a circuit, the LED generally operates with a current of about 20 mA (0.020 A) or less. Assume that an LED is to be connected in a 12 VDC circuit and is to have a current draw of approximately 20 milliamperes. This LED must have a current-limiting resistor connected in series with it. Ohm's law can be used to determine what size resistor should be connected in the circuit.

$$R = \frac{E}{I}$$

$$R = \frac{12}{0.020}$$

$$R = 600 \ \Omega$$

The nearest standard size resistor without going below 600 Ω is 620 Ω. A 620 Ω resistor would be connected in series with the LED, Figure 51–7.

The minimum power rating for the resistor can also be determined using Ohm's law. The LED will have a voltage drop of approximately 1.7 volts. Because the resistor is connected in series with the LED, it will have a voltage drop of 10.3 volts. The power dissipation of the resistor can now be determined.

$$P = E^2/R$$

$$P = \frac{10.3^2}{620}$$

$$P = \frac{106.09}{620}$$

$$P = 0.171 \ watt$$

A ¼-watt resistor can be employed in this circuit. A light-emitting diode is shown in Figure 51–8.

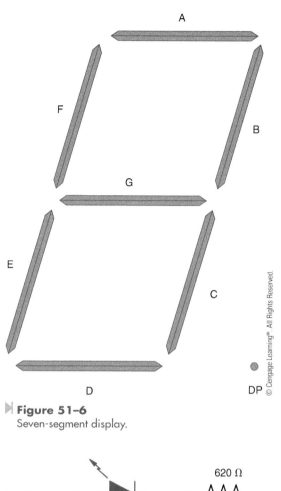

A

F

B

G

E

C

D   DP

▶ **Figure 51–6**
Seven-segment display.

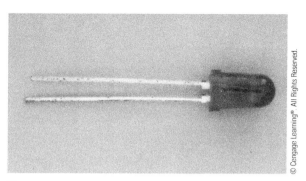

▶ **Figure 51–8**
Light-emitting diode.

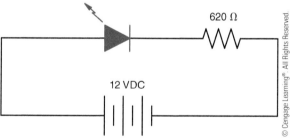

620 Ω

12 VDC

▶ **Figure 51–7**
Current is limited by a resistor connected in series with
the LED.

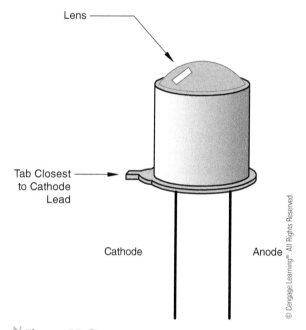

Lens

Tab Closest
to Cathode
Lead

Cathode   Anode

▶ **Figure 51–9**
Photodiode.

PHOTODIODES

The **photodiode** is so named because of its
response to a light source. Photodiodes are housed
in a case that has a window that permits light to
strike the semiconductor material, Figure 51–9.
Photodiodes can be used in two basic ways.

Photovoltaic

Photodiodes can be used as **photovoltaic** devices.
When in the presence of light, they produce a voltage in a manner similar to that of solar cells. The
output voltage is approximately 0.45 volt. The current capacity is small, and use is generally limited
to applications such as operating light-metering
devices. The basic schematic for a photodiode used

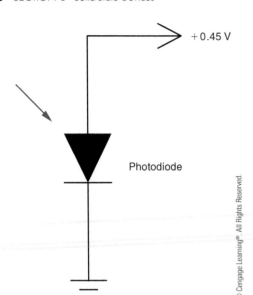

+0.45 V

Photodiode

© Cengage Learning®. All Rights Reserved.

▶ **Figure 51-10**
Photodiode used as a photovoltaic device.

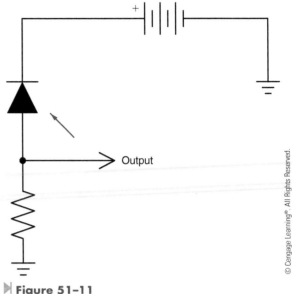

Battery

Output

© Cengage Learning®. All Rights Reserved.

▶ **Figure 51-11**
Photodiode used as a photoconductive device.

as a photovoltaic device is shown in Figure 51–10. Note the symbol used to represent a photodiode. The arrow pointing toward the diode indicates that it must receive light to operate.

Photoconductive

Photodiodes can also be used as **photoconductive** devices. When used in this manner they are connected reverse biased, Figure 51–11. In the presence of darkness, the amount of reverse current flow is extremely small, similar to that of a junction diode connected reverse biased. This current is referred to as the *dark current* (I_D) and is

generally in the range of a few nanoamperes. For most practical purposes, dark current is generally considered to be zero.

When exposed to light, photons enter the depletion region and create electron-hole pairs, increasing conductivity in the reverse direction. The increased conduction may permit several milliamperes of current to flow. This is known as *light current* (I_L). The great advantage of the photodiode over other photoconductive devices, such as the cad cell, is speed of operation. Photodiodes can operate at very high frequencies.

SUMMARY

⊙ The light-emitting diode produces a light when current flows through it.

⊙ The schematic symbol of an LED is a standard diode symbol with an arrow pointing away from the symbol. The arrow indicates that light is being emitted by the device.

⊙ The schematic symbol for a photodiode is a standard diode symbol with an arrow pointing toward the symbol. The arrow indicates that light is received by the device.

⊙ The forward voltage drop of an LED is approximately 1.7 volts.

⊙ Photodiodes are commonly used as photovoltaic devices and as photoconductive devices.

⊙ The greatest advantage of photodiodes over other photoconductive devices is speed of operation.

KEY TERMS

light-emitting diode (LED)
photoconductive
photodiode
photovoltaic

REVIEW QUESTIONS

1. Will the LED rectify an AC voltage into DC voltage?

2. What is the average voltage drop on an LED?

3. How can the anode and cathode leads of an LED be identified?

4. What is the average amount of current permitted to flow through an LED?

5. Can LEDs be tested with most ohmmeters?

6. When used as a photovoltaic device, how much voltage is generally produced by a photodiode?

7. Explain the difference between the schematic symbol used to indicate an LED and the symbol used to indicate a photodiode.

8. What is the greatest advantage of a photodiode used as a photoconductive device as compared with other photoconductive devices such as a cad cell?

UNIT 52 ▷

The Transistor

OBJECTIVES

After studying this unit, the student should be able to:

▷ Discuss the differences between PNP and NPN transistors

▷ Test transistors with an ohmmeter

▷ Identify the leads of standard case-style transistors

▷ Discuss the operation of a transistor

▷ Connect a transistor in a circuit

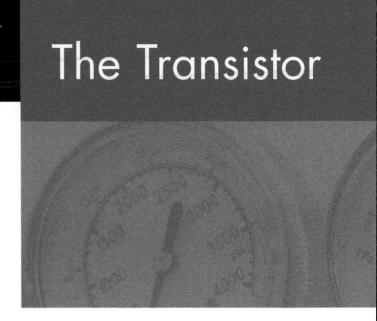

Transistors are made by joining three pieces of semiconductor material together. There are two basic types of transistors, the **NPN** and the **PNP**, Figure 52–1. The schematic symbols for these transistors are shown in Figure 52–2. The difference in these transistors is the manner in which they are connected in a circuit. The NPN transistor must have positive connected to the **collector** and negative connected to the **emitter**. The PNP must have positive connected to the emitter and negative connected to the collector. Notice that the **base** must be connected to the same polarity as the collector to forward bias the transistor. Also notice that the arrows on the emitters point in the direction of conventional current flow.

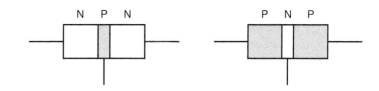

Figure 52–1
Two basic types of transistors.
© Cengage Learning®. All Rights
Reserved.

Figure 52–2
Schematic symbols of
transistors.
© Cengage Learning®. All Rights
Reserved.

TESTING THE TRANSISTOR

The transistor can be tested with an ohmmeter. If the polarity of the output of the ohmmeter leads is known, the transistor can be identified as either NPN or PNP. A transistor will appear to an ohmmeter to be two diodes joined together, Figure 52–3. An NPN transistor will appear to an ohmmeter to be two diodes with their anodes connected together. If the positive lead of the ohmmeter is connected to the base of the transistor, a diode junction should be seen between the base-collector and the base-emitter. If the negative lead of the ohmmeter is connected to the base of an NPN transistor, there should be no continuity between the base-collector and the base-emitter junction.

A PNP transistor appears to be two diodes with their cathodes connected together. If the negative lead of the ohmmeter is connected to the base of the transistor, a diode junction should be seen between the base-collector and the base-emitter. If the positive ohmmeter lead is connected to the base, there should be no continuity between the base-collector or the base-emitter.

The following step-by-step procedure can be used to test a transistor:

1. Using a diode, determine which ohmmeter lead is positive and which is negative. The ohmmeter will indicate continuity through the diode only when the positive lead is connected to the anode and the negative lead is connected to the cathode, Figure 52–4.

2. If the transistor is an NPN, connect the positive ohmmeter lead to the base and the negative lead to the collector. The ohmmeter should indicate continuity. The reading should be about the same as the reading obtained when the diode was tested, Figure 52–5.

3. With the positive ohmmeter lead still connected to the base of the transistor, connect the negative lead to the emitter. The ohmmeter should again indicate a forward diode junction, Figure 52–6.

NOTE: *If the ohmmeter does not indicate continuity between the base-collector or the base-emitter, the transistor is open.*

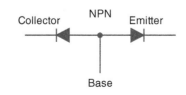
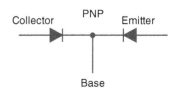

Figure 52–3
Ohmmeter test for transistors.
© Cengage Learning®. All Rights
Reserved.

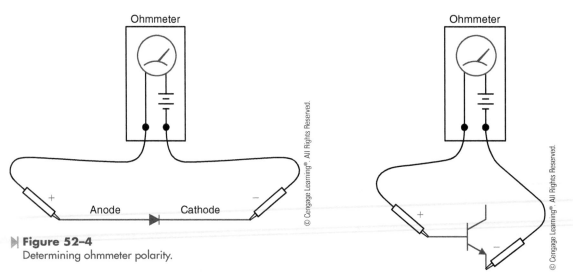

Figure 52–4
Determining ohmmeter polarity.

Figure 52–6
Testing the base-emitter junction.

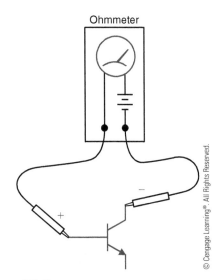

Figure 52–5
Testing the base-collector junction.

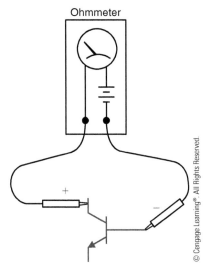

Figure 52–7
Reversing the connection of the base-collector junction.

4. Connect the negative ohmmeter lead to the base and the positive lead to the collector. The ohmmeter should indicate infinity or no continuity, Figure 52–7.

5. With the negative ohmmeter lead connected to the base, reconnect the positive lead to the emitter. There should, again, be no indication of continuity, Figure 52–8.

NOTE: *If a very high resistance is indicated by the ohmmeter, the transistor is "leaky," but it may still operate in the circuit. If a very low resistance is seen, the transistor is shorted.*

6. To test a PNP transistor, reverse the polarity of the ohmmeter leads and repeat the test. When the negative ohmmeter lead is

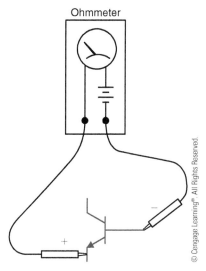

▌ Figure 52–8
Reversing the connection of the base-emitter junction.

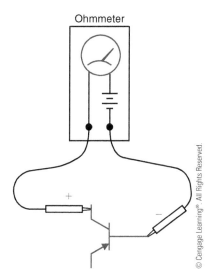

▌ Figure 52–9
Testing the base-collector junction of a PNP transistor.

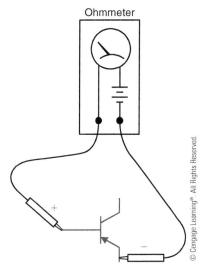

▌ Figure 52–10
Reversing the connection of a PNP transistor.

connected to the base, a forward diode junction should be indicated when the positive lead is connected to the collector or emitter, Figure 52–9.

7. If the positive ohmmeter lead is connected to the base of a PNP transistor, no continuity

should be indicated when the negative lead is connected to the collector or the emitter, Figure 52–10.

TRANSISTOR OPERATION

The simplest way to describe the operation of a transistor is to say it operates like an electric valve. Current will not flow through the collector-emitter until current flows through the base-emitter. The amount of base-emitter current, however, is small when compared to the collector-emitter current, Figure 52–11. For example, assume that when one milliamp (mA) of current flows through the base-emitter junction, 100 mA of current flows through the collector-emitter junction. If this transistor is a linear device, an increase or decrease of base current will cause a similar increase or decrease of collector current. For instance, if the base current is increased to 2 mA, the collector current would increase to 200 mA. If the base current were decreased to 0.5 mA, the collector current would decrease to 50 mA. Notice that a small change in the amount of base current can cause a large change in the amount of collector current. This permits a small amount of signal current to operate a larger device, such as the coil of a control relay.

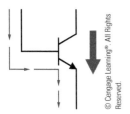

Figure 52-11
A small base current controls a large collector current.

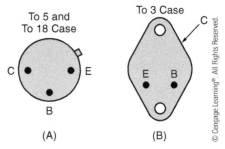

Figure 52-12
Lead identification of transistors.

One of the most common applications of the transistor is that of a switch. When used in this manner, the transistor operates like a digital device instead of an analog device. The term **digital** refers to a device that has only two states such as on or off. An **analog** device can be adjusted to different states. An example of this control can be seen in a simple switch connection. A common wall switch is a digital device. It can be used to either turn a light on or off. If the simple toggle switch is replaced with a dimmer control, the light can be turned on, off, or adjusted to any position between. The dimmer is an example of analog control.

If no current flows through the base of the transistor, the transistor acts like an open switch and no current will flow through the collector-emitter junction. If enough base current is applied to the transistor to turn it completely on, it acts like a closed switch and permits current to flow through the collector-emitter junction. This is the same action produced by the closing contacts of a relay or motor starter, but a relay or motor starter cannot turn on and off several thousand times a second and a transistor can.

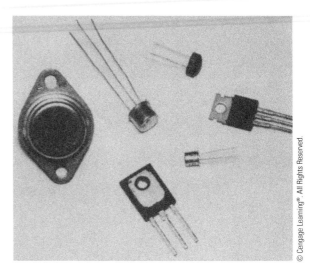

Figure 52-13
Transistors shown in different case styles.

IDENTIFYING TRANSISTOR LEADS

Some **case** styles of transistors permit the leads to be quickly identified. The TO 5 and TO 18 cases, and the TO 3 case are in this category. The leads of the TO 5 or TO 18 case transistors can be identified by holding the case of the transistor with the leads facing you, as shown in Figure 52–12A. The metal tab on the case of the transistor is closest to the emitter lead. The base and collector leads are positioned as shown in Figure 52–12A.

The leads of a TO 3 case transistor can be identified as shown in Figure 52–12B. With the transistor held with the leads facing you and down, the emitter is the left lead and the base is the right lead. The case of the transistor is the collector. Several case styles for the transistor are shown in Figure 52–13.

▷ SUMMARY

- ⊙ Transistors are made by joining three layers of semiconductor material together.
- ⊙ The two basic types of transistors are NPN and PNP.
- ⊙ The three terminal leads of a transistor are the collector, base, and emitter.
- ⊙ A PNP transistor must have a more positive voltage connected to the emitter than the collector.
- ⊙ An NPN transistor must have a more negative voltage connected to the emitter than the collector.
- ⊙ The current flow through the base-emitter of the transistor controls the amount of current flow through collector-emitter.

▷ KEY TERMS

analog	collector	NPN
base	digital	PNP
case	emitter	

▷ REVIEW QUESTIONS

1. What are the two basic types of transistors?

2. Explain how to test an NPN transistor with an ohmmeter.

3. Explain how to test a PNP transistor with an ohmmeter.

4. What polarity must be connected to the collector, base, and emitter of an NPN to make it forward biased?

5. What polarity must be connected to the collector, base, and emitter of a PNP transistor to make it forward biased?

6. Explain the difference between an analog device and a digital device.

The Unijunction Transistor

OBJECTIVES

After studying this unit, the student should be able to:

▷ Discuss the differences between junction transistors and unijunction transistors

▷ Identify the leads of a UJT

▷ Draw the schematic symbol for a UJT

▷ Test a UJT with an ohmmeter

▷ Connect a UJT in a circuit

The **unijunction transistor** is a special transistor that has two bases and one emitter. The unijunction transistor (UJT) is a digital device because it has only two states, on or off, and is generally classified with a group of devices known as **thyristors**. Thyristors are turned completely on or completely off. Thyristors include devices such as the silicon-controlled rectifier (SCR), triac, diac, and the unijunction transistor.

The unijunction transistor is made by combining three layers of semiconductor material as shown in Figure 53–1. The schematic symbol with polarity connections and the base diagram is shown in Figure 53–2.

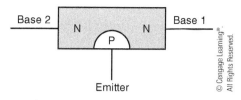

Figure 53–1
Unijunction transistor.

UJT CHARACTERISTICS

The UJT has two paths for current flow. One path is from B2 to B1. The other path is through the emitter and base 1. In its normal state, there is no current flow through either path until the voltage applied to the emitter reaches about 10 volts higher than the voltage applied to base 1. When the voltage applied to the emitter reaches about 10 volts more positive than the voltage applied to base 1, the UJT turns on and current flows through the B1–B2 path and from the emitter through base 1. Current continues to flow through the UJT until the voltage applied to the emitter drops to a point that it is only about 3 volts higher than the voltage applied to B1. When the emitter voltage drops to this point, the UJT turns off and remain turned off until the voltage applied to the emitter again becomes about 10 volts higher than the voltage applied to B1.

CIRCUIT OPERATION

The unijunction transistor is generally connected into a circuit similar to the circuit shown in Figure 53–3. The variable resistor controls the rate of charge time of the capacitor. When the capacitor has been charged to about 10 volts, the UJT turns on and discharges the capacitor through the emitter

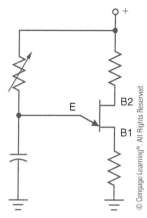

Figure 53–3
Basic UJT connection.

and base 1. When the capacitor has been discharged to about 3 volts, the UJT turns off and permits the capacitor to begin charging again. By varying the resistance connected in series with the capacitor, the amount of time needed for charging the capacitor can be changed, thereby controlling the pulse rate of the UJT ($T = RC$).

The UJT can furnish a large output pulse, because the output pulse is produced by the discharging capacitor, Figure 53–4. This large output pulse is generally used for triggering the gate of an SCR.

The pulse rate is determined by the amount of resistance and capacitance connected to the emitter of the UJT. However, the amount of capacitance that can be connected to the UJT is limited. For instance, most UJTs should not have a capacitor larger than 10 µf connected to them. If the capacitor is too large, the UJT will not be able to handle the current spike produced by the capacitor, and the UJT could be damaged.

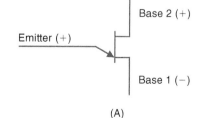

Figure 53–2
The schematic symbol for the unijunction transistor with polarity connections and base diagram.

Emitter (+)

Base 2 (+)

Base 1 (−)

(A)

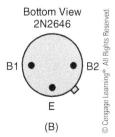

Bottom View
2N2646

B1   B2

E

(B)

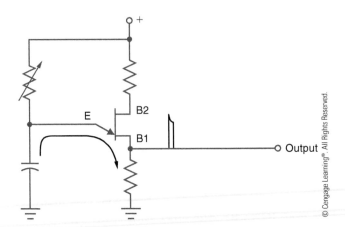

▶ **Figure 53–4**
A large pulse is produced by the capacitor discharge.

TESTING THE UJT

The unijunction transistor can be tested with an ohmmeter in a manner very similar to testing a common junction transistor. The UJT will appear to the ohmmeter to be a connection of two resistors connected to a common junction diode. The common junction point of the two resistors will appear to be at the emitter of the UJT as shown in Figure 53–5. When the positive lead of the ohmmeter is connected to the emitter, a diode junction should be seen from the emitter to base 2 and another diode connection from the emitter to base 1. If the negative lead of the ohmmeter is connected to the emitter of the UJT, no connection should be seen between the emitter and either base.

The following step-by-step procedure can be used to test a unijunction transistor.

1. Using a junction diode, determine which ohmmeter lead is positive and which is negative. The ohmmeter indicates continuity when the positive lead is connected to the anode, and the negative lead is connected to the cathode, Figure 53–6.

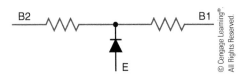

▶ **Figure 53–5**
The UJT appears as two resistors connected to a diode when tested with an ohmmeter.

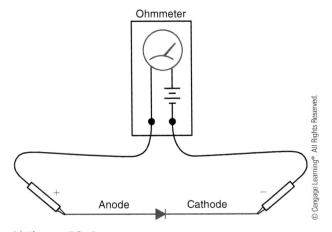

▶ **Figure 53–6**
Determining ohmmeter polarity.

2. Connect the positive ohmmeter lead to the emitter lead and the negative lead to base 1. The ohmmeter should indicate a forward diode junction, Figure 53–7.

3. With the positive ohmmeter lead connected to the emitter, reconnect the negative lead to base 2. The ohmmeter should again indicate a forward diode junction, Figure 53–8.

4. If the negative ohmmeter lead is connected to the emitter, no continuity should be indicated when the positive lead is connected to base 1 or base 2, Figure 53–9.

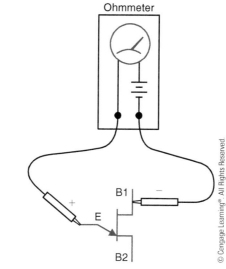

▶ **Figure 53–7**
Testing a UJT.

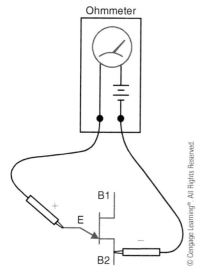

▶ **Figure 53–8**
Testing the emitter-base 2 junction.

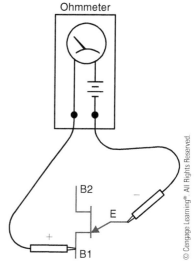

▶ **Figure 53–9**
Reversing the polarity.

SUMMARY

▶ The unijunction transistor has two bases and one emitter.

▶ The unijunction transistor is a member of the thyristor family of components.

▶ Injunction transistors have two states of operation: on or off.

▶ The unijunction transistor operates like a snap action, voltage sensitive switch.

KEY TERMS

thyristors
unijunction transistor

REVIEW QUESTIONS

1. What do the letters UJT stand for?

2. How many layers of semiconductor material are used to construct a UJT?

3. Briefly explain the operation of the UJT.

4. Draw the schematic symbol for the UJT.

5. Briefly explain how to test a UJT with an ohmmeter.

The Silicon-Controlled Rectifier and GTO

OBJECTIVES

After studying this unit, the student should be able to:

▷ Discuss the operation of a silicon-controlled rectifier (SCR) in a DC circuit and an AC circuit

▷ Draw the schematic symbol for an SCR

▷ Discuss phase shifting

▷ Test an SCR with an ohmmeter

▷ Connect an SCR in a circuit

▷ Draw the schematic symbol for a GTO

▷ Discuss the operating differences between an SCR and a GTO

The **silicon-controlled rectifier (SCR)** is often referred to as the **PNPN** junction because it is made by joining four layers of semiconductor material together, Figure 54–1. The schematic symbol for the SCR is shown in Figure 54–2. Notice that the symbol for the SCR is the same as the diode except that a gate lead has been added.

SCR CHARACTERISTICS

The SCR is a member of a family of devices known as thyristors. Thyristors are digital devices in that they have only two states, on or off. The SCR is used when it is necessary for an electronic device to control a large amount of power. Assume an SCR has been connected in a circuit as shown in Figure 54–3. When the SCR is turned off, it will drop the full voltage of the circuit and 200 volts will

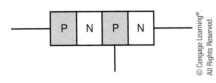

▶ **Figure 54–1**
PNPN junction.

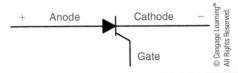

▶ **Figure 54–2**
Schematic symbol of an SCR.

appear across the anode and cathode. Although the SCR has a voltage drop of 200 volts, there is no current flow in the circuit. The SCR does not have to dissipate any power in this condition (200 volts $\times$ 0 amps = 0 watts). When the pushbutton is pressed, the SCR will turn on. When the SCR turns on, it will have a voltage drop across its anode and cathode of about 1 volt. The load resistor limits the circuit current to 2 amps (200 volts/100 ohms = 2 amps). Because the SCR now has a voltage drop of 1 volt and 2 amps of current is flowing through it, it must now dissipate 2 watts of heat (1 volt $\times$ 2 amps = 2 watts). Notice that the SCR is dissipating only 2 watts of power, but is controlling 200 watts.

THE SCR IN A DC CIRCUIT

When an SCR is connected in a DC circuit, as shown in Figure 54–3, the **gate** turns the SCR on but does not turn the SCR off. The gate must be

connected to the same polarity as the anode if it is to turn the anode-cathode section of the SCR on. Once the gate has turned the SCR on, it remains turned on until the current flowing through the anode-cathode drops to a low enough level to permit the device to turn off. The amount of current required to keep the SCR turned on is called the **holding current**, Figure 54–4. Assume resistor R1 has been adjusted for its highest value and resistor R2 has been adjusted to its lowest or 0 value. When switch S1 is closed, no current flows through the anode-cathode section of the SCR because resistor R1 prevents enough current flowing through the gate-cathode section of the SCR to trigger the device. If resistor R1 is slowly decreased in value, current flow through the gate-cathode will slowly increase. When the gate current reaches a certain level, assume 5 mA for this SCR, the SCR fires or turns on. When the SCR fires, current flows through the anode-cathode section and the voltage drops across the device becomes about 1 volt. Once the SCR has turned on, the gate has no more control over the device and could be disconnected from the anode without having any effect on the circuit. When the SCR fires, the anode-cathode becomes a short circuit for all practical purposes and current flow is limited by resistor R3. Now assume that resistor R2 is slowly increased in value. When the resistance of R2 is slowly increased, the current flow through the anode-cathode will slowly decrease. Assume that when the current flow through the anode-cathode drops to 100 mA, the device suddenly turns off and the current flow drops to 0. This SCR requires 5 mA of gate current to turn it on, and has a holding current value of 100 mA.

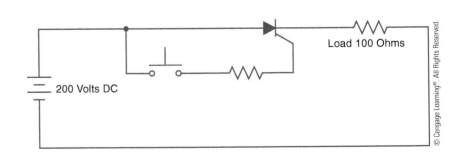

Load 100 Ohms

200 Volts DC

▶ **Figure 54–3**
Gate turns SCR on.

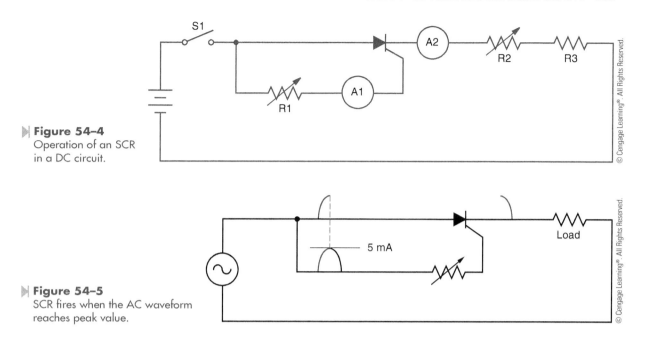

▶ **Figure 54–4**
Operation of an SCR
in a DC circuit.

▶ **Figure 54–5**
SCR fires when the AC waveform
reaches peak value.

THE SCR IN AN AC CIRCUIT

The SCR is a rectifier. When it is connected in an AC circuit, the output is DC. The SCR operates in the same manner in an AC circuit as it does in a DC circuit. The difference in operation is caused by the AC waveform falling back to 0 at the end of each half cycle. When the AC waveform drops to 0 at the end of each half cycle, it permits the SCR to turn off. This means the gate must re-trigger the SCR for each cycle it is to conduct. Refer to the circuit shown in Figure 54–5.

Assume that the variable resistor connected to the gate has been adjusted to permit 5 mA of current to flow when the voltage applied to the anode reaches its peak value. When the SCR turns on, current will begin flowing through the load resistor when the AC waveform is at its positive peak. Current continues to flow through the load until the decreasing voltage of the sine wave causes the current to drop below the holding current level of 100 mA. When the current through the anode-cathode drops below 100 mA, the SCR turns off and all current flow stops. The SCR will remain turned off when the AC waveform goes into the

negative half cycle because it is reverse biased and cannot be fired.

If the resistance connected in series with the gate is reduced, a current of 5 mA will be reached before the AC waveform reaches it peak value, Figure 54–6. This causes the SCR to fire sooner in the cycle. Because the SCR fires sooner, current is permitted to flow through the load resistor for a longer period of time, which causes a higher average voltage drop across the load. If the resistance of the gate circuit is reduced again, as shown in Figure 54–7, the 5 mA of gate current needed to fire the SCR will be reached sooner than before. This permits current to begin flowing through the load sooner than before, which permits a higher average voltage to be dropped across the load.

Notice that this circuit permits the SCR to control only half of the positive waveform. The latest the SCR can be fired in the cycle is when the AC waveform is at 90°, or peak. If a lamp were used as the load for this circuit, it would burn at half brightness when the SCR first turned on. This control would permit the lamp to be operated from half brightness to full brightness, but it could not be operated at a level less than half brightness.

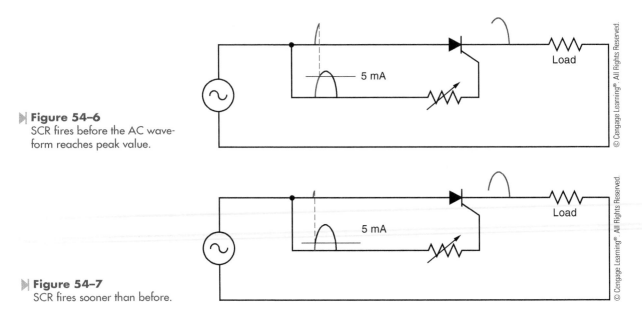

▶| **Figure 54–6**
SCR fires before the AC waveform reaches peak value.

▶| **Figure 54–7**
SCR fires sooner than before.

PHASE SHIFTING THE SCR

If the SCR is to control all of the positive waveform, it must be **phase shifted**. As the term implies, phase shifting means to shift the phase of one thing in reference to another. In this instance, the voltage applied to the gate must be shifted out of phase with the voltage applied to the anode. There are several methods that can be used for phase shifting an SCR, but it is beyond the scope of this text to cover all of them. The basic principles are the same for all of the methods, however, so only one method will be covered.

If the SCR is to be phase shifted, the gate circuit must be unlocked or separated from the anode circuit. The circuit shown in Figure 54–8 will accomplish this. A 24-volt center-tapped transformer has been used to isolate the gate circuit from the anode circuit. Diodes D1 and D2 are used to form a two-diode type of full-wave rectifier to operate the unijunction transistor (UJT) circuit. Resistor R1 is used to determine the pulse rate of the UJT by controlling the charge time of capacitor C1. Resistor R2 is used to limit the current through the emitter of the UJT if resistor R1 is adjusted to 0 ohms. Resistor R3 limits current through the base 1–base 2 section when the UJT turns on. Resistor R4 permits a voltage spike or pulse to be produced across it when the UJT turns on

and discharges capacitor C1. The pulse produced by the discharge of capacitor C1 is used to trigger the gate of the SCR.

Because the pulse of the UJT is used to provide a trigger for the gate of the SCR, the SCR can now be fired at any time, regardless of the voltage applied to the anode. This means the SCR can now be fired as early or late during the positive half cycle as desired, because the gate pulse is now determined by the charge rate of capacitor C1. The voltage across the load can now be adjusted from 0 to the full applied voltage.

TESTING THE SCR

The SCR can be tested with an ohmmeter. To test the SCR, connect the positive output lead of the ohmmeter to the anode and the negative lead to the cathode. The ohmmeter should indicate no continuity. Touch the gate of the SCR to the anode. The ohmmeter should indicate continuity through the SCR. When the gate lead is removed from the anode, conduction may stop or continue, depending on whether the ohmmeter is supplying enough current to keep the device above its holding current level. If the ohmmeter indicates continuity through the SCR before the gate is touched to the anode, the SCR is shorted. If the ohmmeter does not

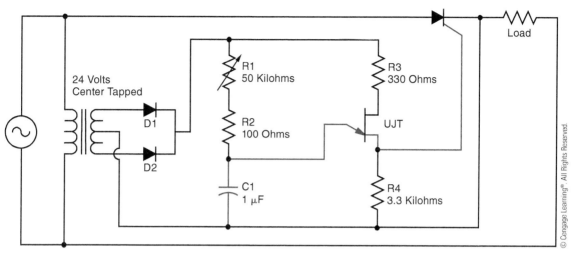

Figure 54–8
UJT phase shift for an SCR.

indicate continuity through the SCR after the gate has been touched to the anode, the SCR is open. The following step-by-step procedure can be used for testing an SCR.

1. Using a junction diode, determine which ohmmeter lead is positive and which is negative. The ohmmeter will indicate continuity only when the positive lead is connected to the anode of the diode and the negative lead is connected to the cathode, Figure 54–9.

2. Connect the positive ohmmeter lead to the anode of the SCR and the negative lead to the cathode. The ohmmeter should indicate no continuity, Figure 54–10.

3. Using a jumper lead, connect the gate of the SCR to the anode. The ohmmeter should indicate a forward diode junction when the connection is made, Figure 54–11.

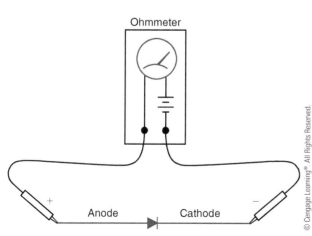

Figure 54–9
Determining ohmmeter polarity.

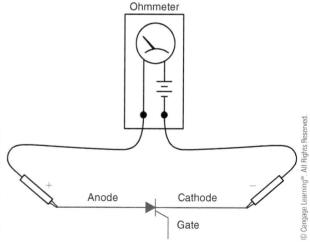

Figure 54–10
There should be no continuity between anode and cathode.

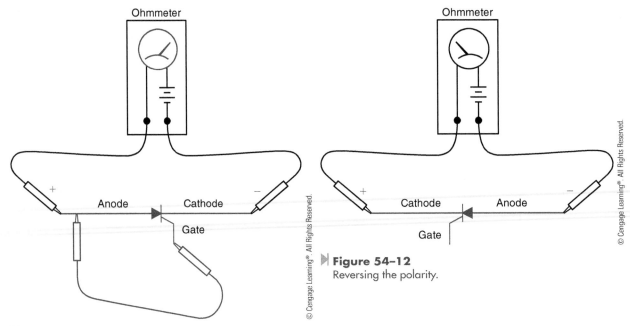

▶ **Figure 54–11**
Shorting the gate and anode causes the SCR to conduct.

▶ **Figure 54–12**
Reversing the polarity.

NOTE: *If the jumper is removed, the SCR may continue to conduct or it may turn off. This will be determined by whether the ohmmeter can supply enough current to keep the SCR above its holding current level.*

4. Reconnect the SCR so that the cathode is connected to the positive ohmmeter lead and the anode is connected to the negative lead. The ohmmeter should indicate no continuity, Figure 54–12.

5. If a jumper lead is used to connect the gate to the anode, the ohmmeter should indicate no continuity, Figure 54–13. SCRs in different case styles are shown in Figure 54–14.

NOTE: *SCRs designed to switch large current (50 amperes or more) may indicate some leakage current with this test. This is normal for some devices.*

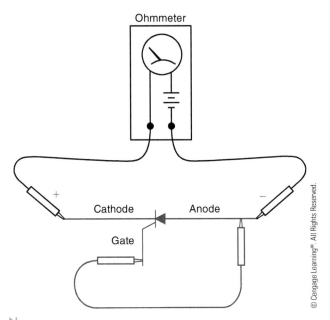

▶ **Figure 54–13**
The SCR will not conduct when the polarity is reversed.

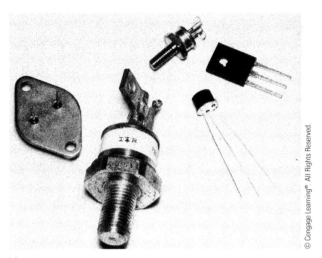

Figure 54–14
SCRs shown in different case styles.

GTO (GATE-TURNOFF) SCR

The **GTO (gate-turnoff) SCR** is also known by the name gate-controlled switch (GCS). The gate-turnoff SCR operates in a similar fashion to a common SCR in that it is turned on by a gate pulse that is positive with respect to the cathode. Unlike the SCR, however, it can be turned off with a gate pulse that is negative with respect to the cathode. A commonly used schematic symbol for the GTO is shown in Figure 54–15. Although the GTO can be turned off with a negative current pulse on the gate it requires from 10 to 20 times more negative gate current than it does to turn it on.

The greatest advantage of the GTO over a common SCR is its ability to control power in a direct current circuit. Because the anode-cathode current does not have to drop below a holding current level to turn it off, it is easier to control than an SCR.

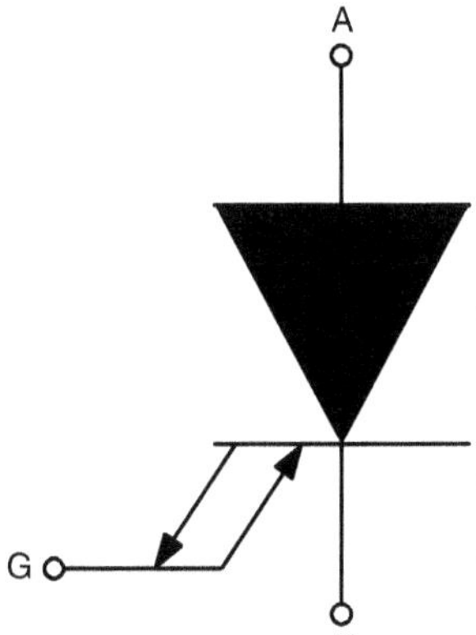

Figure 54–15
Schematic symbol for a GTO.

SUMMARY

▶ The silicon-controlled rectifier (SCR) is often referred to as a PNPN junction.

▶ The SCR is a member of the thyristor family of electronic devices.

▶ The SCR has two states of operation: on or off.

▶ When the SCR is connected in a DC circuit, the gate current controls the turn on, but once the SCR is turned on, the gate cannot turn it off.

▶ Before an SCR can be turned off, the current flow through the anode-cathode section must drop below the holding current level.

▶ When an SCR is connected in an AC circuit, the voltage returning to zero each half cycle turns the SCR off.

▶ The SCR is a rectifier; it will change alternating current into direct current.

▶ In order to gain complete control of the output waveform, the SCR must be phase shifted.

▶ The unijunction transistor is often used to phase shift an SCR.

▶ The GTO can be turned off by applying a current pulse to the gate that is negative with respect to the cathode.

▶ A GTO requires 10 to 20 times more negative gate current to turn it off than it does to turn it on.

KEY TERMS

gate	holding current	silicon-controlled
(GTO) Gate-turnoff	phase shifted	rectifier (SCR)
SCR	PNPN	

REVIEW QUESTIONS

1. What do the letters SCR stand for?

2. How many layers of semiconductor material are joined to form an SCR?

3. SCRs are a member of what family of devices?

4. If an SCR is connected in an AC circuit, is the output AC or DC?

5. Is gate current used to turn the SCR on or off?

6. The amount of current flow through the anode-cathode section needed to keep an SCR turned on is called what?

7. When an SCR is connected in an AC circuit, what must be done to gain complete control of the output waveform?

8. What electronic component is generally used to phase shift an SCR?

9. Explain how a gate-turnoff SCR can be turned off after it has been turned on in a DC circuit.

10. What is the advantage of a GTO over an SCR when used in a direct current circuit?

The Diac and Silicon Bilateral Switch

OBJECTIVES

After studying this unit, the student should be able to:

▷ Draw the schematic symbol for a diac

▷ Discuss the operation of a diac

▷ Connect a diac in a circuit

▷ Draw the schematic symbol for a silicon bilateral switch

▷ Discuss the operation of the silicon bilateral switch

▷ Discuss the differences between a diac switch and a silicon bilateral switch

The **diac** is a special-purpose **bidirectional** diode. The primary function of the diac is to phase shift a triac. The operation of the diac is very similar to that of a unijunction transistor, except the diac is a "bi," or two-directional, device. The diac has the ability to operate in an AC circuit, whereas the UJT is a DC device only.

There are two schematic symbols for the diac, Figure 55–1. Either of these symbols is used in an electronic schematic to illustrate the use of a diac; therefore, you should become familiar with both.

DIAC CHARACTERISTICS

The diac is a voltage-sensitive switch that can operate on either polarity, Figure 55–2. When voltage is applied to the diac, it remains in the turned-off state until the applied voltage reaches

Figure 55-1
Schematic symbols for a diac.

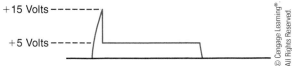

+15 Volts
+5 Volts

Figure 55-3
The diac operates until the applied voltage falls below its conduction level.

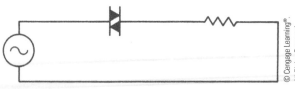

Figure 55-2
The diac can operate on either polarity.

THE SILICON BILATERAL SWITCH (SBS)

The silicon bilateral switch or SBS is another bidirectional device often used to trigger the gate of a triac. The symbol for an SBS is shown in Figure 55–5. The SBS is very similar to the diac in several ways. Both will conduct current in both directions and both exhibit negative resistance. The SBS, however, has a more pronounced negative resistance region than the does the diac. A characteristic voltage curve for both the diac and SBS are shown in Figure 55–6. Notice that the break-back voltage of the SBS is more pronounced than that of the diac. Also, the breakover voltage of an SBS is generally lower than that of a diac. The breakover voltage of a diac will most often range between ± 16 volts and ± 32 volts. The most common breakover voltage for a silicon bilateral switch is ± 8 volts.

The silicon bilateral switch has several other advantages over the diac. SBSs are generally symmetrical to within about 0.3 volt. This means that the positive breakover voltage and the negative breakover will be within about 0.3 volt of each other. The diac is generally symmetrical to within about 1 volt.

a predetermined level. For this example, assume this to be 15 volts. When the voltage reaches 15 volts, the diac turns on, or fires. When the diac fires, it displays a negative resistance, which means it will conduct at a lower voltage than the voltage that was applied to it; assume 5 volts. The diac remains turned on until the applied voltage drops below its conduction level, which in this example is 5 volts. Refer to the waveform shown in Figure 55–3. Because the diac is a bidirectional device, it conducts on either half cycle of the AC applied to it. Refer to the waveform shown in Figure 55–4. Notice that the diac has the same operating characteristic with either half cycle of AC. The simplest way to sum up the operation of the diac is to say it is a voltage-sensitive AC switch.

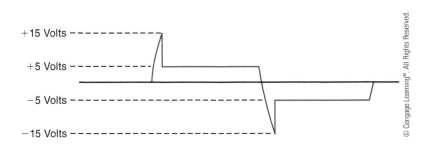

+15 Volts
+5 Volts
-5 Volts
-15 Volts

Figure 55-4
The diac will conduct on either half of the alternating current.

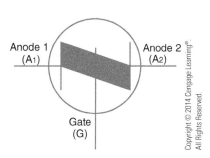

Figure 55–5
Schematic symbol for a silicon bilateral switch.

The SBS also has a gate lead that permits some control over the breakover voltage. If a 3 volt zener diode were to be connected between the gate and anode 2, as shown in Figure 55–7, the positive breakover voltage would be reduced to about 3.6 volts (3 volts for the zener diode plus 0.6 volt needed to turn on a silicon device), but the reverse breakover voltage would be unaffected, as shown in Figure 55–8. This connection could be used if it were desirable to have different forward and reverse breakover values, which is not usually the case. The SBS is most often used with the gate lead not connected.

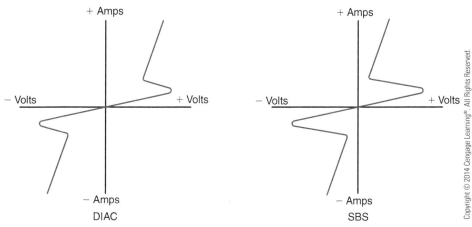

Figure 55–6
Characteristic voltage curves of a diac and silicon bilateral switch.

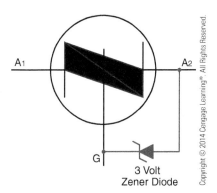

Figure 55–7
A zener diode is used to control the forward breakover voltage.

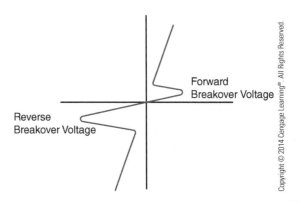

▶ **Figure 55–8**
The forward and reverse breakover
voltages are different.

> **SUMMARY**

- ⊙ The diac is a bidirectional diode.
- ⊙ The primary function of a diac is to phase shift a triac.
- ⊙ The diac operates on AC.
- ⊙ The diac operates like a voltage-sensitive AC switch.
- ⊙ The diac displays a negative resistance characteristic.

> **KEY TERMS**

bidirectional
diac

> **REVIEW QUESTIONS**

1. Briefly explain how a diac operates.
2. Draw the two schematic symbols for the diac.
3. What is the major use of the diac in industry?
4. When a diac first turns on, does the voltage drop, remain at the same level, or increase to a higher level?

The Triac

OBJECTIVES

After studying this unit, the student should be able to:

▷ Draw the schematic symbol for a triac

▷ Discuss the similarities and differences between SCRs and triacs

▷ Discuss the operation of a triac in an AC circuit

▷ Discuss phase shifting of a triac

▷ Connect a triac in a circuit

▷ Test a triac with an ohmmeter

The triac is a PNPN junction connected in parallel with an **NPNP** junction. Figure 56–1 illustrates the semiconductor arrangement of a triac. The triac operates similarly to two SCRs in parallel, facing in opposite directions, with their gate leads connected together, Figure 56–2. The schematic symbol for the triac is shown in Figure 56–3.

When an SCR is connected in an AC circuit, the output voltage is DC. When a triac is connected in an AC circuit, the output voltage is AC. Because the triac operates like two SCRs connected together and facing in opposite directions, it conducts both the positive and negative half cycles of AC current.

When a triac is connected in an AC circuit, as shown in Figure 56–4, the gate must be connected to the same polarity as MT2. When the AC voltage applied to MT2 become positive, the SCR, which

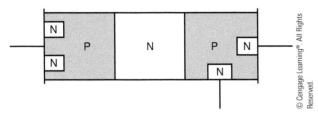

▶ **Figure 56–1**
Semiconductor arrangement of a triac.

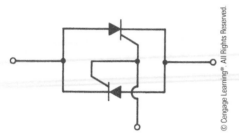

▶ **Figure 56–2**
The triac operates similarly to two SCRs with a common gate.

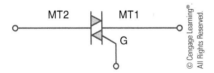

▶ **Figure 56–3**
Schematic symbol of a triac.

is forward biased, will conduct. When the voltage applied to MT2 becomes negative, the other SCR is forward biased and conducts that half of the waveform. Because one of the SCRs is forward biased for each half cycle, the triac conducts AC current as long as the gate lead is connected to MT2.

The triac, like the SCR, requires a certain amount of gate current to turn it on. Once the triac has been triggered by the gate, it will continue to conduct until the current flowing through MT2–MT1 drops below the holding current level.

THE TRIAC USED AS AN AC SWITCH

The triac is a member of the thyristor family and has only two states of operation: on or off. When the triac is turned off, it drops the full applied voltage of the circuit at 0 amps of current flow. When the triac is turned on, it has a voltage drop of about 1 volt, and circuit current must be limited by the load connected to the circuit. The triac has become very popular in industrial circuits as an AC switch. Because it is a thyristor, it has the ability to control a large amount of voltage and current. There are no contacts to wear out; it is sealed against dirt and moisture; and it can operate thousands of times per second. The triac is used as the output of many solid-state relays that are covered later.

THE TRIAC USED FOR AC VOLTAGE CONTROL

The triac can be used to control an AC voltage, Figure 56–5. If a variable resistor is connected in series with the gate, the point at which the gate current reaches a high enough level to fire the triac can be adjusted. The resistance can be adjusted to permit the triac to fire when the AC waveform reaches its peak value. This causes half of the AC voltage to be dropped across the triac and half to be dropped across the load.

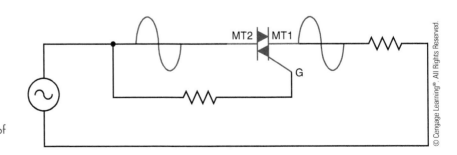

▶ **Figure 56–4**
The triac conducts both halves of the AC waveform.

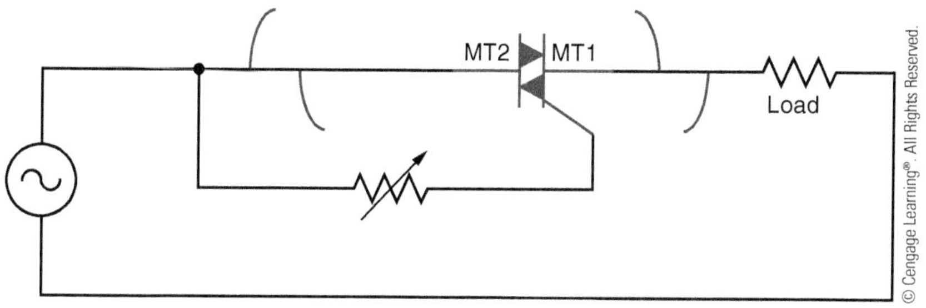

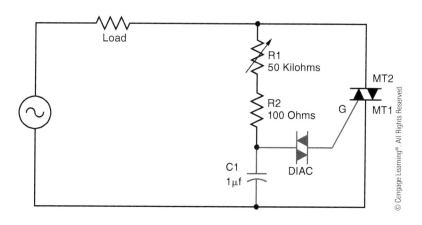

▶ **Figure 56–5**
The triac controls half of the AC
applied voltage.

If the gate resistance is reduced, the amount of gate current needed to fire the triac will be obtained before the AC waveform reaches its peak value. This means that less voltage will be dropped across the triac, and more voltage will be dropped across the load. This circuit permits the triac to control only one half of the AC waveform applied to it. If a lamp is used as the load, it can be controlled from half brightness to full brightness. If an attempt is made to adjust the lamp to operate at less than half brightness, it will turn off.

PHASE SHIFTING THE TRIAC

The triac, like the SCR, must be phase shifted if complete voltage control is to be obtained. There are several methods that can be used to phase shift a triac, but only one is covered in this unit. In this example, a diac is used to phase shift the triac, Figure 56–6. In this circuit, resistors R1 and R2 are connected in series with capacitor C1. Resistor R1 is a variable resistor and is used to control the charge time of capacitor C1. Resistor

R2 is used to limit current if resistor R1 should be adjusted to 0 ohms. Assume the diac connected in series with the gate of the triac will turn on when capacitor C1 has been charged to 15 volts. When the diac turns on, capacitor C1 discharges through the gate of the triac. This permits the triac to fire or turn on.

Once the triac has fired, there is a voltage drop of about 1 volt across MT2 and MT1. The triac remains turned on until the AC voltage drops to a low enough value to permit the triac to turn off. Because the phase shift circuit is connected in parallel with the triac, once the triac turns on, capacitor C1 cannot begin charging again until the triac turns off at the end of the AC cycle. The diac, being a bidirectional device, permits a positive or negative pulse to trigger the gate of the triac.

Notice that the pulse applied to the gate is controlled by the charging of capacitor C1 and not the amplitude of voltage. If the correct values are chosen, the triac can be fired at any point in the AC cycle applied to it. The triac can now control the AC voltage from 0 to the full voltage of the circuit.

▶ **Figure 56–6**
Phase-shift circuit for a triac.

A common example of this type of triac circuit is the light dimmer control used in many homes.

TESTING THE TRIAC

The triac can be tested with an ohmmeter. To test the triac, connect the ohmmeter leads to MT2 and MT1. The ohmmeter should indicate no continuity. If the gate lead is touched to MT2, the triac should turn on and the ohmmeter will indicate continuity through the triac. When the gate lead is released from MT2, the triac may continue to conduct or turn off, depending on whether the ohmmeter supplies enough current to keep the device above its holding current level. This tests one half of the triac. To test the other half of the triac, reverse the connection of the ohmmeter leads. The ohmmeter should again indicate no continuity. If the gate is touched again to MT2, the ohmmeter should indicate continuity through the device. The other half of the triac has been tested. The following step-by-step procedure can be used to test a triac.

1. Using a junction diode, determine which ohmmeter lead is positive and which is negative. The ohmmeter indicates continuity only when the positive lead is connected to the anode and the negative lead is connected to the cathode, Figure 56–7.

2. Connect the positive ohmmeter lead to MT2 and the negative lead to MT1. The ohmmeter

should indicate no continuity through the triac, Figure 56–8.

3. Using a jumper lead, connect the gate of the triac to MT2. The ohmmeter should indicate a forward diode junction, Figure 56–9.

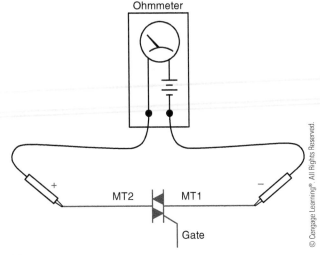

Figure 56–8
No continuity.

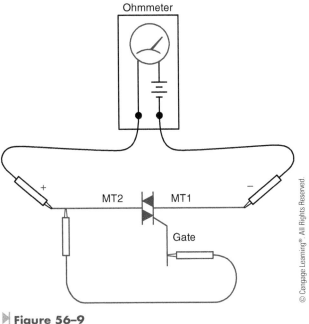

Figure 56–9
The triac conducts when the gate is connected to MT2.

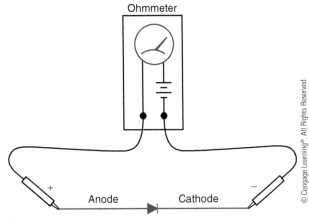

Figure 56–7
Determining ohmmeter polarity.

4. Reconnect the triac so that MT1 is connected to the positive ohmmeter lead and MT2 is connected to the negative lead. The ohmmeter should indicate no continuity through the triac, Figure 56–10.

5. Using a jumper lead, again connect the gate to MT2. The ohmmeter should indicate a forward diode junction, Figure 56–11.

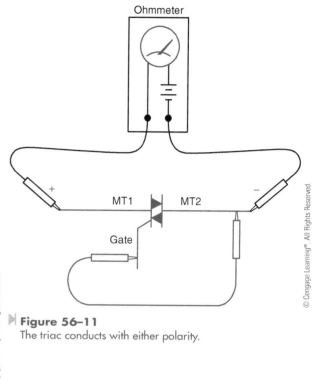

Figure 56–11
The triac conducts with either polarity.

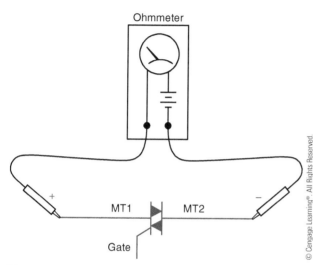

Figure 56–10
Reversing the polarity.

SUMMARY

▶ The triac operates in a manner similar to two SCRs connected in opposite directions.

▶ The triac is a bidirectional device, which means that it operates when connected to AC current.

▶ Triacs are often used as AC switches.

▶ A triac must be phase shifted to gain complete control of the AC waveform.

▶ A diac is often used to phase shift a triac.

KEY TERM

NPNP

REVIEW QUESTIONS

1. Draw the schematic symbol of a triac.

2. When a triac is connected in an AC circuit, is the output AC or DC?

3. The triac is a member of what family of devices?

4. Briefly explain why a triac must be phase shifted.

5. What electronic component is frequently used to phase shift the triac?

6. When the triac is being tested with an ohmmeter, which other terminal should the gate be connected to if the ohmmeter is to indicate continuity?

UNIT 57 ▷

The Operational Amplifier

OBJECTIVES

After studying this unit, the student should be able to:

▷ Discuss the operation of the operational amplifier (op amp)

▷ List the major types of connection for the op amp

▷ Connect a level detector circuit for an op amp

▷ Connect an oscillator, using an op amp

The **operational amplifier** has become another very common component found in industrial electronic circuits. The operational amplifier, or **op amp** as it is generally referred to, is used in hundreds of different applications. There are different types of op amps used, depending on the type of circuit it is intended to operate in. Some op amps use **bipolar transistors** for the input, and others use **field effect transistors**. The advantage of using field effect transistors is their extremely high input impedance, which can be several thousand megohms. The advantage of this extremely high input impedance is that it does not require a large amount of current to operate the amplifier. In fact, op amps, which use FET inputs, are generally considered as requiring no input current.

THE IDEAL AMPLIFIER

Before continuing the discussion of op amps, it should first be decided what an ideal amplifier is. First, the ideal amplifier should have an input impedance of infinity. If the amplifier had an input impedance of infinity, it would require no power drain on the signal source being amplified. Therefore, regardless of how weak the input signal source were, it would not be affected when connected to the amplifier. The ideal amplifier would have 0 output impedance. If the amplifier had 0 output impedance, it could be connected to any load resistance desired and not drop any voltage inside the amplifier. If it had no internal voltage drop, the amplifier would utilize 100% of its gain. Third, the amplifier would have unlimited **gain**. This would permit it to amplify any input signal as much as desired.

741 PARAMETERS

There is no such thing as the ideal or perfect amplifier of course, but the op amp can come close. One of the old reliable op amps, which is still used to a large extent, is the 741. The 741 is used in this description as a typical operational amplifier. Please keep in mind that there are other op amps that have different characteristics of input and output impedance, but the basic theory of operation is the same for all of them.

The 741 op amp uses bipolar transistors for the input. The input impedance is about 2 megohms, and the output impedance is about 75 ohms. Its open loop, or maximum gain, is about 200,000. Actually, the 741 op amp has such a high gain that

it is generally impractical to use, and negative feedback, which will be discussed later, is used to reduce the gain. For instance, assume the amplifier has an output voltage of 15 volts. If the input signal voltage is greater than 1/200,000 of the output voltage or 75 microvolts (15/200,00 = .000075), the amplifier would be driven into saturation, at which point it would not operate.

741 PIN CONNECTION

The 741 operational amplifier is generally housed in an 8-pin in-line IC package, Figure 57–1. Pins 1 and 5 are connected to the **offset null**. The offset null is used to produce 0 volts at the output. What happens is this: The op amp has two inputs called the **inverting input** and the **noninverting input**. These inputs are connected to a differential amplifier that amplifies the difference between the two voltages. If both of these inputs are connected to the same voltage, say by grounding both inputs, the output should be 0 volts. In actual practice, however, there are generally unbalanced conditions in the op amp that cause a voltage to be produced at the output. Because the op amp has a very high gain, a very slight imbalance of a few microvolts at the input can cause several millivolts at the output. The offset nulls are adjusted after the 741 is connected into a working circuit. Adjustment is made by connecting a 10K ohm potentiometer across pins 1 and 5, and connecting the wiper to the negative voltage, Figure 57–2.

Pin 2 is the inverting input. If a signal is applied to this input, the output will be inverted. For instance, if a positive-going AC voltage is applied to the

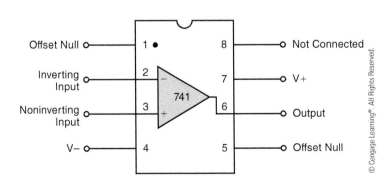

▶ **Figure 57–1**
741 operational amplifier.

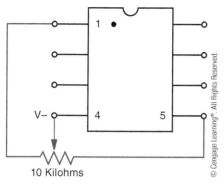

Figure 57–2
Offset null connection.

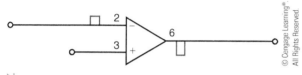

Figure 57–3
Inverted output.

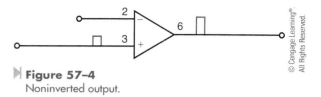

Figure 57–4
Noninverted output.

exception instead of the rule. Pin 4 is connected to the negative- or below-ground voltage, and pin 7 is connected to the positive- or above-ground voltage. The 741 operates on voltages that range from about 4 volts to 16 volts. Generally, the operating voltage for the 741 is 12 to 15 volts plus and minus. The 741 has a maximum power output rating of about 500 milliwatts. Pin 6 is the output and pin 8 is not connected.

NEGATIVE FEEDBACK

As stated previously, the open **loop** gain of the 741 operational amplifier is about 200,000. This amount of gain is not practical for most applications, so something must be done to reduce this gain to a reasonable level. One of the great advantages of the op amp is the ease with which the gain can be controlled, Figure 57–5. The amount of gain is controlled by a negative-feedback loop. This is accomplished by feeding a portion of the output voltage back to the inverting input. Because the output voltage is always opposite in polarity to the inverting input voltage, the amount of output voltage fed back to the input tends to reduce the input voltage. Negative feedback has two effects on the operation of the amplifier. One effect is that it reduces the gain. The other is that it makes the amplifier more stable.

inverting input, the output will produce a negative-going voltage, Figure 57–3.

Pin 3 is the noninverting input. When a signal voltage is applied to the noninverting input, the output voltage will be the same polarity. If a positive-going AC signal is applied to the noninverting input, the output voltage will be positive also, Figure 57–4.

Pins 4 and 7 are the voltage input pins. Operational amplifiers are generally connected to above- and below-ground power supplies. These power supplies produce both a positive and negative voltage as compared to ground. There are some circuit connections that do not require an above- and below-ground power supply, but these are the

Figure 57–5
Negative feedback connection.

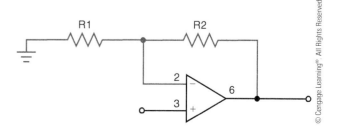

The gain of the amplifier is controlled by the ratio of resistors R2 and R1. If a **noninverting amplifier** is used, the gain is found by the formula (R2 + R1)/R1. If resistor R1 is 1K ohms and resistor R2 is 10K ohms, the gain of the amplifier would be 11 (11,000 ÷ 1000 = 11).

If the op amp is connected as an inverting amplifier, however, the input signal will be out of phase with the feedback voltage of the output. This causes a reduction of the input voltage applied to the amplifier and a reduction in gain. The formula (R2/R1) is used to compute the gain of an **inverting amplifier**. If resistor R1 is 1K ohms and resistor R2 is 10K ohms, the gain of the inverting amplifier would be 10 (10,000 ÷ 1000 = 10).

There are some practical limits, however. As a general rule, the 741 operational amplifier is not operated above a gain of about 100. If more gain is desired, it is generally obtained by using more than one amplifier, Figure 57–6.

As shown in Figure 57–6, the output of one amplifier is fed into the input of another amplifier. The reason for not operating the 741 at high gain is that at high gains it tends to become unstable. Another general rule for operating the 741 op amp is the total feedback resistance (R1 + R2) is usually kept to more than 1000 ohms and less than 100,000 ohms. These general rules apply to the 741 operational amplifier and may not apply to other operational amplifiers.

BASIC CIRCUIT CONNECTIONS

Op amps are generally used in three basic ways. This is not to say that op amps are used in only three circuits, but that there are three basic circuits that are used to build other circuits. One of these basic

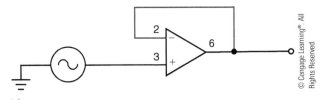

Figure 57–7
Voltage follower connection.

circuits is the **voltage follower**. In this circuit, the output of the op amp is connected directly back to the inverting input, Figure 57–7. Because there is a direct connection between the output of the amplifier and the inverting input, the gain of this circuit is 1. For instance, if a signal voltage of 0.5 volt is connected to the noninverting input, the output voltage will be 0.5 volt also. You may wonder why anyone would want an amplifier that does not amplify. Actually, this circuit does amplify something. It amplifies the input impedance by the amount of the open loop gain. If the 741 has an open loop gain of 200,000 and an input impedance of 2 megohms, this circuit will give the amplifier an input impedance of 200K × 2 megohms, or 400,000 megohms. This circuit connection is generally used for impedance matching purposes.

The second basic circuit is the noninverting amplifier, Figure 57–8. In this circuit, the output voltage is the same polarity as the input voltage. If the input voltage is a positive-going voltage, the output will be a positive-going voltage at the same time. The amount of gain is set by the ratio of resistors R1 + R2/R1 in the negative feedback loop.

The third basic circuit is the inverting amplifier, Figure 57–9. In this circuit, the output voltage is opposite in polarity to the input voltage. If the input

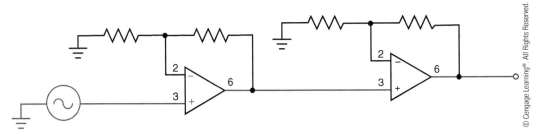

Figure 57–6
Increasing the gain.

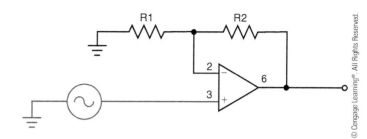

Figure 57–8
Noninverting amplifier connection.

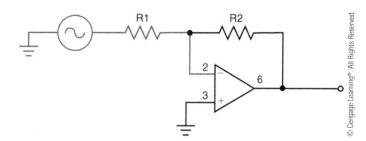

Figure 57–9
Inverting amplifier connection.

signal is a positive-going voltage, the output voltage will be negative-going at the same instant in time. The gain of the circuit is determined by the ratio of resistors R2 and R1.

CIRCUIT APPLICATIONS

The Level Detector

The operational amplifier is often used as a **level detector** or comparator. In this type of circuit, the 741 op amp is as an inverted amplifier to detect when one voltage becomes greater than another. Refer to the circuit shown in Figure 57–10. Notice that this circuit does not use

an above- and below-ground power supply. Instead it is connected to a power supply with a single positive and negative output. During normal operation, the noninverting input of the amplifier is connected to a zener diode. This zener diode produces a constant positive voltage at the noninverting input of the amplifier, which is used as a reference. As long as the noninverting input is more positive than the inverting input, the output of the amplifier will be high. A light-emitting diode, D1, is used to detect a change in the polarity of the output. As long as the output of the op amp remains high, the LED is turned off. When the output of the amplifier is high, the LED has equal voltage applied to both its anode and cathode. Because both the anode and

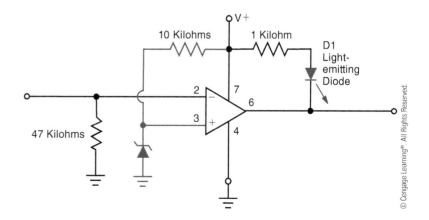

Figure 57–10
Inverting level detector.

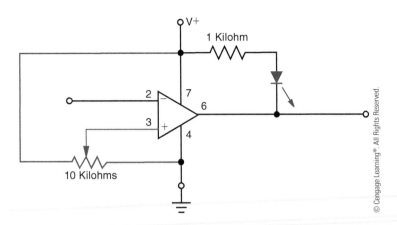

▶ **Figure 57–11**
Adjustable inverting level detector.

cathode are connected to +12 volts, there is no potential difference and therefore no current flow through the LED.

If the voltage at the inverting input becomes more positive than the reference voltage applied to pin 3, the output voltage will go low. The low voltage at the output will be about +2.5 volts. The output voltage of the op amp will not go to 0 or ground in this circuit because the op amp is not connected to a voltage that is below ground. If the output voltage is to be able to go to 0 volts, pin 4 must be connected to a voltage that is below ground. When the output is low, there is a potential of about 9.5 volts (12 − 2.5 = 9.5) produced across R1 and D1, which causes the LED to turn on and indicate that the state of the op amp's output has changed from high to low.

In this type of circuit, the op amp appears to be a digital device in that the output seems to have only two states, high or low. Actually, the op amp is not a digital device. This circuit only makes it appear digital. Notice there is no negative feedback loop connected between the output and the inverting input. Therefore, the amplifier uses its open loop gain, which is about 200,000 for the 741, to amplify the voltage difference between the inverting input and the noninverting input. If the voltage applied to the inverting input becomes 1 millivolt more positive than the reference voltage applied to the noninverting input, the amplifier will try to produce an output that is 200 volts more negative than its high-state voltage (0.001 × 200,000 = 200). The output voltage of the amplifier cannot be driven 200 volts more negative, of course, because there is only 12 volts applied to the circuit, so the output

voltage simply reaches the lowest voltage it can and then goes into saturation. The op amp is not a digital device, but it can be made to act like one.

If the zener diode is replaced with a voltage divider, as shown in Figure 57–11, the reference voltage can be set to any value desired. By adjusting the variable resistor shown in Figure 57–11, the positive voltage applied to the noninverting input can be set for any voltage value desired. For instance, if the voltage at the noninverting input is set for 3 volts, the output of the op amp will go low when the voltage applied to the inverting input becomes greater than +3 volts. If the voltage at the noninverting input is set for 8 volts, the output voltage will go low when the voltage applied to the inverting input becomes greater than +8 volts. Notice that this circuit permits the voltage level at which the output of the op amp will change to be adjusted.

In the two circuits just described, the op amp changed from a high level to a low level when activated. There may be occasions, however, when it is desired that the output be changed from a low level to a high level. This can be accomplished by connecting the inverting input to the reference voltage and connecting the noninverting input to the voltage being sensed, Figure 57–12. In this circuit, the zener diode is used to provide a positive reference voltage to the inverting input. As long as the voltage at the inverting input remains more positive than the voltage at the noninverting input, the output voltage of the op amp will remain low. If the voltage applied to the noninverting input becomes more positive than the reference voltage, the output of the op amp will become high.

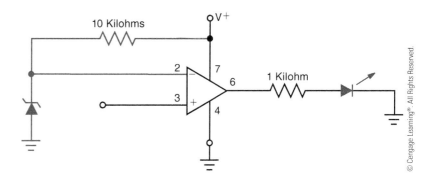

Figure 57–12
Noninverting level detector.

Depending on the application, this circuit could cause a small problem. As stated previously, because this circuit does not use an above- and below-ground power supply, the low output voltage of the op amp is about +2.5 volts. This positive output voltage could cause any other devices connected to the op amp's output to be turned on even if it should be turned off. For instance, if the LED shown in Figure 57–12 were used, it would glow dimly even when the output was in the low state. This problem can be corrected in a couple of different ways. One way would be to connect the op amp to an above- and below-ground power supply, as shown in Figure 57–13.

In this circuit, the output voltage of the op amp is negative, or below ground, as long as the voltage applied to the inverting input is more positive than the voltage applied to the noninverting input. As long as the output voltage of the op amp is negative with respect to ground, the LED is reverse biased and cannot operate. When the voltage applied to the noninverting input becomes more positive than the voltage applied to the inverting input, the output of the op amp becomes positive and the LED turns on.

The second method of correcting the output voltage problem is shown in Figure 57–14. In this circuit, the op amp is connected to a power supply that has a single positive and negative output as before. A zener diode, D2, has been connected in series with the output of the op amp and the LED. The voltage value of diode D2 is greater than the output voltage of the op amp in the low state, but less than the output voltage of the op amp in its high state. For example, assume the value of the zener diode D2 is 5.1 volts. If the output voltage of the op amp in its low state is 2.5 volts, diode D2 is turned off and will not conduct. If the output voltage becomes +12 volts when the op amp switches to its high state, the zener diode will turn on and conduct current to the LED. Notice that the zener diode D2 keeps the LED turned completely off until the op amp switches to its high state and provides enough voltage to overcome the reverse voltage drop of the zener diode.

In the preceding circuits, an LED was used to indicate the output state of the amplifier. Keep in mind that the LED is used only as a detector, and

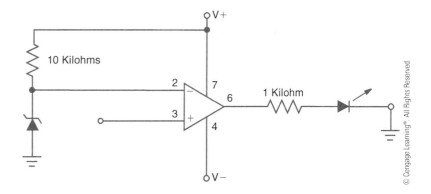

Figure 57–13
Below-ground power connection permits the output voltage to become negative.

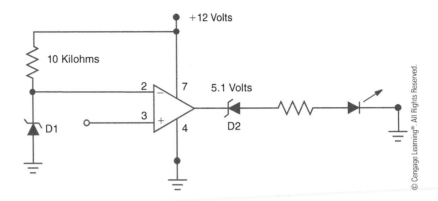

▶ **Figure 57–14**
A zener diode is used to keep the
output turned off.

the output of the op amp could be used to control almost anything. For example, the output of the op amp can be connected to the base of a transistor, as shown in Figure 57–15. The transistor can then control the coil of a relay, which could be used to control almost anything.

The Oscillator

An operational amplifier can be used as an **oscillator**. The circuit shown in Figure 57–16 is a very simple circuit that produces a square wave output. This circuit is rather impractical, however, because it would depend on a slight imbalance in the op amp or random circuit noise to start the oscillator. A slight voltage difference of a few millivolts between the two inputs is all that is needed to cause the output of the amplifier to go high or low. For example, if the inverting input becomes slightly more positive than the noninverting input, the output will go low or negative. When the output becomes negative, capacitor Ct begins to charge

through resistor Rt to the negative value of the output voltage. As soon as the voltage applied to the inverting input becomes slightly more negative than the voltage applied to the noninverting input, the output changed to a high, or positive, value of voltage. When the output becomes positive, capacitor Ct begins charging through resistor Rt toward the positive output voltage. This circuit worked quite well if the op amp has no imbalance and if the op amp is shielded from all electrical noise. In practical application, however, there is generally enough imbalance in the amplifier or enough electrical noise to send the op amp into saturation, which stops the operation of the circuit.

The Hysteresis Loop

The problem with this circuit is that a millivolt difference between the two inputs is enough to drive the amplifier's output from one state to the other. This problem can be corrected by the addition of a **hysteresis loop** connected to the

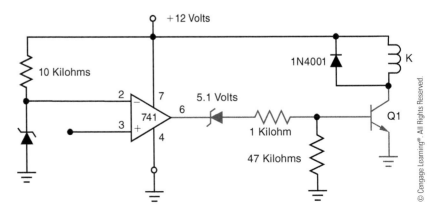

▶ **Figure 57–15**
Controlling a relay with an
op amp.

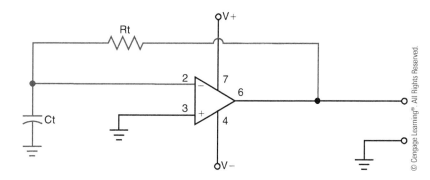

▶ Figure 57–16
Simple square-wave oscillator.

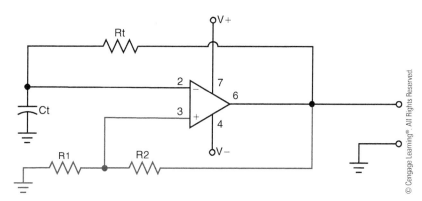

▶ Figure 57–17
Square-wave oscillator using
a hysteresis loop.

noninverting input as shown in Figure 57–17. Resistors R1 and R2 form a voltage divider for the noninverting input. These resistors are generally of equal value. To understand the circuit operation, assume that the inverting input is slightly more positive than the noninverting input. This causes the output voltage to go negative. Also assume that the output voltage is now −12 volts as compared with ground. If resistors R1 and R2 are of equal value, the noninverting input is driven to −6 volts by the voltage divider. Capacitor Ct begins to charge through resistor Rt to the value of the output voltage. When capacitor Ct has been charged to a value slightly more negative than the −6 volts applied to the noninverting input, the op amp's output goes high, or to +12 volts above ground. When the output of the op amp changes from −12 volts to +12 volts, the voltage applied to the noninverting input changes from −6 volts to +6 volts. Capacitor Ct now begins to charge through resistor Rt to the positive voltage of the output. When the voltage applied to the inverting input becomes more positive than the voltage applied to the noninverting input, the output

changes to a low value, or −12 volts. The voltage applied to the noninverting input is driven from +6 volts to −6 volts, and capacitor Ct again begins to charge toward the negative output voltage of the op amp. Notice that the addition of the hysteresis loop has greatly changed the operation of the circuit. The voltage differential between the two inputs is now volts instead of millivolts. The output frequency of the oscillator is determined by the values of Ct and Rt. The period of one cycle can be computed by using the formula ($T = 2RC$).

The Pulse Generator

The operational amplifier can also be used as a **pulse generator**. The difference between an oscillator and a pulse generator is the period of time the output remains on as compared to the period of time it remains low or off. An oscillator is generally considered to produce a waveform that has positive and negative pulses of equal voltage and time, Figure 57–18. Notice that the positive value of voltage is the same as the negative value. Also notice

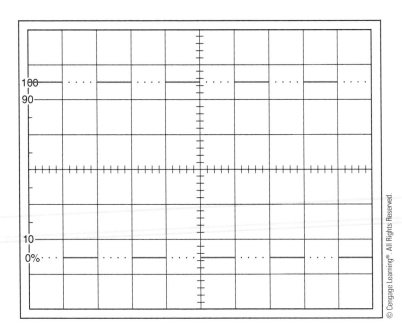

▶ **Figure 57–18**
Output of an oscillator.

that both the positive and negative cycles remain turned on the same amount of time. This waveform is consistent with that which one would expect to see if an oscilloscope is connected to the output of a square-wave oscillator.

If the oscilloscope is connected to a pulse generator, however, a waveform similar to the one shown in Figure 57–19 would be seen. Notice that the positive value of voltage is the same as the negative value just as it was in Figure 57–18. However, the positive pulse is of a much shorter duration than the negative pulse. The device producing this waveform is generally considered to be a pulse generator rather than an oscillator.

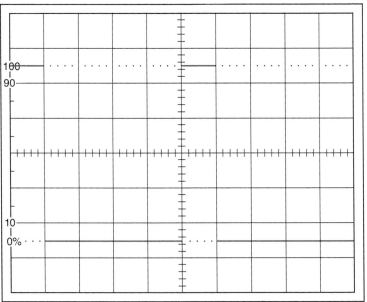

▶ **Figure 57–19**
Output of a pulse generator.

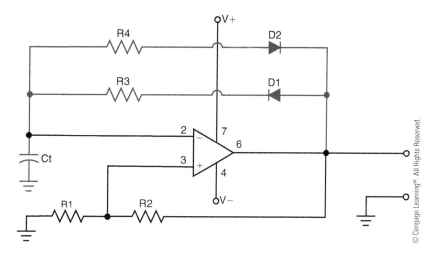

▶ Figure 57–20
Pulse generator circuit.

▶ Figure 57–21
741 operational amplifier in an eight-pin in-line case.

The 741 operational amplifier can easily be changed from a square-wave oscillator to a pulse generator. Refer to the circuit shown in Figure 57–20. This is the same basic circuit as the square-wave oscillator with the addition of resistor R3 and R4, and diodes D1 and D2. This circuit permits capacitor Ct to charge at a different rate when the output is high, or positive, than it does when the output is low, or negative. For instance, assume that the voltage of the op amp's output is low or −12 volts. If the output voltage is negative, diode D1 is reverse biased and no current can flow through resistor R3. Therefore, capacitor Ct must charge through resistor R4 and diode D2, which is forward biased. When the voltage applied to the inverting input becomes more negative than the voltage applied to the noninverting input, the output voltage of the op amp becomes +12 volts. When the output voltage becomes +12 volts, diode D2 is reverse biased and diode D1 is forward biased. Capacitor Ct, therefore, begins charging toward the +12 volts through resistor R3 and diode D1. Notice that the amount of time the output of the op amp remains low is determined by the value of Ct and R4, and the amount of time the output remains high is determined by the value of Ct and R3. The ratio of time the output voltage is high compared to the amount of time it is low can be determined by the ratio of resistor R3 to resistor R4. A 741 operational amplifier is shown in Figure 57–21.

SUMMARY

- ⊙ Some op amps use bipolar transistors for the inputs and other use field effect transistors for the inputs.
- ⊙ The 741 operational amplifier has an input impedance of about 2 megohms.
- ⊙ The 741 op amp has an output impedance of about 75 ohms.
- ⊙ The 741 operational amplifier has an open loop, or maximum gain, of about 200,000.
- ⊙ Op amps have two input connections called the inverting and noninverting inputs.
- ⊙ If the noninverting input is more positive than the inverting input, the output voltage will be positive with respect to ground.
- ⊙ If the inverting input is more positive than the noninverting input, the output voltage will be negative with respect to ground.
- ⊙ Operational amplifiers are generally connected to above- and below-ground power supplies.
- ⊙ Negative feedback is used to reduce the gain of operational amplifiers.
- ⊙ The voltage follower connection produces a gain of 1, but increases the input impedance.
- ⊙ Inverting amplifiers produce an output waveform that is inverted or opposite that of the input.
- ⊙ Noninverting amplifiers produce an output waveform that is the same as the input.

KEY TERMS

bipolar transistors	level detector	operational amplifier
field effect transistor	loop	(op amp)
gain	noninverting amplifier	oscillator
hysteresis loop	noninverting input	pulse generator
inverting amplifier	offset null	voltage follower
inverting input		

REVIEW QUESTIONS

1. When the voltage connected to the inverting input is more positive than the voltage connected to the noninverting input, will the output be positive or negative?
2. What is the input impedance of a 741 operational amplifier?
3. What is the average open loop gain of the 741 operational amplifier?
4. What is the average output impedance of the 741?
5. List the three common connections for operational amplifiers.
6. When the operational amplifier is connected as a voltage follower, it has a gain of one. If the input voltage does not get amplified, what does?
7. Name two effects of negative feedback.

8. Refer to Figure 57–8. If resistor R1 is 200 ohms and resistor R2 is 10K ohms, what is the gain of the amplifier?

9. Refer to Figure 57–9. If resistor R1 is 470 ohms and resistor R2 is 47K ohms, what is the gain of the amplifier?

10. What is the purpose of the hysteresis loop when the op amp is used as an oscillator?

SECTION 9

Solid-State Controls

Programmable Logic Controllers

OBJECTIVES

After studying this unit, the student should be able to:

▷ List the principal parts of a programmable controller

▷ Describe the differences between programmable controllers and other types of computers

▷ Discuss differences among the I/O rack, CPU, and program loader

▷ Draw a diagram of how the input and output modules work

Programmable logic controllers (PLCs) were first used by the automotive industry in the late 1960s. Each time a change was made in the design of an automobile, it was necessary to change the control system operating the machinery. This consisted of physically rewiring the control system to make it perform the new operation. Rewiring the system was, of course, very time consuming and expensive. What the industry needed was a control system that could be changed without the extensive rewiring required to change relay control systems.

DIFFERENCES BETWEEN PLCS AND PCS

One of the first questions generally asked is, "Is a programmable logic controller a computer?" The answer to that question is yes. The PLC is a special type of computer designed to perform a special function. Although the programmable logic controller (PLC) and the personal computer (PC) are both computers, there are some significant differences. Both generally employ the same basic type of computer and memory chips to perform the tasks for which they are intended, but the PLC must operate in an industrial environment. Any computer that is intended for industrial use must be able to withstand extremes of temperature; ignore voltage spikes and drops on the power line; survive in an atmosphere that often contains corrosive vapors, oil, and dirt; and withstand shock and vibration.

Programmable logic controllers are designed to be programmed with schematic or ladder diagrams instead of common computer languages. An electrician who is familiar with ladder logic diagrams can generally learn to program a PLC in a few hours as opposed to the time required to train a person how to write programs for a standard computer.

BASIC COMPONENTS

Programmable logic controllers can be divided into four primary parts:

1. The power supply
2. The central processing unit (CPU)
3. The programming terminal or program loader
4. The I/O (pronounced eye-oh) rack.

The Power Supply

The function of the **power supply** is to lower the incoming AC voltage to the desired level, rectify it to direct current, and then filter and regulate it. The internal logic of a PLC generally operates on 5 to 24 volts DC, depending on the type of controller. This voltage must be free of voltage spikes and other electrical noise and be regulated to within 5% of the required voltage value. Some manufacturers of PLCs build a separate power supply, and others build the power supply into the central processing unit.

The CPU

The **CPU**, or **central processing unit**, is the "brains" of the programmable logic controller. It contains the microprocessor chip and related integrated circuits to perform all the logic functions. The microprocessor chip used in most PLCs is the same as that found in most home and business personal computers.

The central processing unit often has a key located on the front panel, Figure 58–1. This switch must be turned on before the CPU can be programmed. This is done to prevent the circuit from being changed or deleted accidentally. Other manufacturers use a *software switch* to protect the circuit. A software switch is not a physical switch. It is a command that must be entered before the program can be changed or deleted. Whether a physical switch or a software switch is used, they both perform the same function.

Courtesy of Rockwell Automation, Inc.

Figure 58–1
A central processing unit.

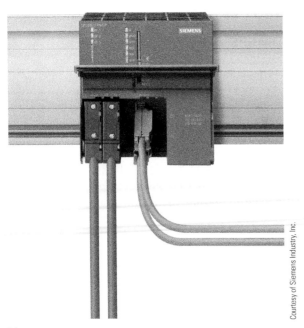

Courtesy of Siemens Industry, Inc.

▶ **Figure 58–2**
Plug connections located on the CPU.

They prevent a program from being accidentally changed or deleted.

Plug connections on the central processing unit provide connection for the programming terminal and I/O racks, Figure 58–2. CPUs are designed so that once a program has been developed and tested, it can be stored on some type of medium such as tape, disk, CD, or other storage device. In this way, if a central processor unit fails and has to be replaced, the program can be downloaded from the storage medium. This eliminates the time-consuming process of having to reprogram the unit by hand.

The Programming Terminal

The **programming terminal**, or **loading terminal**, is used to program the CPU. The type of terminal used depends on the manufacturer and often the preference of the consumer. Some are small handheld devices that use a liquid crystal display or light-emitting diodes to show the program,

Figure 58–3. Some of these small units display one line of the program at a time, and others require the program to be entered in a language called Boolean.

Another type of programming terminal contains a display and keyboard, Figure 58–4. This type of terminal generally displays several lines of the program at a time and can be used to observe the operation of the circuit as it is operating.

Many industries prefer to use a notebook or laptop computer for programming, Figure 58–5. An interface that permits the computer to be connected to the input of the PLC and software program is generally available from the manufacturer of the programmable logic controller.

The terminal is used not only to program the PLC but also to troubleshoot the circuit. When the terminal is connected to the CPU, the circuit can be examined while it is in operation. Figure 58–6 illustrates a circuit typical of those which are seen on the display. Notice that this schematic diagram is different from the typical ladder diagram. All of the line components are shown as normally open or normally closed contacts. There are no NEMA symbols for pushbutton, float switch, limit switches, and so on. The programmable logic controller recognizes only open or closed contacts. It does not know if a contact is connected to a pushbutton, a limit switch, or a float switch. Each contact, however, does have a number. The number is used to distinguish one contact from another.

In this example, coil symbols look like a set of parentheses instead of a circle as shown on most ladder diagrams. Each line ends with a coil and each coil has a number. When a contact symbol has the same number as a coil, it means that the contact is controlled by that coil. The schematic in Figure 58–6 shows a coil numbered 257 and two contacts numbered 257. When coil 257 is energized, the programmable logic controller interprets both contacts 257 to be closed.

A characteristic of interpreting a diagram when viewing it on the screen of most loading terminals is that when a current path exists through a contact or if a coil is energized, either will be highlighted on the display. In the example shown in Figure 58–6, coil 257, both 257 contacts, contact 16, and contact 18 are drawn with dark heavy lines, illustrating

▶ **Figure 58–3**
Small programmable controller and handheld programming unit.

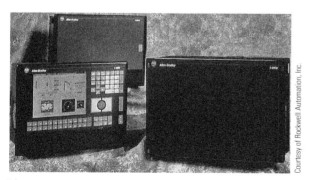

▶ **Figure 58–4**
Programming terminal.

▶ **Figure 58–5**
A notebook computer is often used as the programming terminal for a PLC.

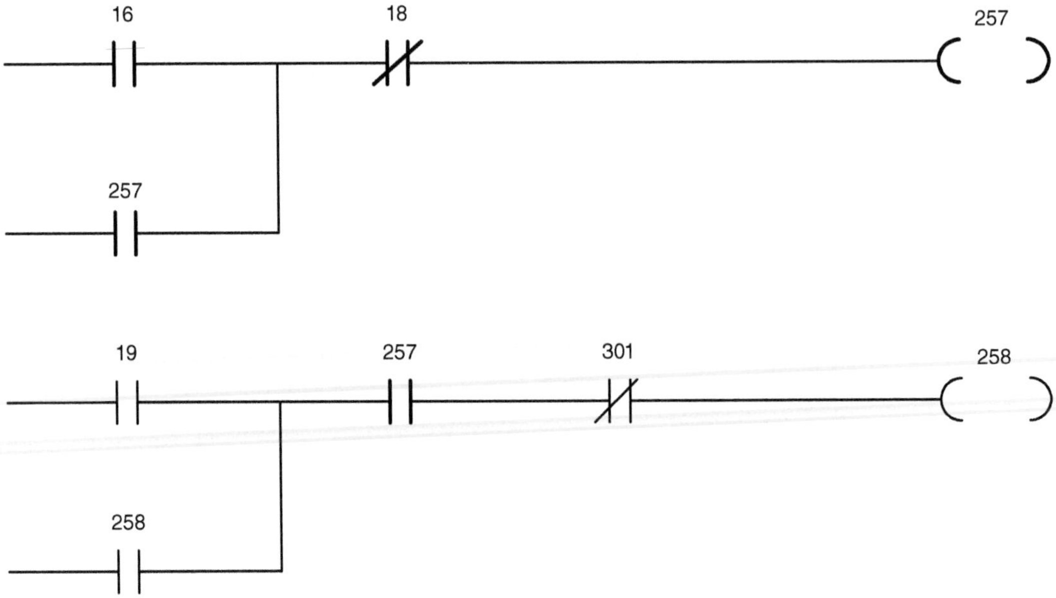

▶ **Figure 58–6**
Analyzing circuit operation with a terminal.

that they are highlighted or **illuminated** on the display. Highlighting a contact does not mean that it has changed from its original state. It means that there is a complete circuit through that contact. Contact 16 is highlighted, indicating that coil 16 has energized, and contact 16 is closed and providing a complete circuit. Contact 18, however, is shown as normally closed. Because it is highlighted, coil 18 has not been energized, because a current path still exists through contact 18. Coil 257 is shown highlighted, indicating that it is energized. Because coil 257 is energized, both 257 contacts are now closed, providing a current path through them.

When the loading terminal is used to load a program into the PLC, contact and coil symbols on the keyboard are used, Figure 58–7. Other keys permit specific types of relays, such as timers, counters, or retentive relays to be programmed into the logic of the circuit. Some keys permit parallel paths, generally referred to as down rungs, to be started and ended. The method employed to program a PLC is specific to the make and model of the controller. It is generally necessary to consult the manufacturer's

literature if you are not familiar with the specific programmable logic controller.

The I/O Rack

The **I/O rack** is used to connect the CPU to the outside world. It contains input modules that carry information from control sensor devices to the CPU and output modules that carry instructions from the CPU to output devices in the field. I/O racks are shown in Figure 58–8A and Figure 58–8B. Input and output modules contain more than one input or output. Any number from 4 to 32 is common, depending on the manufacturer and model of PLC. The modules shown in Figure 58–8A can each handle 16 connections. This means that each input module can handle 16 different input devices such as pushbutton, limit switches, proximity switches, float switches, and so on. The output modules can each handle 16 external devices such as pilot lights, solenoid coils, or relay coils. The operating voltage can be either alternating or direct current, depending on the manufacturer and model of controller, and is generally either 120 or 24 volts.

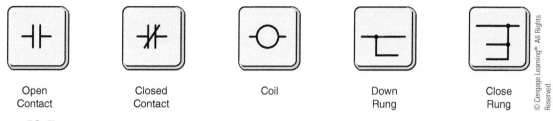

| Open Contact | Closed Contact | Coil | Down Rung | Close Rung |

▶ **Figure 58–7**
Symbols are used to program the PLC.

▶ **Figure 58–8A**
I/O rack with input and output modules.

▶ **Figure 58–9**
Central processor with I/O racks.

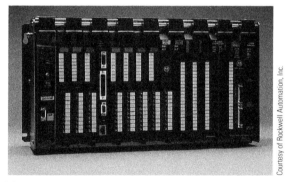

▶ **Figure 58–8B**
I/O rack with input and output modules.

The I/O rack shown in Figure 58–8A can handle 10 modules. Because each module can handle 16 input or output devices, the I/O rack is capable of handling 160 input and output devices. Many programmable logic controllers are capable of handling multiple I/O racks.

I/O CAPACITY

One factor that determines the size and cost of a programmable logic controller is its **I/O capacity**. Many small units may be intended to handle as few as 16 input and output devices. Large PLCs can generally handle several hundred. The number of input and output devices the controller must handle also affects the processor speed and amount of memory the CPU must have. A central processing unit with I/O racks is shown in Figure 58–9.

THE INPUT MODULE

The central processing unit of a programmable logic controller is extremely sensitive to voltage spikes and electrical noise. For this reason, the

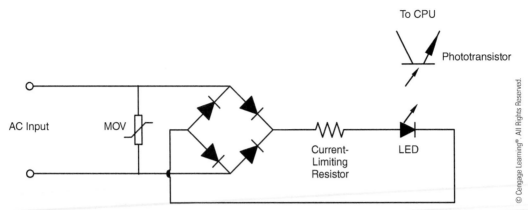

To CPU

Phototransistor

AC Input

MOV

Current-
Limiting
Resistor

LED

© Cengage Learning®. All Rights Reserved.

▷ **Figure 58–10**
Input circuit.

input I/O uses opto-isolation to electrically separate the incoming signal from the CPU. Figure 58–10 shows a typical circuit used for the input. A metal-oxide-varistor (MOV) is connected across the AC input to help eliminate any voltage spikes that may occur on the line. The MOV is a voltage-sensitive resistor. As long as the voltage across its terminals remains below a certain level, it exhibits a very high resistance. If the voltage should become too high, the resistance almost instantly changes to a very low value. A bridge rectifier changes the AC voltage into DC. A resistor is used to limit current to a light-emitting diode. When power is applied to the circuit, the LED turns on. The light is detected by a phototransistor that signals the CPU that there is a voltage present at the input terminal.

When the module has more than one input, the bridge rectifiers are connected together on one side to form a common terminal. On the other side, the rectifiers are labeled 1, 2, 3, and 4. Figure 58–11 shows four bridge rectifiers connected together to form a common terminal. Figure 58–12 shows a limit switch connected to input 1, a temperature switch connected to input 2, a float switch connected to input 3, and a normally open pushbutton connected to input 4. Notice that the pilot devices complete a circuit to the bridge rectifiers. If any switch closes, 120 volts AC will be connected to a bridge rectifier, causing the corresponding light-emitting diode to turn on and signal the CPU that

the input has voltage applied to it. When voltage is applied to an input, the CPU considers that input to be at a high level.

THE OUTPUT MODULE

The output module is used to connect the central processing unit to the load. Output modules provide line isolation between the CPU and the external circuit. Isolation is generally provided in one of two ways. The most popular is with optical isolation very similar to that used for the input modules. In this case, the CPU controls a light-emitting diode. The LED is used to signal a solid-state device to connect the load to the line. If the load is operated by direct current, a power phototransistor is used to connect the load to the line, Figure 58–13. If the load is an alternating current device, a triac is used to connect the load to the line, Figure 58–14. Notice that the central processing unit is separated from the external circuit by a light beam. No voltage spikes or electrical noise can be transmitted to the CPU.

The second method of controlling the output is with small relays, Figure 58–15. The CPU controls the relay coil. The contacts connect the load to the line. The advantage of this type of output module is that it is not sensitive to whether the voltage is AC or DC and can control 120- or 24-volt circuits. The disadvantage is that it does contain moving parts that can wear. In this instance, the CPU is isolated

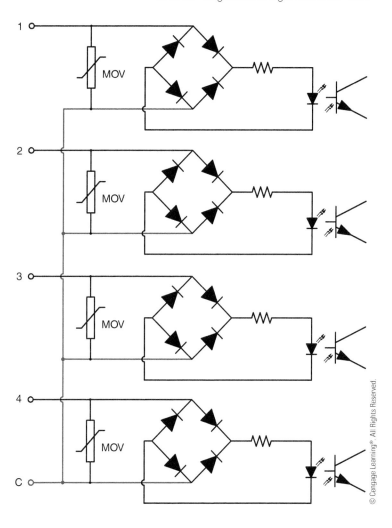

▶ **Figure 58–11**
Four-input module.

from the external circuit by a magnetic field instead of a light beam.

If the module contains more than one output, one terminal of each output device is connected together to form a common terminal similar to a module with multiple inputs, Figure 58–16. Notice that one side of each triac has been connected together to form a common point. The other side of each triac is labeled 1, 2, 3, or 4. If power transistors were used as output devices, the collectors or emitters of each transistor would be connected to form a common terminal. Figure 58–14 shows a relay coil connected to the output of a triac. Notice that the triac is used as a switch to connect the load to the

line. The power to operate the load must be provided by an external source. *Output modules do not provide power to operate external loads.*

The amount of current an output can control is limited. The current rating of most outputs can range from 0.5 to about 3 amperes, depending on the manufacturer and type of output being used. Outputs are intended to control loads that draw a small amount of current such as solenoid coils, pilot lights, and relay coils. Some outputs can control motor starter coils directly and others require an interposing relay. Interposing relays are employed when the current draw of the load is above the current rating of the output.

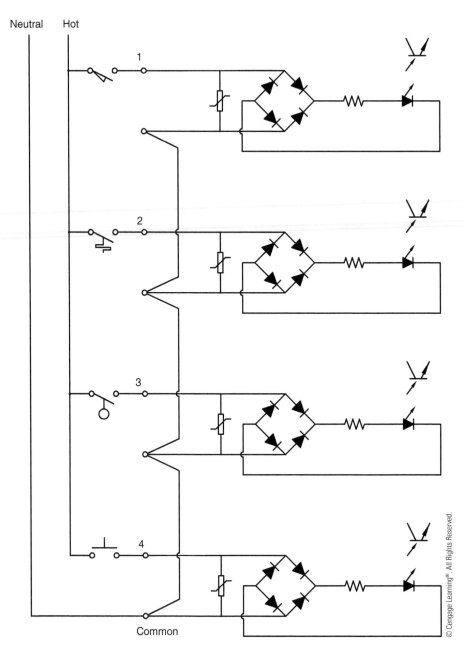

▶ **Figure 58–12**
Pilot devices connected to input modules.

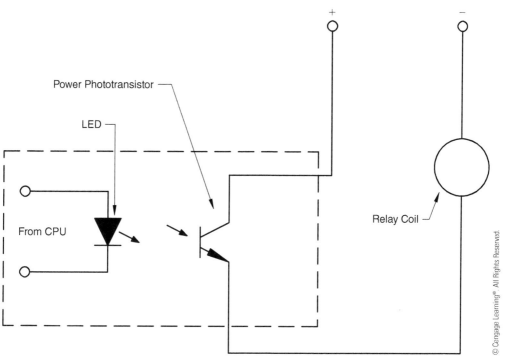

▶ **Figure 58–13**
A power phototransistor connects a DC load to the line.

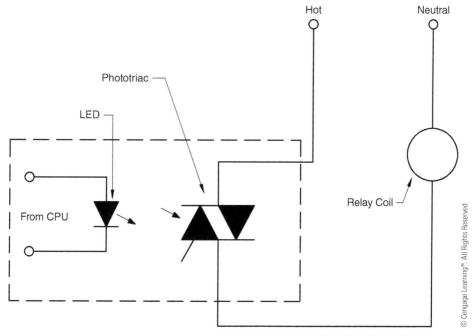

▶ **Figure 58–14**
A triac connects an AC load to the line.

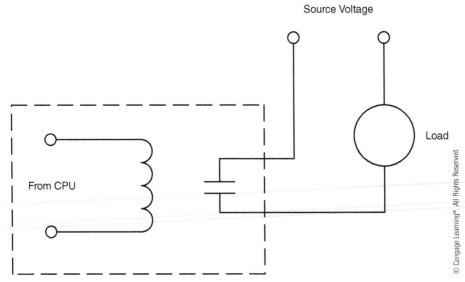

Source Voltage

Load

From CPU

▶ **Figure 58–15**
A relay connects the load to the line.

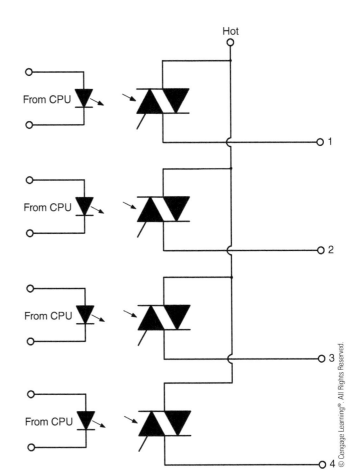

Hot

From CPU

From CPU

From CPU

From CPU

1

2

3

4

▶ **Figure 58–16**
Multiple-output module.

INTERNAL RELAYS

The actual logic of the control circuit is performed by **internal relays**. An internal relay is an imaginary device that exists only in the logic of the computer. It can have any number of contacts from one to several hundred, and the contacts can be programmed normally open or normally closed. Internal relays are programmed into the logic of the PLC by assigning them a certain number. Manufacturers provide a chart that lists which numbers can be used to program inputs and outputs, internal relay coils, timers, counters, and so on. When a coil is entered at the end of a line of logic and is given a number that corresponds to an internal relay, it will act like a physical relay. Any contacts given the same number as that relay will be controlled by that relay.

TIMERS AND COUNTERS

Timers and counters are internal relays also. There is no physical timer or counter in the PLC. They are programmed into the logic in the same manner as any other internal relay, by assigning them a number that corresponds to a timer or counter. The difference is that the time delay or number of counts must be programmed when they are inserted into the program. The number of counts for a counter are entered using numbers on the keys on the load terminal. Timers are generally programmed in 0.1-second intervals. Some manufacturers provide a decimal key and others do not. If a decimal key is not provided, the time delay is entered as 0.1-second intervals.

If a delay of 10 seconds is desired, for example, the number 100 would be entered. One hundred tenths of a second equals 10 seconds.

OFF-DELAY CIRCUIT

Some programmable logic controllers permit a timer to be programmed as on or off delay, but others permit only on-delay timers to be programmed. When a PLC permits only on-delay timers to be programmed, a simple circuit can be used to permit an on-delay timer to perform the function of an off-delay timer, Figure 58–17. To understand the action of the circuit, recall the operation of an off-delay timer. When the timer coil is energized, the timed contacts change position immediately. When the coil is de-energized the contacts remain in their energized state for some period of time before returning to their normal state. In the circuit shown in Figure 58–17, it is assumed that contact 400 controls the action of the timer. Coil 400 is an internal relay coil located somewhere in the circuit. Coil 12 is an output and controls some external device. Coil TO-1 is an on-delay timer set for 100 tenths of a second. When coil 400 is energized, both 400 contacts change position. The normally open 400 contact closes and provides a current path to coil 12. The normally closed 400 contact opens and prevents a circuit from being completed to coil TO-1 when coil 12 energizes. Note that coil 12 turned on immediately when contact 400 closed. When coil 400 is de-energized, both 400 contacts return to their normal position. A current path is maintained to coil 12 by the now closed 12 contact in parallel

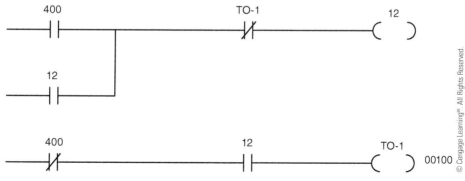

▶ **Figure 58–17**
Off-delay timer circuit.

with the normally open 400 contact. When the normally closed 400 contact returns to its normal position, a current path is established to coil TO-1 through the now closed 12 contact. This starts the time sequence of timer TO-1. After a delay of 10 seconds, the normally closed TO-1 contact opens and de-energizes coil 12, returning the two 12 contacts to their normal position. The circuit is now back in the state shown in Figure 58–16. Note the action of the circuit. When coil 400 was energized, output coil 12 turned on immediately. When coil 400 was de-energized, output 12 remained on for 10 seconds before turning off.

The number of internal relays and timers contained in a programmable logic controller is determined by the memory capacity of the computer. As a general rule, PLCs that have a large I/O capacity have a large amount of memory. The use of programmable logic controllers has steadily increased since their invention in the late 1960s. A PLC can replace hundreds of relays and occupies only a fraction of the space. The circuit logic can be changed easily and quickly without requiring extensive hand rewiring. PLCs have no moving parts or contacts to wear out, and their down time is less than that of an equivalent relay circuit. When replacement is necessary, they can be reprogrammed from a media storage device.

The programming methods presented in this text are general because it is impossible to include

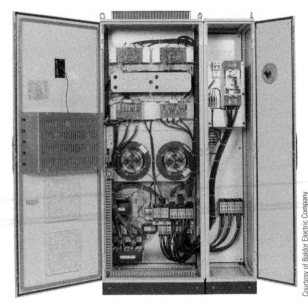

Courtesy of Baldor Electric Company

Figure 58–18
DC drive unit controlled by a programmable logic controller.

examples of each specific manufacturer. The concepts presented in this chapter, however, are common to all programmable controllers. A programmable logic controller used to control a DC drive is shown in Figure 58–18.

SUMMARY

- ⊙ Programmable logic controllers were first used by the automotive industry in the early 1960s.
- ⊙ The major parts of a programmable logic controller are the power supply, central processing unit, programming terminal, and I/O rack.
- ⊙ The power supply changes AC into DC, and then filters and regulates it to the proper voltage.
- ⊙ The central processing unit performs all the logic functions loaded into memory.
- ⊙ The programming terminal is used to load or amend a program in a programmable logic controller.
- ⊙ Laptop computers are often employed as programming terminals.
- ⊙ The I/O rack provides inputs and outputs for the programmable logic controller.
- ⊙ Most programmable logic controllers use opto-isolation in the input and output modules.
- ⊙ Input modules provide information to the central processing unit from the outside circuit.

⊙ Output modules provide information from the central processing unit to the outside circuit.

⊙ Internal relays are relays that exist only in the logic of the computer.

⊙ Some internal relays can be employed as counters and timers.

▷ KEY TERMS

central processing unit (CPU)

illuminated

internal relay

I/O capacity

I/O rack

loading terminal

power supply

programmable logic controller (PLC)

programming terminal

▷ REVIEW QUESTIONS

1. What industry first started using programmable logic controllers?
2. Name two differences between PLCs and common home or business computers.
3. Name the four basic sections of a programmable logic controller.
4. In what section of the PLC is the actual logic performed?
5. What device is used to program a PLC?
6. What device separates the central processing unit from the outside world?
7. What is opto-isolation?
8. If an output I/O controls a DC voltage, what solid-state device is used to connect the load to the line?
9. If an output I/O controls an AC voltage, what solid-state device is used to connect the load to the line?
10. What is an internal relay?
11. What is the purpose of the key switch located on the front of the CPU in many programmable logic controllers?
12. What is a software switch?

Programming a PLC

OBJECTIVES

After studying this unit, the student should be able to:

» Convert a relay schematic to a schematic used for programming a PLC

» Enter a program into a programmable logic controller

In this unit, a **relay schematic** is converted into a diagram used to program a programmable logic controller. The process to be controlled is shown in Figure 59–1. A tank is used to mix two liquids. The control circuit operates as follows:

1. When the start button is pressed, solenoids A and B energize. This permits the two liquids to begin filling the tank.

2. When the tank is filled, the float switch trips. This de-energizes solenoids A and B and starts the motor used to mix the liquids together.

3. The motor is permitted to run for 1 minute. After 1 minute has elapsed, the motor turns off and solenoid C energizes to drain the tank.

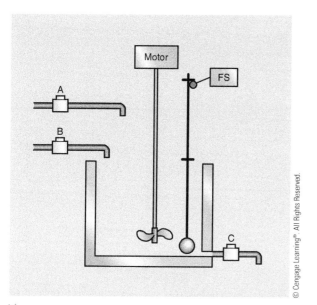

Figure 59-1
Tank used to mix two liquids.

4. When the tank is empty, the float switch de-energizes solenoid C.

5. A stop button can be used to stop the process at any point.

6. If the motor becomes overloaded, the action of the entire circuit will stop.

7. Once the circuit has been energized, it will continue to operate until it is manually stopped.

CIRCUIT OPERATION

A relay schematic that will perform the logic of this circuit is shown in Figure 59–2. The logic of this circuit is as follows:

1. When the start button is pushed, relay coil CR is energized. This causes all CR contacts to close. Contact CR-1 is a holding contact used to maintain the circuit to coil CR when the start button is released.

2. When contact CR-2 closes, a circuit is completed to solenoid coils A and B. This permits

the two liquids that are to be mixed together to begin filling the tank.

3. As the tank fills, the float rises until the float switch is tripped. This causes the normally closed float switch contact to open and the normally open contact to close.

4. When the normally closed float switch opens, solenoid coils A and B de-energize and stop the flow of the two liquids into the tank.

5. When the normally open contact closes, a circuit is completed to the coil of a motor starter and the coil of an on-delay timer. The motor is used to mix the two liquids together.

6. At the end of the 1-minute time period, all of the TR contacts change position. The normally closed TR-2 contact connected in series with the motor starter coil opens and stops the operation of the motor. The normally open TR-3 contact closes and energizes solenoid coil C, which permits liquid to begin draining from the tank. The normally closed TR-1 contact is used to ensure that valves A and B cannot be reenergized until solenoid C de-energizes.

7. As liquid drains from the tank, the float drops. When the float drops far enough, the float switch trips, and its contacts return to their normal positions. When the normally open float switch contact reopens and de-energizes coil TR, all TR contacts return to their normal positions.

8. When the normally open TR-3 contact reopens, solenoid C de-energizes and closes the drain valve. Contact TR-2 recloses, but the motor cannot restart because of the normally open float switch contact. When contact TR-1 recloses, a circuit is completed to solenoids A and B. This permits the tank to begin refilling, and the process starts over again.

9. If the stop button or overload contact opens, coil CR de-energizes and all CR contacts open. This de-energizes the entire circuit.

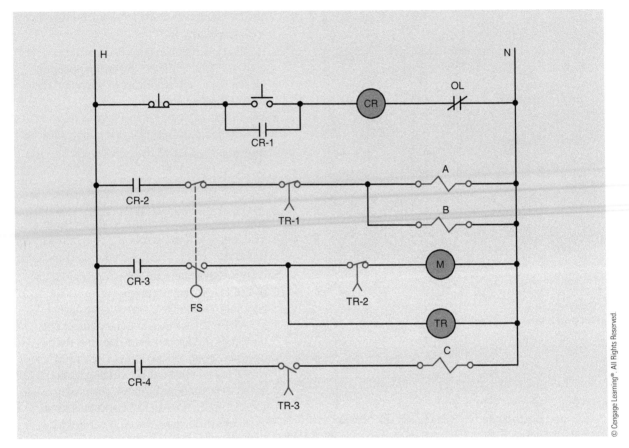

▌**Figure 59–2**
Relay schematic.

DEVELOPING A PROGRAM

This circuit will now be developed into a program that can be loaded into the programmable controller. Figure 59–3 shows a program being developed on a computer. Assume that the controller has an I/O capacity of 32, that I/O terminals 1 through 16 are used as inputs, and that terminals 17 through 32 are used as outputs.

Before a program can be developed for input into a programmable logic controller, it is necessary to assign which devices connect to the **input** and **output** terminals. This circuit contains four input devices and four output devices. It is also assumed that the motor starter for this circuit contains an overload relay that has two contacts instead of one.

One contact is normally closed and will be connected in series with the coil of the motor starter. The other contact is normally open and is used to supply an input to a programmable logic controller. If the motor should become overloaded, the normally closed contacts will open and disconnect the motor from the line. The normally open contacts will close and provide a signal to the programmable logic controller that the motor has tripped on overload. The input devices are as follows:

1. Normally closed stop pushbutton
2. Normally open start pushbutton
3. Normally open overload contact
4. A float switch that contains both a normally open and normally closed contact

Figure 59–3
A program being developed on a programming terminal.

The four output devices are listed here:

1. Solenoid valve A
2. Solenoid valve B
3. Motor starter coil M
4. Solenoid valve C

The connection of devices to the inputs and outputs is shown in Figure 59–4. The normally closed stop button is connected to input 1; the normally open start button is connected to input 2; the normally open overload contact is connected to input 3; and the float switch is connected to input 4.

The outputs for this PLC are 17 through 32. Output 17 is connected to solenoid A; output 18 is connected to solenoid B; output 19 is connected to the coil of the motor starter; and output 20 is connected to solenoid C. Note that the outputs *do not* supply the power to operate the output devices. The outputs simply complete a circuit. One side of each output device is connected to the grounded or neutral side of a 120 VAC power line. The ungrounded or hot conductor is connected to the common terminal of

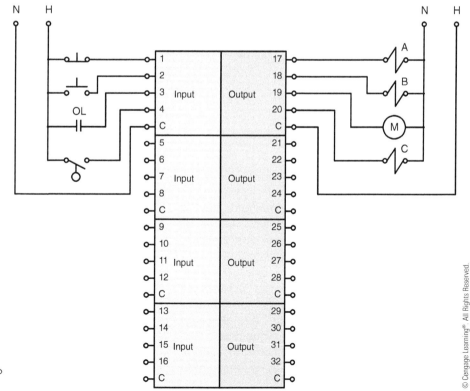

Figure 59–4
Component connection to I/O rack.

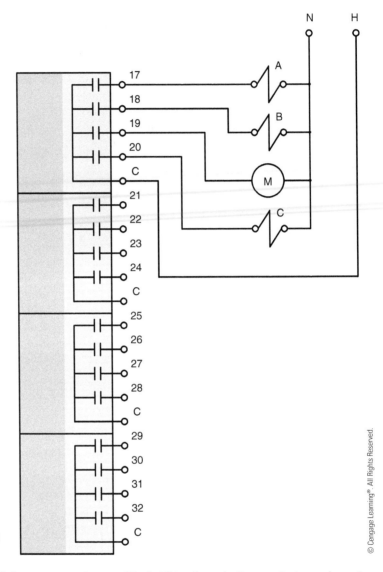

▶ **Figure 59–5**
Output modules complete a circuit to connect
the load to the line.

the four outputs. A good way to understand this is to
imagine a set of contacts controlled by each output,
as shown in Figure 59–5. When programming the
PLC, if a coil is given the same number as one of the
outputs, it will cause that contact to close and connect the load to the line.

Unfortunately, programmable logic controllers
are not all programmed the same way. Almost every
manufacturer employs a different set of coil numbers to perform different functions. It is necessary
to consult the manual before programming a PLC
with which you are not familiar. In order to program
the PLC in this example, refer to the information in

Figure 59–6. This chart indicates that numbers 1
through 16 are inputs. Any contact assigned a number between 1 and 16 will be examined each time
the programmable logic controller scans the program. If an input has a low (0-volt) state, the contact assigned that number will remain in the state it
was programmed. If the input has a high (120-volt)
state, the program will interpret that contact as having changed position. If it was programmed as open,
the PLC will now consider it as closed.

Outputs are 17 through 32. Outputs are treated
as coils by the PLC. If a coil is given the same number as an output, that output will turn on (close

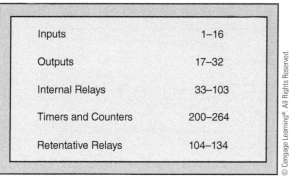

Figure 59–6
Numbers that correspond to specific PLC functions.

Inputs	1–16
Outputs	17–32
Internal Relays	33–103
Timers and Counters	200–264
Retentive Relays	104–134

the contact) when the coil is energized. Coils that control outputs can be assigned internal contacts as well. Internal contacts are contacts that exist in the logic of the program only. They do not physically exist. Because they do not physically exist, a coil can be assigned as many internal contacts as desired, and they can be normally open or normally closed.

The chart in Figure 59–6 also indicates that internal relays number from 33 to 103. Internal relays are like internal contacts. They do not physically exist. They exist as part of the program only. They are programmed into the circuit logic by inserting a coil symbol in the program and assigning it a number between 33 and 103.

Timers and counters are assigned coil numbers 200 through 264, and retentive relays are numbered 104 through 134.

CONVERTING THE PROGRAM

Developing a program for a programmable logic controller is a little different from designing a circuit with relay logic. There are several rules that must be followed with almost all programmable logic controllers.

1. Each line of logic must end with a coil. Some manufacturers permit coils to be connected in parallel and some do not.
2. Generally, coils cannot be connected in series.
3. The program will scan in the order that it is entered.
4. Generally, coils cannot be assigned the same number. (Some programmable logic controllers require reset coils to reset counters and timers. These reset coils can be assigned the same number as the counter or timer they reset.)

The first two lines of logic for the circuit shown in Figure 59–2 can be seen in Figure 59–7. Notice that contact symbols are used to represent inputs instead of logic symbols such as pushbuttons, float switches, and so on. The programmable logic controller recognizes all inputs as open or closed contacts. It does not know what device is connected to which input. This is the reason that you must first determine which device connects to which input before a program can be developed. Also notice that input 1 is shown as a normally open contact. Referring to Figure 59–4, it can be seen that input 1 is connected to a normally closed pushbutton. The input is programmed as normally open because the normally closed pushbutton will supply a high voltage to input 1 in normal operation. Because input 1 is in a high state, the PLC will change the state of the open contact and consider it closed. When the stop pushbutton is pressed, the input voltage changes to low and the PLC changes the contact back to its original open state and causes coil 33 to de-energize.

Referring to the schematic in Figure 59–2, a control relay is used as part of the circuit logic. Because the control relay does not directly cause any output device to turn on or off, an internal relay is used. The chart in Figure 59–6 indicates that internal

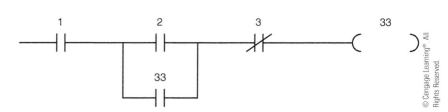

Figure 59–7
Lines 1 and 2 of the program.

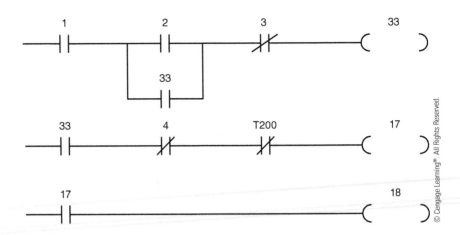

Figure 59–8
Lines 3 and 4 of the circuit are
added.

relays number between 33 and 103. Coil number
33 is an internal relay and does not physically exist.
Any number of contacts can be assigned to this
relay, and they can be open or closed. The number
33 contact connected in parallel with input 2 is the
holding contact labeled CR-1 in Figure 59–2.

The next two lines of logic are shown in
Figure 59–8. The third line of logic in the schematic in
Figure 59–2 contains a normally open CR-2 contact,
a normally closed float switch contact, a normally
closed on-delay timed contact, and solenoid coil A.
The fourth line of logic contains solenoid coil B
connected in parallel with solenoid coil A. Line 3 in
Figure 59–8 uses a normally open contact assigned
the number 33 for contact CR-2. A normally closed
contact symbol is assigned the number 4. Because
the float switch is connected to input number 4, it
controls the action of this contact. As long as input
4 remains in a low state, the contact remains closed.
If the float switch should close, input 4 will become
high, and the number 4 contact will open.

The next contact is timed contact TR-1. The
chart in Figure 59–6 indicates that timers and
counters are assigned numbers 200 through 264.
In this circuit, timer TR will be assigned number
200. Line 3 ends with coil number 17. When coil
17 becomes energized, it will turn on output 17 and
connect solenoid coil A to the line.

The schematic in Figure 59–2 shows that solenoid coil B is connected in parallel with solenoid
coil A. Most programmable logic controllers do not
permit coils to be connected in parallel. Each line of
logic must end with its own coil. Because solenoid

coil B is connected in parallel with A, they both
operate at the same time. This logic can be accomplished by assigning an internal contact the same
number as the coil controlling output 17. Notice
in Figure 59–8 that when coil 17 energizes, it will
cause contact 17 to close and energize output 18 at
the same time.

In Figure 59–9, lines 5 and 6 of the schematic
are added to the program. A normally open contact
assigned number 33 is used as for contact CR-3. A
normally open contact assigned the number 4 is
controlled by the float switch, and a second normally closed timed contact controlled by timer 200
is programmed in line 5. The output coil is assigned
the number 19. When this coil energizes, it turns
on output 19 and connects motor starter coil M to
the line.

Line 6 contains timer coil TR. Notice in Figure 59–2
that coil TR is connected in parallel with contact
TR-2 and coil M. As was the case with solenoid
coils A and B, coil TR cannot be connected in parallel with coil M. According to the schematic in
Figure 59–2, coil TR is actually controlled by contacts CR-3 and the normally open float switch. This
logic can be accomplished as shown in Figure 59–9
by connecting coil T200 in series with contacts
assigned the numbers 33 and 4. Float switches
do not normally contain this many contacts, but
because the physical float switch is supplying a high
or low voltage to input 4, any number of contacts
assigned the number 4 can be used.

The last line of the program is shown in
Figure 59–10. A normally open contact assigned

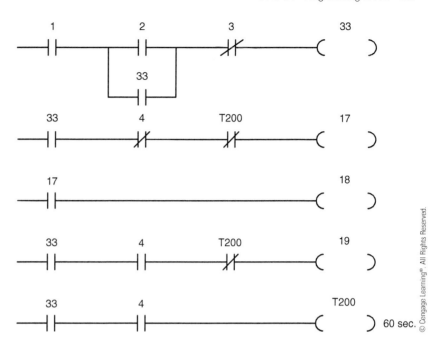

Figure 59–9
Lines 5 and 6 are added to the program.

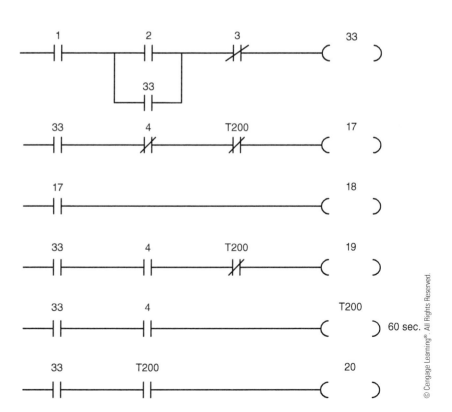

Figure 59–10
Line 7 of the program.

the number 33 is used for contact CR-4, and a normally open contact controlled by timer T200 is used for the normally open timed contact labeled TR-3. Coil 20 controls the operation of solenoid coil C.

The circuit shown in Figure 59–2 has now been converted to a program that can be loaded into a programmable logic controller. The process is relatively simple if the rules concerning PLCs are followed.

ENTERING A PROGRAM

The manner in which a program is entered into the memory of the PLC is specific to the manufacturer and type of programming terminal used. Some programming terminals employ keys that contain contact, coil, and rung symbols to basically draw the program as it is entered. Small programming terminals may require that the program be entered in a language called **Boolean**. Boolean uses statements such as *AND*, *OR*, *NOT*, and *OUT* to enter programs. Contacts connected in series, for example, would be joined by *and* statements, and contacts that are connected in parallel with each other would be programmed with *or* statements. In order to program a contact normally closed instead of normally open, the *not* statement is used. Different PLCs also require the use of different numbers to identify particular types of coils. One manufacturer may use any number between 600 and 699 to identify coils that are used as timer and counters. Another manufacturer may use any number between 900 and 999 to identify coils that can be used as timers and counters. When programming a PLC, it is always necessary to first become familiar with the programming requirements of the model and manufacturer of the programmable logic controller being programmed.

PROGRAMMING CONSIDERATIONS

When developing a program for a programmable logic controller, there are certain characteristics of a PLC that should be considered. One of these is the manner in which a programmable logic controller performs its functions. Programmable logic controllers operate by scanning the program that has been entered into memory. This process is very similar to reading a book. It scans from top to bottom and from left to right. The computer scans the program one line at a time until it reaches the end of the program. It then resets any output conditions that have changed since the previous scan. The next step is to check all inputs to determine whether they are high (power applied to that input point) or low (no power applied to that input point). This information is available for the next scan. The next step is to update the display of the programming terminal if one is connected. The last step is to reset the **"watchdog" timer**. Most PLCs contain a timer that runs continually when the PLC is in the RUN mode. The function of this timer is to prevent the computer from becoming hung in some type of loop. If the timer is not reset at the end of each scan, the watchdog timer will reach zero and all outputs will be turned off. Although this process sounds long, it actually takes place in a few milliseconds. Depending on the program length, it may be scanned several hundred times each second. The watchdog timer duration is generally set for about twice the amount of time necessary to complete one scan.

Scanning can eliminate some of the problems with **contact races** that occur with relay logic. The circuit shown in Figure 59–11 contains two control relay coils. A normally closed contact, controlled by the opposite relay, is connected in series with each coil. When the switch is closed, which relay will turn on and which will be locked out of the circuit? This is called a contact race. The relay that is turned on depends on which one managed to open its normally closed contact first and break the circuit to the other coil. There is no way to really know which relay will turn on and which will remain off. There is not even a guarantee that the same relay will turn on each time the switch is closed.

Programmable logic controllers eliminate the problem of contact races. Because the PLC scans the program in a manner similar to reading a book, if it is imperative that a certain relay turns on before another one, simply program the one that must turn on first ahead of the other one. A similar circuit is shown in Figure 59–12. When contact 1 closes, coil 100 will always be the internal relay that turns on because it is scanned before coil 101.

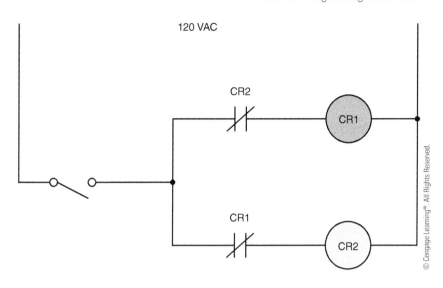

▶ Figure 59–11
A contact race can exist in relay control circuits.

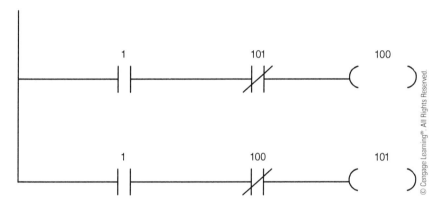

▶ Figure 59–12
Scanning eliminates contact races in PLC logic.

▶ SUMMARY

- As a general rule, the schematic diagrams used for relay logic must be changed before they can be loaded into a programmable logic controller.

- Each line of logic must end with a coil.

- All inputs are assumed to be normally open.

- When a normally closed component is connected to the input of a programmable logic controller, the logic of the program must be reversed for that input.

- Before a program can be developed, the sensor devices such as pushbuttons, limit switches, and float switches must be assigned to an input. Outputs such as solenoid coils, pilot lights, and relay coils must be assigned to an output.

- Outputs do not supply power to operate devices.

KEY TERMS

Boolean	input	relay schematic
contact race	output	watchdog timer

REVIEW QUESTIONS

1. Why are NEMA symbols such as pushbuttons, float switches, and limit switches not used in programmable logic controller schematics?

2. How are such components as coils and contacts identified and distinguished from others in a PLC schematic?

3. Why are normally closed components such as stop pushbuttons programmed normally open instead of normally closed when entering a program into the memory of a PLC?

4. What is an internal relay?

5. Why is the output of a PLC used to energize the coil of a motor starter instead of energizing the motor directly?

6. List four basic rules for developing a program for a PLC.

7. A programmable logic controller requires that times be programmed in 0.1-second intervals. What number should be entered to produce a time delay of 3 minutes?

8. When programming in Boolean, what statement should be used to connect components in series?

9. When programming in Boolean, what statement should be used to connect components in parallel?

10. In a control circuit, it is imperative that a coil energize before another one. How can this be done when entering a program into the memory of the PLC?

11. What is the function of a watchdog timer?

UNIT 60

Analog Sensing for Programmable Controllers

OBJECTIVES

After studying this unit, the student should be able to:

» Describe the differences between analog and digital inputs

» Discuss precautions that should be taken when using analog inputs

» Describe the operation of a differential amplifier

Many of the programmable controllers found in industry are designed to accept analog as well as digital inputs. Analog means continuously varying. These inputs are designed to sense voltage, current, speed, pressure, proximity, temperature, and so on. When an **analog input** is used, such as a thermocouple for measuring temperature, a special module that mounts on the I/O rack is used. These types of sensors are often used with set point detectors that can be used to trigger alarms and turn on or off certain processes. For example, the voltage produced by a thermocouple increases with a change of temperature. Assume that you want to sound an alarm if the temperature of an object reaches a certain level. The detector is preset with a particular voltage. As the temperature of the thermocouple increases, its output voltage increases also. When the voltage of the thermocouple becomes greater than the preset voltage, an alarm sounds.

INSTALLATION

Most analog sensors can produce only very weak signals. Zero to 10 volts or 4 to 20 milliamps is common. In an industrial environment where intense magnetic fields and large voltage spikes abound, it is easy to lose the input signal amid the electrical noise. For this reason, special precautions should be taken when installing the signal wiring between the sensor and input module. These precautions are particularly important when using analog inputs, but they should be followed when using digital inputs also.

KEEP WIRE RUNS SHORT

Try to keep wire runs as short as possible. The longer a wire run is, the more surface area of wire there is to pick up stray electrical noise.

PLAN THE ROUTE OF THE SIGNAL CABLE

Before starting, plan how the signal cable should be installed. Never run signal wire in the same conduit with power wiring. Try to run signal wiring as far away from power wiring as possible. When it is necessary to cross power wiring, install the signal cable so that it crosses at a right angle, as shown in Figure 60–1.

USE SHIELDED CABLE

Shielded cable is generally used for the installation of signal wiring. One of the most common types, Figure 60–2, uses twisted wires with a **Mylar® foil shield**. The ground wire must be grounded if the shielding is to operate properly. This type of shielded cable can provide a **noise reduction ratio** of about 30,000:1.

Another type of signal cable uses a twisted pair of signal wires surrounded by a braided shield. This type of cable provides a noise reduction of about 300:1.

Common coaxial cable should be avoided. This cable consists of a single conductor surrounded by a braided shield. This type of cable offers very poor noise reduction.

GROUNDING

Ground is generally thought of as being electrically neutral or zero at all points. This may not be the case in practical application, however. It is not uncommon to find different pieces of equipment that have ground levels that are several volts apart, Figure 60–3.

One method that is sometimes used to overcome this problem is to use large cable to tie the two pieces of equipment together. This forces them to exist at the same potential. This method is sometimes referred to as the **brute force method**.

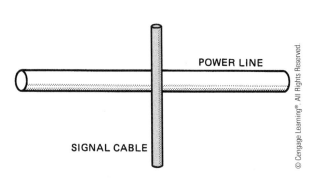

▌**Figure 60–1**
Signal cable crosses power line at right angle.

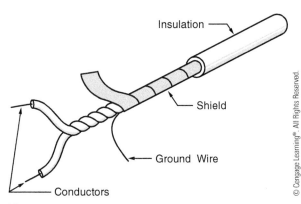

▌**Figure 60–2**
Shielded cable.

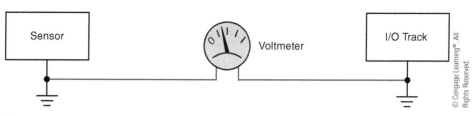

▶ **Figure 60–3**
All grounds are not equal.

Where the brute force method is not practical, the shield of the signal cable is grounded at only one end. The preferred method is generally to ground the shield at the sensor.

THE DIFFERENTIAL AMPLIFIER

An electronic device that is often used to help overcome the problem of induced noise is the **differential amplifier** shown in Figure 60–4. This device detects the voltage difference between the pair of signal wires and amplifies this difference. Because the induced noise level should be the same in both conductors, the amplifier will ignore the noise. For example, assume an analog sensor is producing a 50-millivolt signal. This signal is applied to the input

module, but induced noise is at a level of 5 volts. In this case, the noise level is 100 times greater than the signal level. The induced noise level, however, is the same for both of the input conductors. The differential amplifier therefore ignores the 5-volt noise and amplifies only the voltage difference, which is the 50 millivolts.

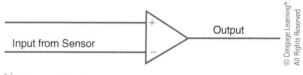

▶ **Figure 60–4**
Differential amplifier detects difference of signal level.

▶ SUMMARY

- ⊙ Analog inputs sense a range of values instead of operating in an on or off mode.
- ⊙ Most analog inputs operate with a standard of 4 to 20 milliamps.
- ⊙ Special precautions should be used when installing analog inputs.
- ⊙ Wire runs should be kept as short as possible. This reduces the surface of the wire that is susceptible to electrical noise.
- ⊙ Signal wiring should never be run close to power wiring.
- ⊙ When a signal wire must cross power wiring, it should cross at a 90° angle.
- ⊙ Signal inputs should be run with shielded cable.
- ⊙ The shield of shielded cable should be grounded, generally at the sensor.
- ⊙ Differential amplifiers are sometimes used to help eliminate electrical noise induced in the signal cable.

KEY TERMS

analog input
brute force method

differential amplifier
Mylar foil shield

noise reduction ratio
shielded cable

REVIEW QUESTIONS

1. Explain the difference between digital inputs and analog inputs.

2. Why should signal-wire runs be kept as short as possible?

3. When signal wiring must cross power wiring, how should the crossing be done?

4. Why is shielded wire used for signal runs?

5. What is the brute force method of grounding?

6. Explain the operation of the differential amplifier.

GLOSSARY

A

AC (alternating current) Current that reverses its direction of flow at regular intervals.

AC electrolytic capacitor Can house a large amount of capacitance in a small case size, is designed for short-term use only, and is used as the starting capacitor on single-phase motors.

across-the-line A method of motor starting that connects the motor directly to the supply line on starting or running. Also known as full voltage starting.

adjustable Describes construction of pressure switches used for commercial or industrial systems that allows service technicians to use the switch on different systems.

alternator A machine used to generate alternating current by rotating conductors through a magnetic field.

ambient air temperature The temperature in the surrounding area of a device.

ammeter A low-impedance device used to measure current in a circuit.

amortisseur Winding named for the set of type "A" squirrel-cage bars in a synchronous motor rotor.

ampacity The maximum current rating of a wire or device.

ampere (amp) A measurement of the actual amount of electricity that flows through a circuit; defined as 1 coulomb per second.

amplifier A device used to increase a signal.

amplitude The highest value reached by a signal, voltage, or current.

analog Device that can be adjusted to different states because it represents continuously changing quantities.

analog input An input sensor designed to sense voltage, current, speed, pressure, proximity, temperature, and other quantities in programmable computers that can trigger alarms or start–stop processes.

analog meter (voltmeter) A voltmeter that uses a meter movement to indicate the voltage value.

anode The positive terminal of an electrical device.

antifreeze protection Feature in some differential thermostats that turns the pump on and circulates warm water through the collector when its temperature is near freezing. Avoids damage to the collector.

anti-short-cycling A control that prevents the compressor from being restarted within a certain time after it has stopped.

apparent power (volt-amps) In an AC circuit, the applied voltage multiplied by the current flow in the circuit.

applied voltage The amount of voltage connected to a circuit or device.

arc-over Spark or illumination that occurs in a gap or breakage in a circuit.

armature The movable arm attracted by the magnetic field of the iron core of a solenoid.

ASA American Standards Association.

atom The smallest part of an element that contains all the properties of that element.

attenuator A device that decreases the amount of signal voltage or current.

auger Device used for boring, forcing, or moving material. Presses ice and excess water out of flaker-type ice maker.

automatic Self-acting; operation by its own mechanical or electrical mechanism.

automatic reset overload Used to protect single-phase motors. Constructed of bimetal strips that snap open contacts when overheated.

autotransformer Transformer that uses only one coil that is tapped to provide the correct voltage.

auxiliary contacts Small contacts located on relays and motor starters, used to operate other control components.

auxiliary limit One of two high-limit contacts connected in series with an automatic gas valve.

AWG (American Wire Gauge) Standard units of measure for wiring.

B

bands of color The color bands of a resistor that indicate its value and tolerance.

base The semiconductor region between the collector and emitter of a transistor. The base controls the current flow through the collector-emitter circuit.

bellows Device that draws in air through a valve and expels it through a tube. Used to operate pressure switches in most air-conditioning units.

bellows-type thermostat A thermostat that uses a refrigerant-filled bellows to operate a set of contacts.

bias A DC voltage applied to the base of a transistor to preset its operating point.

bidirectional Able to move in two directions.

bimetal strip A strip made by bonding two unlike metals together that, when heated, expand at different temperatures. This causes a bending or warping action.

bimetal type One of two basic kinds of overload relays. Uses bimetal strips in construction of unit.

bin thermostat Senses the level of ice in the ice storage bin.

bipolar transistors Dual-poled transistors used by operational amplifiers for input.

Boolean A type of language using statements such as AND, OR, and NOT to program a programmable controller.

branch circuit That portion of a wiring system that extends beyond the circuit protective device, such as a fuse or circuit breaker.

breakdown torque The maximum amount of torque that can be developed by a motor at rated voltage and frequency before an abrupt change in speed occurs.

bridge circuit A circuit that consists of four sections connected in series to form a closed loop.

bridge rectifier A device constructed with four diodes that converts both positive and negative cycles of AC voltage into DC voltage. The bridge rectifier is one type of full-wave rectifier.

brushless Generally refers to a motor that does not contain brushes.

brushless exciter A device for providing the necessary excitation current to the rotor of a synchronous motor that does not require the use of brushers or slip rings.

brute force method When several pieces of equipment have unequal ground levels, the method of using a large cable to tie the equipment together, forcing them to exist at the same potential.

busway An enclosed system used for power transmission that is voltage and current rated.

C

cabinet thermostat One of two thermostats in a refrigerator using a flex tray ice maker that, if opened, interrupts the water fill cycle.

cadmium sulfide cell (CAD cell) A solid-state device that changes its resistance in accordance with the amount of light it is exposed to.

cam A rotating or sliding piece that moves freely on its roller or by notches picked up by pins or gears.

capacitance The electrical size of a capacitor.

capacitive reactance Symbolized by X_C, reactance caused by capacitance.

capacitor A device made with two conductive plates separated by an insulator or dielectric.

capacitor-start capacitor-run motor A type of single-phase motor that employs a capacitor that remains in the circuit at all times.

capacitor-start induction-run motor A single-phase induction motor that uses a capacitor connected in series with the start winding to increase starting torque.

carbon brushes Material connected to slip rings on rotors to create resistance.

case Style of transistor that permits the leads to be quickly identified; sometimes the collector of the transistor.

cathode The negative terminal of an electrical device.

center-tapped transformer A transformer that has a wire connected to the electrical midpoint of its winding. Generally, the secondary winding is tapped.

central processing unit (CPU) The brains of a programmable controller, it contains the microprocessor chip and related integrated circuits to perform all the logic functions.

centrifugal force The law that states a spinning object will pull away from its center point; the faster the spin, the greater the force.

centrifugal switch Used to disconnect start windings from the circuit, configured to move when the centrifugal force spins parts upward.

CFM (condenser fan motor) relay Its coil energizes and starts the condenser fan motor in a flow switch–operated air-conditioning system.

CFS (condenser flow switch) Airflow switch operated by the force of air created by the condenser fan in an air-conditioning circuit.

choke An inductor designed either to present an impedance to AC current or to be used as the current filter of a DC power supply.

circuit breaker A device designed to open under an abnormal amount of current flow. The device is not damaged and may be used repeatedly. Circuit breakers are rated by voltage, current, and horsepower.

circular mils Measurement for the cross-sectional area of wire, equal to 1/1000 (0.001) of an inch.

clamp-on Type of ammeter where the jaw of the meter is clamped around one of the conductors supplying power to the load.

clock timer A time-delay device that uses an electric clock to measure the delay time.

code letter Provided on nameplates of AC motors, it indicates the type of bars used in the rotor.

collapse (of a magnetic field) Occurs when a magnetic field suddenly changes from its maximum value to a zero value.

collector A semiconductor region of a transistor that must be connected to the same polarity as the base.

combination A circuit that contains both series and parallel connections within the same circuit.

common denominator In fractions, the number into which all the denominators will divide.

compact Type of household ice maker.

comparator A device or circuit that compares two like quantities, such as voltage levels.

compressor The component of an air-conditioning or refrigeration system that maintains the difference in pressure between the high and low sides.

compressor relay coil Can be energized only after the condenser fan and evaporator fan relay coils have energized in an air-conditioning unit.

condensing unit The component of an air-conditioning or refrigeration system in which heat is removed from the refrigerant and dissipated to the surrounding air or liquid.

conduction level The point at which an amount of voltage or current will cause a device to conduct.

conductor A device or material that permits current to flow through it easily.

consequent pole motor A multispeed AC motor that changes the motor speed by changing the number of its stator poles.

contact A conducting part of a relay that acts as a switch to connect or disconnect a circuit or component.

contact race A condition that can occur in relay control circuits when coils are energized at the same time.

contactor Similar to relays; although it may contain large-load contacts designed to control large amounts of current, it may also contain auxiliary contacts. Used to connect power to resistance heater banks.

contact section Section of a solder melting–type overload relay that is connected in series with the coil of the motor starter.

continuity A complete path for current flow.

continuous run timer A defrost timer connection (pigtail to terminal 1, terminal 3 to neutral) that permits the motor to operate on a continuous basis.

control connector The connector that is connected to the control system.

control contacts In an oil-pressure failure switch, conducting parts of a relay that act as switches and provide power to the heater of the timer.

control transformers Used to change the value of line voltage to the value needed for the control circuit.

control valve The heart of a gas heating system, it controls the flow of gas to the main burner and pilot light (if used).

conventional current flow theory Assumes the current flows from positive to negative.

cooling anticipator A resistive heating element that operates in an opposite sense to the heat anticipator. While the air-conditioning unit is not running, it slightly heats the thermostat to close the contacts before the ambient temperature closes them.

coulomb A quantity of measurement for electrons at rest on a surface area. One coulomb contains 6.25×10^{18} electrons.

CR (control relay) Electrically operated switch whose coil energizes and closes its contacts, starting other operations in a circuit.

crankcase heater A component of a commercial air-conditioning unit that, as long as power is connected, is energized at all times.

cube-size thermostat Mounted on the evaporator plate of a cube-type ice maker, its closed contacts complete the circuit to the timer motor for ice harvest.

cumulative compressor run timer This defrost timer connection (pigtail to terminal 2, timer contact between terminals 1 and 4) allows the timer motor to operate only when the compressor is in operation and the thermostat is closed.

current The rate of flow of electrons.

current flow The amount of current in a circuit.

current-limiting resistor A component of a time-delay circuit that is center tapped to allow either 240 volts to be connected in series with the heater or 120 volts to be connected to the center tap.

current rating The amount of current flow a device is designed to withstand.

current relay A relay that is operated by a predetermined amount of current flow. Current relays are often used as one type of starting relay for air-conditioning and refrigeration equipment.

current transformer Used with an ammeter to provide multiscale capability.

D

DC (direct current) Current that does not reverse its direction of flow.

DC excitation current Adjusted to change the power factor of the synchronous motor to normal, overexcited, or underexcited, depending on adjustment.

definite time control A method to short out the steps of resistance by using time relays to control when resistance is shorted out of the circuit.

defrost control A heat-pump control that reverses the flow of refrigerant in a system to heat the outside heat exchange unit and remove frost.

defrost heater thermostat One of two thermostats in a flex tray ice maker that when closed permits the water fill cycle to operate.

defrost thermostat Heat sensor located on the outside heat exchanger of a heat pump.

defrost timer The "brain" of frost-free appliances, this disconnects the compressor circuit and connects a resistive heating element located near the evaporator to melt frost at regular time intervals. Operated by a single-phase synchronous motor.

delay-on-break A type of short cycle timer that starts its time delay when the thermostat contacts open.

delay-on-make A type of short cycle timer that begins its time delay when the thermostat contacts close.

delta, delta connection A triangular-shaped circuit, resembling the Greek letter *delta*, for three-phase current transformers and motor windings. A circuit formed by connecting three electrical devices in series to form a closed loop. It is used most often in three-phase connections.

diac A bidirectional diode used to phase shift a triac. It is a voltage-sensitive switch that operates on AC current.

dielectric An electrical insulator.

dielectric stress In a charged capacitor, the electron orbit of the atoms in the dielectric extend and are considered to be in tension, or dielectric stress.

differential amplifier Helps overcome induced noise by detecting voltage difference between pairs of signal wires and amplifying the difference so noise is ignored.

differential pressure switch Used to measure the difference between the suction pressure and the discharge pressure of an oil pump, e.g., device used to measure actual oil pressure in an oil-pressure failure switch.

differential thermostat Found mostly in solar-powered heating systems, uses two separate temperature sensors and is activated by the difference of temperature between them.

digital, digital device A device that has only two states of operation: on or off.

digital logic Circuit elements connected in such a manner as to solve problems using components that have only two states of operation.

digital meter (voltmeter) A voltmeter that uses direct reading numerical display as opposed to a meter movement.

diode A two-element device that permits current to flow through it in only one direction.

disconnecting means (disconnect) Device(s) used to disconnect a circuit or device from its source of supply.

door interlock A switch that permits a heating unit to operate only when the furnace door is closed.

double-acting Describes the circuitry in a thermostat permitting it to be used for both heating and cooling.

double-break contact A contact that breaks connection at two points.

dual-pressure switch Switch that has both high- and low-pressure switches in the same housing.

dynamic braking (1) Using a DC motor as a generator to produce counter torque and thereby produce a braking action. (2) Applying direct current to the stator winding of an AC induction motor to cause a magnetic braking action.

E

eddy current Circular-induced current contrary to the main currents. Eddy currents are a source of heat and power loss in magnetically operated devices.

EFM (evaporator fan motor) relay Its coil energizes and starts the evaporator fan motor.

EFS (evaporator flow switch) Airflow switch operated by the force of air created by the evaporator fan.

E-I cores A type of transformer or inductor core.

ejector blades Blades in ice makers that rotate and push ice cubes out of the mold tray and into a bin.

electrical interlock When the contacts of one device or circuit prevent the operation of some other device or circuit.

electrical pressure The electromotive force (voltage) that pushes electrons through a wire.

electric arc Used to ignite the gas flame in gas-operated appliances. Permits gas to flow but turns the gas off if no flame is soon detected.

electric controller Device(s) used to govern in some predetermined manner the operation of a circuit or piece of electrical apparatus.

electricity The flow of electrons through a completed path.

electric resistance heating element Component in an air conditioner used to provide heat in cool weather.

electromagnetic field Created in windings when DC current is applied to the rotor. This field of the rotor locks in step with the rotating magnetic field of the stator.

electromotive force (EMF) The force that pushes electrons through a wire. Also known as voltage and electrical pressure.

electron One of the three major parts of an atom. The electron carries a negative charge.

electron flow theory Assumes that current flows from negative to positive.

electrostatic charge The energy of a capacitor stored in the dielectric; permits production of extremely high currents under certain conditions.

emitter The semiconductor region of a transistor that must be connected to a polarity different from the base.

enclosure Mechanical, electrical, or environmental protection for components used in a system.

eutectic alloy A metal with a low and sharp melting point used in thermal overload relays.

evaporator An air-conditioning or refrigeration system component that removes heat from the surrounding air or liquid to cause a change of state in the refrigerant.

excitation current The direct current necessary to cause the rotor of a synchronous motor to become a magnet.

expansion Method for sensing temperature determined by the expansion of metal, it considers the type of metal and the amount of heat used.

F

fan limit switch Contains both a fan switch and a high-limit switch in one housing. Its operation causes the burner to turn off and prevent damage to a heating system.

fan switch A single-pole double-throw switch, connected on one side to the thermostat control and on the other to the control voltage.

fan switches (heating) Operated by a bimetal strip that closes a set of contacts when the temperature of the heat exchanger reaches a high enough level so that no cold air is blown into the living area.

farad Rating, or unit of measurement, of a capacitor.

feeder The circuit conductor between the service equipment, or the generator switchboard of an isolated plant, and the branch circuit overcurrent protective device.

field discharge resistor Used across a winding to prevent excessive voltage; helps to reduce the voltage induced into the rotor by the collapsing magnetic field when current is disconnected.

field effect transistors High-input impedance transistors used by op amps that require little current to operate the amplifier.

filter A device used to remove the ripple produced by a rectifier.

"fire eye" A gas flame–sensing device that changes its resistance in the presence of light.

fixed resistor A resistor whose value cannot be changed.

flaker-type Ice maker that produces ice chips or flakes.

"flame rod" A flame–sensing device that operates by using the gas flame as a conductor of electricity.

flex tray Type of ice maker that fills with water and turns at an angle to dump ice cubes into a bin.

flow switch (sail switch) Used in air-conditioning systems to sense the flow of air instead of the flow of liquid. Operates on the principal of a sail.

flow washer Part of an electric solenoid valve in an ice maker that meters the amount of water used.

forward biased A diode is forward biased when a positive voltage is connected to the anode and a negative voltage is connected to the cathode.

four-way valve (reversing valve) A common solenoid valve used to change the direction of refrigerant flow in a heat-pump system. Composed of a main valve and a pilot valve operating together.

FR (fan relay) Controls the fan motor. Designed so that a switch can be used to turn the circuit completely off, operate the fan manually, or permit the fan to be operated by a thermostat.

freezer assembly A hollow tube surrounded by a cylindrical container that is the evaporator in the refrigeration unit.

frequency The number of complete cycles of AC voltage that occur in one second.

FSCR (flow switch control relay) The coil of relay connected to the line when the flow switch closes.

full-load torque The amount of torque necessary to produce the full horsepower of a motor at rated speed.

full-wave rectifier Rectifier that uses two diodes to convert AC to DC. Both halves of the AC waveform are used.

fuse A device used to protect a circuit or electrical device from excessive current. Fuses operate by melting a metal link when current becomes excessive.

fused jumper A device used in troubleshooting a circuit.

G

gain The increase in signal power produced by an amplifier.

gate (1) A device that has multiple inputs and a single output. There are five basic types of gates, the and, or, nand, nor, and inverter. (2) One terminal of some electronic devices, such as SCRs, triacs, and field effect transistors.

germanium Gray-white, hard, and brittle chemical element used in the manufacture of semiconductor materials such as transistors and diodes.

GTO Gate-turnoff SCR An SCR that can be turned off by applying a negative pulse with respect to the cathode to the gate.

gun-type An oil furnace ignited by an electric arc.

H

hair Human hair is used in some humidistats to sense humidity, because it contracts and expands with changes in the amount of humidity in the air.

half-wave rectifier The simplest type of rectifier, it is one diode that changes or converts AC voltage into DC voltage over half an AC waveform.

harmonics Wave forms that become induced in an AC circuit that can cause overheating of devices.

heat anticipator The component of a thermostat that preheats the sensing element and causes the thermostat contacts to open before the room heat has reached the set point of the thermostat.

heater section Section of a solder melting–type overload relay that is connected in series with the motor.

heating element In a solder melting–type of overload relay, it is wound around the tube and is calibrated to produce a certain amount of heat when a predetermined amount of current flows through it.

heat pump A system that uses refrigerant to supply both heating and cooling to a dwelling.

heat sink A metallic device designed to increase the surface area of an electronic component for the purpose of removing heat at a faster rate.

henry Unit of measurement for inductors, or inductance. One henry is present when the current through a coil, changing 1 ampere per second, produces 1 volt across the coil terminals.

hermetic compressor A compressor that is completely enclosed and airtight.

hertz (Hz) The international unit of frequency.

high leg A configuration in a delta connection where one transformer is larger and center tapped.

high-pressure switch Used to sense amount of pressure in air-conditioning and refrigeration units. The bellows in this switch is connected to the discharge side of the compressor via the tube; the high-pressure switch activates if pressure becomes too great.

holding current The amount of current needed to keep an SCR or triac turned on.

holding relay (HR) Used with short-cycling timers, this relay energizes and changes holding contacts during short-cycling recovery time.

holding, sealing, or maintaining contacts Contacts used to maintain continuous current flow to the coil of a relay.

holding switch Functions in an ice maker by changing positions to maintain the circuit until the cam returns to the freeze, or off, position.

horsepower A measure of power for electrical and mechanical devices.

hopscotch method A method of troubleshooting a circuit.

hot-gas solenoid valve In harvest cycle of ice making, opens and permits high-pressure hot gas to be diverted to the evaporator plate.

hot-wire relay A type of starting relay used to disconnect the start windings of a single-phase motor; so named because it uses a length of resistive wire connected in series with the motor to sense motor current.

humidistat A device that can sense the amount of humidity in the air and activate a set of contacts if the humidity should become too high or too low.

hysteresis loop A graphic curve that shows the value of magnetizing force for a particular type of material.

I

illuminate Act or state of being highlighted by light or color. In a PC, illuminated contacts prove they are closed and providing current paths.

impedance The total opposition to current flow in an electrical circuit.

indoor fan relay (IFR) A relay that may control the coils of other relays, which connect the fan motors to the line; may operate several fans at once.

indoor resistance heat Part of a schematic drawing of a heat-pump control system depicting the indoor circuits.

induced current Current produced in a conductor by the cutting action of a magnetic field.

induction motor A motor whose current flow in the rotor is produced by induced voltage from the rotating magnetic field of the stator.

inductive reactance The opposition to current flow in an AC circuit, caused by an inductor. Measured in ohms; its symbol is X_L.

inductor A coil.

inlet Side of a solenoid valve that is connected to the side of the system with the highest pressure (for liquid or gas to flow).

inline Being directly connected into the circuit.

in phase In pure-resistive circuits, the voltage and current are *in phase* when they cross the zero line at the same point and have their peak positive and negative values at the same time.

input The section of a programmable controller where information is supplied to the CPU.

input impedance Resistance or opposition to current flow in an AC or DC circuit.

input voltage The amount of voltage connected to a device or circuit.

insulation Material that inhibits or slows current.

insulator A material used to electrically isolate two conductive surfaces.

interlock, interlocking A device or method used to prevent some action from taking place in a piece of equipment or circuit until some other action has occurred.

internal relay An imaginary device that exists only in the logic of a computer, programmed by assigning a coil a number greater than the I/O capacity.

inverting amplifier Connection in an op amp that renders the input signal out of phase with the feedback voltage of the output, reducing input voltage and gain.

inverting input Inverts the output of an op amp: if positive-going AC is applied, the output produces negative-going voltage.

I/O capacity Measure of input and output ability of a programmable controller. Number of tracks × I/O track (32) = I/O capacity (8 I/O tracks = 256 I/O capacity).

I/O rack Contains input and output modules (typically from 2 to 8 each) used to connect the central processing unit to the outside world.

isolation transformer A transformer whose secondary winding is electrically isolated from its primary winding.

J

jumper A short piece of conductor used to make connection between components or a break in a circuit.

junction diode A diode that is made by joining together two pieces of semiconductor material.

K

kick-back diode A diode used to eliminate the voltage spike induced in a coil by the collapse of a magnetic field.

L

lattice structure The pattern that semiconductor material molecules arrange themselves in when refined into a pure form.

law of charges States that like charges repel each other, and unlike charges attract each other.

leakage current The small amount of current leaked when a very sudden increase of resistance opens a set of contacts. This leakage maintains the temperature of the thermistor and prevents it from returning to a low resistance.

LED (light-emitting diode) A diode that produces light when current flows through it.

legend In a schematic diagram, a list that shows a symbol or notation and gives the definition of that symbol or notation.

level detector Detects when one voltage becomes greater than another; a comparator.

limit switch A mechanically, generally bimetal-operated switch that detects the position or movement of an object.

line isolation The ability of an isolation transformer to physically separate electric circuits.

line voltage thermostat Used to control loads such as blower fans and heating elements without an intervening relay.

load (1) Anything that may draw current from an electrical power source. (2) In a motor schematic, the large motor (M) contacts that are connected in series with the overload heater element and the motor.

load center Generally, the service entrance. A point from which branch circuits originate.

loading terminal Either a handheld device or a cathode ray tube (CRT) used to program the programmable controller. Also known as a program terminal.

locked rotor current The amount of current produced when voltage is applied to a motor and the rotor is not turning.

locked rotor torque The amount of torque produced by a motor at the time of starting.

lock-out relay A mechanical device used to prevent the operation of some other component.

loop A closed electric circuit.

low-pressure switch Used to sense the amount of pressure in an air-conditioning and refrigeration system. This switch is connected to the suction side of the compressor and activates if pressure becomes too low.

low-voltage controls Section of a heat-pump schematic detailing low-voltage circuitry.

low-voltage protection A magnetic relay circuit so connected that a drop in voltage causes the motor starter to disconnect the motor from the line.

M

magnetic contactor A contactor operated electro-mechanically.

magnetic field (1) The space in which a magnetic force exists. (2) Lines of force used to represent magnetic induction.

magnetic flux Lines of magnetic force.

main auxiliary limit A high limit switch located in the low voltage side of the control circuit.

main limit One of two high-limit contacts connected in series with automatic gas valve of a heating unit. May be located in a fan limit switch.

maintaining contact Also known as a holding or sealing contact. It is used to maintain the coil circuit in a relay control circuit. The contact is connected in parallel with the start pushbutton.

major components The basic necessary elements, devices, and connections that make up a unit (switch, fan motor, compressor, etc.) and are described in schematic diagrams.

manual controller A controller operated by hand at the location of the controller.

measuring instruments Devices for measuring values of voltage, current, resistance, and power.

metering device Area in a heat pump where liquid or gas is changed to low-pressure liquid.

microfarad A measurement of capacitance; one-millionth of a farad, symbolized by μf.

microprocessor A small computer. The central processor unit is generally made from a single integrated circuit.

mil-foot Standard measurement of resistance of wire. A mil-foot of wire is 1 circular mil in diameter and 1 foot long.

mode A state or condition.

mold heater Part of an ice maker that warms the mold enough to release ice cubes.

motor A device used to convert electrical energy into rotating motion.

motor controller A device used to control the operation of a motor.

motor starters Contactors (designed to control large amounts of current) with the addition of overload relays. Usually contain auxiliary contacts as well as load contacts.

movable contact A contact that moves to make connection with another contact.

multiple taps A method of connecting to different parts of a transformer winding.

multiranged Ability to use one meter movement to measure several ranges of voltage.

multispeed AC motor A motor that can be operated at more than one speed.

mutual induction Induction that results when there is an interaction of adjacent inductors. When the magnetic field of the primary induces a voltage into the secondary winding wound on the same core.

Mylar foil shield Material used in installations to shield cable for signal wiring to reduce noise ratio.

N

nameplates Plates on electric motors that provide important information describing the characteristics of the motor (horsepower, volts, etc).

negative One polarity of a voltage, current, or charge that has an excess of electrons.

negative feedback Reduces the gain and makes the amplifier more stable in an op amp.

negative temperature coefficient (NTC) In a thermistor that has this type of coefficient, resistance decreases as the temperature increases.

NEMA National Electrical Manufacturers Association.

NEMA ratings Electrical control device ratings of voltage, current, horsepower, and interrupting capability given by NEMA.

neutral The grounded conductor in an AC circuit.

neutron One of the principle parts of an atom. The neutron has no charge and is part of the nucleus.

nichrome A material used in the construction of wire wound resistors and electric heating elements.

node In schematic drawings, the dot in the center of the cross created by connecting wires.

noise reduction ratio Ratio of ability to diminish stray electrical noise or interference in circuits.

nonadjustable Type of pressure switch that must be matched to the refrigerant system, unlike adjustable switches that can be used to switch different systems.

noninductive load An electrical load that does not have induced voltages caused by a coil. Noninductive loads are generally considered to be resistive, but they can be capacitive.

noninverting amplifier One of two input connections in op amps that outputs in phase with the feedback: positive in produces positive out.

noninverting input Signal voltage applied to this input outputs voltage the same polarity: positive in produces positive out.

nonreversing A device that can be operated in only one direction.

normally closed The contact of a relay that is closed when the coil is de-energized.

normally open The contact of a relay that is open when the coil is de-energized.

NPN Transistor with N and P materials in N-P-N order. Must have positive connected to the collector and negative connected to the emitter.

NPNP Rectifier junction made by joining four layers of semiconductor material together in N-P-N-P order.

N-type material Semiconductor material that has been impurified and rendered with excess electrons, resulting in a net negative charge.

nucleus Central part of an atom, containing neutrons and protons.

nylobraid tube Carries ice to the ice storage bin in an ice maker.

nylon Synthetic material used as a sense element in some humidistats.

O

off-delay timer A timer that delays changing its contacts back to their normal position when the coil is de-energized.

offset null Connection in an op amp used and adjusted to produce 0 volts at the output.

ohm The measure of resistance to the flow of current.

ohmmeter A device used to measure resistance.

Ohm's law Current is equal to the voltage divided by the resistance ($I = E/R$). It takes 1 volt to push 1 amp through 1 ohm.

oil-filled capacitor Made with two metal foil plates separated by paper that has been soaked in a special dielectric oil.

oil-pressure failure switch Switch containing several control functions in the same unit, used in a forced-oil system to protect the compressor from insufficient oil pressure.

on-delay timer A timer that delays changing the position of its contacts when the coil is energized.

open-delta system Type of three-phase service that needs only two transformers to provide three-phase voltage.

opens on rise of differential An inlet pressure-regulating valve that responds to changes in the condenser pressure in an adjustable head pressure system.

opens on rise of inlet Provides a bypass around the condenser in an adjustable head pressure system.

operational amplifier (op amp) An integrated circuit used as an amplifier.

opposition Resistance or repelling movement.

opto-isolated Situation when the load side of the relay is optically isolated from the control side of the relay and controlled by a light beam.

optoisolator A device used to connect different sections of a circuit by means of a light beam.

orifice Passage on each side of a four-way reversing valve; each provides a path for a very small amount of refrigerant to flow.

oscillator A device used to change DC voltage into AC voltage; produces a waveform that has positive and negative pulses of equal voltage and time.

oscilloscope A voltmeter that displays a waveform of voltage in proportion to its amplitude with respect to time.

outdoor compressor controls Section of a schematic detailing outdoor circuitry of a heat-pump system.

outlet Side of solenoid valve at which, if reversed and applied to the outlet side of the system, the pressure could cause valve leakage.

out-of-phase The condition in which two components do not reach their positive or negative peaks at the same time.

output The section of the programmable controller where information is supplied to the outside circuits by the CPU.

overload Potentially damaging situation where too much current flows through a circuit not built to sustain the excess, thus overloading or overheating.

overload relay A relay used to protect a motor from damage due to overloads. The overload relay senses motor current and disconnects the motor from the line if the current is excessive for a certain length of time.

P

panelboard A metallic or nonmetallic panel used to mount electrical controls, equipment, or devices.

parallel circuit A circuit that contains more than one path for current flow.

peak-inverse/peak-reverse voltage The rating of a semiconductor device that indicates the maximum amount of voltage in the reverse direction that can be applied to the device.

peak-to-peak voltage The amplitude of AC voltage measured from its positive peak to its negative peak.

peak voltage The amplitude of voltage measured from zero to its highest value.

permanent magnet rotor The rotor of a motor or generator that is constructed using permanent magnets.

permanent split-capacitor motor (PSC) A single-phase induction motor similar to the capacitor start motor except that the start windings and the starting capacitor remain connected in the circuit during normal operation.

phase shift, phase shifted A change in the phase relationship between two quantities of voltage or current (lead and lag), from one to another.

photoconductive A device that changes resistance in accord with the amount of light present.

photodetector Connected to a triac gate to control the output; permits current flow.

photodiode Semiconductor device for detecting and measuring radiant energy (light) by means of its conversion into an electric current.

photovoltaic A device that produces a voltage in the presence of light. Generally called a solar cell.

picofarad Used in extremely small capacitors; one millionth of a microfarad, symbolized by µµf or pf.

pilot device A control component designed to control small amounts of current. Pilot devices are used to control larger control components.

pilot light A small gas flame that burns continuously near the main burner of a gas burner.

pilot valve Operates with the main valve in a four-way reversing valve. Controls the operation of the main valve that controls the flow of refrigerant in a system.

plunger type Type of solenoid that is generally used with relays that use double-break contacts. The coil is surrounded by the iron core that has an opening in it for the shaft of the armature to pass.

pneumatic timer A device that uses the displacement of air in a bellows or diaphragm to produce a time delay.

PN junction An accurate and linear device, or diode, that measures temperature.

PNP Transistor with P and N materials in P-N-P order. Must have positive connected to the emitter and negative connected to the collector.

PNPN Rectifier junction made by joining four layers of semiconductor material together in P-N-P-N order.

polarity The characteristic of a device that exhibits opposite quantities within itself: positive and negative.

positive temperature coefficient (PTC) In a thermistor with this type of coefficient, resistance increases as temperature increases.

potential relay Operates by sensing an increase in the voltage developed in the start winding when the motor is operating.

potentiometer A variable resistor with a sliding contact that is used as a voltage divider.

power connector The connector that is connected to the incoming power.

power factor A comparison of the true power (watts) to the apparent power (volt amps) in an AC circuit.

power rating The rating of a device that indicates the amount of current flow and voltage drop that can be permitted.

power supply Part of a programmable controller used to lower the incoming AC voltage to desired level, rectify it to DC, then filter and regulate it.

power transistor Used to connect the load to the line in a relay designed to control a DC load.

pressure regulator An internal component of many control valves, it maintains a constant pressure to the main burner.

pressure switch A device that senses the presence or absence of pressure and causes a set of contacts to open or close.

primary control The major part of an oil-fired control system, it ensures that when the thermostat calls for heat, the flame will be created within a predetermined amount of time.

primary winding The winding of a transformer connected to power.

printed circuit A board on which a predetermined pattern of printed connections has been made.

programmable logic controllers (PLCs) Comprised of a power supply, central processing unit, program loader or terminal, and I/O rack, these machines are generally designed to be programmed with relay schematic and ladder diagrams, instead of computer languages, for quick, versatile changes and use.

programmable thermostat Can be set to automatically operate at different temperature settings at different times.

programming terminal Used to load or amend a program in a programmable controller. Also known as a loading terminal.

proton One of the three major parts of an atom. The proton has positive charge.

PSCR (pressure switch control relay) Used in a circuit designed to turn off a compressor if the pressure in the system reaches a predetermined (high) level. If power is disconnected, a warning light appears, and the circuit must be manually reset.

psig (pounds per square inch gauge) Standard measuring system to measure pressure.

P-type material Semiconductor material that has only three valence electrons, leaving a hole in its lattice structure, resulting in it having a net positive charge.

pulse generator Op amp that produces a waveform of positive and negative pulses of inconsistent voltage and time (pulse duration changes).

pure-capacitive circuit The current in this type of circuit is limited by the voltage of the charged capacitor.

pure-inductive circuit A circuit in which the current lags behind the voltage by 90°. There is no true power or watts in this circuit.

pure-resistive circuit Similar to a DC circuit in having true power or watts equal to the voltage multiplied by the current. In this circuit, the voltage and current are in phase with each other.

pushbutton A pilot control device operated manually by being pushed or pressed.

R

rapid cooldown Removing heat from an object in a short period of time.

reactance The opposition to current flow in an AC circuit offered by pure inductance or pure capacitance.

reciprocal The opposite of any number found by dividing that number into 1.

rectifier A device or circuit used to change AC voltage into DC voltage.

reed relay A small set of reed contacts connected to the gate of the triac; these contacts are closed by a magnetic field and in turn cause the triac to turn on. Used to control solid-state relays.

regulator A device that maintains a quantity at a predetermined level.

relay A magnetically operated switch that may have one or more sets of contacts.

relay schematic Schematic diagram used for relay logic. These must be adapted to be loaded into a programmable computer.

remote control Controls the functions of some electrical device from a distant location.

resistance The opposition to current flow in an AC or DC circuit.

resistance-start induction-run motor One type of split-phase motor that uses the resistance of the start winding to produce a phase shift between the current in the start winding and the current in the run winding.

resistance temperature detector (RTD) Made of platinum wire, this measures temperature accurately, because the resistance of platinum changes greatly with temperature.

resistive heating element Heating component located near the evaporator of a frost-free appliance, connected at set intervals by a defrost timer to melt any frost on the evaporator.

resistor A device used to introduce some amount of resistance into an electrical circuit.

reversed biased When negative voltage is connected to the anode and positive voltage is connected to the cathode, resulting in an open connection with no current flow.

reversing valve (four-way valve) used to change the direction of flow of refrigerant in a heat-pump system.

reversing valve solenoid A device in a heat-pump system that, if de-energized, indicates the unit is in the heating mode.

rheostat A variable resistor.

RMS value The value of AC voltage that will produce as much power when connected across a resistor as a like amount of DC voltage.

rotating field speed Determined by the number of stator poles and the frequency.

rotating magnetic field The principle of operation for all three-phase motors. A magnetic field that rotates due to either voltages being out of phase, voltages changing polarity, or the arrangement of stator winding in a motor.

rotor A large electromagnet that is the moving part of the alternator.

rotor bars Bars that are part of the rotor in a squirrel-cage induction motor.

rotor slip Condition produced by a weakening magnetic field due to series impedance, eventually causing the motor speed to decrease.

run winding In the stator of a split-phase motor, it is made of large wire and is placed in the bottom of the stator core.

S

sail switch (flow switch) Used in air-conditioning systems to sense the flow of air instead of the flow of liquid. Operates on the principle of a sail.

saturation The maximum amount of magnetic flux a material can hold.

SBS Silicon bilateral switch Often used to phase shift a triac.

schematic An electrical drawing showing components in their electrical sequence without regard for physical location.

SCR (silicon-controlled rectifier) A semiconductor device that can be used to change AC voltage into DC voltage. The gate of the SCR must be triggered before the device will conduct current.

secondary winding The winding of a transformer connected to the load.

self-induction Occurs when a coil induces a voltage into itself.

self-transformer An autotransformer.

semiconductor A material that contains four valence electrons and is used in the production of solid-state devices.

sensing device A pilot device that detects a quantity and converts it into an electrical signal.

sequence timer Used to turn the strip heaters (heating elements) on in stages instead of all at once.

series circuit A circuit that contains only one path for current flow.

service The conductors and equipment necessary to deliver energy from the electrical supply system to the premises served.

service factor An allowable overload for a motor indicated by a multiplier that, when applied to a normal horsepower rating, indicates the permissible loading.

shaded-pole induction motor An AC induction motor that develops a rotating magnetic field by shading part of the stator windings with a shading loop.

shading coil A large loop of copper wire or a copper band wound around one end of a shaded pole.

shading loop A large copper wire or band connected around part of a magnetic pole piece to oppose a change of magnetic flux.

shielded cable Cable that has been wrapped in protective covering to provide noise reduction in signal wiring.

short circuit An electrical circuit that contains no resistance to limit the flow of current.

short cycling The starting and stopping of a compressor in rapid succession.

short-cycling timer A cam-operated, motor driven, on-delay timer that allows time to operate and reset short-cycled contacts to their original positions.

silicon Most common element used to make semiconductor devices due to its ability to withstand heat.

silicon bilateral switch (SBS) A solid-state component in a primary control of an oil burner that, if disconnected, behaves like a diac and will conduct current to the primary control's triac.

silicon-controlled rectifier (SCR) A thyristor, also called the PNPN junction, used when an electronic device is needed to control large amounts of power.

sine-wave voltage A voltage waveform whose value at any point is proportional to the trigonometric sine of the angle of the generator producing it.

single-phase Formed by connecting a single transformer to a three-phase line.

single-pole breaker A circuit breaker used for connecting a 120-volt circuit.

slip The difference in speed between the rotating magnetic field and the speed of the rotor in an induction motor.

slip ring When present on rotor shafts, permits the connection of external resistance to the rotor windings. Always used in pairs.

snap-action The quick opening and closing action of a spring-loaded contact.

solder melting–type An overload relay that uses a heating element to detect overloads.

solenoid A magnetic device used to convert electrical energy into linear motion.

solenoid coil Component of a heat-pump system that, if energized, will change the position of the plunger of the pilot valve.

solenoid valve A valve operated by an electric solenoid used to control the flow of gases or liquids.

solid-state hard starting kit Used to increase the starting torque of a permanent-split capacitor motor; contains a solid-state relay and an AC electrolytic capacitor.

solid-state relay A device or an electronic component constructed from semiconductor material, used to control either AC or DC loads. Advantages are that it has no moving parts, is resistant to vibration, and is sealed against dirt and moisture.

solid-state starting relay Actually, an electronic component known as a thermistor, intended to replace the current-type starting relay.

split-phase motor A type of single-phase motor that uses resistance or capacitance to cause a shift in the phase of the current in the run winding and the current in the start winding. The three primary types of split-phase motors are resistance-start induction-run, capacitor-start induction-run, and permanent split-capacitor.

squirrel-cage induction motor An induction motor whose rotor contains a set of bars that resemble a squirrel cage.

staging thermostat A thermostat that contains more than one set of contacts that operate at different times in accord with the temperature.

star Also known as the wye connection, one of the three-phase wiring configurations.

starter A relay used to connect a motor to the power line.

starting relay Located away from the motor and used to disconnect the start windings when the motor has reached about 75% of its full speed. Used especially in single-phase motors that are hermetically sealed and unable to use a centrifugal switch.

start winding In the stator of a split-phase motor, it is made of small wire and is placed near the top of the stator core.

stationary contact A contact that is fixed and cannot move.

stator The stationary winding of an AC motor.

step-down transformer A transformer that produces a lower voltage at its secondary than is applied to its primary.

step starting Similar to shifting gears in the transmission of an automobile, method used in most large wound motors, as opposed to actual variable resistors. Starts with maximum resistance and with increasing speed shorts out resistance.

step-up transformer A transformer that produces a higher voltage at its secondary than is applied to its primary.

surge A transient variation in the current or voltage at a point in the circuit. Surges are generally unwanted and temporary.

switch A mechanical device used to connect or disconnect a component or circuit.

synchronous condenser A synchronous motor operated at no load and used for power factor correction.

synchronous motor A three-phase motor that is *not* an induction motor; it runs at a constant speed from no load to full load and can correct its own power factor and the power factors of other motors connected to the same line.

synchronous speed The speed of the rotating magnetic field of an AC induction motor.

T

tape wound A type of transformer core.

tapped A circuit that has been cut in on by another circuit or split in its configuration by a circuit.

temperature Affects resistance of wire; resistance increases as temperature increases.

temperature relay A relay that functions at a predetermined temperature. Generally used to protect some other component from excessive temperature.

terminal, terminal board A fitting attached to a device for the purpose of connecting wires to it.

terminal markings Identification letters and symbols assigned to circuitry in schematic drawings to aid in tracing the circuits.

termination temperature Temperature at which the terminals of devices connected to the conductor will be withstood.

test points Labeled on a plate behind the ice maker's front cover, they provide the option of testing different parts of the electrical circuit with a volt- or ohmmeter.

thermistor A resistor that changes its resistance with a change of temperature.

thermocouple Made by joining two dissimilar metals together at one end. The voltage produced is proportional to the types of metals used and the difference in temperature of the two junctions.

thermopile A series connection of thermocouples to permit the voltages to add and produce a higher output voltage.

thermostats Temperature-sensitive switches that employ a variety of methods to sense temperature and can be found with different contact arrangements.

three phase Having three separate voltage waveforms produced by the alternator.

three-phase motors Motors operated by a rotating magnetic field: squirrel-cage induction motor, wound rotor induction motor, and synchronous motor.

three-phase squirrel-cage induction motor A wye-connected motor with no external resistors for the rotor circuit and no DC circuit to excite the rotor.

three-pole breaker A circuit breaker used for connecting a three-phase circuit.

three-wire control circuits More flexible than two-wire control circuits, these are characterized by the fact that they are operated by a magnetic relay or motor starter.

thyristor An electronic component that has only two states of operation: on or off.

time-delay circuit Consists of a current-limiting resistor, a resistance heating element, and a bimetal strip; permits the compressor to operate long enough for oil pressure to build up in a system.

timer clock Device used in place of a defrost timer to activate defrost cycles. Its construction, including a separate timer release solenoid, allows the defrost cycle to be activated during minimum use periods.

timer motor Part of a short-cycle timer that is geared to permit a delay of about 3 minutes before contacts change position to restart a compressor.

tolerance The permissible deviation form a value.

torque The turning force developed by a motor.

transducer A device that converts one type of energy into another type of energy. Example: A solar cell converts light into electricity.

transformer An electrical device that changes one value of AC voltage into another value of AC voltage.

transistor A solid-state device made by combining three layers of semiconductor material together. A small amount of current flow through the base-emitter can control a larger amount of current flow through the collector-emitter.

triac A bidirectional thyristor used to control AC voltage.

troubleshoot To locate and eliminate problems in a circuit.

true power The wattage, or measure of the amount of power that is being used in a circuit.

turns ratio Relationship between the number of turns in the primary of a transformer and the number of turns in the secondary.

two-conductor romex A cable used in circuit breaking containing three wires—black (hot), white (neutral), and bare copper (grounding).

two-phase power Produced by having an alternator with two sets of coils wound 90° out of phase with each other.

two-pole breaker Circuit breaker used for connecting a 240-volt single-phase circuit.

two-speed compressors The systems that maintain the difference in pressure between the high and low sides and that have two speeds.

two-wire control circuit In this circuit, a simple switch is used to control the power applied to a small motor. Switch open means no complete path for current to flow; switch closed supplies power.

U

unijunction transistor A digital transistor that has two bases and one emitter made by combining three layers of semiconductor material.

unity Pure resistance, or having a power factor of 100%.

V

valence electrons Electrons located in the outer orbit of an atom.

valence shell The outer shell of an atom.

variable frequency Frequency that can be changed over a range.

variable resistor A resistor whose resistance value can be varied between its minimum and maximum values.

variable speed A motor or device that permits a change of speed.

variable speed motors Shaded-pole or permanent-split capacitor motors able to change speeds without disconnecting by a switch or relay. Commonly used to operate light loads such as ceiling fans and blower motors.

varistor A resistor that changes its resistance value with a change of voltage.

VARs (volt-amps-reactive) In a pure-inductive circuit, the value of the voltage multiplied by the current. Also known as wattless power.

voltage An electrical measurement of potential difference, electrical pressure, or electromotive-force (EMF).

voltage drop The amount of voltage required to cause an amount of current to flow through a certain resistance.

voltage follower A basic op amp circuit in which the output is connected directly back to the inverting input, producing a gain of 1; used for impedance matching purposes.

voltage rating A rating that indicates the amount of voltage that can be safely connected to a device.

voltage regulator A device or circuit that maintains a constant value of voltage.

volt-amps The apparent power of an AC circuit; the applied voltage multiplied by the current flow.

voltmeter An instrument used to measure a level of voltage.

volt-ohm-milliammeter (VOM) A test instrument so designed that it can be used to measure voltage, resistance, or milliamperes.

volts per turn The amount of voltage induced in each turn of a coil.

W

watchdog timer A time that runs continually to ensure that the program does not become locked in a loop.

water solenoid valve Valve in an ice maker that opens and permits fresh water to flow into the sump.

watt, wattage A measure of true power, or the amount of power being used in a circuit. It is proportional to the amount of voltage and the amount of current flow.

waveform The shape of a wave obtained by plotting a graph of voltage with respect to time.

wire wound A type of resistor.

wiring diagram An electrical diagram used to show components in their approximated physical location with connecting wires.

wound rotor induction motor A three-phase induction motor used for large air-conditioning units. Stator winding is the same as squirrel-cage stator; however, the rotor contains wound coils of wire instead of squirrel-cage rotor bars, and containd as many poles as there are stator poles.

wye, wye connection A three-phase wiring configuration that resembles the letter Y when the three coils are connected and schematically drawn. Also known as star, or the star connection.

Z

zener diode A diode that has a constant voltage drop when operated in the reverse direction. Zener diodes are commonly used as voltage regulators in electronic circuits.

zero switching Occurs when a relay is told to turn off while the AC voltage is in the middle of a cycle. The relay will continue to conduct until the AC voltage drops to a zero level and will then turn off.

zone Generally refers to heating or cooling a certain section of a dwelling.

INDEX

Note: "f" indicates a figure